Psychobiology of Behaviour

Kostas N. Fountoulakis
Ioannis Nimatoudis
Editors

Psychobiology of Behaviour

Springer

Editors
Kostas N. Fountoulakis
3rd Department of Psychiatry, Faculty of Medicine, School of Health Sciences
Aristotle University of Thessaloniki
Thessaloniki
Greece

Ioannis Nimatoudis
3rd Department of Psychiatry, Faculty of Medicine, School of Health Sciences
Aristotle University of Thessaloniki
Thessaloniki
Greece

ISBN 978-3-030-18322-6 ISBN 978-3-030-18323-3 (eBook)
https://doi.org/10.1007/978-3-030-18323-3

This Springer imprint is published by the registered company Springer Nature Switzerland AG
The registered company address is: Gewerbestrasse 11, 6330 Cham, Switzerland

Foreword

Introduction and Guidance for the Study of the Book

This book is a comprehensive textbook of the psychobiology of behavior. It is written primarily for mental health professionals in general, doctors of various specialties outside mental health and students of medicine, psychologists and others who are interested in this topic. The editors apparently had in mind also particularly GPs who are increasingly faced with mental health or psychological or behavioral problems of their patients. Prof. Fountoulakis, who has written three chapters in this book, and the other authors had the vastly different tasks and needs of different groups in mind, considering multiple levels of details and depth.

Since the editors are both academic researchers and active doctors in everyday clinical practice, they could best meet the very demanding challenge of summarising the extensive research results in this field in a way that is digestible by the target group of readers.

A complete textbook, as the current one, needs to present the essence of things, but still the authors need to keep in mind that the era of sole authority has long passed. Any teaching text should provide, together with the essence, links to sources and data. In other words, no author is considered today a source unto himself/herself; he/she is rather the medium who carries knowledge to readers in a comfortable and practical way. Thus, a balance is needed between the inclusion of too many references, which would make the text too technical and difficult to read, and the inclusion of too few references, which would make the reader rely solely on the judgment and expertise of the author. The authors have tried to follow a middle line but seem inclined towards the first option. The approach is that it is much more important to show the reader the field of knowledge no matter how difficult and fuzzy it might be, rather than simply provide him with the conclusion.

This is obviously a difficult task. Currently, knowledge is increasing rapidly, and the sources one needs to scan are growing exponentially. New knowledge emerges constantly concerning neuroanatomy, neurophysiology, neuropharmacology, neuropsychology, behavior psychology, psychosocial research, etc. Still many of these recent developments should be incorporated within the existing theoretical and practical knowledge in the respective fields. The time when theoretical knowledge and keeping in line with recent developments was considered to be a hobby or an extravagant intellectual exercise has long passed. In modern times, this is an important prerequisite for people working in the respective areas mentioned above.

The book follows the principle of an integrated approach. It looks from different angles at the complexity of the psychobiology of human behavior. This goal demands to include all kinds of knowledge, not only the evidence coming from the best designed empirical studies. A meaningful approach on the psychobiology of behavior cannot be restricted to the individual, but has to consider also the environment in all relevant aspects. Also conceptual issues should be discussed encompassing terms like psychological, behavioral, brain related, neurobiological, psychosomatic, etc. as well as the body-soul interaction. Finally, the intended better understanding of the psychobiology of human behavior shall help the mental help professionals and the doctors beyond this field to fulfil their mission which is primarily to help patients and their families.

Psychiatry/mental health have the unique characteristic that multiple perspectives could and should be taken into consideration in order to explain and understand the etiopathogenesis of mental disorders or in order to arrive at meaningful therapeutic tools. It is banal to repeat that humans are shaped by the interaction of their neurobiology with environmental events. What we need in the twenty-first century is to recognise the specific weight of different factors in the etiopathogenesis and the treatment tools. A vague bio-psycho-social model does no longer correspond to the state of the art of psychiatric/mental health research and practice. A better conceptualisation of mental illness is a pressing necessity today, and for this a deep understanding of the psychobiology of human behavior with all its dimensions, empirical results and theoretical conceptualisations is necessary.

Finally, this book deals with something which is indeed maybe the most challenging of all human desires: to understand human nature/human behavior itself. Whether such an understanding is possible is still a matter of philosophical debate; several philosophers argue that it is impossible for the human mind to fully understand itself since you need a higher level of functioning to understand a lower one. In this endeavor, the social and ideological and philosophical and religious philosophy itself is often embarrassing, one-sided and not always helpful for a deeper and meaningful understanding. For example, it was several centuries ago when Rene Descartes (1596–1650) proposed his mind-body dichotomy, better known later as the "Cartesian dualism". With this Descartes provided the philosophical rationale for natural sciences and expelled metaphysical and religious explanations from the study of nature and the physical universe (res extensa), including the human body, but as a compromise he gave the mind (res cogitans) to religious beliefs and metaphysical ideas and preserved the concept concerning the divinity and immortality of the soul. These ideas, although are scientifically obsolete today, are much alive in lay culture and in the belief systems of groups of individuals and society in general, irrespective of specific religious and ideological or political inclinations or affiliations. This resilience of the Cartesian dualism has done much damage both in the progress of psychiatry and in the way mental health and illness is perceived and accepted by the general public. Dualism can be traced at the roots of stigma and rejection as well as in the form of aggression towards patients, since according to such an understanding of the human nature, mental illness is a sign of moral weakness, sin or ill will. A modern and scientifically based psychobiology of human behavior might help to overcome some of these problems. Such a scientific psychobiology of human behavior provides valuable insights for all professionals in the field, but also for the people beyond.

More concretely, this book tries to cover in 13 chapters different aspects of the psychobiology of human behavior. The titles of these chapters are derived from different dimensions. Thus, we find chapters on more general/theoretical topics, for example, the "functional anatomy of the brain" (going much beyond that which is traditionally described as functional anatomy); on "basic vital functions and instincts"; on "genetics and behavior"; on "the contribution of sociology, ethology and other disciplines"; on "psychophysiology and psychosomatics"; on "psychobiology and psychoanalysis"; etc. The other topics that focus on special aspects of behavior alterations are, for example, "temperament-personality-character", "sexuality" "sleep and dreams", "substance abuse", "biological psychiatry and psychopharmacology" and "aging". All the chapters are well written and rich in the thoughtful presentation of empirical data as well as in considering the respective theoretical frameworks. One is impressed how much knowledge is available in the field of neurobiology of human behavior. On the other hand, one understands that in spite of all this knowledge, it is very difficult to condense all this into a unified complex theory. We see primarily facets of a theory of human behavior, but nevertheless this is already quite a lot, and we should be thankful to the authors for their excellent contributions.

To go a bit more in detail, I will briefly mention the first chapter, the one on functional anatomy, written by one of the editors. It offers much more than we would expect reading the short title. It starts with a general introduction and historical perspective, discussing issues like "brain" and "mind", historical findings of descriptions of the brain and brain-related therapeutic interventions, the humoral pathology of Hippocrates, the mind-body dichotomy proposed by Descartes, the phrenology of Gall and others and concepts of the modern brain anatomy. In the following section, a comprehensive description of the basic anatomy and physiology of the brain is presented, describing all the brain lobes and subcortical structures, the cellular structure, the neurophysiology as well the localisation and function of the neurotransmitters (the latter in much detail).

To mention another chapter as an example which describes another world, I select the chapter on ethology, evolutionary psychology, sociobiology and evolutionary psychiatry. Already the title underlines that this chapter deals with a very complex field, involving very different scientific approaches. First some definitions, short descriptions and historical remarks on each of these approaches are presented. Then a special focus is on evolutionary psychology, a very rich section encompassing among others environment of evolutionary adaptiveness, life history theory, evolution of emotion, mating and parenting, evolution of language, personality, etc. Furthermore, two sections describe evolutionary social psychology and sociobiology. Finally, a chapter on Darwinian (evolutionary) psychiatry follows. All this is described in many details, with a deep understanding, but also with a lot of critical reflections.

Altogether a very interesting book, which hopefully will find many readers!

1 May 2018

Hans-Jürgen Möller
Department of Psychiatry
Ludwig–Maximilians University
Munich, Germany

Contents

About the Editors

Kostas N. Fountoulakis, M.D., is Professor of Psychiatry at Aristotle University of Thessaloniki, AHEPA University Hospital, in Thessaloniki, Greece.

General Background

Dr. Fountoulakis received his medical degree (1989), performed his residency in psychiatry (1998) and earned his doctorate in psychiatry (1999) at the Aristotle University of Thessaloniki and served as Research Fellow in the Department of Psychiatry, Division of Neuropsychiatry, at the University of Geneva in Switzerland.

Research and Teaching Topics

Dr. Fountoulakis' areas of clinical and research interest are reflected in the topics that he teaches: general psychiatry, biological psychiatry, psychopharmacology, mood disorders, schizophrenia and personality disorders. He has co-authored more than 400 papers, and more than 250 of them are published in international journals such as the *The Lancet, The BMJ, The American Journal of Psychiatry, British Journal of Psychiatry, Biological Psychiatry, International Journal of Neuropsychopharmacology, Journal of Affective Disorders, Schizophrenia Research, Psychiatry Research, Bipolar Disorders* and the *Annals of General Psychiatry*, among others, with over 10,000 citations and $h = 50$ (Publish or Perish).

He has authored or co-authored a number of chapters in books. He has authored the book *Bipolar Disorders: An Evidence-Based Guide to Manic Depression* (Springer-Verlag 2015) and edited the WPA book *Advances in Psychiatry*.

Dr. Fountoulakis is Editor in Chief of *Annals of General Psychiatry* and has served as Section Editor of *Current Opinion in Psychiatry* as well as Guest Editor in other international journals.

Membership in Scientific Associations

He was the Chair at the ISNP and at times was Chair of several sections of the World Psychiatric Association. He served as Chair of the CINP Credentials and Membership Committee (2010–2012) and is an active Member of a number of national and international professional organisations and Peer Referee for the Cochrane Collaboration.

International Awards

He has received a number of national and international research awards, including the 2012 Kraepelin-Alzheimer Medal of the University of Munich and the 2015 Excellence in Education Award of the WFSBP.

Since 2014, he is Honorary Member of the WPA.

Ioannis Nimatoudis is Professor of Psychiatry at the Aristotle University of Thessaloniki, Greece, and Head of the 3rd Department of Psychiatry at the same university.

He received his medical degree from the Aristotle University of Thessaloniki in 1979 and completed his residency in Psychiatry–Neurology in 1984. Since 1990, he has been an active member in Psychiatric Reform and rehabilitation in Greece, and specially devoted to the "Leros Programme" of the Regulation 815/84 of the EU. During 1991–2003 he served as Scientific Coordinator of rehabilitation units in Thessaloniki under the auspices of the NGO "Society of Mental Health and Social Rehabilitation", in collaboration with the Ministry of Health and the EU.

He has supervised, as part of a three-member committee, nine Ph.D. researches, four of which were completed successfully, as part of a three-member committee. He is currently the main Supervisor in four Ph.D. researches that are in process. He is responsible for organising elective courses in Neuropsychiatry; he participates in education programmes of other medical universities in Greece and is responsible for organising educational workshops in international conferences. His main research interests lie in neuropsychology, psychopharmacology, psychopathology and evaluation of psychiatric services. He served as Member of the organising committee of 13 international and 13 national conferences. He is an active Member of 11 national and international scientific societies. He has participated in 83 Greek and international scientific conferences and in 58 round tables, lectures and seminars as invited speaker. He is Author or Coauthor of more than 250 papers presented in conferences or published in national and international scientific journals.

The Human Connectome: Functional Anatomy of the Brain

1

Kostas N. Fountoulakis, Ioannis Nimatoudis, and Xenia Gonda

1.1 General Introduction and Historical Perspective

Trying to describe the way the human brain functions is a great challenge. Trying to describe how the human mind functions is an even more difficult endeavor. There is a difference between the words "brain" and "mind." The word "brain" is of Germanic origin and probably is analogous to the Greek word *εγκέφαλος* (enkephalos, meaning what is inside the skull). On the other hand, the respected word in the Latin is cerebrum or cerebro which come from the Proto-Indo-European "keres" or "ceres." This root is related to a number of words with different meanings but some common line of concept in their development, including "top of the head," daughter (in Greek *κόρη*), female goddess, cereal, "soft tender and grayish-like cooked cereal," etc. The etymology of the word "mind" comes from the Latin word "mens" which is related to the Greek words *μένος* (menos) and *μήνις* (minis) which both mean anger. It is also related to the verb *μένω* (meno = stay) and thus reflects a condition in which the state of the mind is stalled because of intrusion of anger and intense emotions. Interestingly, in ancient Greek the word *μήνις* also means crescent and thus relates the states of the mind with the phases of the moon. The Greek word for month

K. N. Fountoulakis (✉) · I. Nimatoudis
3rd Department of Psychiatry, Faculty of Medicine, School of Health Sciences, Aristotle University of Thessaloniki, Thessaloniki, Greece

X. Gonda
Department of Psychiatry and Psychotherapy, Semmelweis University, Budapest, Hungary

MTA-SE Neuropsychopharmacology and Neurochemistry Research Group of the Hungarian Academy of Sciences and Semmelweis University, Budapest, Hungary

NAP-A-SE New Antidepressant Target Research Group, Semmelweis University, Budapest, Hungary

K. N. Fountoulakis, I. Nimatoudis (eds.), *Psychobiology of Behaviour*,
https://doi.org/10.1007/978-3-030-18323-3_1

(which corresponds to moon phases) is *μήνας* (minas), while the Latin word is mensis. This group includes also the words "menses" and "menstrual," and thus they possess both a temporal/cyclical and an emotional meaning. The word *μαινάς* (= maenad) used for women who participated in the orgiastic rites of God Dionysus and *μηνίσκος* (meniscus of the knee) have a similar etymology. The equivalent Greek word for mind is *νους* which is of unknown etymology but probably is related to the Latin word.

Essentially the brain is the anatomical formation, the organ in which the processes that constitute and give birth to the mind take place.

1.2 Basic Anatomy and Physiology of the Brain

1.2.1 Gross Morphology

At the macroscopic level, the central nervous system includes the spinal cord, the medulla oblongata (responsible for vital autonomic functions), the pons, the cerebellum (which is responsible for motor function), the midbrain (coordinates visual and auditory reflexes and controls sensory and motor functions), the diencephalon (includes the thalamus and the hypothalamus), and the brain hemispheres. Another description includes three parts of the human brain: the hindbrain (medulla, pons, and cerebellum), the midbrain, and the forebrain (thalamus, hypothalamus, and hemispheres). The midbrain and hindbrain are also called the "brain stem" (Fig. 1.1).

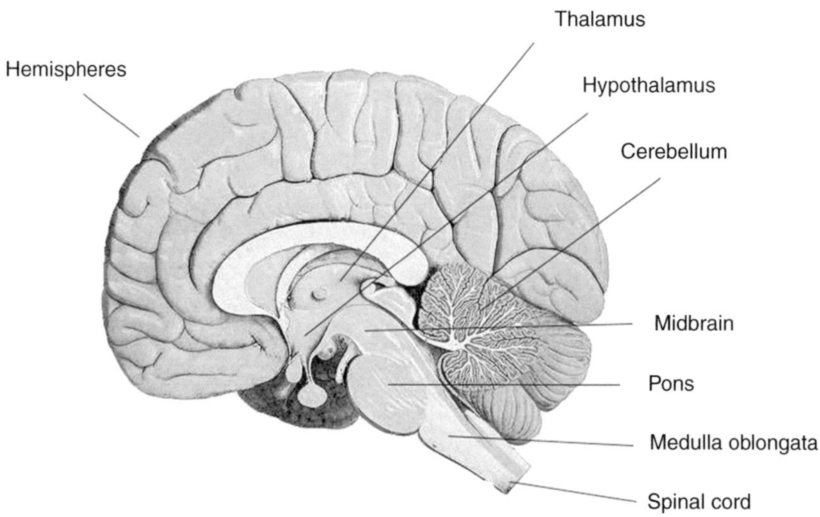

Fig. 1.1 The human brain and its parts

The connections between the brain areas and structures are many and complex but not infinite. Some are but others are not reciprocal, some are uni- and other bi-directional, some are convergent (gather information from many sources to a single point), while others are divergent (distribute information from a single source to many recipients), some are hierarchical (they transfer the information from a to b then to c followed by d), while others are parallel (they transfer information from a1 to b1 and then to c1, and the same information travels from a2 to b2 and then to c2).

There is also a relative differentiation and specialization of specific brain areas to carry out a specific function. The spinal cord receives bodily sensory information and contains motor neurons responsible for both voluntary and reflex movements of the trunk and limbs. There are 31 spinal nerves serving these functions. The brain stem includes functions pertaining to sensory (e.g., hearing and taste) and motor (e.g., balance) systems as well as autonomic function (cardiac output, blood pressure, eye pupil reflex, etc.). There are 12 cranial nerves functionally analogous to the spinal nerves responsible for the sensory information and movement of the head, neck, and face. In the brain stem, there is the reticular formation which influences the arousal level. The medulla participates in the regulation of blood pressure and respiration, and it is also involved in taste, hearing, and maintenance of balance as well as the control of neck and facial muscles. The pontine nuclei relay information about movement and sensation from the cerebral cortex to the cerebellum, and also they are involved in respiration, taste, and sleep. The midbrain provides important linkages between the cerebellum, the basal ganglia, and the cerebral hemispheres. The cerebellum is the part of the brain which contains a far greater number of neurons and synapses than any other part. It receives somatosensory and motor information, and it is important for maintaining posture, coordinating head and eye movements, and fine-tuning the movements of the muscle. It is important in learning motor skills but also in language and other neurocognitive functions. The diencephalon contains the thalamus and hypothalamus. The thalamus determines whether sensory information will reach conscious awareness in the neocortex. It participates in the integration of motor information from the cerebellum and the basal ganglia and subsequently transmits the processed result in the cortex. The hypothalamus regulates several bodily vital behaviors that are essential for homeostasis and reproduction including eating, drinking, maternal behavior, and circadian rhythms.

The cerebral hemispheres constitute the largest region of the human brain and are concerned with perceptual, motor, and higher cognitive functions, including memory, emotion, and decision-making (Ungerleider 1995). They include the cerebral cortex (2–4 mm thick), the underlying white matter, the basal ganglia (responsible for the control of fine movement), the amygdala (where mood is generated), and the hippocampus (responsible for memory). Each hemisphere is divided into four anatomically distinct lobes (frontal, parietal, temporal, and occipital) each hosting different functions. Their surface is full of infoldings whose crests are called gyri, while the grooves are called sulci. These infoldings serve the purpose to increase the surface of the brain (and the gray matter) while keeping the overall

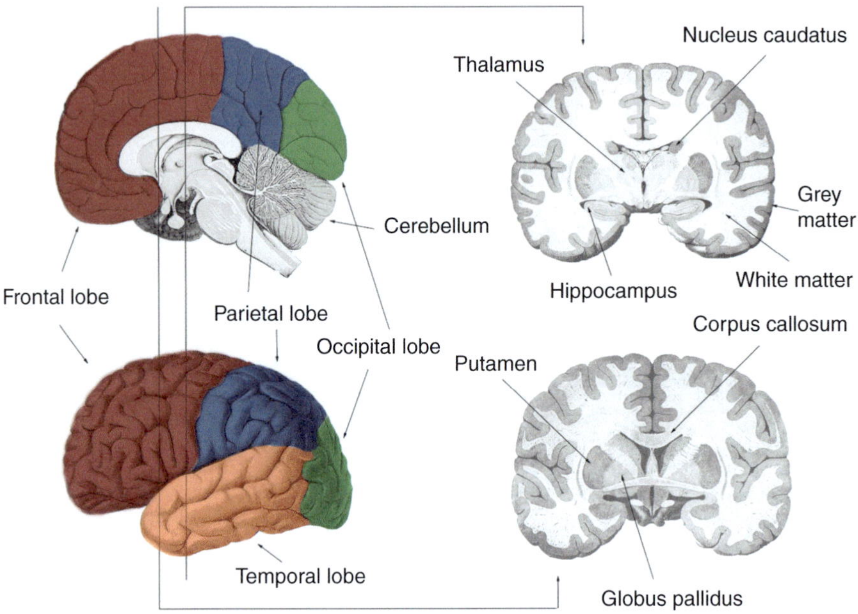

Fig. 1.2 Brain lobes and subcortical structures

volume as small as possible. The major gyri and sulci are stable across persons, and they hold specific names (Fig. 1.2).

An important characteristic of the functioning of the hemispheres is that each hemisphere function is primarily related to sensory and motor function of the opposite (contralateral) half of the body, and hemispheres although they are similar in appearance are not similar or symmetrical in function (Kandel 2000).

The two hemispheres are interconnected by the corpus callosum, which is essentially a thick set of fibers (the largest of the commissures) that connect symmetrical and similar regions of the left and right sides of the brain.

1.2.2 Cellular Structure

The basic unit of the brain is the nerve cell. The brain includes an extraordinary number of nerve cells at the order of 100 billion (10^{11}). These cells are mainly divided into two broad categories: the nerve cells (neurons) and the glial cells (glia). Glial cells outnumber neurons 10–50:1, and their name comes from the Greek word *γλοια* (meaning glue).

They are further divided into microglia and macroglia. Microglia are phagocytes mobilized in case of injury and infection of certain diseases. Macroglia include oligodendrocytes which are located in the brain, Schwann cells which are located in the peripheral nervous system, and astrocytes. Oligodendrocytes provide the brain

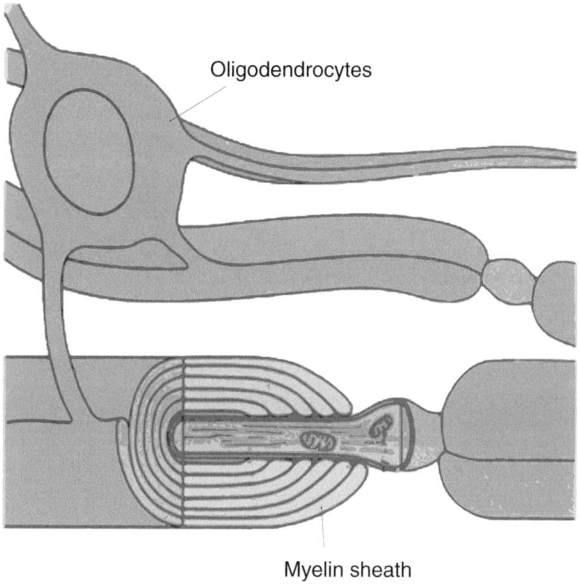

Fig. 1.3 Oligodendrocytes and the myelin sheath (from Bunge 1968)

with structure and support and separate and also insulate neurons with a myelin sheath by wrapping their processes around the axons (Fig. 1.3) (Bunge 1968). Some of glia cells (astrocytes) place end feet on the walls of blood vessels where they cause the formation of tight junctions between endothelial cells, thus creating the blood-brain barrier which isolates blood vessels from the brain and prevents various substances to enter the brain.

The nerve cells share more or less a similar structure throughout the brain.

They include four morphologically defined regions (Fig. 1.4):

(a) The cell body, which contains the nucleus (with the genes and the DNA), the endoplasmic reticulum where proteins are synthesized, and the mitochondria which constitute the power stations of the cell, the places where energy is produced and manipulated.
(b) The dendrites are short processes which arise from the cell body and serve the connection with neighboring cells.
(c) The axon is a single and very long dendrite which extends far away from the cell body. It is the main conducting unit and serves as communication with distant neurons. It can carry signals up to 3 m away. The axons are wrapped with a myelin sheath stemming from oligodendrocytes. This sheath isolates them from one another. In this way the electrical phenomena occurring in one axon do not spread to neighboring axons and cells.

The cerebral cortex is organized into six cell layers (Felleman and Van Essen 1991). Layer I (molecular layer) is an acellular layer occupied by dendrites and axons of the cells located deeper in the cortex. Layer II (external granule cell layer)

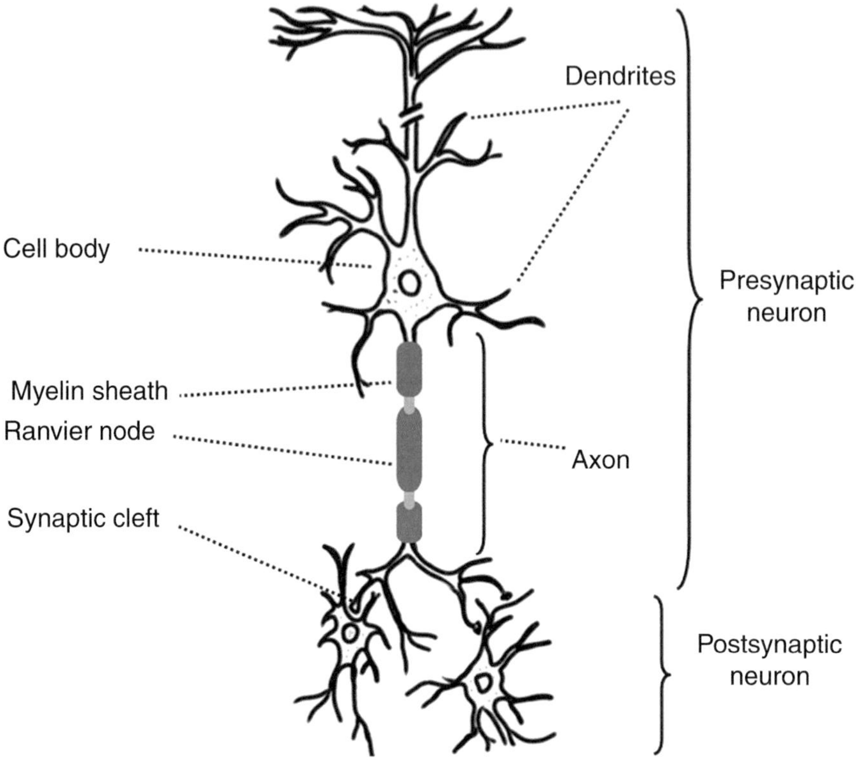

Fig. 1.4 Presynaptic neuron and its components

is comprised of small spherical cells called granule cells. Layer III (external pyramidal cell layer) contains a variety of cell types, many of which are pyramidally shaped; layer IV (internal granule cell layer), like layer II, is made up primarily of granule cells, layer V (internal pyramidal cell layer) contains mainly pyramidally shaped cells that are typically larger than those in layer III, and layer VI (polymorphic or multiform layer) is a heterogeneous layer of neurons.

The projection neurons typically have pyramidally shaped cell bodies and are located mainly in layers III, V, and VI with glutamate being their primary neurotransmitter, while local interneurons are located in all layers and use GABA as their primary neurotransmitter.

1.2.3 Neuronal Action Potential

The communication among neurons is achieved through electrical signals, called "action potentials." These potentials constitute an "all-or-none" phenomenon which does not decay and travels to distant targets and is periodically regenerated.

At rest all neurons maintain a difference in the electrical potential on either side of the cell membrane, which is about −65 mV on average and is called "resting potential." This potential depends on the unequal distribution of ions and particularly of Na^+ and K^+, as well as on the selective permeability of the cell membrane to K^+ alone. This potential is maintained by the Na^+-K^+ pump which pumps Na^+ out of the cell and K^+ in. When the resting potential increases, the phenomenon is called hyperpolarization, and it is considered to be inhibitory since the neuron's ability to produce an action potential is reduced. On the contrary, when the resting potential decreases, then the phenomenon is called depolarization, and since it enhances the ability of the neuron to produce an action potential, it is considered to be excitatory.

The action potentials are initiated at the "hillock" which is the specific region of the cell body where the axon originates and where ion channels are denser. While the action potential is a long-distance signal, the neuron produces also a number of local signals in the form of receptor or synaptic potentials, which typically spread only for a few millimeters. While action potential is an "all-or-none" phenomenon, the local potentials could be graded, based on the amount of chemical neurotransmitter released and the duration these molecules are active in the synaptic cleft. On the contrary, the action potential carries information only in terms of frequency of potential generation and duration.

However, while the amplitude of the action potential is an "all-or-none" phenomenon, the duration is not, and this duration of the action potential determines the amount of Ca^{2+} that flows into the terminal, with more Ca^{2+} flowing into the cell after prolonged action potentials (Smith and Augustine 1988). In this frame, the steady-state Ca^{2+} influx is enhanced by depolarization and decreased by hyperpolarization, and vice versa the action potential is modulated by voltage-sensitive Ca^{2+} channels (Baker et al. 1971).

Intense activity alters synaptic effectiveness since a high-frequency sequence of action potentials is followed by a period during which action potentials produce successively larger postsynaptic potentials. The very high frequency in the production of action potentials (even up to 1000 s^{-1} in some neurons) is called "tetanic stimulation" and leads to a significant increase in the size of the potentials in the postsynaptic neuron, a phenomenon which is called "potentiation." Often this increase persists even after the resolution of the tetanic stimulation, and this perseveration is called "post-tetanic potentiation" and usually lasts several minutes, but it can persist for more than an hour (Erulkar and Rahamimoff 1978).

On the contrary, when one neuron causes a hyperpolarization in the cell body or the dendrites of another, in this way it decreases the likelihood that the postsynaptic cell will fire. This is called "postsynaptic inhibition." When through a contact on the axon terminal of another cell, it reduces the amount of the released transmitter by the second cell (presynaptic neuron) with a third cell as a target (postsynaptic neuron); this action is called "presynaptic inhibition," while when it increases the amount of the neurotransmitter, it is called "presynaptic facilitation." In sensitized neurons, a high-frequency sequence of action potentials could result in tetanic stimulation. The increase in size of the postsynaptic potentials during tetanic stimulation is called potentiation. The increase that persists after tetanic stimulation is called post-tetanic potentiation (Erulkar and Rahamimoff 1978).

One very interesting feature is that presynaptic neurons keep records of their activity in the form of residual Ca^{2+} in their terminals. This eventually leads to the strengthening of the presynaptic connection that persists for many minutes. Thus, post-tetanic potentiation is followed by an even longer-lasting process, which also depends on Ca^{2+} influx and is called "long-term potentiation." This could last from hours to days (Kandel 1981). Similar mechanisms which also involve the free Ca^{2+} concentration in the presynaptic terminal govern the plastic capabilities of the neuron (Dunlap et al. 1995; Klein et al. 1980).

Essentially the calcium ions (Ca^{2+}) are intracellular messengers which translate the electrical activity of depolarization into all the non-electrical activities inside the neuron. Thus these ions serve the output of the system, and any delay in the initiation of these non-electrical activities reflects the time it takes for Ca^{2+} channels to open and for Ca^{2+} to trigger the process.

1.2.4 Neurotransmitters

The term "neurotransmitter" refers to a group of substances with diverse chemical structure but with the common function to serve the synaptic transmission. The term was introduced for the first time by Henry Dale (1875–1968) and George Barger (1878–1939), ironically in an article attacking the concept (Barger and Dale 1910). At that time and for many decades to come, the electrophysiology was considered to be the dominant function of nerve cells and the brain. However, notions of a receptive substance were first outlined by John Newport Langley (1852–1925) in 1878 (Maehle 2004). These ideas were picked up by Alexander Crum Brown (1838–1922) and Sir Thomas Richard Fraser (1841–1920) in Edinburgh, who drew attention to the differences between L-hyoscine and D-hyoscine, the isomers of atropine. However an even more important development came from an unexpected field: The Nobel Laureate Paul Ehrlich (1854–1915) used a variety of dyes to stain different bacteria and tissues. Out of this work came the notion that agents related to the dyes might chemically bind selectively and specifically, and thus Ehrlich conceived the notion of the "magic bullet." This was to become the dominant therapeutic metaphor for the second half of the twentieth century, and it gave early currency to the idea of a receptor. In the same line of thinking, another Nobel Laureate, Hermann Emil Louis Fischer (1852–1919), created the metaphor of the "lock and key" concerning the reaction between messenger substances and receptors, in 1890 (Lemieux and Spohr 1994).

However, the concept that receptors were behind neurotransmission and the effects of medication on the human body was still a matter of debate. At the root of this debate was a resistance to accept a materialistic view of the human body since electrophysiology was leaving space for spirituality, but this was not the case with chemistry. There are authors suggesting that the biological aspect of psychiatry was finally been accepted in part because of the "spiritual" and cultural effects of psychedelic drugs (Gerard 1949, 1955a, b; Gordon 1948).

The concept became popular in the 1930s, after the Nobel Laureate Otto Loewi (1873–1961) demonstrated the release of acetylcholine from vagus terminals in the frog heart and Henry Dale published his work on cholinergic and adrenergic transmission.

The essential function and characteristic of a neurotransmitter is that it is released in the synaptic cleft by the presynaptic neuron and affects the condition of the postsynaptic neuron. In contrast to hormones, neurotransmitters are never released in the bloodstream (to act as neurotransmitters), but instead their target neuron is in contact or very close to the secreting neuron. While their direct effects are short-lived and last from milliseconds to minutes, they might result in long-term changes in the postsynaptic neuron.

Although an important characteristic of neurotransmitters is that their effects are transient, lasting from milliseconds to minutes, neurotransmitter action can result in long-term changes in target cells lasting hours or days.

The modern definition of a neurotransmitter demands the following four criteria to be met:

1. It is synthesized in the presynaptic neuron.
2. It is located in the presynaptic terminal and is released in amounts sufficient to exert a specific action on the postsynaptic neuron.
3. When administered exogenously it mimics the action of the endogenously released transmitter exactly.
4. A specific mechanism exists for removing it from its site of action.

All neurotransmitters are synthesized in the cell body and are subsequently stored in areas called synaptic vesicles, which are located in the presynaptic terminal. Each vesicle contains several thousand molecules of a single specific neurotransmitter (Kelly 1993). The synaptic vesicles tend to cluster at regions of the membrane which are specialized for releasing the neurotransmitter, and they are called active zones (Unwin and Zampighi 1980).

The release of the neurotransmitter into the synaptic cleft is followed by a complex cascade of events, but the principal role is played by Ca^{2+}, which during the discharge of a presynaptic action potential enters the presynaptic terminal through voltage-gated Ca^{2+} channels located at the active zone. This abrupt rise in the concentration of intracellular Ca^{2+} causes the vesicles to fuse with the cell membrane (Bahler and Greengard 1987; Hanson et al. 1997). In this way a process called "exocytosis" permits the release of the neurotransmitter molecules into the synaptic cleft (Lawson et al. 1977). After exocytosis, the vesicle membrane which had been fused with the cellular membrane is retrieved rapidly and is recycled so that its material is used to generate new synaptic vesicles (Schweizer et al. 1995; von Gersdorff and Matthews 1994).

After their release into the synaptic cleft, the neurotransmitter molecules bind to their specific receptors located on the membrane of the postsynaptic neuron. This activates the receptors and causes a cascade of procedures in the postsynaptic neuron. All these steps take time, varying from 0.3 ms to several milliseconds or longer, and this is why chemical synapses are not as fast as electrical synapses. However, chemical transmission has the important property of amplification, since a small presynaptic electrical current can result in the release of thousands of neurotransmitter molecules in the cleft and in turn to the activation of a large number of postsynaptic receptors with a profound effect even on a large postsynaptic neuron. Also, the action of a neurotransmitter in the postsynaptic neuron depends on the properties of the postsynaptic receptors that recognize and bind the specific neurotransmitter rather than on the chemical properties of the transmitter itself.

It is important to note that the storage and subsequently the release of neurotransmitters are done in standardized and fixed amounts (quanta) which are specific for each neurotransmitter and correspond to several thousand molecules. Following this, each vesicle affords to store only one quantum of the specific neurotransmitter and releases its entire contents into the synaptic cleft during exocytosis (Lindau and Almers 1995; Matthews 1996). The only exception identified so far to the quantal mode of neurotransmitter release is the retina. Each action potential in the brain releases only 1–10 quanta in comparison to an average of 150 quanta released at the neuromuscular synapse. Each quantum of transmitter produces a postsynaptic potential of fixed size, called the "quantal synaptic potential" (Liley 1956).

The neurotransmitters can be roughly classified into two big groups:

1. *Small-molecule neurotransmitters or biogenic amines*. This second term is rather imprecise in chemical terms but is being used for decades to label a group of neurotransmitters, including the catecholamines (dopamine, epinephrine, and norepinephrine) and serotonin but often also histamine which is chemically remote from both the catecholamines and the indolamines.

   The catecholamines are synthesized from the essential amino acid tyrosine in a common biosynthetic pathway. During the first step of this pathway, tyrosine hydroxylase converts tyrosine to L-dihydroxy-phenylalanine (L-DOPA). At the next step, L-DOPA is decarboxylated by a decarboxylase giving dopamine and CO_2. Then at the third step, dopamine β-hydroxylase converts dopamine to norepinephrine.

   Serotonin (5-hydroxytryptamine or 5-HT) is synthesized from the essential amino acid tryptophan. They both are indoles with an aromatic structure. Histamine has been identified as important during inflammatory reactions in the body. It is synthesized from histidine by decarboxylation.
2. *Excitatory amino acids and neuroactive peptides*.

   All types of neurotransmitters and neuroactive molecules can coexist in the same neuron. The combination usually consists of one of the small-molecule transmitters and one or more peptides derived from one kind of polyprotein (Kupfermann 1991).

1.2.4.1 Small-Molecule Neurotransmitters

Acetylcholine

Acetylcholine (Ach) is not an amino acid or its direct derivative, but still it is the only low-molecular-weight amine substance considered to be a neurotransmitter. It is synthesized from choline (which derives from diet) in a reaction which includes a single step, catalyzed by choline acetyltransferase with acetyl-coenzyme A (acetyl CoA) as co-substrate. Acetylcholine is catabolized by acetylcholinesterase (AchE) and butyrylcholinesterase (BCHE).

Ach is widely distributed in the mammalian brain, and it is produced in the cell bodies of neurons located mainly in the Meynert nucleus (nucleus basalis magnocellularis). According to the literature (Mesulam 1994; Mesulam et al. 1983), there are eight regions where the cell bodies of cholinergic neurons are located (Ch1–Ch8). The Ch1 group concerns those cholinergic neurons located in the medial septal nucleus (Fujishiro et al. 2006), the Ch2 those in the nucleus of the vertical limb of the diagonal band (Fujishiro et al. 2006), the Ch3 in the lateral portion of the horizontal limb nucleus of the diagonal band (Mesulam et al. 1983), the Ch4 in the Meynert nucleus (Liu et al. 2015), the Ch5 in the pars compacta and pars dissipata of the pedunculopontine nucleus (Manaye et al. 1999), the Ch6 centered around the laterodorsal tegmental nucleus and spreads into the central gray and medial longitudinal fasciculus (Mesulam et al. 1989), the Ch7 in the medial habenula (Dautan et al. 2016), and the Ch8 in the parabigeminal nucleus (Mufson et al. 1986). The Ch1 and Ch2 groups provide the main cholinergic innervation to the hippocampus, the Ch3 to the olfactory bulb, the Ch4 to the hemisphere cortex and the amygdala, the Ch5 and Ch6 to the thalamus, the Ch7 to the interpeduncular nucleus, and the Ch8 to the superior colliculus. There are also projections of lesser importance from the Ch1–Ch4 and Ch8 to the thalamus and from the Ch5–Ch6 to the cortex (Geula and Mesulam 2011).

The main function of cholinergic system is to turn non-cholinergic neurons more sensitive to other excitatory stimuli. There is a bulk of pharmacological data suggesting that the central cholinergic system plays an important role in the acquisition as well as the expression of learned behaviors since it seems that scopolamine and atropine exert a negative effect on this kind of behaviors (Reiner and Fibiger 1994).

Serotonin

Serotonin (5-hydroxytryptamine, 5-HT) (Marsden 1996) is an indoleamine which can be found not only in humans but also in most animals and even in plants. In mammals it can be found in the platelets, the mast cells, and the chromaffin cells of the intestine. In the brain, 5-HT acts as a neurotransmitter and regulates a number of functions including sleep, food intake, thermoregulation, emotions, and psychotic experiences. However its main role is to inhibit behaviors and to reduce impulsivity. Additionally, it seems that the mission of the serotonergic system of the forebrain is to attenuate the impact on human behavior stressful and adverse life events have. In other words its mission is prevention and adjustment to stress (Deakin 2013; Deakin and Graeff 1991).

Serotonin is synthesized in the raphe nucleus of the brain stem. It is divided in the dorsal raphe nucleus (DRN) and median raphe nucleus (MRN). The cells in the DRN project into the same regions of the brain in which the dopaminergic neurons of the ventral tegmental area (VTA) project and especially to the basal ganglia, the amygdala, and the nucleus accumbens. They are implicated mainly in the regulation of locomotor and incentive-motivation activity. On the contrary, the cells of the MRN project to the cortex, the thalamus, and the hippocampus, and they contribute to the process of sensory stimuli. Thus, the first group of neurons is related to the motor part of the response mechanisms and acts mainly through the 5-HT_2, 5-HT_{2A}, and 5-HT_{2D} receptors, while the second group is related to the sensory part and acts possibly through the 5-HT_{1A} receptors (Blue et al. 1988; Mamounas et al. 1991). It is widely accepted that these pathways also regulate the approach-avoidance behaviors (Graeff et al. 1993; Hodges et al. 1987; LeDoux et al. 1990). Particularly high concentrations of 5-HT can be found in the limbic system and imply its relationship with emotions. However all the projections of serotonergic neurons, which are abundant and concern all brain areas, stem from the dorsal and median raphe nuclei. The median provides with thick fibers while the dorsal with more thin, which are probably more sensitive to toxic agents.

Norepinephrine

Norepinephrine is the main catecholamine in the brain and one of the most important neurotransmitters. Its name comes from the Greek name of the adrenal glands which is *επινεφρίδια* (epinefridia, meaning "on top of the kidney"). It is synthesized via the catecholamine synthesis pathway from phenylalanine and tyrosine.

It is abundant in the brain stem and the dorsal nucleus of vagus nerve and plays an important role in the regulation of blood pressure. However the main norepinephrine pathway originates from the cells of the locus coeruleus and projects to the thalamus, the dorsal hypothalamus, the hippocampus, and the cortex. The ventral norepinephric bundle emerges ventral/rostral to the locus coeruleus and reaches the hypothalamus and the subcortical parts of the limbic system. The dorsal norepinephric bundle emerges from the locus coeruleus and projects to the cortex. Both bundles seem to relate with volition, drive, and motivation as well as with reward mechanisms and REM sleep.

Dopamine

The dopaminergic system is significantly more complex in comparison to the norepinephric. Below its classical structure is described, which is considered to be important for educational purposes, although it is currently disputed due to the accumulation of more data. Traditionally, four main dopaminergic pathways are mentioned:

- The nigrostriatal pathway whose fibers originate from nucleus A9 and project to the nucleus caudatus and the putamen of the lentiform nucleus.
- The mesolimbic pathway whose fibers originate from the ventral tegmental area (VTA) and project to the amygdaloid nucleus, the dorsolateral septal nucleus, and the nucleus accumbens. The experience of stress activates this pathway.
- The mesocortical pathway whose fibers originate from the VTA and project to the frontal cortex and areas of the septum and the hippocampus. Again the experience of stress by the person activates this pathway.

- The tuberoinfundibular pathway (A12 region) (Moore and Lookingland 1994), whose fibers originate from the arcuate nucleus in the hypothalamus and project to the median eminence. Essentially they constitute a small minority of dopaminergic neurons; however they release dopamine in the stream of the hypophyseal portal veins, which eventually reaches the anterior pituitary gland where it inhibits the secretion of prolactin. Experiencing stress by the individual could lead to the inhibition of this pathway resulting in an increase in prolactin secretion and subsequently in sexual disorders in males and abnormal lactation and amenorrhea in females.

The dopaminergic neurons of mesolimbic and mesocortical pathways receive regulatory influence from serotonergic and norepinephric neurons located in the brain stem. Some VTA dopaminergic neurons are especially sensitive to input from excitatory amino acids.

The overall conclusion from the study of the dopaminergic system suggests that it does not serve any specific function, but instead it regulates and permits the synthesis of the functions which are hosted in the structures where this system projects (Le Moal 1994). Probably one of the rare but most important specific dopaminergic effects is a dopaminergic involvement in reward systems (dopaminergic theory of reward) (Jacques 1979; Wise 2008).

It is important to note that although low dopaminergic activity seems to relate to low ability to concentrate and reduced performance, also excessive increase is related to deficits in attention and stereotypical behaviors. It has been reported that intrapsychic conflict as a result of ambiguous situations with unclear solution leads to an increase in dopaminergic activity, and in this way they trigger displacement activities which reduce the level of arousal and stabilize the system (Tazi et al. 1986, 1988).

Histamine

Histamine is derived from the decarboxylation of the amino acid histidine by L-histidine decarboxylase. In the brain it is catabolized by histamine-*N*-methyltransferase. The cell bodies of histamine neurons are located in the tuberomammillary nuclei of the thalamus, and they project throughout the brain, including to the cortex, through the medial forebrain bundle. Interestingly, they have a firing pattern which is closely related to wakefulness. During wakefulness the fire frequency is high, while during periods of relaxation or tiredness, the frequency is significantly lower. It seems they completely stop firing during REM and non-REM sleep. When antihistamines (i.e., older-type H1 receptor antagonists) cross the blood-brain barrier, they produce drowsiness. Similarly, destruction of histamine-releasing neurons or inhibition of histamine synthesis leads to an inability to maintain vigilance. On the contrary, H3 receptor antagonists increase wakefulness. Therefore it is believed that histamine increases wakefulness and prevents sleep (Brown et al. 2001). It is also involved in local immune responses and in the function of the gut. During immune responses, histamine is secreted by basophils and mast cells. It increases the permeability of the capillaries to white blood cells and some proteins.

Additionally to its arousing and stimulatory effects, histaminergic activity also protects against the susceptibility to convulsion, drug sensitization, denervation supersensitivity, ischemic lesions, and stress, probably through a suppressive type of effect (Yanai and Tashiro 2007). It has also been suggested that histamine controls the mechanisms by which memories and learning are forgotten (Alvarez 2009).

1.2.4.2 Excitatory Amino Acids and Neuroactive Peptides

Glutamate (Glu)

Glu is the most important and frequently used excitatory amino acid throughout the brain. It is also a major part of a wide variety of proteins, and therefore it is one of the most abundant amino acids in the human body (Meldrum 2000). Normally it is obtained from the diet and there is no need to be synthesized. However it can also be synthesized from alpha-ketoglutaric acid (part of the citric acid cycle which has citrate as starting point). Glu is actively transported through the blood-brain barrier (Shigeri et al. 2004; Vandenberg 1998). As a neurotransmitter it is dominant throughout the cortex and in most subcortical pathways (Stahl 2008). It plays a role in the mechanisms concerning brain plasticity and higher neurocognitive function. It could be said that the most important function of all other neurotransmitters is to regulate Glu activity.

Gamma-Aminobutyric Acid (GABA)

GABA is produced from Glu via the action of the enzyme glutamic acid decarboxylase. It is present at high concentrations throughout the brain and can also be found in other tissues (e.g., the pancreas and the adrenal gland). In the brain it is the major inhibitory neurotransmitter (Paul 1994).

Glycine (Gly)

Gly is the major transmitter in inhibitory interneurons of the spinal cord. It is synthesized from serine, and it is the smallest of the 20 amino acids commonly found in proteins. It bounds to ionotropic receptors, and it is a required co-agonist along with glutamate for NMDA receptors.

Neuroactive Peptides

More than 50 short peptides are active as messengers in the brain. Some of them had been previously identified as hormones (e.g., angiotensin and gastrin) or as products of neuroendocrine secretion (e.g., oxytocin, vasopressin, somatostatin, and thyrotropin-releasing hormone). (A decade of neuropeptides: past, present, and future. Tenth Annual Winter Neuropeptide Conference. Breckenridge, Colorado, January 16–20, 1989.) Most of them play an important role in mechanisms including thermoregulation, food and liquid intake, memory, learning, and reaction to stress and pain.

Most of them do not satisfy all criteria to be considered a neurotransmitter, but several do (Myers 1994). Neurons could contain and release several peptides which come from the same polyprotein, and their release leads to potentially different postsynaptic actions (Fisher et al. 1988).

ATP and its degradation products (e.g., adenosine) act as transmitters at some synapses. Adenine and guanine and their derivatives are called purines.

1.2.5 Neurotransmitter Reuptake

The neurotransmitter effect in the synaptic cleft lasts only for short periods of time, and after this the neurotransmitter molecules should be removed in time. Failure to do so could result in "noise" or even in total inhibition of new signal transmission, in the diffusion to neighboring synapsis resulting in signal contamination, while the signal itself these molecules carry will tend to degrade. Therefore the timely removal of transmitters from the synaptic cleft is critical for the quality of synaptic transmission. Transmitter molecules are removed from the cleft by three mechanisms:

- Diffusion.
- Enzymatic degradation.
- Reuptake.

While diffusion removes some fraction of all chemical messengers, it cannot remove the complete quantity. Enzymatic degradation of transmitter molecules is used primarily by the cholinergic synapses.

Neuropeptides are removed more slowly than small-molecule transmitters, and this probably contributes to their more lasting effects.

The most important mechanism of removal is reuptake, and it is important also because it "salvages" the neurotransmitter and its fragments for recycling. It is mediated by transporter molecules which are classically proteins which use transmembrane ion gradients and electrical potential to transport neurotransmitter across the membrane. There are several transporters for each transmitter (Nelson and Lill 1994; Amara and Arriza 1993).

Fast reuptake and fast recycling of secreted neurotransmitters contribute also to a fast overall information flow and quick reaction to environmental stimuli.

1.2.6 The Synapses

The point where two neurons communicate is called the synapse (Jessell and Kandel 1993). The word "synapse" comes from the Greek verb *συνάπτω* (bring together, join), and it was introduced in 1897 by Michael Foster (1836–1907) at the suggestion of Arthur Woollgar Verrall (1851–1912) which was an English classical scholar (Tansey 1997) (Fig. 1.5).

The neuron which transmits the signal is called "presynaptic," and the neuron which receives it is "postsynaptic." The physiologists, led by John Eccles (1903–1997), argued that all synaptic transmissions are electrical and that the action potential in the presynaptic neuron generates a current that flows passively into the postsynaptic cell. The pharmacologists, led by Henry Dale (1875–1968), argued

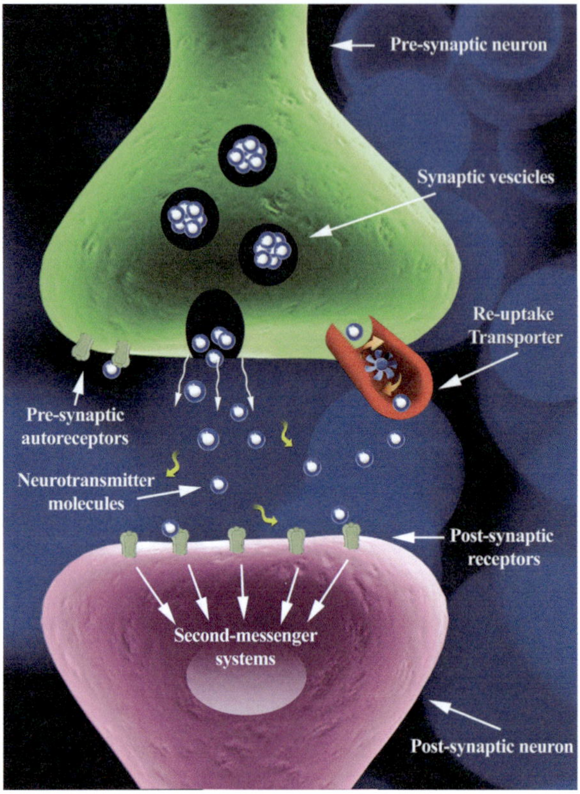

Fig. 1.5 The synapse and its components

that transmission is chemical and that the action potential in the presynaptic neuron leads to the release of a chemical substance that in turn initiates current flow in the postsynaptic cell. When physiological techniques improved in the 1950s and 1960s, it became clear that both forms of transmission exist. Although most synapses use a chemical transmitter, some operate purely by electrical means (Eccles 1976).

So today we know that there are two kinds of synapses:

- Chemical synapses, where the electrical activity in the presynaptic neuron is converted into the release of a neurotransmitter.
- Electrical synapses, where the presynaptic and postsynaptic cell membranes are connected by channels called gap junctions or are in touch through a synaptic cleft that is capable of passing electric current.

It is to be noted that synaptic communication is distinct from ephaptic coupling (from Greek *εφάπτομαι* meaning being in touch), which refers to the exchange of ions between the cells or to the effect of local electric fields. In this way it influences the synchronization and timing of the firing of action potentials in neighboring neurons. It is exactly the phenomenon of ephaptic interactions that myelination prohibits (Arvanitaki 1942).

In chemical synapses, the two neurons are not in physical contact; instead there is an empty space between them, called the "synaptic cleft." The end of the presynaptic neuron is called "presynaptic terminal," and it can end almost anywhere on the postsynaptic neuron but preferentially on dendrites.

In the brain, electrical synaptic transmission is very rapid and rather stereotyped, and it is triggered explosively in an all-or-none manner. In contrast, chemical synapses are capable of more variable signaling and thus can produce more complex behaviors. They can mediate either excitatory or inhibitory actions in postsynaptic cells and produce electrical changes in the postsynaptic cell that last from milliseconds to many minutes. This comes at the price of longer latency, that is, longer time between the presynaptic spike and the postsynaptic potential, because chemical transmission demands several biochemical steps. On the contrary, at electrical synapses the gap-junction channels connect the cytoplasm of the two cells, and in this way they allow a direct ion flow between them. One important characteristic of the communication between neurons is that its strength can be enhanced or diminished by cellular activity, a phenomenon which is called plasticity, and it is crucial to memory and other higher brain functions.

One important principle in the function of neurons and synapses is that they transmit a signal always in one direction (principle of dynamic polarization) from the receiving site to the presynaptic terminal. A second principle is that neurons connect to each other in a way which leads to the formation of neuronal networks which serve some specific function (principle of connectional specificity). However, the complexity of these networks is extremely high. On average, a spinal motor neuron receives 10,000 contacts from other neurons, of which 2000 on the cell body and the rest on its dendrites, while a Purkinje neuron in the cerebellum could receive up to 150,000 contacts.

In chemical synapses and in the membrane of the postsynaptic neuron oriented toward the synaptic cleft, there are molecules called receptors. These receptors serve as docking sites for neurotransmitters. They are a sort of "receivers" of the signal which is transmitted by the presynaptic neuron. One important characteristic of transmission is that it is the receptor that determines whether a particular synapse is excitatory or inhibitory and whether an ion channel will be activated directly by the transmitter or indirectly through a second messenger.

The notion of a receptor was introduced in the late nineteenth century by the German bacteriologist Paul Ehrlich (1854–1915) to explain the selective action of toxins and other pharmacological agents and the great specificity of immunological reactions (lock and key). Chemical neurotransmitters act either directly or indirectly in controlling the opening of ion channels in the postsynaptic cell.

There are two main types of receptors:

- Ionotropic receptors: They include an ion channel in their structure, which opens as a response to the binding of the neurotransmitter (Unwin 1993).
- Metabotropic receptors: They act by stimulation of the production of second messengers, which are small freely diffusible intracellular metabolites such as c-AMP and diacylglycerol. Subsequently these second messengers activate protein

kinases that phosphorylate a variety of proteins, or they mobilize Ca^{2+} ions from intracellular stores (Tanaka and Nishizuka 1994). Often metabotropic receptors gate ion channels, but this happens only indirectly through second messengers. There are two types of metabotropic receptors: the G-protein-coupled receptors and the receptor tyrosine kinases. The G-protein family contains the α- and β-adrenergic receptors, the muscarinic ACh receptors, the $GABA_B$, and others (Gilman 1989, 1990, 1995). Receptor tyrosine kinases are activated by hormones, growth factors, and neuropeptides.

Ionotropic and metabotropic receptors have different functions. The ionotropic receptors produce relatively fast synaptic actions lasting only milliseconds. These are commonly found in neural circuits that mediate rapid behaviors, such as the muscle stretch receptor reflex. The metabotropic receptors produce slower synaptic actions lasting seconds to minutes. These slower actions can modulate behavior by altering the excitability of neurons and the strength of the synaptic connections of the neural circuitry mediating behavior. Such modulatory synaptic pathways often act as crucial reinforcing pathways in the process of learning. Additionally ligand-gated channels function as simple on-off switches triggering an all-or-nothing effect, while on the contrary the metabotropic receptors manifest a complex function with the ability to vary in space and time as well as in the number of intracellular systems affected. While ionotropic excitation can only increase the ion flux through the channel, metabotropic activity can either increase or decrease (Hille 1994).

Essentially metabotropic receptors exert a modulatory effect on the postsynaptic neuron since they are normally not capable of firing an action potential. Their role is mainly to regulate the electrophysiological properties of a cell (e.g., resting potential, action potential duration, repetitive firing characteristics, etc.).

The main receptors for each neurotransmitter are as follows:

1.2.6.1 Acetylcholine

Cholinergic receptors are classified into two main families, the muscarinic and the nicotinic, on the basis of muscarine and nicotine to mimic the effects of Ach.

Nicotinic receptors (Arneric et al. 1994) consist of one alpha (α1–α7) and one beta (β1–β4) subunits. Different combinations of these subunits result in different properties of the receptor. The only documented nicotinic neurotransmission in the brain concerns the communication between the motor neuron and a Renshaw cell in the spinal cord. However a nicotinic type of activity has been recorded in the retina, the hippocampus, the respiratory center in the brain stem, the cortex, the thalamus, the hypothalamus, the substantia nigra, the striatum, and the locus coeruleus. Nicotine interacts with nicotinic receptors and increases the secretion of various neurotransmitters, while the cholinergic system, in general, regulates and governs a number of important functions including attention and concentration, brain regional blood flow, glucose consumption, and the electrical activity of the cortex. Each one of these functions is enhanced by nicotine and attenuated by its antagonists. Additionally nicotine could improve memory and learning, and it has an anxiolytic effect probably through the release of endogenous substances acting on GABA receptors. Nicotine-related disorders are related to depression, but the nature of this relationship is unknown (Edwards et al. 2011; Lyons et al. 2008; Fu et al. 2007; Killen et al. 2003; Cardenas et al. 2002; Breslau et al. 1993).

On the other hand, muscarinic receptors (Ehlert et al. 1994) are widely distributed in the human body. Their activation in the periphery causes drop of heart frequency, vasodilation, constriction of airways, increase of gastric tube mobility, and miosis in the eye pupil, among others. In the brain they are implicated in function including memory, learning, and control of posture.

There have been identified four types of muscarinic receptors with the use of pharmacological methods and five (M1–M5) with molecular biology methods. There is almost complete correspondence between the types identified with these two methods, and all receptors seem to act through their coupling with a second messenger system of a G-protein.

- The M1 type is abundant in the forebrain and the sympathetic ganglia, while its density is lower in the hindbrain. Its highest concentration is in the hippocampus, followed by the cortex, the striatum, the olfactory bulb, and the thalamus, while the concentration is far lower in the hypothalamus. This receptor is denser in the limbic system and the association cortices. Its activation results in an increase in the hydrolysis of phosphoinositol.
- The M2 type is widespread in the body, including the brain, but its density is rather low. It is the only muscarinic receptor located in the myocardium. It is the main muscarinic receptor of smooth muscles, excluding the intestine ones. Its density in the brain is low but the distribution is rather homogenous. It is relatively denser in the cerebellum, the medulla oblongata, and the mesencephalon, followed by the hypothalamus and the thalamus. Its highest density is in the primary sensory and motor areas. Its activation leads to the inhibition of adenylcyclase.
- The M3 type is the main muscarinic receptor of exocrine glands. It also triggers the constriction of smooth muscle fibers in spite of the fact it constitutes a minority among the muscarinic receptors in these fibers. Probably it constitutes the main receptor governing the intestine mobility. It exists in low density throughout the brain, and it is slightly denser in the forebrain. Its activation results in an increase in the hydrolysis of phosphoinositol.
- The M4 type is abundant in the telencephalon, the striatum, and the olfactory bulb. It is denser in the forebrain, and its activation leads to the inhibition of adenylcyclase.
- The M5 type constitutes less than 2% of total muscarinic receptors, and it is found exclusively in the brain. Its activation results in an increase in the hydrolysis of phosphoinositol.

1.2.6.2 Serotonin

A significant number of 5-HT receptors (most recent number mounts to 14) have been identified so far, but for the role of most of them, we know little. The main are the following:

- 5-HT_1 (A, B, D, E, and F): They are negatively linked to the adenylcyclase system. They inhibit the transformation of ATP into c-AMP (in the hippocampus, substantia nigra, gray matter around the Sylvius aqueduct). An increase in the 5-HT_1 activity in the raphe nucleus leads to increased appetite as it reduces serotonergic activity. An increase of the activity of the rest of the 5-HT_1 receptors leads to reduced appetite (Montgomery and Fineberg 1989).
- 5-HT_2 (A, B, and C): They are positively linked to the phosphoinositol system and lead to an increase of intracellular Ca or to protein kinase activation. Animal research has shown that social isolation causes hypersensitivity of 5-HT_{2C}. These same receptors of the amygdala-hippocampus circuit probably lead to the manifestation of aggressive behaviors when danger is confronted, by overcoming the inhibitory effects of 5-HT_1 and 5-HT_{2A} located in the gray matter around the aqueduct. An increase in the activity of these receptors again could lead to a reduction in appetite, especially in bulimic patients which also manifest affective disorders and disordered saturation feeling mediated by cholecystokinin (CCK). In general, 5-HT_2 exert a tonic inhibition on noradrenalin in the hippocampus.
- 5-HT_3: They have a positive link with fast ion channels and increase their permeability. They inhibit Ach activity in the cortex but augment dopaminergic in the corpus striatum and the limbic system.
- 5-HT_4: They have a positive link with the adenylcyclase system. They increase the transformation of ATP into c-AMP. Their relationship to behavior is unknown.
- 5-HT_5 (A and B): It is unknown the exact second messenger system they relate to, and equally unknown is their relationship to behavior.
- 5-HT_6: They have a positive link with the adenylcyclase system. They increase the transformation of ATP into c-AMP. Their relationship to behavior is unknown.
- 5-HT_7: They have positive link with the adenylcyclase system. They increase the transformation of ATP into c-AMP. Their relationship to behavior is unknown. Recently it has been suggested that they mediate the antidepressant effect of lurasidone (Cates et al. 2013).

1.2.6.3 Norepinephrine

Norepinephrine (adrenergic) receptors belong to the G-protein-coupled superfamily. They can be found pre- and postsynaptically on neurons but also on glial cells (Salm and McCarthy 1992; Stone and Ariano 1989).

- Alpha-1 ($\alpha 1$; a, b, c, and d): Postsynaptic excitatory receptors linked to the phosphatidylinositol system (Hieble et al. 1995; Bylund et al. 1994). They can also be found in the muscles of the vessel wall and the smooth muscles of the intestine and the heart.
- Alpha-2 ($\alpha 2$; a, b, c, and d): They can be found presynaptically, and they are inhibitory receptors linked to the adenylcyclase system (Patel et al. 1981; Langer 1974; Aghajanian and VanderMaelen 1982; Anden et al. 1970; Cedarbaum and Aghajanian 1977; Langer and Arbilla 1990; Starke 1971; Svensson et al. 1975; Bylund et al. 1994). Apart from the brain, they can be found in the muscle of the vessel wall, the smooth muscles of the intestine, and the platelet membrane.

Subtypes are unevenly distributed in the brain, with high densities of α2a and α2c in the LC, amygdala, and hippocampus, while α2b is found mainly in the thalamus (Scheinin et al. 1994; Rosin et al. 1996; Talley et al. 1996; Wamsley et al. 1992; Nicholas et al. 1993a).

- Beta-1 (β1): They are excitatory receptors linked to the adenylcyclase system (Frielle et al. 1987; Shorr et al. 1982; Nicoll et al. 1990). They predominate in the cerebral cortex, the dentate gyrus, the CA1, and the medial dorsal hypothalamic nuclei (Nicholas et al. 1993b; Palacios and Kuhar 1982; Rainbow et al. 1984). Apart from the brain, they can be found in the heart.
- Beta-2 (β2): They are excitatory receptors linked to the adenylcyclase system (Tholanikunnel et al. 1999; Ostrowski et al. 1992; O'Dowd et al. 1988; Nicoll et al. 1990). They are more abundant in the cerebellum and reticular, paraventricular, and central thalamic nuclei (Nicholas et al. 1993b; Palacios and Kuhar 1982; Rainbow et al. 1984), but they can also be found both in the smooth and skeletal muscles and the liver. They are also present in the membrane of the lymphocytes where their function is unknown.
- Beta-3 (β3): They can be found in the adipose tissue, and little is known for them (Emorine et al. 1989).

1.2.6.4 Dopamine

The following dopaminergic receptors have been identified so far (Civelli 1994; Mansur and Watson 1994). They can be grossly classified into D1-like (D1 and D5) and D2-like (D2, D3, and D4):

- D1: It is a postsynaptic excitatory receptor linked to the adenylcyclase system through a G-protein. It is abundant in the anterior cingulate gyrus, the cortex of the frontal pole, the insular (Reil cortex) and the olfactory cortex, the nucleus caudatus, the putamen, the nucleus accumbens, and the olfactory bulb and to lower density in the septum, the hypothalamus, and the rest of the cortex. It is absent from the substantia nigra, the VTA, and the hippocampus. It is dominant in the amygdaloid nucleus where D2 is absent. D1 is present solely in the brain and not in the rest of the body.
- D2: It is a presynaptic autoreceptor and inhibits the adenylcyclase system through a G-protein; however there is evidence that at least a subgroup of these receptors acts independently of this second messenger system (Roth and Elsworth 1994). It is the dominant dopamine receptor in the substantia nigra, the VTA, the globus pallidus, and the hippocampus where D1 is absent. It is also abundant in the olfactory cortex, the nucleus caudatus and the putamen, the nucleus accumbens, the olfactory bulb, and the hypothalamus and in lower density in the septum and the rest of the cortex. It is absent from the amygdaloid nucleus where D1 is dominant. Outside the brain it can be found in the cortex of the adrenal glands, the pituitary, and the retina. Possibly it plays a role in the regulation of pain through receptors located in the reticular nucleus of the thalamus and the gigantocellular reticular nucleus (magnocellularis) in the hindbrain.

- D3: It is a presynaptic autoreceptor, and probably it is linked to the same G-protein D2 is linked, but this is not solidly proven. It is located in the same areas with D1 and D2 but with a much lower density. It manifests some selectivity for the limbic system. It is not found outside the brain.
- D4: Probably it is linked to the same G-protein D2 is linked, but this is not solidly proven. It is located in the same areas with D1 and D2 but with a much lower density. It also manifests some selectivity for the limbic system. Probably it is also located in the kidneys and the heart.
- D5: It is an excitatory receptor linked to the adenylcyclase system through a G-protein. It is located solely in the hippocampus, the hypothalamus, and the parafascicular nucleus in the thalamus, and it is implicated in functions concerning affect, pain, and neuroendocrine secretion. It is probably located also in the kidneys and the heart.

1.2.6.5 Histamine

Histamine receptors are classified as H1–H4 and are all bound to a G-protein. It is also believed that histamine activates ligand-gated chloride channels in the brain. These are located in the thalamus and hypothalamus and produce fast inhibitory postsynaptic potentials (Panula et al. 2015).

- H1: It is produced in the tuberomammillary nucleus, and fibers project to the dorsal raphe and the locus coeruleus, among other areas. Activation of this receptor increases wakefulness, while other roles include the regulation of temperature, endocrine function, appetite, and neurocognition (Panula et al. 2015; Blandina et al. 2012).
- H2: Their receptors are located exclusively in the gastrointestinal system and in vascular smooth muscle cells (Wouters et al. 2016; Panula et al. 2015).
- H3: It manifests both autoreceptor and heteroceptor functions and exerts an inhibitory effect on the secretion of histamine, Ach, norepinephrine, and 5-HT (Gilman 1990).
- H4: Found primarily in the basophils, the bone marrow, the thymus, the small intestine, the spleen, and the colon. It plays a role in various mechanisms of inflammation, etc. Its role on cognition is unclear (Panula et al. 2015).

1.2.6.6 Glutamate (Glu)

Glu is excitatory at ionotropic receptors and modulatory at metabotropic receptors.

The *ionotropic receptors* of Glu are classified on the basis of the agonist they bind to, into the following types:

- *N*-methyl-D-aspartate (NMDA): These receptors are essentially a ligand-gated rapidly transmitting calcium ion channel. Glutamate binding causes the channel to open and be excited. NMDA receptors possess unique characteristics that allow them to be part of large macromolecular synaptic signaling complexes. In addition to glutamate, NMDA receptors also require the simultaneous binding of glycine as a co-agonist (Monaghan and Jane 2009; Asztely and Gustafsson

1996; Mayer et al. 1984; Monaghan et al. 1989). They are composed of multiple subunits which may belong to three related families: NR1, NR2, and NR3 (Nakanishi 1992). NMDA receptors with different subunit compositions show distinct distributions in the brain and different properties and regulation (Yamakura and Shimoji 1999). NR1 subunits contain a glycine-binding site, while NR2 subunit a glutamate-binding site (Hirai et al. 1996; Laube et al. 1997). It is important to note that NR1 receptor subunits are distributed ubiquitously in the brain and NR1 subunit mRNA is expressed throughout the different stages of neurodevelopment (Mony et al. 2009; Yamakura and Shimoji 1999; Laurie et al. 1997).

NMDA receptors are located in synaptic, presynaptic, and extrasynaptic sites, but in most neurons NMDA receptor density is higher in dendritic spines within the postsynaptic density (Mony et al. 2009). They are often found co-localized with AMPA receptors and are not activated by single synaptic events like the Glu binding alone (Huang and Bergles 2004). Two simultaneously occurring events are necessary for their activation: presynaptic release and binding of glutamate to NMDA receptors and sufficiently strong postsynaptic membrane depolarization leading to the removal of Mg^{++} blockade which is normally present (Wu and Zhuo 2009). NMDA receptors become fully activated upon extensive stimulation of the synapse, because the repetitive activation of AMPA receptors results in sufficient depolarization of the postsynaptic membrane and thus the lifting of the Mg^{++} block, during a process called use-dependent calcium influx (Monaghan and Jane 2009).

The role of the NMDA receptors includes synaptic plasticity, synaptic development, and function- and experience-related refinement of synaptic connections; thus these receptors play a crucial role in learning and memory formation and consolidation, general cognition, and attention but also in mood and anxiety (VanDongen 2009). However they also mediate a phenomenon called excitotoxicity which leads to neurodegeneration. Excitotoxicity is a pathological process that eventually triggers reckless glutamate activity causing a dangerous opening of the calcium channel allowing too much Ca^{++} to enter and by the activation of intracellular enzymes leading to the formation of free radicals, which exert a toxic action on cellular organelles and membranes killing the cell (Stahl 2008).

- α-Amino-3-hydroxy-5-methyl-4-isoxazolepropionic acid (AMPA) receptor: They are composed of combinations of four subunits (GluA1–GluA4) which combine to form heterotetrameric blocks consisting of two sub-blocks of GluA2 and either GluR1, GluR3, or GluR4 (Mayer 2005; Greger et al. 2007). Their distribution in the brain is similar to that of the NMDA.
- Kainate (KA): They are composed of heteromeric combinations of two related subfamilies of proteins which comprise the receptor subunits, GluR 5–7 and KA1 and KA2, and their variants (Bettler et al. 1990, 1992; Egebjerg et al. 1991; Herb et al. 1992; Sakimura et al. 1992; Werner et al. 1991). They have both presynaptic and postsynaptic actions, they are fewer in comparison to the other ionotropic receptors, and their function is not well understood (Contractor et al. 2011).

The *metabotropic Glu receptors* are coupled with a G-protein system and are divided into three groups and eight subtypes, mGluR1–8 (Pin and Acher 2002). They are causing a wide range of physiological effects on the postsynaptic neuron.

1.2.6.7 Gamma-Aminobutyric Acid (GABA)

There are two main GABA receptors, named A and B (Zorumski and Isenberg 1991):

- $GABA_A$ is a fast ionotropic receptor linked to an ion channel (Cossart et al. 2005). It is made of at least five subunits which are transmembrane and are arranged in a ring creating a pore in the center, which constitutes the ion channel. There are three major types of subunits named α, β, and γ, but also the existence of δ, ε, π, and θ has been reported. The usual combination is two α subunits, two β subunits, and one γ subunit. These three subunits are allosterically connected, which means that binding to any of them modifies the binding ability of the others (Olsen and Sieghart 2009). The presence of a distinct type of subunit, named ρ, determines a subgroup of $GABA_A$ receptors named $GABA_C$ which do not manifest allosteric properties (Enz and Cutting 1998).
- $GABA_B$ is a metabotropic receptor and is linked to a G-protein. It is located presynaptically in the neurons containing GABA or Glu.

Most areas of the brain contain both A and B GABA receptors.

1.2.7 Second Messenger Systems

There are only a few second messengers which are well characterized and studied, in comparison to almost 100 neurotransmitters. Second messenger systems are classified into two broad categories: nongaseous and gaseous.

- The nongaseous second messengers include hydrophobic (water-insoluble) molecules like diacylglycerol and phosphatidylinositols and hydrophilic (water-soluble) molecules like c-AMP, cGMP, IP3, and intracellular Ca^{2+}. Hydrophobic messengers are membrane-associated and diffuse from the plasma membrane into the intermembrane space, while hydrophilic messengers are located within the cytosol (Majerus 1992; Arachidonate related lipid mediators 1990; Needleman et al. 1986). The most studied nongaseous second messenger systems are those of c-AMP, phosphoinositol, arachidonic acid, cGMP, and tyrosine kinase systems. The first three are coupled with the neurotransmitter signal through a G-protein. One interesting property of arachidonic acid and its metabolites is that they are highly lipid soluble, and thus they diffuse through membranes. In this way they can act also in neighboring cells, acting as a type of transcellular synaptic messengers.
- The gaseous second messengers (Mustafa et al. 2009) are highly diffusible and include nitric oxide (NO), carbon monoxide (CO), and hydrogen sulfide (H2S).

They are extremely short-lived. All of them can diffuse both through cytosol and across cellular membranes. The criteria of "gasotransmitter" are the following (Wang 2002, 2004):

- Small molecule of a gas.
- Freely permeable to membranes and its effects do not rely on membrane receptors.
- It is endogenously and enzymatically generated and its production is regulated.
- It has well-defined and specific functions at physiologically relevant concentrations.
- Functions of this endogenous gas can be mimicked by its exogenously applied counterpart.
- Its cellular effects may or may not be mediated by second messengers but should have specific cellular and molecular targets.

Eventually, all second messenger systems produce changes in specific target proteins within the cell, and the phosphorylation mediated by protein kinases and the action of phospholipases are central in the whole process (Tanaka and Nishizuka 1994). Protein kinases can also induce the synthesis of new proteins by altering gene expression. This leads to long-term changes which are important for neuronal development and for long-term memory. To make things more complex, the effects of second messenger systems are broadened through interaction between them with parallel, convergent, and antagonistic actions.

1.3 Functional Organization of the Brain

The human brain has a complex but highly efficient organization (Gazzaniga 1989, 1995). Throughout the brain, neuronal cells are more or less the same both morphologically and also in terms of function. It is their interconnections and the networks that arise which produce the differences in function between brain areas, and they give birth to behavior itself (Mishkin 1993; Changeux and Dehaene 1993; Lewis and Oeth 1995). Thus, this fundamental simplicity of structure supports a great complexity through astronomical numbers and exponential power. The end result is currently beyond our comprehension, even at its middle levels (e.g., coding and storing of external stimuli) not to mention the higher levels (e.g., consciousness).

The basic operationalized conceptualization of the human brain is that of a helmet with spheres at its basis. Both the helmet and the spheres are made of gray matter, while the space between them, that is, the internal space of the helmet, is full of white matter. The helmet corresponds to the cortex and the spheres to the thalamus, the hypothalamus, the amygdala, the hippocampus, and the basal ganglia (nucleus caudatus, putamen, and globus pallidus). Gray matter consists of the cell bodies of neurons, which are the processors where the information is being analyzed and processed, while the white matter corresponds to the axons, which are the cables connecting the various processors. The spheres/basal ganglia deal with more primitive and primary processes (basic emotions, instincts, vital functions, and motor coordination), while the helmet/cortex deals with higher cognitive function especially in the frontal areas (detailed process and analysis, abstract thinking and planning). This whole design is extremely ergonomic; it permits the circulation of information in the fastest, most efficient, and economical way possible. The ventricles serve as the cooling system, and they also assist in the removal of toxic and metabolic waste (Fig. 1.2).

An important feature of the brain organization is that most inputs cross over to the opposite (contralateral) side of the brain or spinal cord. Subsequently the sensory and motor activities are mediated by the opposite-side brain hemisphere. The reason for this is not yet understood. What is however also known is that this asymmetry is accentuated in humans in comparison to primates because of the presence of complex and symbolic thought and behavior. While this is true, it is also a fact that most activities engage both hemispheres but with a different contribution. Classically, the dominant hemisphere is more concerned with detailed thought, abstract thinking, and speech and subsequently with intentional behavior, while the non-dominant hemisphere is more concerned with emotional aspects of behavior and spatial analysis. In some way it could be said that the dominant hemisphere is the "digital," while the non-dominant is the "analog" or better the "fuzzy." In the normal brain, there is communication between the two hemispheres, and there is now evidence that the capacity of one hemisphere to perform a particular task may deteriorate after commissurotomy.

The first component of behavior is the sense of the environment, external but also internal. The human brain utilizes five main senses: sight, hearing, smell, taste, and touch plus pain and the sensation of body movements, which give information both for the external and the internal environments. A number of other senses also provide information concerning the internal environment, which is of course the human body.

Concerning the external environment, specialized apparatus and receptor cells on the body surface receive stimuli and encode information. Essentially they transform external natural phenomena (e.g., sound, light, pressure, etc.) into neuronal phenomena, and in this way they register their intensity and their temporal and spatial characteristics. An important characteristic of these "sensors" is that they tend to focus on and prioritize specific characteristics of the stimuli. For example, sight has high resolution only in a small fraction of the visual field (spatial focus, priority to vision directly in front), while the visual color spectrum is limited (other species have different visual spectra probably as a result of natural selection pressure). Also the senses from the face or the fingers are much more detailed in terms of information in comparison to trunk or legs. As a result, the receptor cells in the sensor organs are arranged in a "topographical" way, reflecting this focus and prioritization, and this topographical arrangement is kept through the successive stages of processing; in this way an orderly neural map of information from the receptive surface is

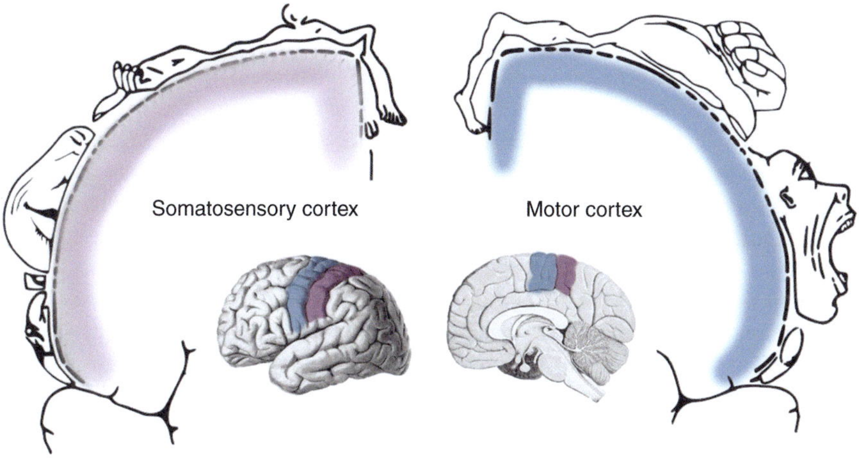

Fig. 1.6 Topographic maps in the somatosensory and motor cortices

retained at each successive level in the brain (Fig. 1.6) (Penfield and Rasmussen 1950).

Sensory information is processed in several parallel pathways each of which handles a different set of information concerning the specific sense. This increases computational power. Central to this processing is the involvement of a number of neuronal circuits which contribute to the coding and processing of the information. There are three such major neural systems.

- The first is the thalamus which is the biggest structure in the diencephalon (Fig. 1.1). Its name comes from the Greek *θάλαμος* meaning "chamber." It is comprised by a number of nuclei. These nuclei are classified into the relay, the specific relay, the association relay, and the diffuse-projection nuclei. One of them, the reticular nucleus, regulates the functional relationship between the thalamic nuclei. The thalamus receives input from the spinal cord with the lateral spinothalamic tract, which carries pain and temperature information, and the anterior (or ventral) spinothalamic tract, which carries crude touch and pressure information. Every sensory system (with the exception of the olfactory system) includes a thalamic nucleus that receives sensory signals and sends them to the associated primary cortical area. The thalamus is manifoldly connected to the hippocampus via the mammillothalamic tract (Carlesimo et al. 2011; Stein et al. 2000), and this connection serves the link between memory mechanisms and new stimuli. The projection to the cerebral cortex is carried via the thalamocortical radiations (Briggs and Usrey 2008). These pathways provide information input to the cortex from the senses, but also the thalamus relays input from the basal ganglia and cerebellum to the cortical motor areas

(Asanuma et al. 1983; Evarts and Thach 1969). To summarize, the thalamus filters and processes both the sensory input to the cortex during the first steps of the process of information coming from the environment, but also it filters and processes the motor responses from the basal ganglia before their entering into the motor cortex during the early stages of response, after the process of the sensory information is completed. With its link to the hippocampus, it allows the brain to keep record of these processes in the form of a type of memory.

- Depending on the author, the limbic system includes the cingulate gyrus, the olfactory bulbs, the hippocampus, the hypothalamus, the amygdala, the fornix, the mammillary body, the septum pellucidum, the habenular commissure, the parahippocampal gyrus, and the limbic midbrain areas (Morgane et al. 2005). Essentially it includes the structures where the production of mood (amygdala), emotion (cingulate gyrus), and memory (hippocampus) is seated. There are a number of connections, many of them reciprocal, between these structures and the thalamus and the frontal association cortex. Information concerning sensory information are projected to the amygdala which adds an emotional component which is essentially a primitive kind of assessment and decision (e.g., fear in the sight of a snake) and the hippocampus, which stores aspects of perception in long-term memory and also provides with feedback from previous experiences and memories. In this way a comprehensive internal representation of the external environment is formulated. The hippocampus is responsible for the formation of long-term memories, but it is not the permanent storage site of memories. If damaged the subject loses the ability to form new memories but keeps old ones. Overall, the role of these connections is to input affect on the information process and the decision-making and to develop and store memories and to selectively retrieve them on the basis of current sensory input and emotional state (Van Hoesen 1993). Because of this complex emotional-mnemonic function, it has been proposed as the seat of the "social brain" (Brotehrs 1990).
- The basal ganglia (Lanciego et al. 2012; Ward et al. 2013; O'Connor 1998) include the nucleus caudatus, the putamen, the globus pallidus, the subthalamic nucleus, and the substantia nigra (Fig. 1.2). The term "striatum" refers to the nucleus caudatus, the putamen, and the globus pallidus, while the term "lentiform" or "lenticular" nucleus refers to the putamen and the globus pallidus. The basal ganglia take part mainly in the motor response, and their function will be described below.

Information is processed and transformed at every step, and at each stage every postsynaptic neuron typically receives inputs from thousands of presynaptic neurons. Eventually the sensory information are projected onto the cortex at areas specific for each sense. As said, most of this information comes through the thalamus which acts as a gatekeeper for information to the cerebral cortex and together with the limbic system prevents or enhances the passage of specific information and the retrieval of relevant memories and storage of new, depending on the behavioral state of the animal. These areas in the cortex are called "primary association areas," and their main characteristic is that they receive input from a single sense, and because

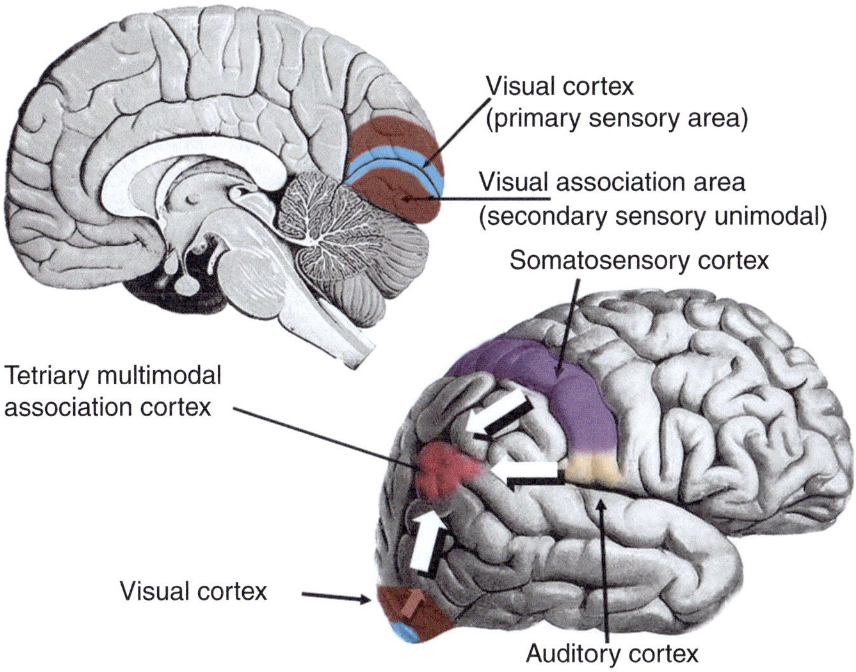

Fig. 1.7 Association areas and the flow of information

of this they are called "unimodal." The areas where each of the five senses projects are well known, and it is the pole of the occipital lobe (calcarine sulcus) for sight and the Heschl's gyri in the temporal lobe for hearing and the posterior central gyrus in the parietal lobe for somatosensory senses. Next to or surrounding the primary association cortex, an adjacent, higher-order area, the secondary association cortex which is also unimodal is located (Fig. 1.7). The concept of "association areas" was introduced by John Hughlings Jackson (1835–1911).

An important characteristic is that the projection into the cortex is done in an orderly manner, thus creating a sort of a topographical map. This map corresponds to the characteristics of the specific sense (e.g., light touch vs. pain on the same area of skin), and it is spatial for sight and somatic senses and oriented toward the analysis of frequency and similar features for sounds whose source is also able to locate (Clark et al. 1988; Kaas et al. 1979; Jenkins et al. 1990; Meldrum 2000). It also corresponds to the differential resolution (and subsequently importance) given to each part of the sensory stimulus by the external sensors. Thus, the topographical map is not an accurate representation of the physical stimulus but rather a distorted one, according to the weight given by the senses. It reflects both the physical topography and the density of the perceived information. For example, in the somatosensory cortex, the face, tongue, and fingers occupy a much larger surface and neurons and subsequently more focus in processing, in comparison to the trunk or legs (Fig. 1.6). In the visual field, the center of vision (the fovea) occupies a much larger area in comparison to the peripheral parts of the visual field (Fox et al. 1987; Penfield and Rasmussen 1950). These maps change with experience (Mogilner et al. 1993; Jenkins et al. 1990; Ramachandran 1993).

All secondary association areas project to one of the three tertiary major multimodal association areas. In these areas the integration of two or more sensory modalities takes place, and a more comprehensive internal neuronal representation of the external environment is developed (Van Hoesen 1993).

Three multimodal association areas are particularly important:

- The posterior association area, at the margin of the parietal, temporal, and occipital lobes. It receives information from several sensory modalities (Fig. 1.7).
- The limbic association area, which includes the anterior pole of the temporal lobe and the medial edge of the cerebral hemisphere. It is concerned with emotion and memory storage.
- The anterior association area (prefrontal cortex) which is located anterior to the postcentral gyrus. It is concerned with the executive functions, judgment, planning for the future, and holding and organizing events from memory for prospective action (working memory) as well as with the planning of movement (Goldman-Rakic 1992, 1996).

A simplistic functional description of the frontal lobe could consider it as a trilateral pyramid with the base toward the parietal lobe and with its top corresponding to the frontal pole. One of its lateral surfaces is horizontal and corresponds to the orbital cortex, while the other two correspond to the outer and the medial surfaces of the lobe. The two lateral surfaces corresponding to the orbital and the medial surfaces are joined in a 90° angle (Fig. 1.8). The outer surface (dorsal prefrontal association area) corresponds to what is closer to RAM in modern computers and hosts working memory. The medial surface (medial prefrontal cortex) concerns the production of affect, while the orbital surface (ventral orbitofrontal cortex) hosts the decision circuits.

After the decision is made, projections to the premotor cortex are triggered. Every decision implies the initiation of a behavior which almost always has a motor part, either to act or to stand still. A topographical map exists also in the

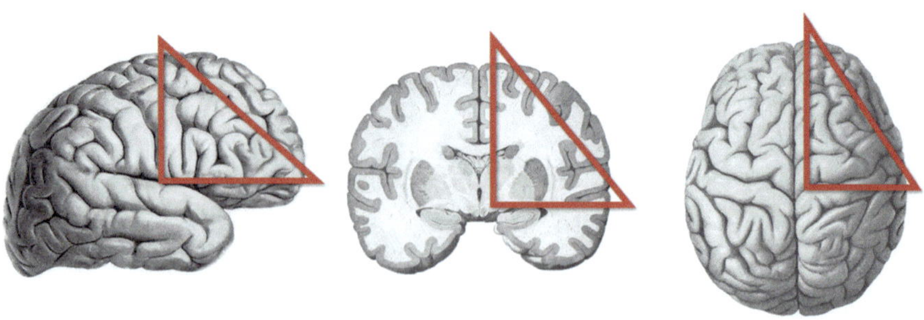

Fig. 1.8 The simplified pyramidal shape of the frontal lobe

motor cortex (Fig. 1.6). This map, like the sensory maps, does not represent every part of the body equally, but it puts specific emphasis on individual body parts reflecting the fineness of control required for the motor function of these particular body parts.

The primary motor cortex probably contains the "software" of movement and is located rostral to the central sulcus and is associated with the motor systems of the spinal cord. The human corticospinal tract consists of about one million axons, and about 40% of them originate in the motor cortex, and the entire projection is also called the "pyramidal tract."

In this process, there is input to the motor cortex from the basal ganglia which constitute key structures for motor planning. There are two major pathways with projections from and to the basal ganglia:

- The "direct" pathway, which carries projections from cortical cells to the striatum via the SNr-GPi complex, which in turn projects to the thalamus through the inhibitory ansa lenticularis pathway, and eventually the thalamus projects to the motor cortex which in turn projects to the brain stem and ultimately mobilizes muscle fibers via the lateral corticospinal tract.
- The "indirect" pathway also originates from the striatum, and after stimulation by the cortex, it projects to the globus pallidus externa, which in turn projects to the subthalamic nucleus which also in turn projects to the SNr-GPi complex which inhibits the thalamus. This indirect pathway regulates the inhibition of the motor cortex by the thalamus. Both pathways are based on a serial sequence of inhibitory and excitatory projections which produce a fine-tuned end output. Overall, the direct and the indirect pathways are antagonistic in their functions, and this antagonism is modulated by the substantia nigra pars compacta. The antagonism serves balance and fine-tuning of motor movement.

It is important for the long-term adaptation and survival that those decisions, choices, and responses with a positive outcome for the individual be ranked higher in the consideration of future resolutions and given more weight and speed in situations of ambiguity. In other words, behaviors which are successful are also rewarding, and this is registered not only in terms of essential mnemonic registrations but also at a lower unconscious level. In general, dopaminergic neurons in the midbrain mediate these rewarding aspects of behavior.

The basal ganglia circuit probably serves the selection and enforcement of behaviors that lead to reward by using past experience to predict which patterns of input from the neocortex and the thalamus will lead to reward. In order to execute this function, the cells in the globus pallidus and the substantia nigra fire tonically producing a constant inhibition of neurons in the thalamus and superior colliculus. In case this inhibition pauses, then the neurons in these areas are released and respond to excitatory inputs that would otherwise be subthreshold. The central location of the globus pallidus in the basal ganglia formation implies also a central role in the functioning. It constitutes the inner component of the lentiform nucleus, and with the putamen it forms a cone-like structure, with its tip directed medially. The majority of projections from the neostriatum (which receives input from the neocortex, the thalamus, and the substantia nigra) are to the globus pallidus which serves as an intermediate processing center (Wilson 2004). A role for the nucleus accumbens and the VTA is particularly well established. Numerous things that people find rewarding, including addictive drugs, good-tasting food, and sex, have been shown to elicit activation of the ventral tegmental area dopamine system.

This above network probably tests a variety of input concerning both the interpretation and also the response options on the basis of templates and maybe error memories provided by the prefrontal cortex to choose those that systematically and reliably elicit reward reactions in the brain. The prefrontal cortex provides the templates to test against but on the other hand also inhibits or modifies the "resolutions" of the basal ganglia. This is why although mood, affect, and initial tendencies to act are not under volitional control and cannot be influenced, the final output (externalized behavior) is under volitional control through effortful regulation.

A comprehensive but simplistic representation of the complete process from sensory input to the manifestation of motor response is shown in Fig. 1.9.

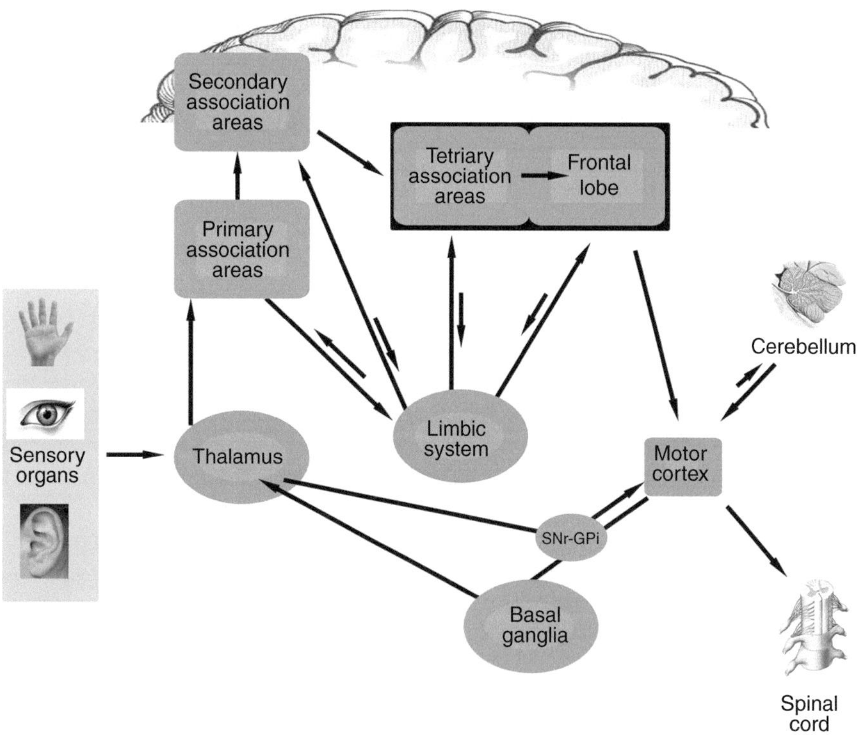

Fig. 1.9 Abstract representation of the flow of information and its process through different brain areas

1.4 Thought and Intelligence

Thought is the end result of brain function and constitutes what is conceptualized with the term "mind." Philosophical and analytical approaches of this issue are beyond the scope of the current book chapter.

What is important to consider as an extension of all that has been mentioned and described so far in this book chapter are some basic macroscopic characteristics of the thought process and the concept of intelligence.

One astonishing characteristic of the human brain is that it is capable of performing complex process of information in extremely short time but is incapable of performing much less demanding process of information. It is as if it is calibrated to perform on high demands. For example, the human brain, apart from its ability to comprehend and analyze abstract concepts and meanings, in everyday life is able, e.g., to perform the complex visuospatial and motor calculations necessary to throw a spear to hit a moving target at a distance or to perform complex body acrobatics, in which modern engineering and computing find it difficult if not impossible to fully simulate, while on the contrary the human brain is unable, e.g., to perform a multiplication between two five-digit numbers, which is one of the easiest tasks for electronic-based calculators even at credit card size. This is because the essential architecture and the basic principles on which the brain function is based are completely different from that of electronic computers and specifically of what is called "Turing machines." Again this topic is beyond the scope of the current chapter; however it is useful to note that only recently we have started understanding the principles of the so-called non-Turing machines whose main characteristic is that their functions are not deterministic and of quantum computers which utilize a probabilistic system instead of a binary code.

One probable characteristic of some of these non-deterministic and non-binary computational systems is their complex coding at input and that they are extremely relying on templates and pre-acquired information in order to process novel input. This is also a characteristic of the brain, and at least some of these pre-acquired templates emerge in the form of archetypical thoughts and impulses as well as in the form of cognitive biases.

The concept of cognitive biases was introduced by Amos Tversky and Daniel Kahneman in 1972 (Kahneman and Shane 2002). Kahneman received the Nobel Prize in Economic Sciences in 2002, and his overall work challenges the assumption of human rationality prevailing in modern economic theory and decision-making.

Cognitive biases refer to a systematic pattern of deviation from norm or rationality in judgment. This leads to inferences about the environment and to events in an illogical fashion (Haselton et al. 2005). This way of thinking leads individuals to create their own subjective version of the world and especially of the social reality, and very often the pre-existing biases rather than the objective information input are what determines the response and overall behavior (Morewedge and Kahneman 2010; Kahneman and Klein 2009; Kahneman and Frederick 2007; Kahneman 2003; Kahneman and Tversky 1996, 1982; Tversky and Kahneman 1981; Gigerenzer and Goldstein 1996). Although some of these biases are the unwanted result of limitations in the process capacity of the brain, in principle the role of biases is profoundly adaptive since they enable faster decisions when timeliness are more valuable than accuracy, as illustrated in heuristics (information-processing shortcuts which are simple for the brain to compute but sometimes introduce severe and systematic errors; comes from the Greek word *ευρίσκω*, that is, "to find"). The problem is that for most of biases, their basic principles were embedded in the human brain when humans were living in a much different environment than today, with different everyday needs, and therefore today in modern human society, they might not be always able to play an adaptive role since they can activate stereotypes and inaccurate judgments of others. It is important to note that cognitive biases can be controlled and are influenced by training and education, but often it seems that intelligent and educated persons use their knowledge and skills only to further strengthen and support their biases.

On the other hand, there is a concept which refers to the "objective" capacity to solve specific problems. This is called "intelligence" and has been defined in many different ways, including the capacity for logic, understanding, self-awareness, learning, emotional knowledge, planning, creativity, and problem-solving. Various approaches to human intelligence have been adopted, and its definition is controversial. The word derives from the Latin *intelligere* (to comprehend or perceive). As a scientific topic, it is the result of research on anthropological differences between humans which started in the late eighteenth and early nineteenth centuries. Further elaboration of the matter is beyond the scope of the current chapter.

1.5 Mood, Affect, and Emotions

The modern approach distinguishes between:

- "Mood" which is the long-lasting internal emotional tone and largely characteristic of the individual
- "Affect" which is the general emotional status during the last few days or weeks and is observable through the individual's behavior
- "Emotion" which corresponds to the transient emotional state which also manifests itself through motor behaviors (face mimics, body movements, complex behaviors, etc.)

Mood has an enduring nature, tends to be unfocused and diffused, involves expectation of the future, and is manifested in subtle ways, while, in contrast, emotions tend to be short-lived and to have a clear focus.

In terms of etymology, the word "mood" comes from Proto-Germanic "modaz" and is related to the old high German "muot" (in German: Mut) and the old Saxon mod (in Dutch: moed). Similar words exist in Scandinavian languages, and they mean "anger" and "emotion." The ancient Greek word *μῶθαι* (mothai) and the Latin *mos* (= mores) come for the same Indo-European root. The word "affect" comes from the middle French word "affecter" which is in turn derived from the Latin *afficere* (*ad* + *facere* which means “to act upon,” “influence,” “attack with disease”). It seems that it had entered the English language in the fourteenth century, and it is first recorded in Geoffrey Chaucer’s “Troilus and Cressida.” “Emotion” comes from the French word “émouvoir” which is based on the Latin *emovere* (e-movere means “without move”). Thus the original meanings of these terms relate to anger and to the tendency to act or remain still.

Aristotle’s concept of affect is “that which leads one’s condition to become so transformed that his judgment is affected, and which is accompanied by pleasure and pain” (Aristotle, “Rhetoric” 6). While he was the first to elaborate on human affective states and various terms can be found in his works, in Greek the word for mood is *διάθεση* (diathesi = disposition, tendency, availability), for affect is *συναίσθημα* (synaesthima = complex or combined feelings, sentiment), and for emotion is *συγκίνηση* (synkinisi = complex set of movements as a response).

Affects and emotions serve two main aims.

- The first concerns the internal functioning of the individual and provides the individual with fast decisions which serve the survival of the individual but also of the species. Some of these decisions are easy to understand (e.g., fear of animals), but others are incomprehensible in principle (aesthetics and attraction to the opposite sex). In the same frame, emotions provide feedback concerning the behavior of the individual, and in this way they enhance the expression of the specific behavior or preclude its future manifestations. For example, sadness constitutes the emotional response to loss, defeat, disappointment, or other adversities. Its adaptive function includes permitting withdrawal to conserve resources and asking for support from significant others, and the autonomic arousal which might be present facilitates the search for the lost object or an appropriate substitute.
- The second aim is to communicate the internal emotional state of the individual to others, and this is achieved with facial expressions, gestures, bodily moves and posture, and voice verbal and nonverbal elements. These ways of communicating emotions vary between cultures, but most of the repertoire is universal for human beings. They constitute a main source for the interaction with others, since the emotions of an individual influence the emotions, thoughts, and behaviors of others, produce positive or negative feedback, and give birth to circles of future interactions and reciprocal influence.

The first modern neurobiological theory of emotions was independently developed in the 1880s by William James (1842–1910) and Carl Lange (1834–1900). That theory proposed that emotions are the result of changes in the physiology of the body caused by a stimulus and not the result of the stimulus directly.

A modern approach to the issue goes through a basic approach to brain function which could suggest that there are two distinct mental processes: logical thinking and emotions. While emotions are present also in animals, logical thinking is present primarily in humans, while some elements are also evident in the behavior of primates.

Traditionally, the left hemisphere is considered to be the site of logical thinking, while the right hemisphere serves the creation of emotions. Although this assumption is supported by some neuropsychological data, especially in neurosurgical patients and head injury, it is almost certain that complete and opposite lateralization of logic and emotions does not exist. A consequence of the above assumption is that the dysfunctions leading to schizophrenia are supposed to be localized in the left hemisphere, while those leading to depression are localized in the right.

A very simplified neurobiological model may propose that "mood" derives probably from processes largely taking part in the amygdala and the insula, while "emotion" is generated mainly in the anterior cingulate cortex (ACC) and more specifically in area 25. However, its effortful regulation is likely to implicate area 24 and the dorsolateral prefrontal cortex (DLPFC). In the middle between "mood" and "emotion" lies the "affect" which is at least partially generated in the ACC and partially in other brain areas including the prefrontal cortex (PFC) (Phillips et al. 2003; Fountoulakis et al. 2008).

Emotional processes are evolutionary older and are characterized by speed and dominance. They lead to fast decision-making, on the basis of predetermined strong assumptions concerning the gross characteristics of the situation. For example, fear is triggered immediately and almost before conscious recognition of the stimuli, and it leads to the fast manifestation of adaptive behavior (fight or flight). A snake will always trigger fear, no matter whether it is poisonous or not. On the contrary, logical thinking is slow, requires the conscious elaboration on the stimuli, demands concentration and effort, and is not as strong as emotion concerning the effect on behavior. Emotion is biased toward the triggering of those behaviors that serve the survival of the individual and the species, while logical thinking aims toward an "objective" assessment of the situation. In the language of artificial intelligence, the closest description which can be made today is that of a "fuzzy" vs. "digital" systems.

The database of assumptions emotions use is of unknown origin, probably partially inherited and partially acquired through experience, and possibly it is characteristic of the species. Logical thinking is based mainly on training. Decisions based on emotions are stronger than those based on logical thinking, and when they collide, the person faces a difficult dilemma, since it is very difficult for logical thinking to override emotional pressure.

The two processes, although independent in principle, interact and influence each other. The emotional status causes bias in logical thinking, and logical analysis triggers emotions depending on the positive or negative outcome. This interaction is likely to happen at multiple levels (e.g., selective memory recall, reinforcement through new analysis, biased selection of possible solutions, etc.).

Interest in emotions from an evolutionary perspective was triggered by the publication of the book *The Expression of the Emotions in Man and Animals* by Charles Darwin (1809–1882) in 1872. In that book, Darwin stresses the universal nature of emotions and the connection of mental states to the neurological organization of movement. Central to his understanding was a shared human and animal ancestry in sharp contrast to the contemporary claims that there were divinely created human muscles to express uniquely human feelings. Darwin's original suggestion was that emotions evolved via natural selection and therefore have cross-culturally universal counterparts, a proposal confirmed almost a century later by the works of Paul Ekman (Ekman 1965, 1980, 1992a, b, 1993, 1994, 2003, 2009, 2016; Ekman and Friesen 1967, 1971; Ekman et al. 1969, 1987). Furthermore, animals undergo emotions comparable to those of humans.

1.6 Sleep

The old and lay understanding concerning sleep suggested that the awake state is actively maintained by sensory stimulation, and when fatigue along with a relative lack of sensory stimulation appears, sleep starts as a phase of recovery from the labor of the daytime.

There is a circadian periodicity of a little more than 24 h (close to 25 h) for sleep and wakefulness. A major role in this wake-sleep cycle is played by the reticular formation whose rostral portion contributes to wakefulness, but it is inhibited by the ventral portion which thus induces sleep. External timing cues called zeitgebers (time givers) serve as anchors to adapt the rhythm to the environment. The major external cue is sunlight, which acts through the retino-hypothalamic tract on the major internal clock of the suprachiasmatic nucleus in the anterior hypothalamus.

In terms of description, sleep has two distinct phases; the first is characterized by rapid eye movements (REM sleep), while in the second there are no rapid eye movements (non-REM sleep). During sleep, the two phases alternate cyclically in a highly structured pattern (Aserinsky and Kleitman 1953, 2003).

While its precise function and the reason all animals including humans need sleep (although exceptions and important variations do exist) are unknown, its universal presence in animals suggest an underlying importance. This importance is also shown by the fact that there is a rebound of sleep after sleep deprivation and a rebound of slow-wave or REM sleep after selective deprivation of these stages. Even more impressive is the functional impairments after sleep loss which might lead to death in extreme cases.

It is interesting that total sleep time seems to remain fairly stable from day to day even under widely varying conditions. While the lay belief is that fatigue (both mental and physical) causes a need for prolonged sleep, this has not been supported by data. The only factor that beyond doubt increases sleep is prior sleep loss (sleep deficit).

Although during sleep deprivation, food intake is increased and adipose tissue formation increases, there is no data suggesting that energy conservation is a function of sleep since the metabolism during sleep is only around 15% lower in comparison to wakefulness, but again winter hibernation in some mammals suggests otherwise. On the other hand, increased food intake and adipose tissue formation could be explained as a consequence of the co-occurring significant stress in periods of sleep deprivation. There are some data regarding sleep having thermoregulatory and cooling functions especially in mammals, but this cannot explain the fact that such functions (including somatic rest) could be performed without leaving the animal vulnerable in a state of defense failure. It is essential to note that rest without sleep leaves the individual sleepy. After several days of sleep deprivation, there are no significant physiological changes in the human body, but interestingly there is an intellectual impairment. Even more interesting is the fact that this impairment can be reversed by drugs or motivation. This points to the possibility that the impairment could be due to a pressure to sleep rather than because of a physiological deficit in mental functioning because of prolonged lack of "housekeeping," but again there are no convincing data.

Dreams are another mental function during sleep which has an unknown usefulness for the brain. They are organized thematically and perceptually as separate short stories. They are far more likely to be recalled when subjects were awakened from REM sleep (3/4 or more of awakenings) than from non-REM sleep (less than 10% of awakenings). Non-REM dreams might be shorter, but otherwise there seems to be no difference between dreams in relationship to sleep phase (Rechtschaffen 1978, 1998; Frank 2006; Dement and Wolpert 1958).

Sleep and dreams will be the specific focus of another chapter in this book.

1.7 Consciousness and Human Experience

The word "consciousness" comes from the Latin conscious, meaning "knowing, being aware" which in turn comes from conscire (con meaning "together" + scire meaning "know"). Probably this is a loan translation of the Greek word *συνείδηση* (sinidisi) which comes from the verb *σύνοιδα* (meaning "I know well"). From the Latin scire also come the words "sense" and "science."

While consciousness has been the focus of philosophy in the frame of mind, self, and morality, and most philosophers had adopted the dualistic approach (brain vs. soul/mind) which was suggested for the first time by Rene Descartes (1596–1650) in his books *The Description of the Human Body* (1647) and *Passions of the Soul* (1649) (Descartes et al. 1984), modern advances in neuroscience point to the conclusion that it is fundamentally a function of the brain. However so far there was no success in the efforts to identify neural mechanisms that give rise to consciousness (De Sousa 2013).

A basic approach to consciousness is its conceptualization as a state of awareness. Modern philosophers of mind such as John Searle (1932–) and Thomas Nagel (1937–) ascribe three main features to awareness: subjectivity, unity, and intentionality.

- Subjectivity poses the greatest scientific challenge since it could be in direct conflict with the principles of conducting research and conceptualizing in a scientific way.

- Unity refers to the fact that experiences surface into consciousness as a unified whole and all the various senses are merged into a single conscious experience of reality.
- Intentionality refers to the attribution of meaning beyond the moment of experience.

There is a debate whether the human brain can completely understand itself and subsequently whether consciousness could be scientifically studied. While some philosophers, e.g., Colin McGinn (1950–), believe that there are limits to human cognitive capacities which preclude the comprehension of what consciousness really is, Searle and Nagel disagree. A third approach suggested by the Nobel Laureate Gerald Maurice Edelman (1929–2014) is that consciousness is simply the outcome of the computational workings of the association areas of the brain, and in this frame it could be the focus of scientific research (Edelman 1978, 1989, 1992; Edelman and Changeux 2001).

The great practical problem is the astronomical complexity of the brain's functional anatomy and physiology. A measure of this complexity can be shown with the following example. Let's take not the whole human brain but only one of the two optic nerves. This part of the nervous system contains between 770,000 and 1.7 million nerve fibers (Jonas et al. 1992), and to make calculations easy, let's assume it is one million. Now if each of these neurons can take only two distinct states (e.g., yes/no, fire/silent, etc.) which is a very simplistic approach if one takes into consideration the neuronal function as described earlier, then the total number of states the optic nerve can take is $2^{1,000,000}$. This number, if transformed in power of 10, is larger than $10^{250,000}$ which should be compared with 10^{80}–10^{97} which correspond to estimations of the number of all elementary particles of the visible universe.

This scale of complexity, considered together with the limited number of experimental trials we can conduct and the limited number of subjects and observations we can gather, makes impossible the study of these phenomena with the standard reductionist approach to events, which science normally follows.

Currently we lack the tools both theoretical and also operational to dig deeper into these mechanisms and to understand them, far less to extrapolate subjective properties (consciousness) from the properties of objects (interconnected nerve cells).

While eventually we have a general idea of how the brain works, from the atom and molecule level to the level of social behavior and abstract expression of ideas, we lack the precise knowledge that bridges different steps and levels in this system. We lack the knowledge of precisely how the incoming sensory information is broken, coded, and stored and how a non-deterministic process leads to a decision. We lack the knowledge of how electric currents and chemical reactions through a highly organized super-complexity give rise to consciousness, to sense of individuality, and even more important, in humans, to the realization of the inevitable future death. This realization and subsequent fear are, according to many philosophers, the driving force of creativity.

References

A decade of neuropeptides: past, present, and future. Tenth Annual Winter Neuropeptide Conference. Breckenridge, Colorado, January 16–20 (1989) Proceedings (1990). Ann N Y Acad Sci 579:1–280
Aghajanian GK, VanderMaelen CP (1982) Alpha 2-adrenoceptor-mediated hyperpolarization of locus coeruleus neurons: intracellular studies in vivo. Science 215(4538):1394–1396
Alvarez EO (2009) The role of histamine on cognition. Behav Brain Res 199(2):183–189. https://doi.org/10.1016/j.bbr.2008.12.010
Amara SG, Arriza JL (1993) Neurotransmitter transporters: three distinct gene families. Curr Opin Neurobiol 3(3):337–344
Anden NE, Corrodi H, Fuxe K, Hokfelt B, Hokfelt T, Rydin C, Svensson T (1970) Evidence for a central noradrenaline receptor stimulation by clonidine. Life Sci 9(9):513–523
Arachidonate related lipid mediators (1990) Methods Enzymol 187:1–628
Arneric S, Sullivan J, Williams M (1994) Neuronal nicotinic acetylcholine receptors: novel targets for CNS therapeutics. In: Bloom F, Kupfer D (eds) Psychopharmacology: the fourth generation of progress. Raven Press, New York, NY, pp 95–110
Arvanitaki A (1942) Effects evoked in an axon by the activity of a contiguous one. J Neurophysiol 5(2):89–108
Asanuma C, Thach WT, Jones EG (1983) Cytoarchitectonic delineation of the ventral lateral thalamic region in the monkey. Brain Res 286(3):219–235
Aserinsky E, Kleitman N (1953) Regularly occurring periods of eye motility, and concomitant phenomena, during sleep. Science 118(3062):273–274
Aserinsky E, Kleitman N (2003) Regularly occurring periods of eye motility, and concomitant phenomena, during sleep. 1953. J Neuropsychiatry Clin Neurosci 15(4):454–455. https://doi.org/10.1176/jnp.15.4.454
Asztely F, Gustafsson B (1996) Ionotropic glutamate receptors. Their possible role in the expression of hippocampal synaptic plasticity. Mol Neurobiol 12(1):1–11
Bahler M, Greengard P (1987) Synapsin I bundles F-actin in a phosphorylation-dependent manner. Nature 326(6114):704–707. https://doi.org/10.1038/326704a0
Baker PF, Hodgkin AL, Ridgway EB (1971) Depolarization and calcium entry in squid giant axons. J Physiol 218(3):709–755
Barger G, Dale H (1910) β-Imidazolylethylamine, a depressor constituent of intestinal mucosa. J Physiol 41:499–503
Bettler B, Boulter J, Hermans-Borgmeyer I, O'Shea-Greenfield A, Deneris ES, Moll C, Borgmeyer U, Hollmann M, Heinemann S (1990) Cloning of a novel glutamate receptor subunit, GluR5: expression in the nervous system during development. Neuron 5(5):583–595
Bettler B, Egebjerg J, Sharma G, Pecht G, Hermans-Borgmeyer I, Moll C, Stevens CF, Heinemann S (1992) Cloning of a putative glutamate receptor: a low affinity kainate-binding subunit. Neuron 8(2):257–265
Blandina P, Munari L, Provensi G, Passani MB (2012) Histamine neurons in the tuberomammillary nucleus: a whole center or distinct subpopulations? Front Syst Neurosci 6:33. https://doi.org/10.3389/fnsys.2012.00033
Blue ME, Yagaloff KA, Mamounas LA, Hartig PR, Molliver ME (1988) Correspondence between 5-HT2 receptors and serotonergic axons in rat neocortex. Brain Res 453(1–2):315–328
Breslau N, Kilbey MM, Andreski P (1993) Nicotine dependence and major depression. New evidence from a prospective investigation. Arch Gen Psychiatry 50(1):31–35
Briggs F, Usrey WM (2008) Emerging views of corticothalamic function. Curr Opin Neurobiol 18(4):403–407. https://doi.org/10.1016/j.conb.2008.09.002
Brotehrs L (1990) The social brain: a project for integrating primate behavior and neurophysiology in a new domain. Concepts Neurosci 1:27–51
Brown RE, Stevens DR, Haas HL (2001) The physiology of brain histamine. Prog Neurobiol 63(6):637–672
Bunge RP (1968) Glial cells and the central myelin sheath. Physiol Rev 48(1):197–251

Bylund DB, Eikenberg DC, Hieble JP, Langer SZ, Lefkowitz RJ, Minneman KP, Molinoff PB, Ruffolo RR Jr, Trendelenburg U (1994) International Union of Pharmacology nomenclature of adrenoceptors. Pharmacol Rev 46(2):121–136
Cardenas L, Tremblay LK, Naranjo CA, Herrmann N, Zack M, Busto UE (2002) Brain reward system activity in major depression and comorbid nicotine dependence. J Pharmacol Exp Ther 302(3):1265–1271
Carlesimo GA, Lombardi MG, Caltagirone C (2011) Vascular thalamic amnesia: a reappraisal. Neuropsychologia 49(5):777–789. https://doi.org/10.1016/j.neuropsychologia.2011.01.026
Cates LN, Roberts AJ, Huitron-Resendiz S, Hedlund PB (2013) Effects of lurasidone in behavioral models of depression. Role of the 5-HT(7) receptor subtype. Neuropharmacology 70:211–217. https://doi.org/10.1016/j.neuropharm.2013.01.023
Cedarbaum JM, Aghajanian GK (1977) Catecholamine receptors on locus coeruleus neurons: pharmacological characterization. Eur J Pharmacol 44(4):375–385
Changeux J, Dehaene S (1993) Formal neuronal models for cognitive functions associated with the prefrontal cortex. In: Poggio T, Glaser D (eds) Exploring brain functions – models in neuroscience. John Wiley and Sons, Chichester, pp 249–267
Civelli O (1994) Molecular biology of the dopamine receptor subtypes. In: Bloom F, Kupfer D (eds) Psychopharmacology: the fourth generation of progress. Raven Press, New York, NY, pp 155–161
Clark SA, Allard T, Jenkins WM, Merzenich MM (1988) Receptive fields in the body-surface map in adult cortex defined by temporally correlated inputs. Nature 332(6163):444–445. https://doi.org/10.1038/332444a0
Contractor A, Mulle C, Swanson GT (2011) Kainate receptors coming of age: milestones of two decades of research. Trends Neurosci 34(3):154–163. https://doi.org/10.1016/j.tins.2010.12.002
Cossart R, Bernard C, Ben-Ari Y (2005) Multiple facets of GABAergic neurons and synapses: multiple fates of GABA signalling in epilepsies. Trends Neurosci 28(2):108–115. https://doi.org/10.1016/j.tins.2004.11.011
Dautan D, Hacioglu Bay H, Bolam JP, Gerdjikov TV, Mena-Segovia J (2016) Extrinsic sources of cholinergic innervation of the striatal complex: a whole-brain mapping analysis. Front Neuroanat 10:1. https://doi.org/10.3389/fnana.2016.00001
De Sousa A (2013) Towards an integrative theory of consciousness: part 1 (neurobiological and cognitive models). Mens Sana Monogr 11(1):100–150. https://doi.org/10.4103/0973-1229.109335
Deakin J (2013) The origins of '5-HT and mechanisms of defence' by Deakin and Graeff: a personal perspective. J Psychopharmacol 27(12):1084–1089. https://doi.org/10.1177/0269881113503508
Deakin JF, Graeff FG (1991) 5-HT and mechanisms of defence. J Psychopharmacol 5(4):305–315. https://doi.org/10.1177/026988119100500414
Dement W, Wolpert EA (1958) The relation of eye movements, body motility, and external stimuli to dream content. J Exp Psychol 55(6):543–553
Descartes R, Cottingham J, Stoothoff R, Murdochtrans D (1984) The philosophical writings of rené descartes (1641). Cambridge University Press, Cambridge
Dunlap K, Luebke JI, Turner TJ (1995) Exocytotic Ca2+ channels in mammalian central neurons. Trends Neurosci 18(2):89–98
Eccles J (1976) From electrical to chemical transmission in the central nervous system. Notes Rec R Soc Lond 30(2):219–230
Edelman G (1978) The mindful brain: cortical organization and the group-selective theory of higher brain function. MIT Press, Cambridge
Edelman G (1989) The remembered present: a biological theory of consciousness. Basic Books, New York, NY
Edelman G (1992) Bright air, brilliant fire. Penguin, London
Edelman G, Changeux J (2001) The brain. Transaction Publishers, Piscataway, NJ
Edwards AC, Maes HH, Pedersen NL, Kendler KS (2011) A population-based twin study of the genetic and environmental relationship of major depression, regular tobacco use and nicotine dependence. Psychol Med 41(2):395–405. https://doi.org/10.1017/S0033291710000589

Egebjerg J, Bettler B, Hermans-Borgmeyer I, Heinemann S (1991) Cloning of a cDNA for a glutamate receptor subunit activated by kainate but not AMPA. Nature 351(6329):745–748. https://doi.org/10.1038/351745a0
Ehlert F, Roeske W, Yamamura H (1994) Molecular biology, pharmacology and brain distribution of subtypes of the muscarinic receptor. In: Bloom F, Kupfer D (eds) Psychopharmacology: the fourth generation of progress. Raven Press, New York, NY, pp 111–124
Ekman P (1965) Differential communication of affect by head and body cues. J Pers Soc Psychol 2(5):726–735
Ekman P (1980) Asymmetry in facial expression. Science 209(4458):833–834
Ekman P (1992a) Are there basic emotions? Psychol Rev 99(3):550–553
Ekman P (1992b) Facial expressions of emotion: an old controversy and new findings. Philos Trans R Soc Lond Ser B Biol Sci 335(1273):63–69. https://doi.org/10.1098/rstb.1992.0008
Ekman P (1993) Facial expression and emotion. Am Psychol 48(4):384–392
Ekman P (1994) Strong evidence for universals in facial expressions: a reply to Russell's mistaken critique. Psychol Bull 115(2):268–287
Ekman P (2003) Emotions inside out. 130 years after Darwin's "the expression of the emotions in man and animal". Ann N Y Acad Sci 1000:1–6
Ekman P (2009) Darwin's contributions to our understanding of emotional expressions. Philos Trans R Soc Lond Ser B Biol Sci 364(1535):3449–3451. https://doi.org/10.1098/rstb.2009.0189
Ekman P (2016) What scientists who study emotion agree about. Persp Psychol Sci 11(1):31–34. https://doi.org/10.1177/1745691615596992
Ekman P, Friesen WV (1967) Head and body cues in the judgment of emotion: a reformulation. Percept Mot Skills 24(3):711–724. https://doi.org/10.2466/pms.1967.24.3.711
Ekman P, Friesen WV (1971) Constants across cultures in the face and emotion. J Pers Soc Psychol 17(2):124–129
Ekman P, Sorenson ER, Friesen WV (1969) Pan-cultural elements in facial displays of emotion. Science 164(3875):86–88
Ekman P, Friesen WV, O'Sullivan M, Chan A, Diacoyanni-Tarlatzis I, Heider K, Krause R, LeCompte WA, Pitcairn T, Ricci-Bitti PE et al (1987) Universals and cultural differences in the judgments of facial expressions of emotion. J Pers Soc Psychol 53(4):712–717
Emorine LJ, Marullo S, Briend-Sutren MM, Patey G, Tate K, Delavier-Klutchko C, Strosberg AD (1989) Molecular characterization of the human beta 3-adrenergic receptor. Science 245(4922):1118–1121
Enz R, Cutting GR (1998) Molecular composition of GABAC receptors. Vis Res 38(10):1431–1441
Erulkar SD, Rahamimoff R (1978) The role of calcium ions in tetanic and post-tetanic increase of miniature end-plate potential frequency. J Physiol 278:501–511
Evarts EV, Thach WT (1969) Motor mechanisms of the CNS: cerebrocerebellar interrelations. Annu Rev Physiol 31:451–498. https://doi.org/10.1146/annurev.ph.31.030169.002315
Felleman DJ, Van Essen DC (1991) Distributed hierarchical processing in the primate cerebral cortex. Cereb Cortex 1(1):1–47
Fisher JM, Sossin W, Newcomb R, Scheller RH (1988) Multiple neuropeptides derived from a common precursor are differentially packaged and transported. Cell 54(6):813–822
Fountoulakis KN, Giannakopoulos P, Kovari E, Bouras C (2008) Assessing the role of cingulate cortex in bipolar disorder: neuropathological, structural and functional imaging data. Brain Res Rev 59(1):9–21. https://doi.org/10.1016/j.brainresrev.2008.04.005
Fox PT, Miezin FM, Allman JM, Van Essen DC, Raichle ME (1987) Retinotopic organization of human visual cortex mapped with positron-emission tomography. J Neurosci 7(3):913–922
Frank MG (2006) The mystery of sleep function: current perspectives and future directions. Rev Neurosci 17(4):375–392
Frielle T, Collins S, Daniel KW, Caron MG, Lefkowitz RJ, Kobilka BK (1987) Cloning of the cDNA for the human beta 1-adrenergic receptor. Proc Natl Acad Sci U S A 84(22):7920–7924
Fu Q, Heath AC, Bucholz KK, Lyons MJ, Tsuang MT, True WR, Eisen SA (2007) Common genetic risk of major depression and nicotine dependence: the contribution of antisocial traits in a United States veteran male twin cohort. Twin Res Hum Genet 10(3):470–478. https://doi.org/10.1375/twin.10.3.470

Fujishiro H, Umegaki H, Isojima D, Akatsu H, Iguchi A, Kosaka K (2006) Depletion of cholinergic neurons in the nucleus of the medial septum and the vertical limb of the diagonal band in dementia with Lewy bodies. Acta Neuropathol 111(2):109–114. https://doi.org/10.1007/s00401-005-0004-1
Gazzaniga MS (1989) Organization of the human brain. Science 245(4921):947–952
Gazzaniga MS (1995) Principles of human brain organization derived from split-brain studies. Neuron 14(2):217–228
Gerard RW (1949) Physiology and psychiatry. Am J Psychiatry 106(3):161–173. https://doi.org/10.1176/ajp.106.3.161
Gerard RW (1955a) Biological roots of psychiatry. Science 122(3162):225–230
Gerard RW (1955b) The biological roots of psychiatry. Am J Psychiatry 112(2):81–90. https://doi.org/10.1176/ajp.112.2.81
von Gersdorff H, Matthews G (1994) Dynamics of synaptic vesicle fusion and membrane retrieval in synaptic terminals. Nature 367(6465):735–739. https://doi.org/10.1038/367735a0
Geula C, Mesulam MM (2011) Brainstem cholinergic systems. In: Mai J, Paxinos G (eds) The human nervous system. Academic, Cambridge, MA, pp 456–470
Gigerenzer G, Goldstein DG (1996) Reasoning the fast and frugal way: models of bounded rationality. Psychol Rev 103(4):650–669
Gilman AG (1989) The Albert Lasker Medical Awards. G proteins and regulation of adenylyl cyclase. JAMA 262(13):1819–1825
Gilman AG (1990) Regulation of adenylyl cyclase by G proteins. Adv Second Messenger Phosphoprotein Res 24:51–57
Gilman AG (1995) Nobel lecture. G proteins and regulation of adenylyl cyclase. Biosci Rep 15(2):65–97
Goldman-Rakic PS (1992) Working memory and the mind. Sci Am 267(3):110–117
Goldman-Rakic PS (1996) Regional and cellular fractionation of working memory. Proc Natl Acad Sci U S A 93(24):13473–13480
Gordon WW (1948) Cerebral physiology and psychiatry. J Ment Sci 94(394):118–132
Graeff FG, Silveira MC, Nogueira RL, Audi EA, Oliveira RM (1993) Role of the amygdala and periaqueductal gray in anxiety and panic. Behav Brain Res 58(1–2):123–131
Greger IH, Ziff EB, Penn AC (2007) Molecular determinants of AMPA receptor subunit assembly. Trends Neurosci 30(8):407–416. https://doi.org/10.1016/j.tins.2007.06.005
Hanson PI, Heuser JE, Jahn R (1997) Neurotransmitter release – four years of SNARE complexes. Curr Opin Neurobiol 7(3):310–315
Haselton MG, Nettle D, Andrews PW (2005) The evolution of cognitive bias. In: Buss D (ed) The handbook of evolutionary psychology. John Wiley & Sons Inc, Hoboken, NJ, pp 724–746
Herb A, Burnashev N, Werner P, Sakmann B, Wisden W, Seeburg PH (1992) The KA-2 subunit of excitatory amino acid receptors shows widespread expression in brain and forms ion channels with distantly related subunits. Neuron 8(4):775–785
Hieble JP, Bylund DB, Clarke DE, Eikenburg DC, Langer SZ, Lefkowitz RJ, Minneman KP, Ruffolo RR Jr (1995) International Union of Pharmacology. X. Recommendation for nomenclature of alpha 1-adrenoceptors: consensus update. Pharmacol Rev 47(2):267–270
Hille B (1994) Modulation of ion-channel function by G-protein-coupled receptors. Trends Neurosci 17(12):531–536
Hirai H, Kirsch J, Laube B, Betz H, Kuhse J (1996) The glycine binding site of the N-methyl-D-aspartate receptor subunit NR1: identification of novel determinants of co-agonist potentiation in the extracellular M3-M4 loop region. Proc Natl Acad Sci U S A 93(12):6031–6036
Hodges H, Green S, Glenn B (1987) Evidence that the amygdala is involved in benzodiazepine and serotonergic effects on punished responding but not on discrimination. Psychopharmacology 92(4):491–504
Huang YH, Bergles DE (2004) Glutamate transporters bring competition to the synapse. Curr Opin Neurobiol 14(3):346–352
Jacques S (1979) Brain stimulation and reward: "pleasure centers" after twenty-five years. Neurosurgery 5(2):277–283

Jenkins WM, Merzenich MM, Ochs MT, Allard T, Guic-Robles E (1990) Functional reorganization of primary somatosensory cortex in adult owl monkeys after behaviorally controlled tactile stimulation. J Neurophysiol 63(1):82–104
Jessell TM, Kandel ER (1993) Synaptic transmission: a bidirectional and self-modifiable form of cell-cell communication. Cell 72(Suppl):1–30
Jonas JB, Schmidt AM, Muller-Bergh JA, Schlotzer-Schrehardt UM, Naumann GO (1992) Human optic nerve fiber count and optic disc size. Invest Ophthalmol Vis Sci 33(6):2012–2018
Kaas JH, Nelson RJ, Sur M, Lin CS, Merzenich MM (1979) Multiple representations of the body within the primary somatosensory cortex of primates. Science 204(4392):521–523
Kahneman D (2003) A perspective on judgment and choice: mapping bounded rationality. Am Psychol 58(9):697–720. https://doi.org/10.1037/0003-066X.58.9.697
Kahneman D, Frederick S (2007) Frames and brains: elicitation and control of response tendencies. Trends Cogn Sci 11(2):45–46. https://doi.org/10.1016/j.tics.2006.11.007
Kahneman D, Klein G (2009) Conditions for intuitive expertise: a failure to disagree. Am Psychol 64(6):515–526. https://doi.org/10.1037/a0016755
Kahneman D, Shane F (2002) Representativeness revisited: attribute substitution in intuitive judgment. In: Gilovich T, Griffin D, Kahneman D (eds) Heuristics and biases: the psychology of intuitive judgment. Cambridge University Press, Cambridge, pp 51–52
Kahneman D, Tversky A (1982) Variants of uncertainty. Cognition 11(2):143–157
Kahneman D, Tversky A (1996) On the reality of cognitive illusions. Psychol Rev 103(3):582–591. discusion 592–586
Kandel ER (1981) Calcium and the control of synaptic strength by learning. Nature 293(5835):697–700
Kandel E (2000) The brain and behavior. In: Kandel E, Schwarts J, Jessell T (eds) Principles of neural science. McGraw-Hill, New York, NY, pp 5–18
Kelly RB (1993) Storage and release of neurotransmitters. Cell 72(Suppl):43–53
Killen JD, Fortmann SP, Schatzberg A, Hayward C, Varady A (2003) Onset of major depression during treatment for nicotine dependence. Addict Behav 28(3):461–470
Klein M, Shapiro E, Kandel ER (1980) Synaptic plasticity and the modulation of the Ca2+ current. J Exp Biol 89:117–157
Kupfermann I (1991) Functional studies of cotransmission. Physiol Rev 71(3):683–732
Lanciego JL, Luquin N, Obeso JA (2012) Functional neuroanatomy of the basal ganglia. Cold Spring Harb Perspect Med 2(12):a009621. https://doi.org/10.1101/cshperspect.a009621
Langer SZ (1974) Presynaptic regulation of catecholamine release. Biochem Pharmacol 23(13):1793–1800
Langer SZ, Arbilla S (1990) Presynaptic receptors on peripheral noradrenergic neurons. Ann N Y Acad Sci 604:7–16
Laube B, Hirai H, Sturgess M, Betz H, Kuhse J (1997) Molecular determinants of agonist discrimination by NMDA receptor subunits: analysis of the glutamate binding site on the NR2B subunit. Neuron 18(3):493–503
Laurie DJ, Bartke I, Schoepfer R, Naujoks K, Seeburg PH (1997) Regional, developmental and interspecies expression of the four NMDAR2 subunits, examined using monoclonal antibodies. Brain Res Mol Brain Res 51(1–2):23–32
Lawson D, Raff MC, Gomperts B, Fewtrell C, Gilula NB (1977) Molecular events during membrane fusion. A study of exocytosis in rat peritoneal mast cells. J Cell Biol 72(2):242–259
Le Moal M (1994) Mesocorticolimbic dopaminergic neurons: functional and regulatory roles. In: Bloom F, Kupfer D (eds) Psychopharmacology: the fourth generation of progress. Raven Press, New York, NY, pp 283–293
LeDoux JE, Cicchetti P, Xagoraris A, Romanski LM (1990) The lateral amygdaloid nucleus: sensory interface of the amygdala in fear conditioning. J Neurosci 10(4):1062–1069
Lemieux RU, Spohr U (1994) How Emil Fischer was led to the lock and key concept for enzyme specificity. Adv Carbohydr Chem Biochem 50:1–20
Lewis D, Oeth K (1995) Functional neuroanatomy. In: Kaplan H, Sadock B (eds) Comprehensive textbook of psychiatry, 6th edn. Williams and Wilkins, Baltimore, MD, pp 4–24
Liley AW (1956) The quantal components of the mammalian end-plate potential. J Physiol 133(3):571–587

Lindau M, Almers W (1995) Structure and function of fusion pores in exocytosis and ectoplasmic membrane fusion. Curr Opin Cell Biol 7(4):509–517
Liu AK, Chang RC, Pearce RK, Gentleman SM (2015) Nucleus basalis of Meynert revisited: anatomy, history and differential involvement in Alzheimer's and Parkinson's disease. Acta Neuropathol 129(4):527–540. https://doi.org/10.1007/s00401-015-1392-5
Lyons M, Hitsman B, Xian H, Panizzon MS, Jerskey BA, Santangelo S, Grant MD, Rende R, Eisen S, Eaves L, Tsuang MT (2008) A twin study of smoking, nicotine dependence, and major depression in men. Nicotine Tobacco Res 10(1):97–108. https://doi.org/10.1080/14622200701705332
Maehle AH (2004) "Receptive substances": John Newport Langley (1852–1925) and his path to a receptor theory of drug action. Med Hist 48(2):153–174
Majerus PW (1992) Inositol phosphate biochemistry. Annu Rev Biochem 61:225–250. https://doi.org/10.1146/annurev.bi.61.070192.001301
Mamounas LA, Mullen CA, O'Hearn E, Molliver ME (1991) Dual serotoninergic projections to forebrain in the rat: morphologically distinct 5-HT axon terminals exhibit differential vulnerability to neurotoxic amphetamine derivatives. J Comp Neurol 314(3):558–586. https://doi.org/10.1002/cne.903140312
Manaye KF, Zweig R, Wu D, Hersh LB, De Lacalle S, Saper CB, German DC (1999) Quantification of cholinergic and select non-cholinergic mesopontine neuronal populations in the human brain. Neuroscience 89(3):759–770
Mansur A, Watson S (1994) Dopamine receptor expression in the central nervous system. In: Bloom F, Kupfer D (eds) Psychopharmacology: the fourth generation of progress. Raven Press, New York, NY, pp 207–220
Marsden C (1996) The neuropharmacology of serotonin in the central nervous system. In: Feighner J, Boyer W (eds) Selective serotonin re-uptake inhibitors. John Wiley and Sons, Chichester, pp 1–35
Matthews G (1996) Synaptic exocytosis and endocytosis: capacitance measurements. Curr Opin Neurobiol 6(3):358–364
Mayer ML (2005) Glutamate receptor ion channels. Curr Opin Neurobiol 15(3):282–288. https://doi.org/10.1016/j.conb.2005.05.004
Mayer ML, Westbrook GL, Guthrie PB (1984) Voltage-dependent block by Mg2+ of NMDA responses in spinal cord neurones. Nature 309(5965):261–263
Meldrum BS (2000) Glutamate as a neurotransmitter in the brain: review of physiology and pathology. J Nutr 130(4S Suppl):1007S–1015S
Mesulam M (1994) Structure and function of cholinergic pathways in the cerebral cortex, limbic system, basal ganglia and thalamus of the human brain. In: Bloom F, Kupfer D (eds) Psychopharmacology: the fourth generation of progress. Raven Press, New York, NY, pp 135–146
Mesulam MM, Mufson EJ, Wainer BH, Levey AI (1983) Central cholinergic pathways in the rat: an overview based on an alternative nomenclature (Ch1-Ch6). Neuroscience 10(4):1185–1201
Mesulam MM, Geula C, Bothwell MA, Hersh LB (1989) Human reticular formation: cholinergic neurons of the pedunculopontine and laterodorsal tegmental nuclei and some cytochemical comparisons to forebrain cholinergic neurons. J Comp Neurol 283(4):611–633. https://doi.org/10.1002/cne.902830414
Mishkin M (1993) Cerebral memory circuits. In: Poggio T, Glaser D (eds) Exploring brain functions – models in neuroscience. John Wiley and Sons, Chichester, pp 113–127
Mogilner A, Grossman JA, Ribary U, Joliot M, Volkmann J, Rapaport D, Beasley RW, Llinas RR (1993) Somatosensory cortical plasticity in adult humans revealed by magnetoencephalography. Proc Natl Acad Sci U S A 90(8):3593–3597
Monaghan DT, Jane DE (2009) Pharmacology of NMDA receptors. In: VanDongen AM (ed) Biology if the NMDA receptor. CRC Press, Boca Raton, FL, pp 257–283
Monaghan DT, Bridges RJ, Cotman CW (1989) The excitatory amino acid receptors: their classes, pharmacology, and distinct properties in the function of the central nervous system. Annu Rev Pharmacol Toxicol 29:365–402
Montgomery SA, Fineberg N (1989) Is there a relationship between serotonin receptor subtypes and selectivity of response in specific psychiatric illnesses? Br J Psychiatry Suppl 8:63–69

Mony L, Kew JN, Gunthorpe MJ, Paoletti P (2009) Allosteric modulators of NR2B-containing NMDA receptors: molecular mechanisms and therapeutic potential. Br J Pharmacol 157(8):1301–1317
Moore K, Lookingland K (1994) Dopaminergic neuronal systems in the hypothalamus. In: Bloom F, Kupfer D (eds) Psychopharmacology: the fourth generation of progress. Raven Press, New York, NY, pp 245–256
Morewedge CK, Kahneman D (2010) Associative processes in intuitive judgment. Trends Cogn Sci 14(10):435–440. https://doi.org/10.1016/j.tics.2010.07.004
Morgane PJ, Galler JR, Mokler DJ (2005) A review of systems and networks of the limbic forebrain/limbic midbrain. Prog Neurobiol 75(2):143–160. https://doi.org/10.1016/j.pneurobio.2005.01.001
Mufson EJ, Martin TL, Mash DC, Wainer BH, Mesulam MM (1986) Cholinergic projections from the parabigeminal nucleus (Ch8) to the superior colliculus in the mouse: a combined analysis of horseradish peroxidase transport and choline acetyltransferase immunohistochemistry. Brain Res 370(1):144–148
Mustafa AK, Gadalla MM, Snyder SH (2009) Signaling by gasotransmitters. Sci Signal 2(68):re2. https://doi.org/10.1126/scisignal.268re2
Myers RD (1994) Neuroactive peptides: unique phases in research on mammalian brain over three decades. Peptides 15(2):367–381
Nakanishi S (1992) Molecular diversity of glutamate receptors and implications for brain function. Science 258(5082):597–603
Needleman P, Turk J, Jakschik BA, Morrison AR, Lefkowith JB (1986) Arachidonic acid metabolism. Annu Rev Biochem 55:69–102. https://doi.org/10.1146/annurev.bi.55.070186.000441
Nelson N, Lill H (1994) Porters and neurotransmitter transporters. J Exp Biol 196:213–228
Nicholas AP, Pieribone V, Hokfelt T (1993a) Distributions of mRNAs for alpha-2 adrenergic receptor subtypes in rat brain: an in situ hybridization study. J Comp Neurol 328(4):575–594. https://doi.org/10.1002/cne.903280409
Nicholas AP, Pieribone VA, Hokfelt T (1993b) Cellular localization of messenger RNA for beta-1 and beta-2 adrenergic receptors in rat brain: an in situ hybridization study. Neuroscience 56(4):1023–1039
Nicoll RA, Malenka RC, Kauer JA (1990) Functional comparison of neurotransmitter receptor subtypes in mammalian central nervous system. Physiol Rev 70(2):513–565
O'Connor WT (1998) Functional neuroanatomy of the basal ganglia as studied by dual-probe microdialysis. Nucl Med Biol 25(8):743–746
O'Dowd BF, Hnatowich M, Regan JW, Leader WM, Caron MG, Lefkowitz RJ (1988) Site-directed mutagenesis of the cytoplasmic domains of the human beta 2-adrenergic receptor. Localization of regions involved in G protein-receptor coupling. J Biol Chem 263(31):15985–15992
Olsen RW, Sieghart W (2009) GABA A receptors: subtypes provide diversity of function and pharmacology. Neuropharmacology 56(1):141–148. https://doi.org/10.1016/j.neuropharm.2008.07.045
Ostrowski J, Kjelsberg MA, Caron MG, Lefkowitz RJ (1992) Mutagenesis of the beta 2-adrenergic receptor: how structure elucidates function. Annu Rev Pharmacol Toxicol 32:167–183. https://doi.org/10.1146/annurev.pa.32.040192.001123
Palacios J, Kuhar MJ (1982) Beta adrenergic receptor localization in rat brain by light microscopic autoradiography. Neurochem Int 4(6):473–490
Panula P, Chazot PL, Cowart M, Gutzmer R, Leurs R, Liu WL, Stark H, Thurmond RL, Haas HL (2015) International Union of Basic and Clinical Pharmacology. XCVIII. Histamine receptors. Pharmacol Rev 67(3):601–655. https://doi.org/10.1124/pr.114.010249
Patel S, Patel U, Vithalani D, Verma SC (1981) Regulation of catecholamine release by presynaptic receptor system. Gen Pharmacol 12(6):405–422
Paul S (1994) GABA and Glycine. In: Bloom F, Kupfer D (eds) Psychopharmacology: the fourth generation of progress. Raven Press, New York, NY, pp 87–94
Penfield W, Rasmussen T (1950) The cerebral cortex of man: a clinical study of localization of function. Macmillan, New York, NY

Phillips ML, Drevets WC, Rauch SL, Lane R (2003) Neurobiology of emotion perception II: implications for major psychiatric disorders. Biol Psychiatry 54(5):515–528
Pin JP, Acher F (2002) The metabotropic glutamate receptors: structure, activation mechanism and pharmacology. Curr Drug Targets CNS Neurol Disord 1(3):297–317
Rainbow TC, Parsons B, Wolfe BB (1984) Quantitative autoradiography of beta 1- and beta 2-adrenergic receptors in rat brain. Proc Natl Acad Sci U S A 81(5):1585–1589
Ramachandran VS (1993) Behavioral and magnetoencephalographic correlates of plasticity in the adult human brain. Proc Natl Acad Sci U S A 90(22):10413–10420
Rechtschaffen A (1978) The single-mindedness and isolation of dreams. Sleep 1(1):97–109
Rechtschaffen A (1998) Current perspectives on the function of sleep. Perspect Biol Med 41(3):359–390
Reiner P, Fibiger C (1994) Functional heterogeneity of central cholinergic systems. In: Bloom F, Kupfer D (eds) Psychopharmacology: the fourth generation of progress. Raven Press, New York, NY, pp 147–153
Rosin DL, Talley EM, Lee A, Stornetta RL, Gaylinn BD, Guyenet PG, Lynch KR (1996) Distribution of alpha 2C-adrenergic receptor-like immunoreactivity in the rat central nervous system. J Comp Neurol 372(1):135–165. https://doi.org/10.1002/(SICI)1096-9861(19960812)372:1<135::AID-CNE9>3.0.CO;2-4
Roth R, Elsworth J (1994) Biochemical pharmacology of midbrain dopamine neurons. In: Bloom F, Kupfer D (eds) Psychopharmacology: the fourth generation of progress. Raven Press, New York, NY, pp 227–243
Sakimura K, Morita T, Kushiya E, Mishina M (1992) Primary structure and expression of the gamma 2 subunit of the glutamate receptor channel selective for kainate. Neuron 8(2):267–274
Salm AK, McCarthy KD (1992) The evidence for astrocytes as a target for central noradrenergic activity: expression of adrenergic receptors. Brain Res Bull 29(3–4):265–275
Scheinin M, Lomasney JW, Hayden-Hixson DM, Schambra UB, Caron MG, Lefkowitz RJ, Fremeau RT Jr (1994) Distribution of alpha 2-adrenergic receptor subtype gene expression in rat brain. Brain Res Mol Brain Res 21(1–2):133–149
Schweizer FE, Betz H, Augustine GJ (1995) From vesicle docking to endocytosis: intermediate reactions of exocytosis. Neuron 14(4):689–696
Shigeri Y, Seal RP, Shimamoto K (2004) Molecular pharmacology of glutamate transporters, EAATs and VGLUTs. Brain Res Brain Res Rev 45(3):250–265. https://doi.org/10.1016/j.brainresrev.2004.04.004
Shorr RG, Strohsacker MW, Lavin TN, Lefkowitz RJ, Caron MG (1982) The beta 1-adrenergic receptor of the turkey erythrocyte. Molecular heterogeneity revealed by purification and photoaffinity labeling. J Biol Chem 257(20):12341–12350
Smith SJ, Augustine GJ (1988) Calcium ions, active zones and synaptic transmitter release. Trends Neurosci 11(10):458–464
Stahl SM (2008) Stahl's essential psychopharmacology: neuroscientific basis and practical applications, 3rd edn. Cambridge University Press, Cambridge
Starke K (1971) Influence of alpha-receptor stimulants on noradrenaline release. Naturwissenschaften 58(8):420
Stein T, Moritz C, Quigley M, Cordes D, Haughton V, Meyerand E (2000) Functional connectivity in the thalamus and hippocampus studied with functional MR imaging. AJNR Am J Neuroradiol 21(8):1397–1401
Stone EA, Ariano MA (1989) Are glial cells targets of the central noradrenergic system? A review of the evidence. Brain Res Brain Res Rev 14(4):297–309
Svensson TH, Bunney BS, Aghajanian GK (1975) Inhibition of both noradrenergic and serotonergic neurons in brain by the alpha-adrenergic agonist clonidine. Brain Res 92(2):291–306
Talley EM, Rosin DL, Lee A, Guyenet PG, Lynch KR (1996) Distribution of alpha 2A-adrenergic receptor-like immunoreactivity in the rat central nervous system. J Comp Neurol 372(1):111–134. https://doi.org/10.1002/(SICI)1096-9861(19960812)372:1<111::AID-CNE8>3.0.CO;2-6
Tanaka C, Nishizuka Y (1994) The protein kinase C family for neuronal signaling. Annu Rev Neurosci 17:551–567. https://doi.org/10.1146/annurev.ne.17.030194.003003

Tansey EM (1997) Not committing barbarisms: Sherrington and the synapse, 1897. Brain Res Bull 44(3):211–212

Tazi A, Dantzer R, Mormede P, Le Moal M (1986) Pituitary-adrenal correlates of schedule-induced polydipsia and wheel running in rats. Behav Brain Res 19(3):249–256

Tazi A, Dantzer R, Le Moal M (1988) Schedule-induced polydipsia experience decreases locomotor response to amphetamine. Brain Res 445(2):211–215

Tholanikunnel BG, Raymond JR, Malbon CC (1999) Analysis of the AU-rich elements in the 3′-untranslated region of beta 2-adrenergic receptor mRNA by mutagenesis and identification of the homologous AU-rich region from different species. Biochemistry 38(47):15564–15572

Tversky A, Kahneman D (1981) The framing of decisions and the psychology of choice. Science 211(4481):453–458

Ungerleider LG (1995) Functional brain imaging studies of cortical mechanisms for memory. Science 270(5237):769–775

Unwin N (1993) Neurotransmitter action: opening of ligand-gated ion channels. Cell 72(Suppl):31–41

Unwin PN, Zampighi G (1980) Structure of the junction between communicating cells. Nature 283(5747):545–549

Van Hoesen GW (1993) The modern concept of association cortex. Curr Opin Neurobiol 3(2):150–154

Vandenberg RJ (1998) Molecular pharmacology and physiology of glutamate transporters in the central nervous system. Clin Exp Pharmacol Physiol 25(6):393–400

VanDongen AM (ed) (2009) Biology of the NMDA receptor. Front Neurosci. CRC Press, Boca Raton, FL

Wamsley JK, Alburges ME, Hunt MA, Bylund DB (1992) Differential localization of alpha 2-adrenergic receptor subtypes in brain. Pharmacol Biochem Behav 41(2):267–273

Wang R (2002) Two's company, three's a crowd: can H2S be the third endogenous gaseous transmitter? FASEB J 16(13):1792–1798. https://doi.org/10.1096/fj.02-0211hyp

Wang R (2004) Signal transduction and the gasotransmitters: NO, CO and H2S in biology and medicine. Humana, Totowa, NJ

Ward P, Seri S, Cavanna AE (2013) Functional neuroanatomy and behavioural correlates of the basal ganglia: evidence from lesion studies. Behav Neurol 26(4):219–223. https://doi.org/10.3233/BEN-2012-120264

Werner P, Voigt M, Keinanen K, Wisden W, Seeburg PH (1991) Cloning of a putative high-affinity kainate receptor expressed predominantly in hippocampal CA3 cells. Nature 351(6329):742–744. https://doi.org/10.1038/351742a0

Wilson C (2004) Basal ganglia. In: Shepherd G (ed) The synaptic organization of the brain. Oxford University Press, Oxford, pp 361–413

Wise RA (2008) Dopamine and reward: the anhedonia hypothesis 30 years on. Neurotox Res 14(2–3):169–183. https://doi.org/10.1007/BF03033808

Wouters MM, Vicario M, Santos J (2016) The role of mast cells in functional GI disorders. Gut 65(1):155–168. https://doi.org/10.1136/gutjnl-2015-309151

Wu LJ, Zhuo M (2009) Targeting the NMDA receptor subunit NR2B for the treatment of neuropathic pain. Neurotherapeutics 6(4):693–702

Yamakura T, Shimoji K (1999) Subunit- and site-specific pharmacology of the NMDA receptor channel. Prog Neurobiol 59(3):279–298

Yanai K, Tashiro M (2007) The physiological and pathophysiological roles of neuronal histamine: an insight from human positron emission tomography studies. Pharmacol Ther 113(1):1–15. https://doi.org/10.1016/j.pharmthera.2006.06.008

Zorumski CF, Isenberg KE (1991) Insights into the structure and function of GABA-benzodiazepine receptors: ion channels and psychiatry. Am J Psychiatry 148(2):162–173. https://doi.org/10.1176/ajp.148.2.162

Learning and Memory

2

Stavroula I. Bargiota, Vasilis P. Bozikas,
and Ioannis Nimatoudis

2.1 Introduction

According to the ancient Greek mythology, Mnemosyne was the goddess of memory and the mother of the nine Muses of the arts and science (Dudai and Carruthers 2005). Learning and memory are closely related fundamental higher brain processes that allow individuals to adapt to the environment, create, and widen not only their personal history but also the population culture (Benfenati 2007).

Learning helps people to enquire and encode information making them able to adopt new behaviors. Learning is a whole lifetime process that intervenes in almost all occasions of people's social living. As a result the acquired experiences can alter an individual's behavior (Mazur 2015).

Memory is connected with learning. The initial information is encoded, preserved over time, and used when any need occurs. Any damage in the encoding, the storage, and the retrieval of the information disturbs the process of memory. In fact, memory is the result of learning, but these two procedures are intermingled and essential for the survival of not only humans but also all living creatures (Engel 1999).

S. I. Bargiota · V. P. Bozikas (✉)
2nd Department of Psychiatry, Faculty of Medicine, School of Health Sciences, Aristotle University of Thessaloniki, Psychiatric Hospital of Thessaloniki, Thessaloniki, Greece

I. Nimatoudis
3rd Department of Psychiatry, Faculty of Medicine, School of Health Sciences, Aristotle University of Thessaloniki, Thessaloniki, Greece

University Hospital AHEPA, Thessaloniki, Greece

K. N. Fountoulakis, I. Nimatoudis (eds.), *Psychobiology of Behaviour*,
https://doi.org/10.1007/978-3-030-18323-3_2

2.2 Learning

Learning is the possession of knowledge and skill through systemic study or through experience and error. There are two types of learning under the interest of psychobiology: associative learning and nonassociative learning. The associative learning is the procedure by which an individual finds out the connection between two stimuli or between an action and an event. The nonassociative learning is a simple form of learning that describes a relatively stable alteration in the power of response to a single stimulus due to repeated exposure to this same type of stimulus (Mazur 2015; Rescorla 2014).

2.2.1 Associative Learning

Some typical forms of associative learning are complex enough, and they are presented as classical conditioning, operant conditioning, and observational learning.

2.2.1.1 Classical Conditioning

In classical conditioning subjects learn to form an association between two events happening in sequence. In particular, a neutral stimulus produces a certain response to another stimulus. In the beginning of the twentieth century, the Russian physiologist Ivan P. Pavlov discovered the classical conditioning accidentally during experiments upon digestion. Pavlov gave food to a dog, and he watched the secretion of saliva, while the dog was consuming the food. In particular, he presented a stimulus like the sound of a metronome (conditioned stimuli), and afterward he supplied the food (unconditioned stimuli). He repeated this procedure a few times, and as a result he noticed that the dog started to salivate in response to the sound of the metronome. Pavlov came up with the conclusion that if a specific stimulus was present while the dog was given food, then that particular stimulus could be connected with food and could provoke salivation on its own. He made the dog to repeat this procedure, and as a result he realized that the dog was able to begin salivating before eating. This particular type of learning was named Pavlovian conditioning or classical conditioning (Pavlov et al. 1929).

Later, Pavlov and his colleagues introduced the terms acquisition, extinction, generalization, and discrimination in an attempt to clarify classical conditioning. They also introduced the rules of temporal contiguity and contingency. According to their studies, acquisition is the initial phase when a dog is trained, for example, to salivate at the sound of a bell. A conditioned stimulus should precede the unconditioned stimulus by several seconds interval (temporal contiguity). Then, a certain behavioral response, the conditioned response is produced. Conditioned stimulus must forecast the unconditioned stimulus (contingency). Extinction is a phenomenon that describes the gradual weakness and fading of conditioned response after the repetitive appliance of conditioned stimulus when the unconditioned stimulus is absent. For example, in extinction when salivation of the dog is the conditioned response, it will gradually decline if the bell is ringing repeatedly without supplying any food (unconditioned event). Generalization is the procedure that an individual already conditioned to a specific event will probably respond to a similar event without the occurrence of training to the new event. For example, if a particular tone makes a dog to salivate, tones of higher and lower frequency will also lead to salivation. On the other hand, during discrimination the dog is taught to salivate in a different way to two different stimuli tones.

Classical conditioning was postulated to be related with some clinical presentations like phobias, but it was also used to treat some certain pathological behaviors like addiction (Mackintosh 1983).

2.2.1.2 Operant Conditioning

Edward L. Thorndike and B. F. Skinner are two American psychologists who researched operant conditioning, another form of associative learning which is also called instrumental conditioning. Operant conditioning is a process which includes certain behaviors that a subject needs to perform in order to receive a reward or a punishment. The subject not only answers to the stimuli but also by its behavior causes changes to the environment (Mackintosh 1983).

In the 1890s, Edward L. Thorndike studied operant conditioning on different species of animals like dogs, cats, and chickens (Thorndike 1898). In particular, he placed the animal in a box called the operant box. In case the animal acted according to the right way, the box would open, and the animal would be able to get out and find food that was left outside the box. In the beginning, it was a difficult and long-lasting procedure for the animal to open the box and earn the reward. However, after placing the same animal in the box repeatedly, it started to learn and act correctly in a shorter period of time. After these observations, Thorndike formulated the first formal theory of learning, called law of effect. According to this principle, any behavior connected with pleasant results is likely to be repeated, and any response that produces an unpleasant effect is less likely to occur again (Thorndike 1927).

Later, during the 1930s, B. F. Skinner investigated and stated some significant principles of operant conditioning. He used rats and pigeons to create a way of learning based on reward and punishment. He experimented by training animals inside the known Skinner boxes. Inside the box, the animals were able to get food by using a small lever or by pecking at a food well. He observed the responses of the animal with a device designed to record the impact of food delivery on a subject's response. As a result, a positive reinforcement or a reward increases a behavior by adding a pleasant stimulus, while on the other hand, negative reinforcement or punishment decreases the probability of a certain behavior to happen again. There are a lot of rules that control the frequency and the timing of the reinforcers, like continuous or fixed-ratio schedule, called reinforcement schedules (Skinner 1990).

According to Skinner, punishments reduce the chance of the occurrence of a response and are divided to positive and negative punishments. A positive punishment is the decline of a behavior applying an unpleasant stimulus in case the behavior occurs, while negative punishment means the decline of a behavior by taking away a pleasant stimulus the time the behavior occurs. Shaping is a method of learning animal and human behaviors that they have never met before, by reinforcing the behavior. Extinction is a process of eliminating a trained response by diminishing the reinforcer. For example, if the reward of food is not delivered by pressing a lever in rats, then the behavior will be eliminated. It is noticeable that if individuals are taught to perform a behavior in one occasion, then they are able to behave in the same way and in other similar occasions. For example, the greeting "congratulations" is used to others' happy events. Discrimination refers to the ability to learn when a response is possible to be followed by a reward or not in different circumstances. In the previous example, humans learn not to say "congratulations" to someone in a bad situation since it may lead to negative results (Skinner 1953).

2.2.1.3 Observational Learning

In the early 1960s, Albert Bandura, a Canadian-American psychologist, introduced his social learning theory. It is an unquestionable fact that living organisms can also learn through observation. Bandura conducted several studies to investigate how observational/imitation learning impacts children's behavior (Bandura 1969). According to his theory, observational learning is composed of attention, retention, reproduction, and motivation. At the beginning an individual pays attention to the behavior of others; then the learner retains the information he observed. Afterward, he should mimic to perform the same behavior (reproduction), and finally he should have the motivation to mimic the behavior (Bandura and Walters 1977).

To sum up, learning by observation means that the subject should observe and then imitate the behavior of others. People are able to contact other people by using different languages. These languages have not only been officially taught to them, but they have also been learned through observation. In addition, they develop their personality and their social habits and abilities through experience they gain by observing other people. The mirror neurons that can be found in the ventral premotor cortex and inferior parietal lobe are considered to contribute to imitation. It has been suggested that mirror neurons explain many sides of social cognition, like the ability to realize the actions of others, to "read their mind," and to communicate by gestures and speech (Ramachandran 2000).

Although the precise substrate mechanism of action is still unspecified, observational learning plays an important role in social behaviors and communication through media and educational process.

2.2.2 Nonassociative Learning

The other simpler type of learning is the nonassociative learning, and it consists of several types like habituation and sensitization.

2.2.2.1 Habituation

Habituation is a process that decreases the behavioral response to a repeated and innocent stimulus. Several examples of habituation exist in everyday life. When someone hears a very loud sound for the first time, he may be annoyed and even frightened. But if the sound continues for a certain period of time, his annoyance and shock get reduced which is a typical case of habituation (Pinsker et al. 1970).

The Austrian-American neuropsychiatrist Eric Richard Kandel investigated the neural mechanisms of habituation examining the sea hare *Aplysia californica* because of its simplicity and its relatively large size of the underlying neural circuitry. *Aplysia californica* is a large shell-less sea snail or sea slug. The experiment was to apply a gentle touch to the siphon of *Aplysia* leading to a gill-withdrawal reflex. *Aplysia* hides inside the mantle shelf. Nevertheless, if the siphon is disturbed repeatedly, then the withdrawal reflex shows a gradual decrease. At the end, no response is observed. This steadily diminishment of the reflex after applying repeatedly a mechanical stimulus is called habituation. Kandel and his colleagues proved that habituation in *Aplysia* gill-withdrawal reflex (GWR) was due to a reduction on the synaptic transmission between sensory neurons of the siphon and motor neurons of the gill. The magnitude of the excitatory postsynaptic potential in motor neurons was reduced progressively after stimulating repeatedly the siphon (Kandel et al. 1976).

Habituation exists in short- and long-term form. In *Aplysia*, habituation can last for several weeks and is called long-term habituation (LTH) (Carew et al. 1972). It has been found that by this form of habituation, the sensorimotor pathway is suppressed and the presynaptic terminals and branches of the sensory neurons of the siphon are being retracted (Castellucci et al. 1970). However, the exact signaling pathways which after activation trigger long-term cellular changes are still unclear. Some recent studies suggested that LTH of the GWR relies on protein synthesis and activation of protein phosphatases 1 and 2A and α-amino-3-hydroxy-5-methyl-4-isoxazolepropionic acid (AMPA) and *N*-methyl-D-aspartate (NMDA) receptors. The role of RNA synthesis, calcineurin activity, and l-type Ca^{2+} channel activation in LTH in *Aplysia* has also been tested (Esdin et al. 2010; Ezzeddine and Glanzman 2003).

To conclude, despite the fact that habituation is one of the simplest types of learning, its neurobiological basis remains obscure.

2.2.2.2 Sensitization

Sensitization is the opposite of habituation. The term refers to a nonassociative learning type that increases the behavioral response to repeated and innocent stimuli. Sensitization contributes to the learning of noxious and threatening stimuli in animals. If the animal comes across a threatening and fearful event, then it is able to respond more robustly to other stimuli even if they are not harmful or frightening (Landsteiner and Jacobs 1935).

Except for habituation, other mechanisms of learning were also investigated using *Aplysia californica* (Castellucci and Kandel 1976; Pinsker et al. 1973). As observed in habituation, repeated disturbance of the siphon leads to no GWR in *Aplysia*. However, after a period of rest, the response showed quick recovery. The scientists applied an electric shock one time to the tail of the slug, and the gill-withdrawal reflex was restored. This process of the facilitation of a habituated response after the presentation of a strong tactile stimulus to another part of the animal is also called dishabituation (Pinsker et al. 1970).

Eric Kandel and his group studied the neural mechanisms underlying sensitization by experimenting upon *Aplysia californica*. According to their studies, the appliance of electric shock to the tail generates a strong gill withdrawal. The gill withdrawal exhibits a form of plasticity called sensitization (Kandel et al. 1976). Sensitization like habituation is a result of changes in synaptic transmission of only a few different types of neurons. These neurons are the sensory neurons that innervate the tail and the siphon, motor neurons that stimulate muscles in the gill and interneurons that receive information from various sensory neurons. The electric shock to the tail triggers the sensory neurons that innervate the tail. Afterward, these neurons stimulate modulatory interneurons that secrete serotonin on the presynaptic terminals of siphons' sensory neurons. The synaptic excitation of the motor neurons increases, and a receptor, called stimulatory G protein-coupled, is stimulated by the binding serotonin. This stimulatory receptor binds to a transmembrane protein which once activated is going to stimulate a G protein. Afterward, the enzyme adenylyl cyclase is activated and converts ATP into cyclic AMP (cAMP). Cyclic AMP (cAMP) plays the role of the second messenger and triggers cAMP-dependent protein kinase (PKA). Protein kinase phosphorylates K+ channels, and the action potentials increase as a consequence to the ceasing of K+ channels. Then, the influx of Ca^{2+} raises, and as a result the glutamate release from the presynaptic siphon's neuron to gill's motor neuron is increased. This augmentation of glutamate release is the mechanism that turns *Aplysia*'s gill-withdrawal reflex back to normal and ends habituation. Apart from *Aplysia* sensitization occurs also in other animals, and a similar procedure takes place (Bristol et al. 2004; Squire and Kandel 2000).

2.2.3 Features of Learning

Since learning is a composite process, learning ability differs due to reasons like experience, age, and mental state. Learning can happen in all ages of a living organism, but the process is not the same. For example, aged people find it more difficult than children to learn a new language. Some people are able to speak fluently two languages because they are exposed to them naturally from their infant age. In addition, experience helps people to learn easily new things relevant to already known information. Adults are also more capable to understand abstract thinking related to adolescents.

Learning ability is influenced by developmental disorders. People with attention deficit hyperactivity disorder (ADHD) or with the diagnosis of autism find it difficult to learn and develop social skills. On the other hand, degenerative diseases like Alzheimer's disease can affect the ability of learning as well as memory.

Different levels of intelligence result to different academic achievements, or there are talented people with flair to music or mathematics. Learning differs among individuals because of differences in motivation and emotion. Additionally, motivation can enhance the ability of people, in particular students, to learn (Mazur 2015).

2.3 Memory

Memory is the process according to which information is initially encoded, then retained, stored, and finally retrieved and recalled in the brain. All the living organisms need memory to exist. In particular, memory gives people the ability to remember the past, to acquire experience, and finally to learn (Arnold 2013). A lot of scientists, thinkers, and philosophers have talked and written about memory. "Our memory is a more perfect world than the universe: it gives back life to those who no longer exist" are some words of the French writer Guy de Maupassant (1903).

Memory consists of three distinct systems: the sensory memory, the short-term memory, and the long-term memory. Each one of these categories includes subcategories (Squire 2004).

2.3.1 Sensory Memory

Sensory memory refers to the representation of the stimulation of human sense organs for a brief period of time. Iconic memory means a brief recording of visual information, while echoic memory stands for the representation of auditory stimulation. Sensory memory is also represented in other sensory modalities and shows modality specificity (Anderson 2000).

2.3.1.1 Iconic Memory

The American cognitive psychologist George Sperling was the first to document the iconic memory, using the partial-report model. Specifically, his experiment was to present briefly a 3 × 3 or 3 × 4 array of letters to some observers. He saw that people were able to report all the letters in any indicated row (partial report) if this happened directly after the visual presentation. Moreover, if the individuals were asked to recall all the letters (whole report), they were found able to report only four to five. The probability of the presentation of each row was equal, which means that the observer had access to all the letters at the end of the display. The fact that partial-report shows an advantage compared to whole-report phenomenon led the scientists to conclude that there is a rapid-decaying iconic memory that holds at least 9–12 items (Sperling 1960). The exact duration of iconic memory has been calculated to be about 300–500 ms for young adults (Coltheart 1980; di Lollo 1977; Dick 1974; Mewhort et al. 1981; Neisser 1967).

2.3.1.2 Echoic Memory

Guttman and Julesz (1963) designed a creative experiment in order to measure echoic memory. A computer was used to create repeating patterns of white noise. White noise sounds like "shhh," cannot be memorized or described, and consists of all frequencies mixed together at random. The computer was putting together a repeating segment of white noise with no gap between them. The subjects were instructed by the investigators to wear headphones, listen carefully to the noise, and write down what they heard. The subjects did not know that a sound was being repeated. In case the repeating part of white noise endured longer than a few seconds, the subjects were not able to recognize the repetition. They could only hear a whooshing sound with no pattern. On the other hand, if the segment lasted shorter than 2 s, the subjects could realize that they heard a repeated sound even if they were not able to describe it.

Echoic memory is the system used to preserve the exact copy of sound for a few seconds by giving individuals the capacity to detect a repeating pattern of frequencies.

2.3.2 Short-Term Memory

Short-term memory refers to the system that holds essential information in mind temporarily, in an available and active state for further cognitive activities (James 1950).

Short-term memory has the feature to decay rapidly. In case no practice of the information is conducted, it is preserved for a limited time, no longer than 30 s. The retaining of information for longer periods requires periodical rehearsal. During this process, the information is kept for greater time through reentrance to the short-term store.

Another feature of short-term memory is its limited amount of capacity which when talking about humans is around seven items plus or minus two. These seven items are called "magical number seven." The term "word length effect" refers to the fact that fewer words are recalled when their pronunciation is longer. According to the "phonological similarity effect," fewer words can be recalled when they have similar pronunciation, while a greater number of phonologically dissimilar words can be recalled. On the other side, if the words are familiar or belong to the same category, they can be retrieved more easily (Miller 1956).

Chunking is a property that increases short-term memory capacity and especially the amount of recalled items. The ability of a person to remember information can be improved if every unit is placed in a meaningful word or phrase. For example, it is easier for someone to keep in mind a telephone number if the digits are chunked into groups rather than try to recall ten digits in a row (Gobet et al. 2001).

The cerebral cortex seems to be responsible for the most types of memory and short-term memory as well. Different parts of the body, like the eyes or ears, send sensory information to the cerebral cortex where they are stored for a few seconds. The non-attended information are thrown away, while the attended ones will be stored in the sensory areas of the cortex or will be moved to the hippocampus and will be kept into the long-term memory (Jonides et al. 2008).

2.3.3 Working Memory

Working memory was introduced by Baddeley and Hitch in 1974 (Baddeley and Hitch 1974) and is related to short-term memory although it refers to a distinct procedure. This memory system is used to plan and carry out behavior and is based to the dorsolateral prefrontal cortex. Some individual uses working memory when trying to recall the partial results during the solution of a mathematic problem without paper or to prepare a cake without using the same ingredient twice. Thus, working memory is a mental workbench which not only keeps but also processes information. Opposite to the passive nature of short-term memory, working memory is an active procedure. Working memory is also linked by some scientists with intelligence (Baddeley 2003).

2.3.4 Long-Term Memory

Long-term memory is a brain system that deals with information from the past and stores it for a great period of time. Long-term memory appears capable to keep countless information, from something learned recently to long-lasting memories.

There are different aspects describing the engagement of information in long-term memory. According to one aspect, in the beginning information is stored in short-term memory, and afterward it is processed and transmitted to long-term memory. On the other side, there is the belief that information is processed separately and enters short-term and long-term memory at the same time (Goelet et al. 1986).

Long-term memory is divided in psychobiology into declarative and non-declarative memory.

2.3.4.1 Declarative Memory

Declarative memory is linked with the hippocampus and related structures and gives individuals the ability of conscious recall and declaration of the information. Endel Tulving divided the declarative memory into episodic and semantic memory, in 1972 (Tulving 1972).

Episodic memory refers to the process involved in recalling specific experiences throughout a person's life. In this type of autobiographical memory, an event is related to a context during a specific time in a specific place, answering the questions "where" and "when." For example, episodic memory is used to remember a holiday in a beautiful island (where) with family 2 years ago (when). People are not only able to remember these events, but they talk about them, declaring them. As a result, episodic memory belongs to the category of declarative memory (Tulving 2002).

Semantic memory refers to general knowledge, like information about things that we know or about the world in general. The fact that someone knows that the Big Ben is in London and the Parthenon is in Athens represents an example of semantic memory. This memory system, unlike episodic memory, is not linked with the specific time and place that a memory was created. An individual does not have to keep in mind when and where he learned about Big Ben and Parthenon (Squire and Zola 1998).

2.3.4.2 Non-declarative Memory

Non-declarative memory can be recalled unconsciously without effort, is not declared, and is expressed through performance. Non-declarative memory system includes procedural memory, priming, habits (conditioning), and nonassociative learning (Squire and Zola-Morgan 1988).

Procedural memory, which involves visuomotor learning, includes all the skills that some individuals hold, for instance, playing a guitar, driving a car, or writing. The difference between procedural and declarative memory systems relies on the difference between procedural and declarative learning. Procedural learning refers to how to learn something, whereas declarative learning refers to what someone learns. Procedural memory is expressed through behaviors and can be easily influenced by practice. For example, driving ameliorates over time, and a guitarist can improve his skills by practicing repeatedly.

Priming is a well-studied phenomenon of non-declarative memory that refers to perceptual detection of words and objects. During priming, exposure to one stimulus affects the response to another stimulus by the activation of certain representations or associations in memory, just before executing an action or task. Priming reveals the underlying mechanism of perceptual tasks that is formed without awareness and has been implicitly memorized. For example, if a subject sees the word "red," it will be easier to find the word "heart," since red and heart are closely connected in memory. Priming is assessed through perceptual and conceptual techniques like word-stem completion and word association, respectively. The first task is the ability of representation of words such as medication with the three letters med. The second task incorporates, for example, the free association of the word ship with the related words sea and port (Squire and Zola 1996).

Habits, which occur in conditioning, as well as nonassociative learning, have been already described in this chapter.

2.3.4.3 Explicit and Implicit Memory

The American psychologist William McDougal divided long-term memory into explicit and implicit memory.

Explicit memory is the process that facilitates the recall of past experiences and corresponds to declarative memory. Explicit memory is needed for someone to remember his holiday in a mountain.

In contrast to explicit memory, implicit memory retrieves stored information without any conscious awareness and is supposed to correspond to non-declarative memory. The fact that people are imitating the behaviors of people they have met or unconsciously sing a song that they heard before is attributed to implicit memory. This category of memory also includes priming effect (McDougall 1924).

2.3.4.4 Neural, Regional, Synaptic, and Molecular Mechanisms of Long-Term Memory

Neural

One of the basic aims of neuroscience is to find out how memories are encoded and stored in the brain. Richard Semon was the first to introduce the term "engram"—a memory trace, which consists of neurons that represent memory physically (Semon 1921). Later, the American psychologist Karl Lashley conducted his famous search by training rats. Specifically, he made them traverse a maze to gain a food reward, and after the experiment or before, he surgically removed various lesions of different cortical regions. His investigations were published to his famous "search of the engram" 30 years later (Lashley 1950).

Regional

Nowadays, a lot of brain regions are supposed to interact in order to encode, store, and retrieve distinct information for each separate memory process. Brenda Milner and her colleagues studied the role of the *hippocampus* in episodic memory through the observation of the case of their famous patient H.M. In 1953, an operation was performed to the 27-year-old young man H.M. with the aim of treating his intractable epileptic seizures. A bilateral mesial temporal lobectomy has been conducted including the hippocampus, amygdala, and surrounding cortices, leading to the decrease of the frequency and the severity of the seizures. However, a new amnesic syndrome occurred, and the patient was unable to remember anything that happened to him following his operation, whereas his perceptual and intellectual skills remained intact, as he carried out normally a wide range of tasks. He was able to achieve some retention of simple visual and tactual mazes, tasks that were not long enough for his short-term memory capacity though the rate of acquisition was extremely slow. After investigating the case of H.M. almost for 40 years, the scientists concluded that bilateral lesions of the hippocampus and parahippocampal gyrus cause a severe and enduring disorder in episodic memory, a memory system that relates events with specific places and specific time, and they shed light to the realization that the components of the limbic system are bottleneck structures through which information needs to pass in order to be kept long-term (Milner et al. 1968, 1998; Smith and Milner 1981). The left (dominant) hemisphere has been found to control verbal memory, while a specialization has been found for the right hemisphere for spatial processing, a phenomenon called hemispheric asymmetry of memory (De Renzi et al. 1977; Habib et al. 2003; Ojemann and Dodrill 1985).

Except for the hippocampus, other brain regions are also involved in the function of memory. In 1937, when Papez proposed his well-known emotional circuit, the Papez circuit (hippocampal formation (subiculum) → fornix → mammillary bodies → mammillothalamic tract → anterior thalamic nucleus → cingulum → entorhinal cortex → hippocampal formation) for controlling the emotional expression, he also showed that this circuit plays an important role in the transfer of information into long-term memory (Markowitsch 2005; Papez 1937). Modern brain imaging procedures provided information about the participation of brain regions like the cerebellum, striatum, amygdala, and other motor or sensory systems in non-declarative memory (Markowitsch 2005).

Synaptic Molecular

The distinctive property of the nervous system which is called plasticity was first proposed by Santiago Ramon y Cajal (1852–1934) as the potential of the brain to adapt to the environment, in a congress in Rome in 1894 (DeFelipe 2006). Donald

Hebb introduced a theory to explain the associative learning in his book in 1949. The Hebbian learning refers to the simultaneous activation of neurons which increases the synaptic strength between those neurons in a way that activity in one results to an activity in the other (Hebb 2005). Later in 1966, the Norwegian Terje Lomo was working with the hippocampus of anesthetized rabbits, studying the consequences of triggering the perforant path to dentate granule cells when he observed that high-frequency tetanic stimuli could lead to an enduring increase in efficiency of transmission at the perforant path-granule cell synapses (Lømo 2003).

In 1968, Terje Lomo and Tim Bliss followed up the preliminary results from 1966 and did the experiments that resulted in the discovery and fully description of long-term potentiation (LTP)—a cellular model of memory and learning in 1973 (Bliss and Lømo 1973). LTP represents an artificial form of plasticity and is measured both as an increase in the efficiency of synaptic transmission at the perforant path synapses and as the increase in the postsynaptic cell population spike (Bliss and Lømo 1973). LTP occurs in both excitatory and inhibitory neural synapses, and aside from the hippocampus, it has been investigated in the cerebral cortex, the amygdala, the cerebellum, and the spinal cord. However, since the hippocampus controls declarative memory, and LTP was induced in the hippocampus, it was concluded that LTP is a cellular model of the memory function (Lynch 2004).

Nowadays, it is widely known that several structural changes lie behind synaptic plasticity and memory formation. In a few milliseconds, glutamate is released from presynaptic neurons and activates the AMPA receptors and depolarizes the postsynaptic neuron. As a result, Mg^{2+} induces the removal of the NMDA receptor inhibition leading to the influx Ca^{2+} through the ion channel of the NMDA receptor. Moreover, voltage-gated calcium channels are also activated by the aforementioned depolarization, leading to further increase of Ca^{2+} supply. A few minutes since the beginning of the process, the influx of the calcium into the synapse leads to the activation of kinases able to modulate their substrates' activity. The substrates are responsible for local alterations at the synapse, like morphological change through cytoskeletal regulation. They also regulate transcription factors and lead to the induction of the transcription of RNA. Finally, in a few hours, the translation of the transcribed RNA into proteins will have taken place. Activated synapses capture these proteins contributing to the stabilization of synaptic changes (Lamprecht and LeDoux 2004).

2.3.5 Forgetting and Other Phenomena Related to Memory

2.3.5.1 Forgetting

Forgetting is a term described by Hermann Ebbinghaus, defined as the loss or modification of information already stored in long-term memory. During his experiments, he created lists of three-syllable meaningless words, consisting of a vowel between two consonants, like KET or SIP. He practiced a lot, and he was able to remember the list incorruptly by heart. He was aware of the duration and the times that he tried in order to learn the list. Afterward, he checked his remembering of the list intermittently. He observed that rapid forgetting took place initially and then a steady increase in forgetting followed. Forgetting is believed to occur because new information interferes as time passes. Although it could cause some trouble, it is considered to naturally clean and exclude outdated and useless data (Ebbinghaus 2013; Murdock 1985).

2.3.5.2 Other Memory Phenomena

Flashbulb memory is a category of memory linked with significant emotional incidents in an individual's lifetime. For example, someone is able to remember where and when he/she heard the bad news of the death of one's beloved person, since a flashbulb memory has been formed (Winograd and Neisser 2006).

Déjà vu describes the feeling of the physical presence in a place that the person has never been before or the feeling of the experience of an event that the person has not experienced before. Déjà vu creates an implicit familiarity of an unrecognized stimuli, as it unconsciously evokes a former situation and as a result provokes a familiar sense (Brown 2003).

Jamais vu is the opposite of déjà vu, as individuals have the feeling that they have never experienced a situation before, even though they know that they did (Cleary 2008).

Tip-of-the-tongue state is a state when a well-known or familiar word cannot be immediately recalled (Brown 2012).

The method of loci, also known as *the memory palace* or *mind palace technique*, is an ancient mnemonic device adopted in Roman and Greek rhetorical essays. Through spatial memory, it helps a person to enhance serial remembering. The trained individuals imagine themselves to walk through in a well-known environment, and they place what they need to remember in specific locations. In order to retrieve specific memories, the person reimagines navigating the environment "searching" for the placed items in order. A lot of mnemonists use this method to recall faces, digits, and lists of words (Legge et al. 2012).

2.3.5.3 Memory and Sleep

Sleep has been found to play a significant role in memory processes. The sleep-related mechanisms of neural plasticity contribute to the consolidation of memory and learning. Rapid eye movement and non-rapid eye movement sleep are important for learning and memory, and a sleep cycle controls memory consolidation through multiple ways, but the actual processes remain unknown (Born and Wilhelm 2012; Maquet et al. 2003).

The relationship between sleep and academic performance is of a great interest. It is widely known that sleep enhances learning and memory processes, while sleep deprivation worsens these functions influencing academic performance. Students with more regular sleep-wake patterns, such as fewer night awakenings, shorter sleep latency, and earlier rise times on weekends, have been found to gain higher grades compared to students that were sleepy during the day due to disturbed night sleep (Vatthauer 2009).

2.3.5.4 Memory Distortion

In 1932, Frederick Bartlett revealed the existence of a phenomenon called memory distortion. Memory distortion means that a memory report is different from the real experience. Bartlett investigated the course of memory, and he came up with the idea that memory transforms over time including omissions, deletions, and distortions. In particular, during his experiments, he asked the subjects to read a folktale the "War of Ghosts." According to the plot, there is a battle between two enemy tribes. He used the method of serial reproduction known to people by the game "telephone," when children sit in a queue and each one tries to reproduce the same initial story, having been heard by each previous player. Likewise, he asked the subjects to remember the tale they heard from their immediate predecessor with the most possible details and try to narrate it to the next one. He revealed that memory of the original story undergoes huge distortion after only a small number of repetitions (Bartlett 1932).

Since then, a lot of work has been done by scientist trying to identify the brain regions that are responsible for true and false memories. The medial temporal lobe has been suspected for false recognition. The prefrontal cortex plays a role in memory monitoring errors. Other studies suggest that true and false memories activate the same brain regions (Schacter and Slotnick 2004).

2.4 Disorders of Learning and Memory

2.4.1 Age-Associated Memory Impairment

It is well known among scientists that aging leads to a decline in the function of memory among healthy individuals. Kral, in 1958, was the first to use the term "benign senescent forgetfulness" (BSF) to separate adults with mild memory impairment from those with severe changes or normal memory systems (Kral 1958). Later, the term age-associated memory impairment was constructed, and criteria were presented with the aim of facilitating the communication among scientists about the memory loss that may occur in healthy elderly individuals in the later decades of life (Crook et al. 1986). In the mid-1990s, the term mild cognitive impairment was introduced by Peterson and his colleagues (Petersen et al. 1999). It refers to a transitional stage between age-associated impairment and dementia presenting with subjective memory complaints, lower scores on memory tasks, absence of dementia, and good levels of quality in activities.

During aging a lot of changes take place in the brain. Recent MRI studies have shown that atrophy of the medial temporal lobe and the hippocampus is a usual phenomenon in elderly persons and might lead to memory decline (episodic) (Golomb et al. 1993). Frontal lobes present with the greatest reduction in brain volume during aging (Hänninen and Soininen 1997); they have been found to underlie in several memory systems such as the working memory, the retrieval process which needs conscious mental effort, the temporal organization of memory, and source memory.

2.4.2 Amnesias

2.4.2.1 Introduction

The term amnesia refers to a situation characterized by the impaired ability of a person to learn new details and recall information from the past. Common reasons for amnesia not only include brain injuries, neurological deficits, and vascular incidents like a stroke but neurodegenerative and psychological disorders as well (Snodgrass and Corwin 1988). Amnesia can happen with or without the existence of other cognitive deficits.

Memory consists of multiple different systems, but not all of them are impaired in amnesia. While amnesic humans are not able to acquire information about facts and events (explicit memory), they keep the ability for several types of learning like conditioning and habit learning and skill learning. Explicit—declarative—memory relies on the structures of the medial temporal lobe and diencephalon unlikely implicit memory that depends on brain structures that remain intact in amnesia (Kopelman 2002).

2.4.2.2 Anterograde and Retrograde Amnesias and Memory Consolidation

Anterograde amnesia is called the failure to learn new information. The patient finds it difficult to remember new persons, situations, and words following the onset of amnesia. Anterograde amnesia is more severe when more locations of the medial temporal lobe are defected. On the other hand, retrograde amnesia refers to the damaged ability of an individual to recall events that took place soon before amnesia occurred. Thus, memories estimated in the time period before amnesia are lost, while old memories, like childhood events, remain. Retrograde amnesia is met alone or in combination with anterograde amnesia (Winocur 1990).

During the process of memory consolidation, after the initial encoding of the information, cortical and neural processes and reformation take place, leading to the permanent storage of memory. Retrograde amnesia supports the idea of memory consolidation proving that what has been learned is not immediately made stable. Memory consolidation starts after information which has been encoded in the neocortex are linked with a memory trace in the hippocampus and related structures in the medial temporal lobes as well as the diencephalon, i.e., the posterior part of the forebrain that includes the thalamus, hypothalamus, and ventral thalamus. This initial procedure engages a short-term consolidation process, which ends within seconds or minutes. Afterward, long-term consolidation begins. Even though the hippocampus and related structures store and retrieve the information in the beginning, as consolidation proceeds, the neocortex alone becomes capable of keeping the permanent memory trace and mediating its retrieval (Nadel and Moscovitch 1997).

2.4.2.3 Amnesias of Dementias

Dementias are disorders that have taken their name from their profound symptoms characterized of deficits in mentation. The pathophysiological reasons for dementias are considered to be cortical and subcortical. For example, Alzheimer's disease is a cortical disorder but includes decline of cholinergic neurons in subcortical regions, as well. On the other hand, vascular dementia is a subcortical disease presented also with atrophy of the frontal cortex. The predominating impairment in most dementias is amnesia which depends on the stage of the pathology as well as the participating brain locations (Economou et al. 2006).

Alzheimer's Disease

Alzheimer's disease refers to a chronic neurodegenerative illness that usually starts gradually and deteriorates over time and leads to the majority of cases of dementia. The cause of Alzheimer's disease is believed to be 70% genetic with many genes involved. The disease process is associated with plaques and tangles in the brain. In particular, the neurodegeneration in Alzheimer's disease (AD) may be attributed to the deposition of amyloid β (Aβ) peptide in plaques in brain tissue (Hardy and Selkoe 2002). It most often affects people over 65 years of age, although 4–5% of cases are early-onset Alzheimer's disease (Hardy and Higgins 1992).

Episodic memory impairment is the first and main deficit in Alzheimer's disease, while deficits in semantic memory follow (Perry et al. 2000). In fact, patients in the early stages of the disorder indicate a decline in retaining new information. They are not able to remember the names of other persons, appointments, conversations, and where they have left their bag or keys. This episodic memory decline does not regard only verbal context, but the known information about visuospatial context is restricted (Barbeau et al. 2004).

Patients with Alzheimer's disease also suffer from semantic memory impairment which is proven from their impairment on various tasks like basic word retrieval, naming objects, and recalling examples of a semantic category (Nebes 1989). They also find it difficult to define concepts and separate them from one another (Alathari et al. 2004). Alzheimer's disease is not related with perceptual implicit memory decline though there are some scientists that support the opposite. Habit learning is preserved in Alzheimer's disease, while fear conditioning is often impaired.

Medications for Alzheimer's disease improve cognitive deficits and the affective and behavioral presentations. Donepezil, rivastigmine, and galantamine have been proved to ameliorate cognition in patients with Alzheimer's disease. The mechanism of action of acetylcholinesterase inhibitors relies on the inhibition of acetylcholinesterase, leading to an increase in the available acetylcholine, stimulating postsynaptic cholinergic receptors, and reducing the deficit in cholinergic transmission attributed to the degeneration of basal forebrain nuclei (Birks 2006).

Vascular Dementias

There are several types of vascular dementias with different clinical symptoms depending on the etiology of damage and the region of the brain that is affected. Vascular dementias include stroke-related dementias (single-infract and multi-infract dementia), subcortical dementia, and mixed dementia (along with Alzheimer's disease). Single-infract dementia, caused by a single infract, is capable of leading to amnesia, affecting cortical or subcortical areas responsible for memory functions (Szirmai et al. 2002). The most prevalent subcategory of vascular dementias is the subcortical ischemic dementia, characterized by impairment in the procedural memory system due to deficits in subcortical nuclei such as the caudate, putamen, and substantia nigra (Cummings 1994).

The treatment of vascular dementias depends on the understanding and management of the cause. Manipulating risk factors such like hypertension, diabetes, and smoking could help to prevent the illness. The techniques of carotid endarterectomy and atrial fibrillation using anticoagulant therapy are secondary preventive methods. Cholinesterase inhibitors have been found to improve the cognitive symptoms of the disease (Román 2003).

Frontotemporal Dementia

Frontotemporal dementia describes subjects with dementia associated with focal atrophy of the orbitomedial frontal and anterior temporal lobes. As a result, it is divided into temporal and frontal variant. The temporal variant leads to semantic amnesia, whereas the frontal variant is connected with behavioral disorders as well as deficits in working memory and retrieval of information (Englund et al. 1994).

The frontal variant of frontotemporal dementia is also called behavioral variant. Patients with behavioral variant frontotemporal dementia are characterized by insidious changes in personality, interpersonal behavior, and emotional variety. These changes show progressive collapse of the neural circuits that take part in social cognition, emotion regulation, motivation, and decision-making. In addition, apathy characterized by reduced interest in previous activities and social isolation coexists with disinhibition that leads to impulsive actions along with socially embarrassing behavior (Piguet et al. 2011; Rascovsky et al. 2007).

The temporal variant of frontotemporal dementia is also represented by the term semantic amnesia. It was introduced in 1975 by Warrington and refers to the loss or the inaccessibility of events and ideas that have been pieces of an individual's personal knowledge. Recent memories are spared, and episodic memories along with the possession of new information are not noticeably affected (Warrington 1975). Taking into account that the most semantic memory are verbally coded, semantic amnesia occurs as difficulties in finding specific words, naming objects, recalling and linking facts, and understanding the meaning of already known words and concepts. The syndrome of semantic amnesia is caused by bilateral temporal lobe lesions located posteriorly, inferiorly, and laterally but not medially from the poles. As a result, the hippocampus and neighbor limbic structures stay intact (Squire and Zola 1998).

Lewy Body-Related Dementias

This category includes two dementia syndromes: the synonymous dementia in which Lewy bodies are cortical in location and Parkinson's disease, in which the pathology is primarily in the basal ganglia. The clinical presentation of both syndromes deteriorates over time, and the patients tend to present with identical cognitive impairments.

Parkinson's Disease

In patients with dementia due to Parkinson's disease, attentional problems are more severe than memory deficits. The cognitive decline must follow established Parkinson's disease, and there must be a gradual development. Working memory and visual-perceptual process are impaired. Decline in procedural learning also occurs, making these patients unable to perform mainly motor learning tasks. As a result, patients with Parkinson's disease become dependent on their caregivers, since not only their motor dysfunction but also working memory deficits prevent them from performing daily living activities.

Dementia with Lewy Bodies

Patients with dementia with Lewy bodies present with fluctuating cognition and pronounced variations in attention and alertness. In particular, this disorder includes early changes in complex attention and executive function.

Treatment of Lewy body-related dementias involves dopamine replacement by L-dopa and dopaminergic agonists in order to improve cognitive function rising psychomotor speed. Acetylcholinesterase inhibitors are also effective because they reduce visual hallucinations, apathy, anxiety, and sleep disturbances caused by cholinergic deficits of these patients (Papanicolaou 2005).

Huntington's Disease

Huntington's disease is an illness that presents with unequivocal, extrapyramidal motor abnormalities (choreoathetosis) along with progressive cognitive impairment. It is diagnosed either in family members with a history of the disease or in individuals that show an expansion in the HTT gene on chromosome 4 and atrophy in the caudate nucleus. Impairment in psychomotor skills, speed of processing information, and initial memory functions like procedural memory are the main deficits in the early stages of Huntington's disease, due to its frontostriatal etiology, characterized by a progressive nature. Deficits in visuospatial and semantic memory show no progressive decline.

Pharmacological treatments help to manage the symptoms as well as the social, physical, and occupational deficits of this progressive disease. First-generation antipsychotics are used to improve choreoathetosis and psychiatric symptomatology, but due to their implications, they have been replaced by the newer second-generation antipsychotic agents (Butters et al. 1985).

2.4.2.4 Traumatic Amnesia

A traumatic brain injury refers to any insult to the brain caused by a physical external force leading to alteration in consciousness and affecting cognitive function.

The cognitive impairments in people who survive after a serious traumatic brain injury are divided into four stages according to the time of occurrence related to the injury. The first two phases last some days and a post-traumatic delirium is included. The first stage is the period of an altered state of consciousness or coma. The second stage consists of different abnormalities in cognition and behavior, psychomotor activity as well as an inability to recall events and/or acquire new information, which is also called post-traumatic amnesia. Cognitive functions rehabilitate rapidly after 6–12 months (third stage). The fourth phase may include permanent damages in processes like attention and vigilance, memory and new learning, verbal skills, self-regulation of mood, executive functions, emotional reactions, and awareness of one's limitations (Kosmidis et al. 2006).

2.4.2.5 Transient Global Amnesia

Transient global amnesia was initially introduced by Fisher and Adam in the 1950s, and it describes an unknown pathophysiology well-recognized clinical syndrome. Patients present with a sudden onset of both anterograde and retrograde amnesia, without any other cognitive and neurological abnormalities. Transient global amnesia has a total duration of a few hours until 1 day and is observed mainly in middle-age to elderly persons (Fisher and Adams 1964).

Patients who develop transient global amnesia have memory deficits, but they retain their full consciousness and awareness. As a result their ability to carry on normal activities remains intact. Disorientation to time and place is the basic clinical feature of this entity. Memory for events during the acute phase is permanently damaged. The inner world of the patients remains to a specific point of time in the past, while the outer world occurs in the present time. The pathogenesis of transient global amnesia remains unclear. Some scientists support that focal ischemic lesions, migraine headaches, and brain tumors can be some causes. The medial temporal lobe is suspected (Szabo 2014). Other studies have suspected ischemic dysfunction in the medial temporal lobe, especially the hippocampus, to underlay in transient global amnesia (Webb and Rothwell 2013).

No semantic or procedural memory impairments occur during the episode of transient global amnesia (Papanicolaou 2005).

2.4.2.6 Amnesic Syndrome

The amnesic syndrome is defined as permanent and total disorder of memory attributed to organic brain dysfunction. The amnesic syndrome presents alone, without any other perceptual or cognitive disturbance. It is a mental disorder that affects short- and long-term memory with anterograde and sometimes retrograde amnesia. This decline occurs in a normal state of consciousness. The amnesic syndrome is associated with different causes that lead to damage to certain structures in the median temporal lobe and the diencephalon. It may result from close-head injury, penetrating head injury, subarachnoid hemorrhage, cerebral infraction, hypoglycemia, hypoxia, tumor, and thiamine deficiency as the result of heavy alcoholism (Wernicke's disease-Korsakoff's syndrome), carbon monoxide poisoning, and herpes simplex encephalitis (Parkin and Leng 1993).

The clinical manifestations of the amnesic syndrome involve disorientation, confabulation, and a lack of insight into the memory deficit. The term confabulation was introduced by Whitlock in 1981 and refers to "false statements that are not made to deceive, are typically more coherent than thoughts produced during delirium, and do not reflect underlying psychopathology" (Johnson et al. 1997). These manifestations are related to frontal lobe dysfunction produced by damage to neural links in a thalamo-frontal network (Johnson et al. 1997).

2.4.2.7 Psychogenic Amnesia

Psychogenic amnesias or functional amnesias are retrograde and present with reversible amnesia of autobiographical events. This type of memory loss is supposed to have psychological reasons since no obvious structural abnormalities are involved. Nevertheless, there are functional deficits in frontotemporal region that reverse after amnesia resolves (Savvidou et al. 2006).

Dissociative amnesia is subsequent to a traumatic experience, and according to DSM-5, it is defined as the inability of the individual to recall important autobiographical information of a traumatic or stressful nature, inconsistent with ordinary forgetting. The specifier for dissociative fugue refers to an apparently purposeful travel or bewildered wandering that is associated with amnesia for identity or for other important autobiographical information (APA 2013).

2.5 Conclusion

Learning and memory are closely connected brain processes. The application of the underlying biological mechanisms to the observation of the behavior is the aim of the field of psychobiology. Describing this aspect of memory and learning functions can be a useful tool for every scientist, leading to a deeper and more complete understanding of daily tasks and events. Memory is considered to be a big, rich, and awesome phenomenon, and it must be studied as such (Cahill et al. 2001).

A lot of light has been shed into understanding how the brain learns, forms, preserves, or loses memory. Conclusions should be derived from carefully investigating behaviors, taking under consideration the previous findings in the history of science. Actually, forgetting past lessons and experiences can be nothing more than a memory disorder.

References

Alathari L, Trinh Ngo C, Dopkins S (2004) Loss of distinctive features and a broader pattern of priming in Alzheimer's disease. Neuropsychology 18:603

Anderson JR (2000) Learning and memory: an integrated approach, 2nd edn. John Wiley & Sons Inc, Hoboken, NJ

APA (2013) DSM 5. American Psychiatric Association, Philadelphia, PA

Arnold MB (2013) Memory and the brain. Psychology Press, London

Baddeley A (2003) Working memory: looking back and looking forward. Nat Rev Neurosci 4:829–839

Baddeley AD, Hitch G (1974) Working memory. Psychol Learn Motiv 8:47–89

Bandura A (1969) Principles of behavior modification. Holt, Rinehart and Winston, Inc., New York, NY

Bandura A, Walters RH (1977) Social learning theory. Prentice Hall, Englewood Cliffs, NJ

Barbeau E, Didic M, Tramoni E, Felician O, Joubert S, Sontheimer A, Ceccaldi M, Poncet M (2004) Evaluation of visual recognition memory in MCI patients. Neurology 62:1317–1322

Bartlett FC (1932) Remembering: an experimental and social study. Cambridge University, Cambridge

Benfenati F (2007) Synaptic plasticity and the neurobiology of learning and memory. Acta Biomed 78:58–66
Birks JS (2006) Cholinesterase inhibitors for Alzheimer's disease. Cochrane Database Syst Rev 1:CD005593
Bliss TV, Lømo T (1973) Long-lasting potentiation of synaptic transmission in the dentate area of the anaesthetized rabbit following stimulation of the perforant path. J Physiol 232:331–356
Born J, Wilhelm I (2012) System consolidation of memory during sleep. Psychol Res 76:192–203
Bristol AS, Sutton MA, Carew TJ (2004) Neural circuit of tail-elicited siphon withdrawal in aplysia. I Differential lateralization of sensitization and dishabituation. J Neurophysiol 91:666–677
Brown AS (2003) A review of the deja vu experience. Psychol Bull 129:394
Brown AS (2012) The tip of the tongue state. Taylor & Francis, Abingdon
Butters N, Wolfe J, Martone M, Granholm E, Cermak LS (1985) Memory disorders associated with Huntington's disease: verbal recall, verbal recognition and procedural memory. Neuropsychologia 23:729–743
Cahill L, McGaugh JL, Weinberger NM (2001) The neurobiology of learning and memory: some reminders to remember. Trends Neurosci 24:578–581
Carew TJ, Pinsker HM, Kandel ER (1972) Long-term habituation of a defensive withdrawal reflex in aplysia. Science 175:451–454
Castellucci V, Kandel E (1976) Presynaptic facilitation as a mechanism for behavioral sensitization in aplysia. Science 194:1176–1178
Castellucci V, Pinsker H, Kupfermann I, Kandel ER (1970) Neuronal mechanisms of habituation and dishabituation of the gill-withdrawal reflex in aplysia. Science 167:1745–1748
Cleary AM (2008) Recognition memory, familiarity, and déjà vu experiences. Curr Dir Psychol Sci 17:353–357
Coltheart M (1980) Iconic memory and visible persistence. Percept Psychophys 27:183–228
Crook T, Bartus RT, Ferris SH, Whitehouse P, Cohen GD, Gershon S (1986) Age-associated memory impairment: proposed diagnostic criteria and measures of clinical change—report of a national institute of mental health work group. Dev Neuropsychol 2:261–276
Cummings JL (1994) Vascular subcortical dementias: clinical aspects. Dement Geriatr Cogn Disord 5:177–180
De Maupassant G (1903) The complete short stories of Guy de Maupassant. WJ Black, New York, NY
De Renzi E, Faglioni P, Previdi P (1977) Spatial memory and hemispheric locus of lesion. Cortex 13:424–433
DeFelipe J (2006) Brain plasticity and mental processes: cajal again. Nat Rev Neurosci 7:811–817
Dick A (1974) Iconic memory and its relation to perceptual processing and other memory mechanisms. Percept Psychophys 16:575–596
Dudai Y, Carruthers M (2005) The Janus face of Mnemosyne. Nature 434:567–567
Ebbinghaus H (2013) Memory: a contribution to experimental psychology. Ann Neurosci 20:155
Economou A, Papageorgiou SG, Papanicolaou AC (2006) Amnesias associated with the dementias. In: Papanicolau AC (ed) The amnesias. The clinical textbook of memory disorders. Oxford University Press, New York, NY, pp 75–110
Engel S (1999) Context is everything: the nature of memory. WH Freeman, New York, NY
Englund B, Brun A, Gustafson L, Passant U, Mann D, Neary D, Snowden J (1994) Clinical and neuropathological criteria for frontotemporal dementia. J Neurol Neurosurg Psychiatry 57:416–418
Esdin J, Pearce K, Glanzman DL (2010) Long-term habituation of the gill-withdrawal reflex in aplysia requires gene transcription, calcineurin and L-type voltage-gated calcium channels. Front Behav Neurosci 4:181
Ezzeddine Y, Glanzman DL (2003) Prolonged habituation of the gill-withdrawal reflex in aplysia depends on protein synthesis, protein phosphatase activity, and postsynaptic glutamate receptors. J Neurosci 23:9585–9594
Fisher CM, Adams RD (1964) Transient global amnesia. Acta Neurol Scand 40(Suppl 9):1

Gobet F, Lane PC, Croker S, Cheng PC, Jones G, Oliver I, Pine JM (2001) Chunking mechanisms in human learning. Trends Cogn Sci 5:236–243
Goelet P, Castellucci VF, Schacher S, Kandel ER (1986) The long and the short of long-term memory: a molecular framework. Nature 322(6078):419–422
Golomb J, de Leon MJ, Kluger A, George AE, Tarshish C, Ferris SH (1993) Hippocampal atrophy in normal aging: an association with recent memory impairment. Arch Neurol 50:967–973
Guttman N, Julesz B (1963) Lower limits of auditory periodicity analysis. J Acoust Soc Am 35:610–610
Habib R, Nyberg L, Tulving E (2003) Hemispheric asymmetries of memory: the HERA model revisited. Trends Cogn Sci 7:241–245
Hänninen T, Soininen H (1997) Age-associated memory impairment. Drugs Aging 11:480–489
Hardy JA, Higgins GA (1992) Alzheimer's disease: the amyloid cascade hypothesis. Science 256:184
Hardy J, Selkoe DJ (2002) The amyloid hypothesis of Alzheimer's disease: progress and problems on the road to therapeutics. Science 297:353–356
Hebb DO (2005) The organization of behavior: a neuropsychological theory. Psychology Press, London
James W (1950) The principles of psychology: in 2 volumes. Dover Publications, Mineola, NY
Johnson MK, O'Connor M, Cantor J (1997) Confabulation, memory deficits, and frontal dysfunction. Brain Cogn 34:189–206
Jonides J, Lewis RL, Nee DE, Lustig CA, Berman MG, Moore KS (2008) The mind and brain of short-term memory. Annu Rev Psychol 59:193
Kandel E, Brunelli M, Byrne J, Castellucci V (1976) A common presynaptic locus for the synaptic changes underlying short-term habituation and sensitization of the gill-withdrawal reflex in Aplysia. In: Cold Spring Harbor symposia on quantitative biology. Cold Spring Harbor Laboratory Press, Cold Spring Harbor, NY, pp 465–482
Kopelman MD (2002) Disorders of memory. Brain 125:2152–2190
Kosmidis MH, Giazkoulidou A, Bozikas VP, Papanikolaou AC (2006) Traumatic amnesia. In: Papanicolau AC (ed) The amnesias. The clinical textbook of memory disorders. Oxford University Press, New York, NY, p 156
Kral V (1958) Senescent memory decline and senile amnestic syndrome. Am J Psychiatr 115:361–362
Lamprecht R, LeDoux J (2004) Structural plasticity and memory. Nat Rev Neurosci 5:45–54
Landsteiner K, Jacobs J (1935) Studies on the sensitization of animals with simple chemical compounds. J Exp Med 61:643–656
Lashley KS (1950) In search of the engram. In: Society for Experimental Biology. Physiological mechanisms in animal behavior. Academic, Oxford
Legge EL, Madan CR, Ng ET, Caplan JB (2012) Building a memory palace in minutes: equivalent memory performance using virtual versus conventional environments with the method of loci. Acta Psychol 141:380–390
di Lollo V (1977) Temporal characteristics of iconic memory. Nature 267:241–243
Lømo T (2003) The discovery of long-term potentiation. Philos Trans R Soc Lond B Biol Sci 358:617–620
Lynch M (2004) Long-term potentiation and memory. Physiol Rev 84:87–136
Mackintosh NJ (1983) Conditioning and associative learning. Clarendon Press, Oxford
Maquet P, Smith C, Stickgold R (2003) Sleep and brain plasticity. Oxford University Pres, Oxford
Markowitsch HJ (2005) 10 The neuroanatomy of memory. In: Halligan PW, Wade DT (eds) The effectiveness of rehabilitation for cognitive deficits, vol 105. Oxford University Press, Oxford
Mazur JE (2015) Learning and behavior. Psychology Press, London
McDougall W (1924) Outline of psychology. Psychology Press, London
Mewhort D, Campbell A, Marchetti FM, Campbell JI (1981) Identification, localization, and "iconic memory": an evaluation of the bar-probe task. Mem Cogn 9:50–67
Miller GA (1956) The magical number seven, plus or minus two: some limits on our capacity for processing information. Psychol Rev 63:81

Milner B, Corkin S, Teuber H-L (1968) Further analysis of the hippocampal amnesic syndrome: 14-year follow-up study of HM. Neuropsychologia 6:215–234
Milner B, Squire LR, Kandel ER (1998) Cognitive neuroscience and the study of memory. Neuron 20:445–468
Murdock BB (1985) The contributions of Hermann Ebbinghaus. J Exp Psychol Learn Mem Cogn 11:469
Nadel L, Moscovitch M (1997) Memory consolidation, retrograde amnesia and the hippocampal complex. Curr Opin Neurobiol 7:217–227
Nebes RD (1989) Semantic memory in Alzheimer's disease. Psychol Bull 106:377
Neisser U (1967) Cognitive psychology. Appleton-Century-Crofts.[aAC] Nelson, K.(2003). Self and social functions: individual autobiographical memory and collective narrative. Memory 11:12536
Ojemann GA, Dodrill CB (1985) Verbal memory deficits after left temporal lobectomy for epilepsy: mechanism and intraoperative prediction. J Neurosurg 62:101–107
Papanicolaou AC (2005) The amnesias: a clinical textbook of memory disorders. Oxford University Press, Oxford
Papez JW (1937) A proposed mechanism of emotion. Arch Neurol Psychiatr 38:725–743
Parkin AJ, Leng NR (1993) Neuropsychology of the amnesic syndrome. Psychology Press, London
Pavlov IP, Gantt WH, Volborth G, Cannon W (1929) Lectures on conditioned reflexes. J Philos 26(10):275–277
Perry RJ, Watson P, Hodges JR (2000) The nature and staging of attention dysfunction in early (minimal and mild) Alzheimer's disease: relationship to episodic and semantic memory impairment. Neuropsychologia 38:252–271
Petersen RC, Smith GE, Waring SC, Ivnik RJ, Tangalos EG, Kokmen E (1999) Mild cognitive impairment: clinical characterization and outcome. Arch Neurol 56:303–308
Piguet O, Hornberger M, Mioshi E, Hodges JR (2011) Behavioural-variant frontotemporal dementia: diagnosis, clinical staging, and management. Lancet Neurol 10:162–172
Pinsker H, Kupfermann I, Castellucci V, Kandel E (1970) Habituation and dishabituation of the GM-withdrawal reflex in Aplysia. Science 167:1740–1742
Pinsker HM, Hening WA, Carew TJ, Kandel ER (1973) Long-term sensitization of a defensive withdrawal reflex in Aplysia. Science 182:1039–1042
Ramachandran VS (2000) Mirror neurons and imitation learning as the driving force behind "the great leap forward" in human evolution. Edge, New York, NY
Rascovsky K, Hodges JR, Kipps CM, Johnson JK, Seeley WW, Mendez MF, Knopman D, Kertesz A, Mesulam M, Salmon DP (2007) Diagnostic criteria for the behavioral variant of frontotemporal dementia (bvFTD): current limitations and future directions. Alzheimer Dis Assoc Disord 21:S14–S18
Rescorla RA (2014) Pavlovian second-order conditioning (psychology revivals): studies in associative learning. Psychology Press, London
Román GC (2003) Vascular dementia: distinguishing characteristics, treatment, and prevention. J Am Geriatr Soc 51:S296–S304
Savvidou I, Bozikas VP, Papanicolaou AC (2006) Psychogenic amnesias. In: Papanicolau AC (ed) The amnesias. The clinical textbook of memory disorders, vol 214. Oxford University Press, New York, NY
Schacter DL, Slotnick SD (2004) The cognitive neuroscience of memory distortion. Neuron 44:149–160
Semon RW (1921) The mneme. G. Allen & Unwin Limited, Crows Nest, NSW
Skinner BF (1953) Science and human behavior. Simon and Schuster, New York, NY
Skinner BF (1990) The behavior of organisms: an experimental analysis. BF Skinner Foundation, Binghamton, NY
Smith ML, Milner B (1981) The role of the right hippocampus in the recall of spatial location. Neuropsychologia 19:781–793
Snodgrass JG, Corwin J (1988) Pragmatics of measuring recognition memory: applications to dementia and amnesia. J Exp Psychol Gen 117:34

Sperling G (1960) The information available in brief visual presentations. Psychol Monogr Gen Appl 74:1
Squire LR (2004) Memory systems of the brain: a brief history and current perspective. Neurobiol Learn Mem 82:171–177
Squire LR, Kandel ER (2000) Memory: from mind to molecules. Macmillan, London
Squire LR, Zola SM (1996) Structure and function of declarative and nondeclarative memory systems. Proc Natl Acad Sci 93:13515–13522
Squire LR, Zola SM (1998) Episodic memory, semantic memory, and amnesia. Hippocampus 8:205–211
Squire LR, Zola-Morgan S (1988) Memory: brain systems and behavior. Trends Neurosci 11:170–175
Szabo K (2014) Transient global amnesia. In: Szabo K, Hennerici MG (eds) The hippocampus in clinical neuroscience. Karger Publishers, Berlin, pp 143–149
Szirmai I, Vastagh I, Szombathelyi É, Kamondi A (2002) Strategic infarcts of the thalamus in vascular dementia. J Neurol Sci 203:91–97
Thorndike EL (1898) Animal intelligence: an experimental study of the associative processes in animals. Psychol Rev Monogr Suppl 2:i-109
Thorndike EL (1927) The law of effect. Am J Psychol 39:212–222
Tulving E (1972) Episodic and semantic memory 1. Organ Memory 381:382–404
Tulving E (2002) Episodic memory: from mind to brain. Annu Rev Psychol 53:1–25
Vatthauer, K. E. (2009). Sleep and academic performance
Warrington EK (1975) The selective impairment of semantic memory. Q J Exp Psychol 27:635–657
Webb AJS, Rothwell PM (2013) Transient global amnesia associated with bilateral restricted diffusion in the lateral hippocampus. J Neurol Neurosurg Psychiatry. https://doi.org/10.1136/jnnp-2012-304542
Winocur G (1990) Anterograde and retrograde amnesia in rats with dorsal hippocampal or dorsomedial thalamic lesions. Behav Brain Res 38:145–154
Winograd E, Neisser U (2006) Affect and accuracy in recall: studies of 'flashbulb' memories. Cambridge University Press, Cambridge

Basic Vital Functions and Instincts

3

Panagiotis Ferentinos, Nikolaos Kokras,
and George Konstantakopoulos

3.1 Basic Vital Functions

Basic vital functions are supposed to be what is common in all living organisms and what distinguishes them from nonliving things. However, there is no unequivocal definition of the phenomenon of life, which represents a challenge for biologists and philosophers (Cleland and Chyba 2002). Therefore, a variable number of basic vital functions have been proposed in the literature, including nutrition, metabolism, respiration, excretion, homeostasis, growth and development, organization and compartmentalization, adaptability, sensitivity to environmental stimuli, locomotion, reproduction, and evolution (Koshland 2002; Chodasewicz 2014). We focus on three core vital functions recognized by many authors as the defining features of life: *nutrition (feeding), interaction with the environment (social interaction)*, and *reproduction*. Here follows an overview from a psychobiological perspective of the first two core vital functions, with a specific emphasis on humans, while the third is discussed in Chap. 6.

P. Ferentinos (✉)
2nd Department of Psychiatry, School of Medicine, Attikon General Hospital,
National and Kapodistrian University of Athens, Athens, Greece

Institute of Psychiatry, Psychology & Neuroscience, Social, Genetic and Developmental Psychiatry Center, King's College London, London, UK
e-mail: pferentinos@med.uoa.gr

N. Kokras
1st Department of Psychiatry, School of Medicine, Eginition Hospital,
National and Kapodistrian University of Athens, Athens, Greece

Department of Pharmacology, School of Medicine, National and Kapodistrian University of Athens, Athens, Greece

G. Konstantakopoulos
1st Department of Psychiatry, School of Medicine, Eginition Hospital,
National and Kapodistrian University of Athens, Athens, Greece

Institute of Psychiatry, Psychology & Neuroscience, Section of Cognitive Neuropsychiatry, King's College London, London, UK

K. N. Fountoulakis, I. Nimatoudis (eds.), *Psychobiology of Behaviour*,
https://doi.org/10.1007/978-3-030-18323-3_3

3.1.1 Nutrition (Feeding)

Feeding provides the energy that is essential for survival and therefore is subject to intense regulation by the human brain. Adequate nutrition is ensured by a complex brain system regulating the levels of various nutrients in the blood and in the body stores. The hypothalamus is the center of the network of control on food intake and metabolism in response to peripheral signals that reflect the feeding state and energy reserve, i.e., homeostatic control.

Hunger is associated with discomfort providing a strong drive for feeding and satiety and is accompanied with satisfaction preventing further consumption of food. However, the rewarding nature of food goes beyond the feelings of hunger and satiety. Modern humans often eat without being hungry, and nowadays obesity is a serious public health problem. Hedonic eating, i.e., eating based on pleasure rather than energy needs, is controlled by complex neural mechanisms associated with reward. The insular cortex, orbitofrontal cortex, nucleus accumbens (NAc), amygdala, and ventral tegmental area (VTA) have a key role in the control of feeding behavior in response to the hedonic aspects of food.

Over the last two decades, our knowledge of neural circuits and molecules involved in *homeostatic* and *hedonic control* of food intake has improved substantially, in large part due to the findings of research in experimental animals and functional neuroimaging in humans. These findings also provide insight into the mechanisms underlying obesity and abnormal feeding behavior in neuropsychiatric disorders. Only the main aspects of the current knowledge on mechanisms controlling feeding behavior can be emphasized here.

3.1.1.1 Homeostatic Control of Food Intake

The hypothalamic network that regulates food intake and energy balance consists of interconnected neurons located in the arcuate (infundibular in humans) nucleus (ARC), ventromedial nucleus (VMH), paraventricular nucleus (PVN), dorsomedial nucleus (DMH), and lateral hypothalamic area (LHA). Peripheral signals that stimulate (orexigenic) or inhibit (anorexigenic) food intake are received by neurons in the medial zones of the hypothalamus, including signals from circulating nutrients (glucose, fatty acids), hormones (insulin, leptin, ghrelin), and gastrointestinal peptides (cholecystokinin and peptide YY_{3-36}) (Adan et al. 2008; Chaudhri et al. 2006; Benarroch 2010). The dorsomedial and lateral hypothalamic neurons receive circadian influences and interact with neural circuits for thermoregulation and arousal (Saper et al. 2002; Benarroch 2010). The integration between orexigenic and anorexigenic signals is accomplished via complex interactions between the hypothalamic nuclei mediated by a variety of neurotransmitters (Meister 2007). The hypothalamic network exerts control on food intake and peripheral metabolism acting via projections to sympathetic and parasympathetic nuclei (nucleus of the solitary tract, area postrema, dorsal motor nucleus of the vagus, and locus coeruleus) on the endocrine glands and the gastrointestinal system (Williams et al. 2001; Gao and Horvath 2007). Cognitive and emotional aspects of food intake relay on reciprocal connections of the hypothalamus with cortical and mesolimbic circuits and the hippocampus (Gao and Horvath 2007; Kampe et al. 2009).

In the following we present the main peripheral and central signals and hypothalamic pathways related to feeding behavior, which are also briefly displayed in Table 3.1.

Table 3.1 Main signals and mechanisms for homeostatic control of food intake

Signal	Source	Target (receptors)	Effect	Mechanisms of action
Peripheral (hormones)				
Insulin	Pancreas	Hypothalamus (insulin receptors, IR)	↓ Food intake	Activation of POMC neurons in ARC Activation of BDNF neurons in VMH Inhibition of LHA neurons
Cholecystokinin Peptide YY_{3-36}	Gut	Hypothalamus via vagus nerve (CCK-1, Y2)	↓ Food intake	Stimulation of vagus nerve—signals via NTS and PBN projections to POMC neurons in ARC
Leptin	Adipose tissue	Hypothalamus (leptin receptors, OB-R)	↓ Food intake ↑ Metabolism	Activation of POMC neurons in ARC Activation of BDNF neurons in VMH Inhibition of LHA neurons
Ghrelin	Stomach	Hypothalamus (GHR1)	↑ Food intake	Activation of neuropeptide Y, Agouti-related peptide, and GABA neurons in ARC
Central				
α- and β-MSH	ARC	Hypothalamus (MC4R)	↓ Food intake	Agonists of MC4R in PVN and VMH
Agouti-related peptide	ARC	Hypothalamus (MC4R)	↑ Food intake ↓ Metabolism	Inverse agonist of MC4R in PVN
Neuropeptide Y	ARC	Hypothalamus (Y1, Y2, Y5)	↑ Food intake ↓ Metabolism	Direct activation of PVN Inhibition of POMC neurons in ARC
BDNF	VMH	Hypothalamus (Tropomyosin receptor kinase B, TrkB)	↓ Food intake	Agonist of TrkB and MC4R in PVN and VMH
Melanin-concentrating hormone	LHA	Hypothalamus, VTA (MCH1 and MCH2)	↑ Food intake ↓ Metabolism	Agonist of MCH receptors in hypothalamus and VTA
Orexin/hypocretin	LHA	Hypothalamus (OX1 and OX2)	↑ Food intake	Agonist of OX1 and OX2 in PVN (short-term regulation of energy balance)
Endocannabinoids	LHA	Hypothalamus (cannabinoid-1 receptors, CB1)	↑ Food intake ↓ Metabolism	Inhibition of anorexigenic signals via CB1

ARC arcuate (infundibular in humans) nucleus, *BDNF* brain-derived neurotrophic factor, *GABA* γ-aminobutyric acid, *LHA* lateral hypothalamic area, MC4R melanocortin 4 receptor, *MSH* melanocyte-stimulating hormone, *NTS* nucleus of the solitary tract, *PBN* parabrachial nucleus, *POMC* pro-opiomelanocortin, *PVN* paraventricular nucleus, *VMH* ventromedial nucleus, *VTA* ventral tegmental area

Central Regulation of Feeding and Energy Balance

The ARC is a key regulator of food intake and energy balance containing a group of neurons that synthesize two neuropeptides, *α- and β-melanocyte-stimulating hormone (MSH)*, derived from *pro-opiomelanocortin (POMC)*, and another group of neurons synthesizing *neuropeptide Y (NPY)*, *Agouti-related protein (AgRP)*, and *γ-aminobutyric acid (GABA)*. α- and β-MSH decrease food intake and increase energy expenditure acting on melanocortin 4 receptors (MC4R) in the PVN and VMH (Hillebrand et al. 2006). By contrast, NPY via Y1, Y2, and Y5 receptors and AgRP acting as an inverse agonist of MC4R in the PVN increase food intake and reduce energy expenditure (Meister 2007). Moreover, the same group of neurons can inhibit POMC neurons in the ARC via GABA and NPY projections (Benarroch 2010). Thus, the ARC mediates both orexigenic and anorexigenic signals from periphery and regulates feeding and energy metabolism integrating these mutually opposing influences.

Neurons in the VMH that synthesize *brain-derived neurotrophic factor (BDNF)* receive signals from POMC neurons of the ARC, and they also respond to glucose and leptin reducing food intake and increasing energy metabolism (King 2006). Groups of neurons in the PVN receiving signals from the ARC synthesize hormones with anorexigenic effects—corticotropin-releasing hormone (CRH), thyrotropin-releasing hormone (TRH), and oxytocin (Benarroch 2010).

The LHA has also a key role in the regulation of feeding and metabolism, integrating signals from the periphery (i.e., glucose, leptin, ghrelin) and interacting with other hypothalamic areas and the mesolimbic system (Williams et al. 2001; Dietrich and Horvath 2009; Berridge 2009). A group of neurons in the LHA synthesizing *orexin* (or *hypocretin*) plays a significant role in the short-term regulation of energy balance. Orexin neurons are inhibited by glucose and stimulated during fasting, and they promote food intake acting on specific receptors (OX1 and OX2) in the PVN (Tsujino and Sakurai 2009). Moreover, they are involved in food reward as we will present below. Another group of LHA neurons synthesizes *melanin-concentrating hormone (MCH)*, and they act on specific receptors (MCH1 and MCH2) increasing food intake and decreasing energy metabolism (Guyon et al. 2009). The function of LHA on food intake is related to sleep-wake cycle: the MCH neurons are active during slow-wave sleep, while the orexin neurons are activated in wakefulness (Benarroch 2010).

Peripheral Factors Regulating Food Intake and Metabolism

Gut peptides (cholecystokinin, peptide YY_{3-36}*)* are released after a meal and suppress food intake and meal size activating via vagal afferents the nucleus of the solitary tract, which signals fullness to the hypothalamus and other brain regions, initiating satiety and resulting in meal termination (Chaudhri et al. 2006; Dietrich and Horvath 2009). Another gut peptide under investigation with similar effects on food intake and a significant role in the control of glucose and energy homeostasis is *glucagon-like peptide-1* (Williams and Elmquist 2012). On the other hand, *ghrelin* is the hormone that is released from the stomach during fasting and provokes hunger and meal initiation. Ghrelin, acting on the growth hormone secretagogue receptor

(GHSR) in the ARC, stimulates NPY, AgRP, and GABA neurons (Kageyama et al. 2010). *Leptin* is a hormone synthesized in the adipose tissue that circulates at levels proportional to the amount of fat. Leptin, acting on specific receptors in the ARC, stimulates POMC neurons and inhibits the release of NPY and AgRP, thus contributing in long-term weight and glucose homeostasis (Farooqi and O'Rahilly 2008; Williams and Elmquist 2012). It also produces anorexigenic effect stimulating BDNF neurons in VMH while inhibiting LHA neurons (Farooqi and O'Rahilly 2008; Benarroch 2010). *Insulin*, the hormone released by beta cells in the pancreas and regulating glucose homeostasis, has also anorexigenic effects possibly through similar mechanisms of action as those of leptin (Konner et al. 2009; Williams and Elmquist 2012).

3.1.1.2 Hedonic Control of Feeding Behavior

Many aspects of human behavior, like seeking for pleasant food, cooking, or obesity, indicate that feeding is not controlled solely by homeostatic mechanisms but is also influenced by the rewarding nature of food.

Gustatory Regulation of Feeding

Food reward is associated with palatability qualities, particularly taste and smell. Animals consume sweet and salty food beyond their homeostatic needs and avoid sour or bitter food even if they are hungry. In the human brain, taste information passes via the nucleus of the solitary tract and parabrachial nucleus to the thalamus, the lateral frontal cerebral cortex, the central nucleus of amygdala, and several hypothalamic areas, including LHA. Although gustatory thalamus is critical for hedonic aspects of taste, other subcortical areas also mediate the motivational qualities of palatable food cues (Saper et al. 2002).

Reward System for Feeding

The pleasure of palatable food is associated with activation of many areas of the brain reward system, including the VTA dopaminergic system, NAc, ventral pallidum, and amygdala (Berridge 2009; Kampe et al. 2009). Dopamine release in the NAc mediates the motivational aspects of food intake, especially the drive to eat food that is hedonically desirable ("wanting") (Lutter and Nestler 2009). As yet, the mechanisms by which food stimulates dopamine release are not well understood. It has been found that food can stimulate dopamine signalling independent of the processing of taste information (de Araujo et al. 2008).

Release of *orexin* during feeding directly stimulates dopamine neurons in the VTA increasing dopamine release in the NAc (Harris and Aston-Jones 2006). Other hypothalamic neuropeptides may also play a role in hedonic regulation of feeding influencing dopamine release. The *cocaine- and amphetamine-regulated transcript (CART)* which is found in several hypothalamic areas decreases food intake possibly inhibiting dopaminergic neurons in VTA. However, the anorexigenic effect of CART is associated with its multiple actions in hedonic and homeostatic regulating systems, which are not clear yet (Rogge et al. 2008). By contrast, *galanin* stimulates food intake, in particular the intake of fat, possibly acting on specific receptors in the PVN. However, it still remains unknown which of the multiple central and peripheral effects of galanin might be related with this effect (Fang et al. 2011).

The hedonic reaction per se to the pleasure of food reward ("liking") is regulated by *endogenous opioids* and *endocannabinoids* acting via μ-type opioid receptors and CB1 receptors, respectively, within the shell of the NAc and possibly within the ventral pallidum (Berridge 2009). Although "liking" and "wanting" are needed together for complete food reward, they are mediated by interacting but partially independent neural substrates.

3.1.1.3 Interactions of Homeostatic and Hedonic Regulatory Mechanisms

The stability of body weight over adult life in spite of the availability of highly palatable and energy dense food, as well as the discrepancies from normal eating, e.g., overweight, obesity, and eating disorders, indicates an interface between the metabolic and hedonic drives of eating. Therefore, the possible neural circuits and mechanisms that underlie interactions between homeostatic and hedonic regulation of feeding have been a focus of research during the last two decades.

The NAc plays a key role in the integration of homeostatic, hedonic, and cognitive aspects of food intake via its connections with the prefrontal cortex, amygdala, and lateral hypothalamus (Berridge 2009; Kelley et al. 2005). There are also multiple functional connections between hypothalamic, cortical, and mesolimbic circuits mediated by POMC, orexin, and MCH that may play a role in homeostatic–hedonic control interactions (Kampe et al. 2009). Hormones involved in homeostatic regulation of feeding, such as leptin, insulin, and ghrelin, also exert effects on motivation to obtain food through their influence on mesolimbic dopamine signalling, especially on the dopaminergic neurons in the VTA (Murray et al. 2014). Leptin decreases the firing rate of the VTA dopaminergic neurons. Insulin increases dopamine release and the firing rate of dopaminergic neurons but reduces dopamine levels in the VTA probably by upregulation of the dopamine active transporter (DAT). Ghrelin enhances signalling from the VTA to the NAc increasing the activation of dopamine D_1 and D_2 receptors and dopamine levels.

Like ghrelin, other factors involved in meal-to-meal regulation of feeding may also affect food reward in a way that even highly palatable food may be unpleasant after satiation. There is evidence that the rewarding effects of food are potently modulated by indicators of satiety, such as peptide YY_{3-36} that was found to elicit a switch of activation from the hypothalamus to the orbitofrontal cortex and diminished orbitofrontal activation in response to the rewarding aspects of food (Batterham et al. 2007). The main pathways related to hedonic control of feeding behavior are briefly displayed in Table 3.2.

3.1.1.4 Cognitive and Emotional Control of Feeding Behavior

Homeostatic and hedonic mechanisms controlling feeding behavior described above only partially operate outside awareness. However, there is also a "top-down" control of human feeding behavior: interactions between cognitive and emotional processes could lead to different responses to food cues and changes in food intake

Table 3.2 Main signals and mechanisms for hedonic control of eating behavior

Signal	Source	Target (receptors)	Effect	Mechanisms of action
Peripheral (hormones)				
Leptin	Adipose tissue	VTA (leptin receptors, OB-R)	↓ Food intake ↑ Metabolism	Inhibition of dopaminergic neurons in VTA
Insulin	Pancreas	VTA (insulin receptors, IR)	↓ Food intake	Reduction of dopamine levels in VTA probably by upregulation of DAT
Central				
Ghrelin	ARC	VTA (GHR1)	↑ Food intake	Activation of dopaminergic neurons in VTA Increase of the activation of dopamine D1 and D2 receptors and dopamine levels in NAc
Orexin/hypocretin	LHA	VTA (OX1 and OX2)	↑ Food intake	Activation of dopaminergic neurons in VTA
Endocannabinoids	Local	Nucleus accumbens (cannabinoid-1 receptors, CB1)	↑ Food intake ↓ Metabolism	Enhancement of dopamine effect on nucleus accumbens
Endogenous opioids	Local	Nucleus accumbens (μ-opioid receptors)	↑ Food intake	Increase of dopamine release in nucleus accumbens
CART	ARC, LHA	Hypothalamus, Mesolimbic system	↓ Food intake	Unknown
Galanin	ARC	Hypothalamus, especially PVN (GALR)	↑ Food intake	Unknown

ARC arcuate (infundibular in humans) nucleus, *CART* cocaine- and amphetamine-regulated transcript, *DAT* dopamine active transporter, *LHA* lateral hypothalamic area, *NAc* nucleus accumbens, *PVN* paraventricular nucleus, *VTA* ventral tegmental area

(Berthoud 2011). Thus, humans can voluntarily inhibit their drive to eat or develop involuntary changes in their appetite and body weight related to emotional states.

Cognitive control of feeding behavior involves integration of peripheral signals related to energy status of the body, food-related signals in the form of sensory and environmental cues, and memory of past feeding experiences (Benarroch 2010). The insular, orbitofrontal, and anterior cingulate cortical areas have a key role in the processing of interoceptive and food-related information and participate in motivational aspects of feeding behavior (Craig 2009; Shin et al. 2009; Saper et al. 2002).

There is now evidence from preclinical studies that emotional factors influence both hedonic and homeostatic aspects of food intake, altering the activation of many mediators such as ghrelin, orexin, and leptin. For example, chronic stress may influence feeding and body weight independent of palatability of food or energy status of the individual (Lutter and Nestler 2009). This is more obvious in human behavior, since changes in appetite and body weight are frequent symptoms and one of the core diagnostic features of major depressive disorder. Furthermore, the association rate between mood disorders and obesity is about 25% (Simon et al. 2006). Influences of mood on hedonic and homeostatic control of feeding may be partially mediated by the serotonergic system, e.g., action of serotonin on POMC neurons in ARC via $5HT_{2C}$ receptors (Heisler et al. 2002). Aside from depression, serotonergic system dysfunction is also implicated in the pathophysiology of eating disorders, i.e., anorexia nervosa and bulimia nervosa (von Hausswolff-Juhlin et al. 2015).

3.1.2 Interaction with the Environment (Social Interaction)

Humans form organized groups of conspecifics and live within societies. To properly function in such context, we display instinctive social behaviors aimed at facilitating the formation of social networks of support, competing for resources, avoiding or facing conflicts, assessing social interactions and intentions of others, conveying and exchanging emotional and social cues, bonding and creating romantic relationships, nurturing and teaching offspring, etc. Such complex behaviors have their basis in evolutionary preserved neurobiological mechanisms, which can be studied in humans or in other social animal species. For several years, our understanding of this neurobiology was based on neuro-structural and neurochemical-related parameters. Indeed, for example the *superior temporal sulcus* and the *fusiform gyrus* are known to be involved in the processing of perceived social stimuli. Then, the *amygdala* and the *prefrontal cortex* designate an emotional value to this perception (Ordonana et al. 2013). However, over the last decade, there was a significant progress on the role of the pro-social peptides, such as *oxytocin* and *arginine vasopressin* (AVP), which also appear remarkably conserved across species. Overall, the modulation of social behavior by these peptides has been confirmed by several experimental and clinical studies and in various contexts of social behavior (Keverne and Curley 2004; Heinrichs and Domes 2008; Meyer-Lindenberg et al. 2011; Insel 2010). In this chapter, we focus mainly on findings from human studies. However, it should be noted that such studies are sometimes limited by the fact that studying central oxytocin and AVP systems would ideally require direct access to the brain, which can only be done easily when using experimental rodents. In vivo imaging in humans remains impractical for studying the neurobiology of social instincts, as human subjects are constrained by the limitations of imaging techniques (Crockford et al. 2014). Although some would wrongly argue about the ability of rodents and other experimental animals to experience emotions in social contexts, such research has led to significant advances into our understanding of social instincts and emotional behavior (Panksepp 2016). Additionally, much of our understanding about how pro-social peptides act is by studying not the systems themselves but the effects of the exogenous administration of peptides, as it will be discussed further on. Thus, oxytocin is involved in fear response, anxiety, and generally in the stress response. It is thought that oxytocin is involved in processes that attenuate stress related to social interaction and therefore facilitates interaction between humans. Moreover, several studies have shown that oxytocin can decrease activity in the amygdala, a brain region responsible for fear regulation. In addition, oxytocin further modulates human emotional, social, and affiliative behaviors, such as adult bonding, mother-infant relation, and generally social attachments. On the other hand, AVP is known to play an important role in fluid homeostasis and in the regulation of cardiovascular and autonomic systems. Therefore, by regulating those systems and functions, AVP may also play an important role in attention, arousal, competition, and aggression. It is therefore believed that oxytocin and AVP may have opposing actions. Oxytocin is more involved in regulating pro-social behaviors, such as social bonding, romantic relationships, and parenthood. Moreover, it is thought that oxytocin may even inhibit actions of AVP, during certain periods, such as pregnancy and lactation (Galbally et al. 2011). Finally, it should be noted that both AVP and oxytocin present significant sex differences, as evidenced by experimental and human studies and discussed in this chapter (Kokras et al. 2011; Carter 2007).

3.1.2.1 Facial Expression and Impressions

As humans are social beings, we continuously form impressions of conspecifics, based on prior learning, appearance, and behavioral cues of people that we interact with. This human instinctive behavior does not follow objective criteria but instead may present significant bias. A well-studied example is that people experiencing fear or stress tend to assign hostile characteristics to neutral facial expressions that are presented to them. Recognition of faces and extrapolation of useful social and emotional cues from the facial expression are of paramount importance for humans, perhaps in contrast to other mammalian species, which rely on other systems and cues, such as odor and pheromones, for recognition of conspecifics. Several studies have demonstrated that in this process the *dorsomedial prefrontal cortex (DMPFC)* is crucially involved. Interestingly, whereas the hippocampus is thought to be involved in learning and memory functions, specifically for social cues and impressions, it appears that the dorsomedial prefrontal cortex plays a key role. In fact, individuals with hippocampal lesions apparently preserve their ability to form impressions of other people based on facial expressions (Lass-Hennemann et al. 2011). On the other hand, the *amygdala* is a key brain region involved in complex social behaviors. The amygdala activates in the presence of perceived threats but also activates while experiencing emotions, when processing faces, and interestingly, while making social judgments. Decreased activation of the amygdala in humans is linked to decreased social fear and thus resulting into more pro-social behaviors. On the contrary, it is well-known that increased amygdala activity relates to phobias and social avoidance (Ordonana et al. 2013). As humans have to continuously interact socially, the ability to exchange information on emotions is critical. Bonding between members of a social group depends on mutual recognition, for which most mammals rely on odor and pheromones. In humans, however, recognition of conspecifics and bonding relies heavily on vision and facial recognition. Moreover, the communication of our feelings, such as joy or sadness, facilitates others to understand our needs, and inversely by extrapolating the emotional status of people around us, we can obtain valuable information, such as linking fear in others with the notion that we might as well be in danger. Women are generally more capable than men in decoding emotional cues, and aside verbal communication, emotions can be conveyed through facial expressions. The amygdala is the brain region heavily involved in this process, as persons with lesions in that brain structure are incapable of correctly interpreting facial expressions (Somerville et al. 2011). Several brain imaging studies support this finding as activation of the amygdala happens during both positive and negative emotional cues, facial expressions included (Costafreda et al. 2008). It has been demonstrated that oxytocin can attenuate the activation of the amygdala, which is rich in oxytocin receptors, thus reducing fear responses when confronted with facial expressions (Kis et al. 2013). AVP plays an important role in recognizing each other's emotions, as administration of exogenous AVP facilitates men's pro-social behavior and processing of social information. However other studies showed that in men but not in women AVP deteriorated recognition of facial emotions and promoted aggressive responses (Thompson et al. 2006). It appears that AVP increases attention generally to social stimuli with the potential of facilitating either a pro-social or an aggressive response depending on the context. Moreover, AVP mediates many of its behavioral effects through the AVP receptor 1a. It appears that genetic variations of AVP1a affect the behavioral response and bonding of men, but not women, as evidenced by decreased probability of being married, and increased chance of marital problems and divorcing. More sex differences have been observed after intranasal administration of oxytocin. Women display increased brain activity when seeing angry or fearful facial expressions after exogenous oxytocin administration. On the contrary, men show decreased activity in the involved brain regions, such as the amygdala and the temporal lobe (Zink and Meyer-Lindenberg 2012). In this context, it has been proposed that oxytocin and AVP act toward a divergent stress response in men and women. Certainly, a significant overlap exists in the typical "fight or flight" stress response. However, under certain situations and due to the influence of sex-dependent actions of oxytocin and AVP, women tend to adopt an affiliative ("tend-and-befriend") response significantly more often than men (Taylor et al. 2000; Kokras et al. 2011).

3.1.2.2 Competition and Aggression

Competition plays an important role in human societies, as individuals confront each other for primary resources, such as food, and secondary resources, such as employment, money, etc., which in turn improve access to primary resources. Such confrontation is dependent on the characteristics of each social group and accepted social norms and can have many forms, from pure aggression to elaborate political maneuvers in order to successfully compete and prevail. A common characteristic, however, is that such competition obeys rules of social order, whose purpose is to finally resolve and settle competitive confrontation. Several lines of evidence have linked modulation, usually reduction, of *testosterone* levels to those processes that control the evolvement and then mitigation of social competition. However, the complexity of human behavior while competing with conspecifics, together with methodological problems, creates a challenge for testosterone-related research in this field, given that testosterone levels have been found increased, decreased, or unaffected, depending on the involved competitive task, effort invested in obtaining the resources, coping styles, and many other confounding factors. Furthermore, it is now known that women are no less competitive than men, particularly in modern societies that engage women at many levels. However, research in the psychobiology of female competition is dragging behind, and the role of testosterone in modulating aspects of female social competition is still unclear. In any case, social competition, whether it results in dominant and subordinate individuals in experimental studies or more elaborate outcomes in human societies (such as socioeconomic status), is long known to affect the biology of individuals. Parameters that have been studied and found affected by outcomes of social competition are cardiovascular indices, body/organ weight and size, sperm quality, autonomic and endocrine function, as well as brain function as expressed by neurotransmitters, receptors, and neurogenesis (Salvador and Costa 2009). Social competition may also lead to aggression. It has been long known that testosterone levels relate to aggressive behavior; however, this relation is apparently true only for social expressions of violence and not for aggressive behavior in general. Moreover, it seems that only high testosterone levels, or at best considerable increases, induce such effect. Testosterone rather primes males for anger, feeling threatened and challenged and thus more reactive to responding aggressively (Goetz et al. 2014). High testosterone levels also impede the amygdala-prefrontal cortex connectivity, reducing the conscious control over unconstrained social behavior in males (Gettler 2014). Such testosterone-primed aggressive behavioral response may be adaptive or not, depending on the social context. Moreover, in experimental studies, other steroids, such as *estrogen* and of course peptides, such as *AVP*, have been found to fluctuate in conjunction with testosterone, depending on whether confronting a male or a female conspecific and also depending on social context (mating, territorial behavior, social status, etc.) (Bos et al. 2012b). Experiments with intranasal AVP administration in healthy volunteers showed that AVP may facilitate aggression by predisposing persons to respond aggressively to ambiguous social cues, which by a presumed AVP-mediated action are perceived as threatening. Such actions however are sexually dimorphic. Indeed, whereas in men AVP apparently decreases a perception of friendly social cues, in women AVP administration seems to facilitate a friendly, non-threatening perception of ambiguous social cues, such as unfamiliar faces. Genetic studies also support the involvement of AVP receptors, specifically subtype 1A. Polymorphisms in this receptor modulate social bonding in men but not in women. Thus, it seems that centrally acting AVP modulates social human communication in a sex-dependent manner (Heinrichs et al. 2009). Without doubt, many simultaneous physiological systems, neurotransmitters, hormones, and centrally acting peptides are activated, as many researchers have argued (Neumann and Landgraf 2012; Taylor 2006). In addition, AVP and testosterone actions that prime humans for competition and eventually aggression also interact with *oxytocin's* actions that promote social bonding, as it will be discussed further on. Thus, it appears that high oxytocin and low testosterone levels promote nurturing behavior and gentle affiliations, whereas high oxytocin and high testosterone promote loyal comradeship, risky and dangerous group actions that individuals would not undertake alone (such as warfare) (Van Anders et al. 2011; De Dreu 2012). Finally, it should be noted that those neuroendocrinological mechanisms are also thought to be involved in the termination and settlement of aggressive and competitive behaviors. A remarkable example is face blushing. Blushing and gaze aversion are considered, from an evolutionary point of view, as displays of conciliation, thus facilitating the resolution of conflict and aggression between humans (Stein 2015).

3.1.2.3 Social Bonding and Support

Despite humans forming complex and sophisticated social bonds, those functions are based on brain circuits that apparently are conserved through most mammalian species. Thus, substantial research has been devoted into decoding the involvement of specific neuronal circuits affecting social bonding and close relationships. Romantic love between partners and parental love were successfully linked by imaging studies to the *anterior cingulate gyrus (ACG)*, *medial insula*, *caudate nucleus* of the striatum, and *ventral tegmental area* (VTA). Whereas those brain areas are seemingly activated, others such as frontal, parietal, and temporal cortical brain areas, including the amygdala, appear to deactivate. Deactivation of cortical areas in close relationships (which also appears to happen in sexual arousal) would justify common saying that during love, sex, and other highly rewarding conditions, lapses in judgment frequently arise. Moreover, it is now supported with reasonable certainty that the dopaminergic projections coming from the VTA and crucially involved in reward in general are also involved in attachment and social bonding. Furthermore, pro-social neuropeptides (oxytocin and AVP) facilitate this dopaminergic neurotransmission, as in the case of breast feeding, thus enhancing the subjective rewarding feeling and in turn strengthening social bonding and close relationships (Stein and Vythilingum 2009). Specifically for oxytocin, its involvement in the formation of a romantic relationship and more generally in the expression of affectionate behavior is now well documented. However, several studies now support the notion that oxytocin's actions are not universal but instead context-specific and with considerable interactions with other factors, such as whether expression of affectionate behavior is directed toward next of kin individuals or other more distantly familiar individuals (Gettler 2014). Similarly to what was mentioned previously, peptide actions with regard to bonding present significant sex differences. In humans, plasma oxytocin correlated with romantic relationship-related perceived distress in women but not men, while plasma AVP correlated with such distress in men but not women (Taylor et al. 2010). But seeking rewarding interpersonal interactions is not limited to the formation of close romantic relationships only. Forming a social network and perceiving social support are other social instinctive behaviors aimed at improving human health and chances of survival. In fact, the degree of perceived social support by an individual has been linked to reduced stress and autonomic activation, such as reduced blood pressure, heart rate, reduced cortisol and adrenaline levels, etc. Remarkably, similar effects in stress reactivity are observed independently of whether social support is offered or received by an individual, and therefore, the dampening of the HPA axis response is now considered a key mechanism in promoting pro-social behavior (Ditzen and Heinrichs 2014). Several brain regions are implicated in this process in the ACG and the prefrontal cortex. Specifically, the *ventromedial prefrontal cortex (VMPFC)* appears to activate in the presence of a familiar social supporter, providing reassurance and safety and ultimately increasing pain threshold. Interestingly, there are sex differences in the effects and benefits of social support. Men seem to benefit most from verbal forms of support, including practical advice and instructions, whereas women benefit more from nonverbal expressions of support, such as touching, holding hands, smiling, etc. (Eisenberger et al. 2011). Moreover, differences also appear not only due to sex but also because of different coping styles of individuals. Indeed, subjects who have as a primary coping mechanism their reliance on social support display higher levels of salivary cortisol in response to conflict than those who rely on other coping styles. Finally, oxytocin is also considered a key peptide in regulating pro-social behavior, and several studies with intranasal administration of exogenous oxytocin evidenced its role in attachment during adulthood, ability to trust, generosity, altruism, pro-social motivation, empathy, and positive communication between couples (Hostinar et al. 2014). Whereas no mechanism acts independently, the combined action of all those presented neurobiological mechanisms increases the chances of social contact, facilitates repetitive social interactions, and promotes reward when socializing, and this ultimately results in the formation and maintenance of stable social bonds between members of a group (Crockford et al. 2014).

3.1.2.4 Parental Behavior

Parental behavior extends to a varied repertoire of instinctive behaviors aimed at securing the survival of the offspring. Human infants are born more immature, have significantly more needs, and require more attention compared to other mammals. Therefore, a close bond between human parents and offspring is paramount to their survival, and several mechanisms are involved to secure this bonding. Early in the postpartum, levels of hormones such as *estradiol*, *progesterone*, and *cortisol* correlate with maternal feelings of nurturing, which are enhanced. Young mothers are vigilant for sounds of distress from infants (such as crying), displaying accelerated heart rate and cortisol increases. Learned maternal behavioral responses and the reward from successful nurturing in that period appear linked to such fluctuations of cortisol levels (Olazabal et al. 2013). Moreover, genetic polymorphisms in the dopaminergic neurotransmission system, such as in the DRD1/2, DRD4, and the DAT1 genes (Kaitz et al. 2010; Mileva-Seitz et al. 2012), appear to influence how the mother responds and interacts with the infant but with significant gene by environment (or more specifically gene by social context) interactions. Such complex interactions highlight the difficulties encountered in genetic research of social instincts but also difficulties in genetic research in general (Duncan and Keller 2011; Hewitt 2012). In addition to genetic polymorphisms, the mesolimbic dopamine reward circuit of the nurturing mother is also strongly modulated by prolactin and oxytocin, thus ensuring that the young mother experiences reward by nurturing her offspring. *Oxytocin* has been consistently associated with delivery, lactation, nurturing, and other complex parenting behaviors. Increased oxytocin during pregnancy promotes instinctive maternal behavior and facilitates mother-infant bonding. During pregnancy, both mother and fetus produce oxytocin, which is thought to influence antenatal bonding. Oxytocin has a significant role postnatally as well, as it can moderate amygdala activity, dampen hypothalamic-pituitary-adrenal (HPA) axis activity, reduce the stress and anxiety associated with caring of an infant, and, in addition, promote rewarding feelings. Indeed, less anxious mothers are more efficient in recognizing and responding to nonverbal infant cues. During the postpartum, oxytocin levels remain high by stimuli received during lactation but also by other infant-derived stimuli, such as vocal calls, infant facial expressions, and mother-infant physical contacts. Oxytocin is released rapidly in response to such stimuli and thus promotes maternal caregiving behavior and attachment (Galbally et al. 2011). In the human species, usually mothers and fathers cooperatively raise offspring, although the degree of cooperation may differ across cultures and social situations. The neurobiology of paternal behavior is relatively less studied, but fathers with higher oxytocin levels engage more often in playing with their children, and inversely playing with children resulted in more pronounced spikes of oxytocin. Like mothers, fathers with higher oxytocin displayed closer attachment to their infants. However, there have been other conflicting results; hence it is still not clear whether peripheral oxytocin levels accurately depict the parenting behavior or whether there are context-specific parenting situations and related emotional and behavioral responses that are facilitated by oxytocin. *Testosterone* is also involved in expression of paternal behavior. Fathers involved in raising offspring were found with attenuated levels of testosterone. It is hypothesized that from an evolutionary point of view, human fathers displaying reduced testosterone levels would be primed to less aggressive behavior and would be more attentive and more prone to display cooperative behaviors. However this phenomenon appears linked to the degree of paternal investment and commitment; hence cultural differences in the levels of father involvement and cooperation with the mother are expected to influence the degree of testosterone attenuation (Gettler 2014). Moreover, several lines of evidence support the involvement of *prolactin* not only in the well-studied mechanism of lactation in women but generally into shaping the maternal behavior in close interaction with oxytocin, AVP, and importantly the dopaminergic neurotransmission (Rilling and Young 2014). Interestingly, prolactin also seems to play an important role in influencing father behavior. Experimental studies in rodents and other species show that increased prolactin correlates well with paternal behavior, this correlation is absent from species that do not exhibit significant paternal care, and pharmacological suppression of prolactin in specific periods may inhibit the appearance of paternal behavior (Wynne-Edwards 2001). Although human research in this field is not conclusive, prolactin was also shown in human males to increase just before birth, higher prolactin levels associated with stronger paternal emotional responses, and a link with prolactin was found in expectant fathers' sympathetic pregnancy ("couvade") (Storey et al. 2000).

3.1.2.5 Conclusion

Humans possess but also develop complex social instinctive behaviors, which are destined to support social accession and continuous adaptation to varying social stimuli, thus ultimately ensuring survival and promoting personal well-being. Such complex social behaviors rely on appropriate and equally complex neurobiological mechanisms, involving several brain regions, major neurotransmitter systems, stress, and sex hormones, as well as peptides such as oxytocin and AVP, all forming coordinated networks in our brain and body (Skuse and Gallagher 2009). A simplified representation of these systems is shown in Fig. 3.1 (Baribeau and Anagnostou 2015; Meyer-Lindenberg et al. 2011; Boccia et al. 2013). Genetic polymorphisms and epigenetic mechanisms during several critical periods, such as the neonatal, early childhood, and adolescence further carve the neurobiology of social behavior and contribute to high individual variability. Moreover, it should be noted that observing, understanding, predicting, and responding appropriately in the social context may be innate but also learned behavior. In fact, important aspects of the psychobiology of social behavior are not simply hardwired but prone and receptive to social learning. Indeed, innate but untrained

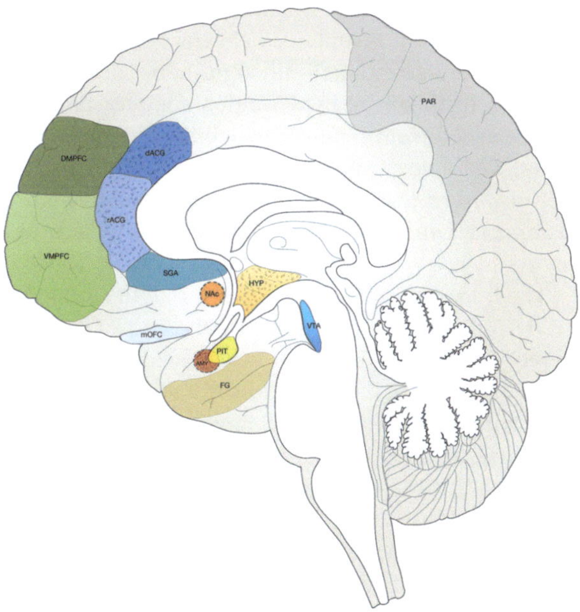

Fig. 3.1 Brain areas significantly involved in social behavior. *DMPFC* dorsomedial prefrontal cortex, *VMPFC* ventromedial prefrontal cortex, *dACG* dorsal anterior cingulate gyrus, *rACG* rostral anterior cingulate gyrus, *SGA* subgenual anterior cingulate cortex, *mOFC* medial orbitofrontal cortex, *NAc* nucleus accumbens, *AMY* amygdala, *PIT* pituitary gland, *HYP* hypothalamus, *FG* fusiform gyrus, *VTA* ventral tegmental area, *PAR* parietal lobe. Of those areas, AMY, HYP and ACG are thought to be particularly rich in oxytocin receptors (shaded areas); however, despite numerous experimental studies, there is inconclusive evidence about the distribution in the human brain of vasopressin 1a and 1b receptors (Baribeau and Anagnostou 2015; Meyer-Lindenberg et al. 2011; Boccia et al. 2013)

neural networks at the time of birth provide later in infancy and childhood the neurobiological basis of learned social skills such as language and imitated behavior. Finally, at the cornerstone of social psychobiology is the understanding of several and important sex differences, as it is clearly the case in aggression, social support, bonding, and parenting.

3.2 Instincts

An instinct is the inherent tendency of a living organism to exhibit a particular complex pattern of behavior. Instinctive behaviors are therefore the manifestation of innate biological factors, are not based on learning or prior experience, and do not implicate consciousness or cognition. Instincts are displayed by most members of the species and can be thought of as the ability possessed by a living organism to perform, without anticipating their benefits, acts which secure the development and preservation of the individual or the species. However, the extent to which instincts determine the behavior of animals varies across species. The higher a species stands on the evolutionary ladder, the more developed its nervous system is, and education/learning plays a greater role than instinct in shaping behavior. Instinctive behaviors escape the individual's volitional control (i.e., they are automatic) although their expression can be modifiable by social learning. Instincts are by definition complex patterns of behavior and should be distinguished from *reflexes*, which are simple behaviors in response to specific stimuli (e.g., deep tendon reflexes) as well as *taxis* (locomotion of organism or cell toward or away from a specific external stimulus, e.g., negative phototropism) and *kinesis* (rate of movement is determined by some feature of the environment).

The term instinct was introduced into the field of psychology by W. Wundt in the 1870s (Wundt 1874), and by the end of the nineteenth century, almost all stereotyped animal behavior was considered as instinctive. Freud (1920, 1923) further elaborated on the notion of instinctive drives ("Trieb") and incorporated them into his topographical model. He described two kinds of unconscious instincts, the life instinct (libido, "eros") and the death instinct (aggression, "thanatos"), which are the major sources of psychic energy and are stored in the id. Both of these are in continuous need of satisfaction, leading to a buildup of tension/excitation; their fulfillment results in discharge of psychic energy and the alleviation of tension (see also Chap. 11 for further details). Instinct as a concept came out of fashion with the advent of behaviorism in the 1920s; most behaviors were then considered as the result of learning, conditioning, and reinforcement.

Interest in innate behaviors was revived again in the 1950s with the work of ethologists Lorenz (1952) and Tinbergen (1951). According to their theory, instinctive behaviors consist of one or more *fixed action patterns* (FAPs), stereotyped behavior sequences exhibited invariantly by all members of the species, which are elicited in response to a specific external sensory stimulus (called the *sign stimulus* or *releaser*) and almost inevitably run to completion once started. FAPs are supposed to be "hardwired" and produced by a neural network (called the *innate releasing mechanism*) described schematically in Lorenz's (1950) *psychohydraulic model* (Fig. 3.2). The model incorporates a reservoir within which action-*specific energy* gradually accumulates over time (representing the individual's state of internal motivation); energy flow is regulated/impeded by a valve on which two very different forces act additively (*law of heterogeneous summation*), i.e., the pressure of the energy building up within the reservoir and the weight exerted by external stimuli (releasers). When the sum of these two forces overcomes the restraining capacity of the valve, energy is discharged to produce a FAP. A period of time follows when a stimulus is less likely to produce a FAP (*behavioral quiescence*). A notable implication of the model is that a FAP can occur even in the absence of external stimuli (*vacuum activity*) following an excessive buildup of energy within the reservoir.

Despite its attractive simplicity, the major limitation of Lorenz's hydraulic model is that it did not include any mechanism for learning, i.e., there was no way that the consequences of the animal's behavior feedback into the system to modify subsequent behaviors. Yet, Lorenz highlighted the effect of developmental phase-sensitive learning (called *imprinting*, first described in geese learning their mother's identity during a critical period). However, he believed that the mechanisms producing instinctive behaviors (FAPs) were largely innate, i.e., solely determined by the effect of genes, and subsequently triggered by environmental stimuli. More recent

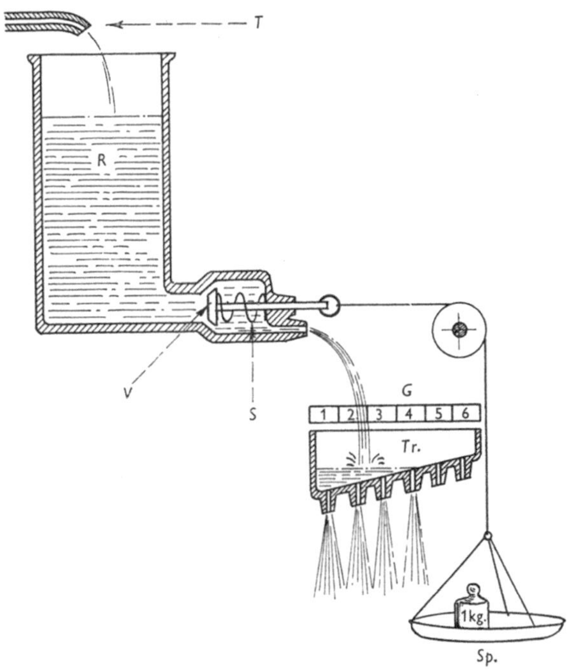

Fig. 3.2 Lorenz's psychohydraulic model of instinctive behavior. *T* tap, *R* reservoir, *V* valve, *S* spring, *Sp* scale-pan with weight, *Tr.* trough, *G* graded scale. Reproduced with permission from Lorenz (1950), courtesy of the Society for Experimental Biology

evidence calls for a revision of Lorenz's model. We now know that since the beginning of ontogenesis environmental stimuli are in constant interplay with genes at several developmental stages to shape the neural substrate in a process called *epigenesis* (Fuller 1954; Sokolowski and Corbin 2012).

Here follows a psychobiological overview of the instinct of aggression, while Chap. 6 covers aspects related to sexuality.

3.2.1 Aggression

Aggression is defined as overt behavior that involves threat or physical damage to another individual in the context of competition/conflicting interests, defense, or emotional arousal. Human aggression is associated with a considerable toll on society. More than one million people die each year as a result of violent acts (homicide, wars); males and young people are particularly overrepresented among the victims of violence (WHO 2010). Patients with neuropsychiatric disorders (psychotic, mood and anxiety disorders, personality disorders, impulse control and substance use disorders, dementia, organic brain damage) are at increased risk of manifesting aggression.

3.2.1.1 Classifications of Aggression

Various classifications of aggression have been proposed, e.g., by the nature of the aggressor (individual or collective), context (defensive or predatory, maternal, territorial, sex-related, dominance, or intermale), the target of aggressive acts (self-directed or other-directed), the mode of its expression (physical or verbal, direct or indirect), or its causes (psychiatric, medical, etc.). Aggression might also be classified according to social acceptability, depending on cultural context. Socially accepted aggression might comprise sports, hunting, corporal punishment, or warfare; socially prohibited aggression includes homicide, rape, child abuse, torture, and terrorism. Normative levels of aggression can facilitate survival, self-defense, or competition; however, when aggression is exaggerated in intensity, persistent, disproportionate to its eliciting stimuli, or out of social context, it is considered as pathological or maladaptive.

However, the most widely used and neurobiologically founded taxonomy of aggression in humans is that of *reactive* (or impulsive or affective or defensive or "hotheaded") versus *proactive* (or premeditated or instrumental or predatory or controlled or "cold-blooded") aggression (Gelles 1994; Barratt et al. 1999). Reactive aggression is considered to be an impulsive response to a perceived threat, associated with anger and autonomic arousal and probably underlies most incidents of aggression encountered in our society; proactive aggression is considered to be more purposeful, often not associated with emotional or autonomic arousal and accounting for higher profile crimes (genocides, mass assassinations). A phenomenological comparison of these two subtypes of aggression is presented in Table 3.3. On the basis of this classification, two quite distinct populations of aggressive individuals can be recognized: those who exhibit mostly reactive aggression and those

Table 3.3 Phenomenological side-by-side comparison of reactive and proactive aggression

Reactive aggression	Proactive aggression
Impulsive, unplanned ("hotheaded")	Planned, premeditated ("cold-blooded")
Responding to perceived threat, provocation, or frustration	Started by the offender, not provoked
Associated with high levels of emotional excitement (frustration, rage, fear) and autonomic arousal	Not associated with emotional or autonomic arousal
Not aiming at self-benefit (only defensive or reactive to rage)	Aiming at self-benefit or gain
Exposure to risk of physical self-harm or damage of own property	Protection against self-harm during aggression
Loss of control of aggressive behavior	Ability to control aggressive behavior
Public display of aggressive behavior	Hidden expression of aggression
May be followed by remorse	Feeling proud to be aggressive
Associated with neuropsychiatric disorders (psychotic, mood, personality and substance use disorders, IED, PTSD, organic brain damage)	Associated with psychopathy and violent offenses

IED intermittent explosive disorder, *PTSD* post-traumatic stress disorder

who exhibit mostly proactive and some reactive aggression; it follows, therefore, that the two subtypes are partly correlated (Poulin and Boivin 2000).

The two subtypes also have distinct pathophysiological underpinnings and neurobiological signatures. Reactive aggression seems to be a conserved pattern of behavior across all mammalian species; it is part of a graded response to increasingly threatening stimuli, starting from freezing toward low-level threats and culminating in flight/escape and reactive aggression toward medium- and high-level threats, respectively (Blanchard et al. 1977). Reactive aggression is largely mediated by hyperresponsiveness in a subcortical (hypothalamic and limbic) network or hyporesponsiveness in higher cortical regulatory areas or both. Proactive aggression in humans has been suggested to parallel predatory aggression in omnivorous/carnivorous animals and is thought to be much more dependent on higher cortical circuits. Much more data is available on reactive aggression, which will be the main focus of the remaining chapter.

3.2.1.2 Functional Neuroanatomy of Aggression

Reactive Aggression

Lesion (ablation) and electrical or, more recently, optogenetic stimulation studies in experimental rodents have highlighted the significance of the so-called subcortical rage circuit in the manifestation of reactive aggression (Falkner et al. 2016; Lin et al. 2011). This circuit relays perceived threatening sensory stimuli either directly via the thalamus (thalamoamygdalar pathway) or indirectly via the cortex (primary and secondary association cortices) to the *medial amygdaloid areas*, then to the *anterior* and *ventromedial hypothalamus (VMH)*, the *bed nucleus of the stria terminalis* and *lateral septal areas*, and from there to the *dorsal half of the periaqueductal gray* (Panksepp 1998; Nelson and Trainor 2007). As expected, lesions of all components of this circuit reduce aggression. However, a hierarchical organization is evident, since, for example, stimulation of the periaqueductal gray elicits a rage response even in the absence of an intact amygdala but not vice versa. Moreover, depending on the social context, different subnuclei of the aforementioned structures may be activated.

Brain lesion and brain imaging studies have in large part provided evidence (although less conclusive) that the subcortical rage circuit is important in mediating reactive aggressive expressions in humans and nonhuman primates, as well (Gregg and Siegel 2001; Davidson et al. 2000). However, in primates (as well as in other mammals but to a much lesser extent) prefrontal cortical areas exert central regulatory surveillance on the subcortical network by evaluating the salience of the threatening stimuli, planning actions, adjusting them to environmental feedback, anticipating their consequences, and assessing their potential reward value (Bachevalier et al. 2011; Anderson et al. 1999; Coccaro et al. 2011). Lesions of the *orbitofrontal cortex (OFC)* are generally associated with increased aggression (de Bruin et al. 1983; Machado and Bachevalier 2008), exemplified in the classic case of Phineas Gage (Damasio et al. 1994). The medial division of the OFC receives feedback on endogenous parameters (e.g., arousal, muscle readiness), while its lateral division receives sensory input about exogenous parameters of the perceived threat (e.g., its size or its proximity); both divisions are reciprocally connected with the amygdala (Ghashghaei and Barbas 2002). It seems, therefore, that the OFC is involved in the interpretation of interpersonal cues and the inhibition of inappropriate behavioral responses. The *anterior cingulate gyrus (ACG)* has been the focus of intensive research in recent years; it receives input both from the limbic system and the prefrontal cortex and serves as a conflict processor by assessing the social appropriateness of emotionally laden aggressive impulses. The infracallosal part of the ACG (often called subgenual ACG) may be more linked to predatory aggression, while the supracallosal part of the ACG is possibly involved in restraining inappropriate social behaviors and reactive aggressive impulses. The *ventromedial prefrontal cortex (VMPFC)*, in concert with the *dorsomedial (DMPFC)* and *ventrolateral (VLPFC) prefrontal cortices*, is involved in the detection of contingency changes of motor plans (i.e., of whether planned actions continue achieving their goals) and in the implementation of alternative plans according to social cues of conspecifics ("social response reversal") (Budhani et al. 2007). Failure to do so results in a buildup of frustration thought to underlie reactive aggression in individuals with psychopathic tendencies (Budhani and Blair 2005; Gorrindo et al. 2005; Dickstein et al. 2010). The *dorsolateral prefrontal cortex (DLPFC)* acts in synergy with the VMPFC and the ACG to consciously plan motor actions and anticipate their consequences before they are executed in the supplementary motor area and premotor and motor cortices. Lastly, the *temporal lobe* is also implicated in reactive aggression responses, especially in patients with organic temporal lesions or temporal lobe epilepsy (Ito et al. 2007). Concludingly, in primates the *medial amygdala* lies at the intersection of subcortical (unconscious) and cortical (partly conscious) circuits mediating reactive aggression; it is, therefore, actively involved in the assessment of the necessity of an aggressive response on the basis of sensory input and social cues and not just a releaser of a stereotyped aggressive behavior (Coccaro et al. 2007).

Interestingly, the hypothalamus, amygdala, septal areas, and prefrontal cortex provide sites of monoaminergic modulation of aggressive responses via their connections with mesolimbic dopaminergic projections stemming from the ventral tegmental area and serotonergic projections from the raphe nuclei, which result in disinhibiting and dampening, respectively, of aggressive behaviors (Seo et al. 2008; Siever 2008). Moreover, oxytocin and AVP exert opposing effects on the amygdala, hypothalamus, and possibly prefrontal cortical areas (Huber et al. 2005).

Proactive Aggression

Predatory aggression in animals is mediated by a circuit including the *dorsolateral hypothalamus* and the *ventral half of the periaqueductal gray* (Gregg and Siegel 2001). However, the suggested association of human proactive aggression with animal predatory aggression has been criticized mainly because the latter is not displayed toward conspecifics and is little influenced by learning, unlike the former.

Proactive aggression is probably dependent on the same cortical circuits involved in any purposeful behavior. A maladaptive proactive aggressive behavior will be selected by the individual when the neural systems involved in moral socialization and current decision-making are dysfunctional. Recent literature suggests that the core deficit in individuals exhibiting mainly proactive aggression is a decrease in the responsiveness of the *amygdala* to fearful and sad facial expressions of others, which is necessary for developing empathy through reinforcement learning and a prerequisite for moral socialization by parents and peers (Dadds et al. 2006; Dolan and Fullam 2006; Marsh and Blair 2008). As reinforcement information is normally represented in the medial OFC, it is no surprise that psychopathic individuals display reduced amygdala responses but also reduced *rostral ACG/medial OFC* activation in response to emotional tasks (Kiehl et al. 2001), while youths with psychopathic/callous-unemotional traits display reduced amygdala-OFC/*VMPFC* functional connectivity when processing fearful expressions (Marsh et al. 2008) or moral dilemmas (Marsh et al. 2011).

3.2.1.3 Neurochemistry of Aggression

There is gross evidence from both animal and human studies about the involvement of several neurotransmitters, neuropeptides, and steroid hormones in the modulation of aggressive behaviors (Table 3.4).

Neurotransmitters and Neuropeptides

Serotonin (5-HT) innervates both subcortical (hypothalamus, amygdala, septal areas, periaqueductal gray, hippocampus) and prefrontal cortical areas implicated in aggression (Frankle et al. 2005). Several lines of evidence point to 5-HT's role in modulating aggressive responses. In specific, low 5-HT levels are associated with higher levels of impulsivity and aggression as suggested by reduced cerebrospinal fluid (CSF) concentrations of the 5-HT metabolite 5-HIAA (5-hydroxyindoleacetic

Table 3.4 Major neurotransmitters, neuropeptides, and steroid hormones modulating aggression

Biological effector molecule	Effect on aggression
(a) *Neurotransmitters and neuropeptides*	
Serotonin (5-HT)	
SSRIs Tryptophan loading 5-HT_{1A} or 5-HT_{1B} agonists 5-HT_{2A} antagonists (atypical antipsychotics) $5HTT^{-/-}$ knockout mice	↓
Reduced 5-HT levels, reduced 5-HIAA Tryptophan depletion $5\text{-HT}_{1B}{}^{-/-}$ knockout mice	↑
Dopamine (DA)	
D_2 blockers (antipsychotics) $D_2{}^{-/-}$ knockout mice	↓
D_2 agonists (ropinirole, pramipexole) $DAT^{-/-}$ knockout mice	↑
Norepinephrine (NE)	
β-Blockers (propranolol) DA β-hydroxylase $(DBH)^{-/-}$ knockout mice	↓
α_2 agonists (clonidine)	↑↓ (Dose-dependent)
γ-aminobutyric acid (GABA)	
GABA reuptake inhibitors (tiagabine) $GABA_A$ agonists (muscimol)	↓
$GABA_A$ allosteric modulators (benzodiazepines, barbiturates)	↑↓ (Dose-dependent)
$GABA_A$ antagonists (flumazenil)	↑
Arginine vasopressin (AVP)	
AVP microinjection in mice Increased AVP levels	↑
Intranasal AVP	↑↓ (Sex-specific)
AVP receptor antagonists $AVP1b^{-/-}$ knockout mice	↓
Oxytocin (OXT)	
Intranasal OXT OXT microinjection in mice Increased OXT levels	↓
OXT receptor antagonists $OXT^{-/-}$ knockout mice	↑
(b) *Steroid hormones*	
Testosterone	
Increased testosterone levels Exogenous testosterone	↑
Castration, androgen replacement AR mutant mice	↓
Cortisol	
Reduced basal cortisol levels Adrenalectomy in mice Increased acute HPA axis activity Increased HPA axis reactivity	↑
Exogenous cortisol	↑ (Females)

AR androgen receptor, *DAT* DA transporter, *GAD* glutamic acid decarboxylase, *SSRI* selective serotonin reuptake inhibitor, *5-HIAA* 5-hydroxyindoleacetic acid, *5HTT* 5-HT transporter acid) (Brown et al. 1982), reduced 5-HT transporter binding sites both in platelets and in cortical areas (ACG, OFC) (Coccaro et al. 1996; Frankle et al. 2005), increased 5-HT_{2A} receptor binding sites in platelets and in the hippocampus (Coccaro et al. 1997b; Soloff et al. 2007), as well as blunted prolactin responses (Coccaro et al. 1997a) and reduced OFC, ACG, and VMPFC activation (Siever et al. 1999; Soloff et al. 2000) following challenge with serotonergic probes (d,l-fenfluramine) in aggressive individuals or violent suicide attempters. Furthermore, lowering 5-HT neurotransmission through depletion of tryptophan (a 5-HT precursor) increases impulsivity and aggression (Wood et al. 2006). Conversely, increasing 5-HT neurotransmission with tryptophan loading, selective serotonin reuptake inhibitors (SSRIs) or 5-HT_{1A}, and 5-HT_{1B} receptor agonists (termed "serenics") reduces aggression (Marsh et al. 2002; Miczek et al. 2001; Knutson et al. 1998); interestingly, 5-HT_{2A} antagonists (including various atypical antipsychotics) also produce the same aggression-dampening effect (Krakowski et al. 2006). Genetic manipulation of experimental rodents has also been extensively used to study 5-HT's role in aggression; knockout mice for the gene encoding the 5-HT transporter display reduced aggression (Holmes et al. 2002), while mice lacking the gene for the 5-HT_{1B} receptor are more impulsive and aggressive (Saudou et al. 1994). Further studies are warranted to clarify the specific role of various 5-HT receptors and their interactions in modulating impulsive aggression.

Dopamine (DA) is considered necessary for the manifestation of all kinds of aggressive behaviors via its mesocorticolimbic projections, although its specific role in modulating aggression is unclear. D_2 agonists increase aggression (Siegel et al. 1999), while D_2 blockers (antipsychotics) have antiaggressive effects, especially in psychotic patients, although these could also be attributed to non-specific sedative, motor and overall antipsychotic effects, as well as concomitant 5-HT_{2A} antagonism. D_2 receptor knockout mice display reduced aggression (Vukhac et al. 2001), while DA transporter (DAT) knockout mice have increased extracellular DA concentrations and exhibit increased levels of aggression (Rodriguiz et al. 2004).

Norepinephrine (NE) is crucially involved in both central and peripheral functions that subserve the individual's adaptive response to threat, by promoting vigilance, cardiovascular, and muscle adaptation. However, evidence for NE's role in aggression is mostly indirect through pharmacological studies. Propranolol, a β-receptor blocker, reduces aggression in both rodents and humans (Silver et al. 1999). Clonidine, an α2-receptor agonist, increases irritability and aggression at low doses but may have useful antiaggressive effects in hyperactive or autistic children (Hazell and Stuart 2003). Finally, knockout mice for DA β-hydroxylase, the enzyme involved in NE synthesis, cannot produce NE and display reduced aggression (Marino et al. 2005).

Pharmacological interventions on *γ-aminobutyric acid (GABA)*, a ubiquitous inhibitory neurotransmitter in the brain, have various effects on aggression. Tiagabine, a GABA reuptake inhibitor, has been shown to mitigate aggressive behaviors (Lieving et al. 2008). $GABA_A$ receptor agonists (e.g., muscimol) reduce aggression, while antagonists (e.g., flumazenil) have the opposite effect (Siegel et al. 1999). However, allosteric modulators of $GABA_A$ receptors, such as benzodiazepines, barbiturates, and alcohol, evoke paradoxical aggression in some patients at low-moderate doses and only reduce aggression at high doses (Miczek et al. 2002), suggesting that individual patient traits also contribute to variability of responses.

Various neuropeptides have also been implicated in aggression, the most important of which are the pro-social nonapeptides *AVP* and *oxytocin*, both key regulators of social interaction. Knockout mice for AVP1b receptors display less aggression between conspecifics (Wersinger et al. 2007), and microinjection of an AVP receptor antagonist into the anterior hypothalamus of male mice has the same effect (Ferris and Potegal 1988). Conversely, oxytocin gene knockout mice show exaggerated aggression (Ragnauth et al. 2005), and microinjection of an oxytocin antagonist increases aggression in female rats (Lubin et al. 2003). CSF AVP concentrations correlated with lifetime aggression in subjects with personality disorders (Coccaro et al. 1998), while CSF oxytocin concentrations correlated inversely with lifetime aggression and suicidality (Lee et al. 2009; Jokinen et al. 2012). Therefore, AVP and oxytocin have opposing complementary effects on aggression; more details about their actions, interactions, and underlying mechanisms can be found in Sect. 3.1.2.

Steroid Hormones

Androgens have both perinatal organizational and pubertal activational effects on the brain and its preparation for aggression. Androgen receptors are abundant within the basic threat system (Hamson et al. 2004); therefore, unsurprisingly, higher testosterone levels are associated with increased reactive aggression in several animals, exogenous testosterone increases aggression, and male castration and androgen replacement reduce aggressive behaviors, while androgen receptor mutant mice are less aggressive (Siegel et al. 1999; Archer 1991). Interestingly, however, the well-established effect of testosterone on aggression in nonhuman animals depends on genetic background, is context-specific (applying more clearly to social aggression), and modulated by interactions with pro-social peptides (AVP, oxytocin), monoamine neurotransmitter systems, and other steroids (estradiol, cortisol). Evidence in humans about the effect of testosterone on aggression is less compelling, with only a weak positive relationship recorded in a meta-analysis of 45 studies (Book et al. 2001). Apart from the aforementioned reasons, inconsistencies in the human literature may be explained by the fact that testosterone levels display age-related and situational variation and circadian and seasonal fluctuations and are affected by past experience and social rank. Furthermore, methodological discrepancies in published studies (self-report aggression measures or task-based design, testosterone sampling method) may be important. In fact, studies using a provocation methodology (rather than relying on self-reported aggression measures) record most often a positive relationship between testosterone and human reactive aggression (Pope et al. 2000; Hermans et al. 2008; Bos et al. 2012a).

Basal *cortisol* concentrations are generally lower in both animals and humans (especially prepubertal children) with higher levels of aggression (McBurnett et al. 2000; van Goozen et al. 2007), and experimentally induced adrenalectomy in mice increases aggression (Haller et al. 2001). Conversely, acute HPA axis activity and reactivity correlate positively (although less consistently) with aggressive responses in both animals and humans (Lopez-Duran et al. 2009). Furthermore, exogenous glucocorticoid administration enhances aggression in human females (Bohnke et al. 2010). Cortisol has, additionally, been shown to moderate testosterone's effect on aggression; testosterone was positively correlated with aggression and activation of subcortical (amygdala and hypothalamic) areas in response to angry faces only in subjects with low serum cortisol (Popma et al. 2007; Hermans et al. 2008), giving rise to the "high testosterone/cortisol ratio" hypothesis of aggression (Terburg et al. 2009). However, the interplay of testosterone and cortisol is actually complex and likely depends on additional factors, such as gender, age, trait aggression, and psychopathy.

3.2.1.4 Genetics of Aggression

Family, adoption, and twin studies suggest genetic effects on aggression; a meta-analysis of 24 twin and adoption studies using dimensional measures of aggression calculated a mean heritability estimate of 50%, which increased with age (Miles and Carey 1997). Of note, no distinction between reactive and proactive aggression is unfortunately made in earlier studies. However, more recent twin studies have shown that psychopathy, and associated callous/unemotional traits are highly heritable (67% heritability) in 7-year-olds (Viding et al. 2005).

More recently, candidate gene association studies in general population or patient samples have reported links of aggressive behavior to several common single nucleotide polymorphisms (SNPs) in various genes, selected on the basis of neurochemistry data (Table 3.5). A recent meta-analysis of 132 association studies of 225 independent SNPs in 11 genes with categorical or continuous outcomes of aggression with at least two replications (i.e., at least three separate samples per SNP) found no significant findings (Vassos et al. 2014). No genome-wide association study (GWAS) has been published yet. Here follows a synopsis of associations for the two most interesting candidate genes, importantly also involved in gene by environment interactions (GxE), accompanied by supporting preclinical and translational data.

The X-linked gene encoding monoamine oxidase A (*MAOA*), an enzyme catabolizing 5-HT, NE, and DA, was one of the first to be implicated in the genetic regulation of aggression, as a missense point mutation (resulting in no MAOA expression) was identified in several males of a Dutch pedigree with mild mental retardation and high levels of impulsive aggression, including arson, attempted rape, and exhibitionism (Brunner et al. 1993). Although very rare and, therefore, unlikely to explain much of the variation in human aggression, this mutation moved the focus of aggression-related genetic research to *MAOA*. Curiously, *MAOA* knockout mice were subsequently shown to have increased brain levels of 5-HT, NE, and DA and displayed increased aggression (Cases et al. 1995). A more common variable number tandem repeat (VNTR) polymorphism in *MAOA* promoter region has four common alleles with 3–5 repeats of a 30-bp sequence; alleles 2 and 3 are associated with increased MAOA expression (MAOA-H) while alleles 1 and 4 with lower (MAOA-L) (Sabol et al. 1998). Although MAOA-VNTR has inconsistently been associated with human aggression (MAOA-L genotypes most often associated with increased lifetime aggression), it has more robustly been shown to interact with the severity of early life stress (child maltreatment) in predicting adult antisocial behavior (Kim-Cohen et al. 2006; Caspi et al. 2002), in one of the most

Table 3.5 Major candidate genes associated with human aggression

Gene (chromosome)	Polymorphism	Risk allele → effect on expression/activity	Associated behavioral phenotype
Catecholaminergic neurotransmission genes			
MAOA (Xp)	VNTR (3–5 30-bp repeats) in promoter region	Alleles 1 (3R) and 4 (5R) → lower MAOA expression (MAOA-L)	– Impulsive aggression, conduct disorder, adult antisocial behavior (males only) – MAOA-L∗ (child maltreatment) → adult antisocial behavior (males only)
	Rare point mutation in exon 8	C936T → premature stop codon (missense) → no MAOA expression	Cross-generational violence (impulsive aggression, arson, rape) and borderline IQ in several males of Dutch kindred
COMT (22q)	SNP in exon 3	G1947A (Val158Met) → lower COMT expression	Inconsistent association with aggression in different populations
DRD4 (11p)	VNTR (2–11 48-bp repeats) in exon 3	7R allele → lower DRD4 affinity for DA	– Physical other-directed aggression in schizophrenia patients – DRD4-7R∗ (prenatal stress) → trait aggression in adulthood
DAT1 (SLC6A3) (5p)	VNTR (3–13 40-bp repeats) in 3′-UTR	10R allele → lower DAT expression	Impulsive aggression, violent delinquencies, conduct disorder, poor inhibitory control
Serotonergic neurotransmission genes			
SLC6A4 (17q)	5HTTLPR (s,l alleles with 14 or 16 22-bp repeats) in promoter region	Short (s) allele → lower 5HTT expression	– Inconsistent association with aggression and impulsivity in different populations – 5HTTLPR∗ (child maltreatment) → adult antisocial behavior
TPH1 (11p)	SNPs in intron 7	A779C/A218C → reduced TPH activity	Impulsive aggression, unprovoked anger
HTR1B (6q)	SNP in single gene exon	G861C (synonymous) → lower HTR1B expression	Impulsive aggression
HTR2A (13q)	SNP in promoter region	A1438G → higher HTR2A expression (?)	Aggression, impulse control disorders

bp base pair, *COMT* catecholamine-*O*-methyltransferase gene, *DA* dopamine, *DAT* DA transporter, *HTR1B* 5-HT receptor 1B type, *HTR2A* 5-HT receptor 2A type, *MAOA* monoamine oxidase A, *R* repeat, *SNP* single nucleotide polymorphism, *TPH* tryptophan hydroxylase, *UTR* untranslated region, *VNTR* variable number tandem repeat, *5HTT* 5-HT transporter cited gene by environment (GxE) interaction paradigms. Finally, functional imaging studies show that MAOA-L carriers exhibit increased amygdalar reactivity to angry and fearful faces and negatively valenced memories retrieval as well as decreased prefrontal reactivity in inhibitory control tasks (Meyer-Lindenberg et al. 2006; Passamonti et al. 2006), thereby being more prone to aggressive behaviors than MAOA-H individuals.

The gene for the 5-HT transporter (5HTT or SERT), named *SLC6A4*, harbors an insertion/deletion polymorphism in its promoter region (5HTTLPR), which has been extensively studied in various internalizing psychiatric phenotypes and in aggression. The polymorphism consists of two common variants, the short (s) and long (l) alleles, containing 14 or 16 copies of a 20- to 23-bp sequence and associated with decreased and increased 5HTT expression, respectively (Lesch et al. 1996; Heils et al. 1996). 5HTTLPR has been inconsistently associated with human aggression, with studies most often reporting higher aggression in s allele carriers. However, 5HTTLPR was shown to interact with acute stress to produce laboratory-assessed aggressive expressions (Verona et al. 2006) as well as with child maltreatment (GxE) in predicting adult antisocial behavior (Cicchetti et al. 2012; Reif et al. 2007). Finally, functional imaging studies have shown that s allele carriers have increased amygdala reactivity to fear-inducing stimuli (Hariri et al. 2002, 2005) and reduced functional connectivity of amygdala to perigenual ACG compared to l/l homozygotes (Pezawas et al. 2005).

3.2.1.5 Integration and Conclusions

Neurobiological evidence seems to substantiate the distinction between reactive and proactive aggression, which manifest in different psychopathological contexts. In reactive aggression, sensory (e.g., hearing or vision) deficits or distortions (caused by drugs, alcohol, or metabolic disturbances) can initially affect sensory processing of the provocative stimulus, increasing the likelihood that it is perceived as threatening; cognitive appraisal of the stimulus in limbic or higher cortical areas can then be modulated by coexisting cognitive impairment or disordered reality testing (e.g., in dementia or psychosis), current mood or anxiety disorders, past emotional conditioning, and affective dysregulation induced by early trauma (e.g., in personality disorders) as well as by cultural and social factors (Siever 2008). In any of these contexts, susceptibility to aggression can be understood as an imbalance between "bottom-up" proaggressive drives arising in limbic subcortical areas (amygdala, hypothalamus) and "top-down" regulatory/suppressing control circuits nested in limbic (OFC, ACG) or higher (VMPFC, DMPFC, VLPFC) cortical areas, which evaluate the salience of the threatening stimuli, adjust planned actions to social cues and environmental feedback, and anticipate their consequences or potential reward value. Connectivity between these subcortical and cortical circuits seems to be developmentally modulated by interactions between genetic susceptibility (e.g., MAOA or 5HTT gene polymorphisms) and early trauma. Finally, aggressive diathesis is further moderated by a continuous complex interplay of catecholaminergic neurotransmitter systems with pro-social neuropeptides (AVP, oxytocin) and steroid hormones (testosterone, cortisol), which is gender-, context-, and developmental stage-specific.

On the other hand, less well-studied proactive aggression manifests mainly in the context of psychopathy. Recent evidence suggests that the core deficit in individuals exhibiting mainly proactive aggression is a hyporesponsiveness of the amygdala to fearful and sad facial expressions of others, which is necessary for developing empathy and a prerequisite for moral socialization by parents and peers. Psychopathy and youths' callous/unemotional traits are actually associated with a reduced responsiveness of both amygdala and limbic cortical (ACG, OFC) centers and a reduced connectivity between amygdala and VMPFC in emotional or moral tasks.

Concludingly, substantial progress has been achieved in understanding the neurobiological underpinnings of aggression, especially the reactive subtype. Further exploring the neurobiological signature of aggression will hopefully contribute to the development of clinically useful biomarkers and effective treatment strategies.

References

Adan RA, Vanderschuren LJ, la Fleur SE (2008) Anti-obesity drugs and neural circuits of feeding. Trends Pharmacol Sci 29(4):208–217. https://doi.org/10.1016/j.tips.2008.01.008

Anderson SW, Bechara A, Damasio H, Tranel D, Damasio AR (1999) Impairment of social and moral behavior related to early damage in human prefrontal cortex. Nat Neurosci 2(11):1032–1037. https://doi.org/10.1038/14833

de Araujo IE, Oliveira-Maia AJ, Sotnikova TD, Gainetdinov RR, Caron MG, Nicolelis MA, Simon SA (2008) Food reward in the absence of taste receptor signaling. Neuron 57(6):930–941. https://doi.org/10.1016/j.neuron.2008.01.032

Archer J (1991) The influence of testosterone on human aggression. Br J Psychol 82(Pt 1)):1–28

Bachevalier J, Machado CJ, Kazama A (2011) Behavioral outcomes of late-onset or early-onset orbital frontal cortex (areas 11/13) lesions in rhesus monkeys. Ann N Y Acad Sci 1239:71–86. https://doi.org/10.1111/j.1749-6632.2011.06211.x

Baribeau DA, Anagnostou E (2015) Oxytocin and vasopressin: linking pituitary neuropeptides and their receptors to social neurocircuits. Front Neurosci 9:335

Barratt ES, Stanford MS, Dowdy L, Liebman MJ, Kent TA (1999) Impulsive and premeditated aggression: a factor analysis of self-reported acts. Psychiatry Res 86(2):163–173. https://doi.org/10.1016/s0165-1781(99)00024-4

Batterham RL, ffytche DH, Rosenthal JM, Zelaya FO, Barker GJ, Withers DJ, Williams SC (2007) PYY modulation of cortical and hypothalamic brain areas predicts feeding behaviour in humans. Nature 450(7166):106–109. https://doi.org/10.1038/nature06212

Benarroch EE (2010) Neural control of feeding behavior: overview and clinical correlations. Neurology 74(20):1643–1650. https://doi.org/10.1212/WNL.0b013e3181df0a3f

Berridge KC (2009) 'Liking' and 'wanting' food rewards: brain substrates and roles in eating disorders. Physiol Behav 97(5):537–550. https://doi.org/10.1016/j.physbeh.2009.02.044

Berthoud HR (2011) Metabolic and hedonic drives in the neural control of appetite: who is the boss? Curr Opin Neurobiol 21(6):888–896. https://doi.org/10.1016/j.conb.2011.09.004

Blanchard RJ, Blanchard DC, Takahashi T, Kelley MJ (1977) Attack and defensive behaviour in the albino rat. Anim Behav 25:622–634. https://doi.org/10.1016/0003-3472(77)90113-0

Boccia M, Petrusz P, Suzuki K, Marson L, Pedersen CA (2013) Immunohistochemical localization of oxytocin receptors in human brain. Neuroscience 253:155–164

Bohnke R, Bertsch K, Kruk MR, Richter S, Naumann E (2010) Exogenous cortisol enhances aggressive behavior in females, but not in males. Psychoneuroendocrinology 35(7):1034–1044. https://doi.org/10.1016/j.psyneuen.2010.01.004

Book AS, Starzyk KB, Quinsey VL (2001) The relationship between testosterone and aggression: a meta-analysis. Aggress Violent Behav 6(6):579–599. https://doi.org/10.1016/s1359-1789(00)00032-x
Bos PA, Hermans EJ, Ramsey NF, van Honk J (2012a) The neural mechanisms by which testosterone acts on interpersonal trust. NeuroImage 61(3):730–737. https://doi.org/10.1016/j.neuroimage.2012.04.002
Bos PA, Panksepp J, Bluthé R-M, van Honk J (2012b) Acute effects of steroid hormones and neuropeptides on human social–emotional behavior: a review of single administration studies. Front Neuroendocrinol 33(1):17–35
Brown GL, Ebert MH, Goyer PF, Jimerson DC, Klein WJ, Bunney WE, Goodwin FK (1982) Aggression, suicide, and serotonin: relationships to CSF amine metabolites. Am J Psychiatry 139(6):741–746. https://doi.org/10.1176/ajp.139.6.741
de Bruin JP, van Oyen HG, Van de Poll N (1983) Behavioural changes following lesions of the orbital prefrontal cortex in male rats. Behav Brain Res 10(2–3):209–232
Brunner HG, Nelen M, Breakefield XO, Ropers HH, van Oost BA (1993) Abnormal behavior associated with a point mutation in the structural gene for monoamine oxidase A. Science 262(5133):578–580. https://doi.org/10.1126/science.8211186
Budhani S, Blair RJ (2005) Response reversal and children with psychopathic tendencies: success is a function of salience of contingency change. J Child Psychol Psychiatry 46(9):972–981. https://doi.org/10.1111/j.1469-7610.2004.00398.x
Budhani S, Marsh AA, Pine DS, Blair RJ (2007) Neural correlates of response reversal: considering acquisition. NeuroImage 34(4):1754–1765. https://doi.org/10.1016/j.neuroimage.2006.08.060
Carter CS (2007) Sex differences in oxytocin and vasopressin: implications for autism spectrum disorders? Behav Brain Res 176(1):170–186
Cases O, Seif I, Grimsby J, Gaspar P, Chen K, Pournin S, Muller U, Aguet M, Babinet C, Shih JC et al (1995) Aggressive behavior and altered amounts of brain serotonin and norepinephrine in mice lacking MAOA. Science 268(5218):1763–1766. https://doi.org/10.1126/science.7792602
Caspi A, McClay J, Moffitt TE, Mill J, Martin J, Craig IW, Taylor A, Poulton R (2002) Role of genotype in the cycle of violence in maltreated children. Science 297(5582):851–854. https://doi.org/10.1126/science.1072290
Chaudhri O, Small C, Bloom S (2006) Gastrointestinal hormones regulating appetite. Philos Trans R Soc Lond Ser B Biol Sci 361(1471):1187–1209. https://doi.org/10.1098/rstb.2006.1856
Chodasewicz K (2014) Evolution, reproduction and definition of life. Theor Biosci 133(1):39–45. https://doi.org/10.1007/s12064-013-0184-5
Cicchetti D, Rogosch FA, Thibodeau EL (2012) The effects of child maltreatment on early signs of antisocial behavior: genetic moderation by tryptophan hydroxylase, serotonin transporter, and monoamine oxidase A genes. Dev Psychopathol 24(3):907–928. https://doi.org/10.1017/S0954579412000442
Cleland CE, Chyba CF (2002) Defining 'life'. Origins of life and evolution of the biosphere. J Int Soc Stud Origin Life 32(4):387–393
Coccaro EF, Kavoussi RJ, Sheline YI, Lish JD, Csernansky JG (1996) Impulsive aggression in personality disorder correlates with tritiated paroxetine binding in the platelet. Arch Gen Psychiatry 53(6):531–536
Coccaro EF, Kavoussi RJ, Cooper TB, Hauger RL (1997a) Central serotonin activity and aggression: inverse relationship with prolactin response to d-fenfluramine, but not CSF 5-HIAA concentration, in human subjects. Am J Psychiatry 154(10):1430–1435. https://doi.org/10.1176/ajp.154.10.1430
Coccaro EF, Kavoussi RJ, Sheline YI, Berman ME, Csernansky JG (1997b) Impulsive aggression in personality disorder correlates with platelet 5-HT2A receptor binding. Neuropsychopharmacology 16(3):211–216. https://doi.org/10.1016/S0893-133X(96)00194-7
Coccaro EF, Kavoussi RJ, Hauger RL, Cooper TB, Ferris CF (1998) Cerebrospinal fluid vasopressin levels: correlates with aggression and serotonin function in personality-disordered subjects. Arch Gen Psychiatry 55(8):708–714

Coccaro EF, McCloskey MS, Fitzgerald DA, Phan KL (2007) Amygdala and orbitofrontal reactivity to social threat in individuals with impulsive aggression. Biol Psychiatry 62(2):168–178. https://doi.org/10.1016/j.biopsych.2006.08.024
Coccaro EF, Sripada CS, Yanowitch RN, Phan KL (2011) Corticolimbic function in impulsive aggressive behavior. Biol Psychiatry 69(12):1153–1159. https://doi.org/10.1016/j.biopsych.2011.02.032
Costafreda SG, Brammer MJ, David AS, Fu CH (2008) Predictors of amygdala activation during the processing of emotional stimuli: a meta-analysis of 385 PET and fMRI studies. Brain Res Rev 58(1):57–70
Craig AD (2009) How do you feel—now? The anterior insula and human awareness. Nat Rev Neurosci 10(1):59–70. https://doi.org/10.1038/nrn2555
Crockford C, Deschner T, Ziegler TE, Wittig RM (2014) Endogenous peripheral oxytocin measures can give insight into the dynamics of social relationships: a review. Front Behav Neurosci 8(68):68. https://doi.org/10.3389/fnbeh.2014.00068
Dadds MR, Perry Y, Hawes DJ, Merz S, Riddell AC, Haines DJ, Solak E, Abeygunawardane AI (2006) Attention to the eyes and fear-recognition deficits in child psychopathy. Br J Psychiatry 189(3):280–281. https://doi.org/10.1192/bjp.bp.105.018150
Damasio H, Grabowski T, Frank R, Galaburda AM, Damasio AR (1994) The return of Phineas Gage: clues about the brain from the skull of a famous patient. Science 264(5162):1102–1105
Davidson RJ, Putnam KM, Larson CL (2000) Dysfunction in the neural circuitry of emotion regulation—a possible prelude to violence. Science 289(5479):591–594. https://doi.org/10.1126/science.289.5479.591
De Dreu CK (2012) Oxytocin modulates cooperation within and competition between groups: an integrative review and research agenda. Horm Behav 61(3):419–428
Dickstein DP, Finger EC, Brotman MA, Rich BA, Pine DS, Blair JR, Leibenluft E (2010) Impaired probabilistic reversal learning in youths with mood and anxiety disorders. Psychol Med 40(7):1089–1100. https://doi.org/10.1017/S0033291709991462
Dietrich MO, Horvath TL (2009) Feeding signals and brain circuitry. Eur J Neurosci 30(9):1688–1696. https://doi.org/10.1111/j.1460-9568.2009.06963.x
Ditzen B, Heinrichs M (2014) Psychobiology of social support: the social dimension of stress buffering. Restor Neurol Neurosci 32(1):149–162. https://doi.org/10.3233/RNN-139008
Dolan M, Fullam R (2006) Face affect recognition deficits in personality-disordered offenders: association with psychopathy. Psychol Med 36(11):1563. https://doi.org/10.1017/s0033291706008634
Duncan LE, Keller MC (2011) A critical review of the first 10 years of candidate gene-by-environment interaction research in psychiatry. Am J Psychiatr 168(10):1041–1049
Eisenberger NI, Master SL, Inagaki TK, Taylor SE, Shirinyan D, Lieberman MD, Naliboff BD (2011) Attachment figures activate a safety signal-related neural region and reduce pain experience. Proc Natl Acad Sci 108(28):11721–11726
Falkner AL, Grosenick L, Davidson TJ, Deisseroth K, Lin D (2016) Hypothalamic control of male aggression-seeking behavior. Nat Neurosci 19(4):596–604. https://doi.org/10.1038/nn.4264
Fang PH, Yu M, Ma YP, Li J, Sui YM, Shi MY (2011) Central nervous system regulation of food intake and energy expenditure: role of galanin-mediated feeding behavior. Neurosci Bull 27(6):407–412. https://doi.org/10.1007/s12264-011-1841-7
Farooqi IS, O'Rahilly S (2008) Mutations in ligands and receptors of the leptin-melanocortin pathway that lead to obesity. Nat Clin Pract Endocrinol Metab 4(10):569–577. https://doi.org/10.1038/ncpendmet0966
Ferris CF, Potegal M (1988) Vasopressin receptor blockade in the anterior hypothalamus suppresses aggression in hamsters. Physiol Behav 44(2):235–239
Frankle WG, Lombardo I, New AS, Goodman M, Talbot PS, Huang Y, Hwang DR, Slifstein M, Curry S, Abi-Dargham A, Laruelle M, Siever LJ (2005) Brain serotonin transporter distribution in subjects with impulsive aggressivity: a positron emission study with [11C]McN 5652. Am J Psychiatry 162(5):915–923. https://doi.org/10.1176/appi.ajp.162.5.915
Freud S (1920) Beyond the pleasure principle. Interaationaler Psycho Analytischer Verlag, Vienna
Freud S (1923) The ego and the Id. Interaationaler Psycho Analytischer Verlag, Vienna

Fuller JL (1954) Nature and nurture: a modern synthesis. Doubleday, New York, NY
Galbally M, Lewis AJ, Ijzendoorn M, Permezel M (2011) The role of oxytocin in mother-infant relations: a systematic review of human studies. Harv Rev Psychiatry 19(1):1–14. https://doi.org/10.3109/10673229.2011.549771
Gao Q, Horvath TL (2007) Neurobiology of feeding and energy expenditure. Annu Rev Neurosci 30:367–398. https://doi.org/10.1146/annurev.neuro.30.051606.094324
Gelles RJ (1994) Aggression – its causes, consequences, and control – Berkowitz, L. Contemp Sociol 23(4):575–576. https://doi.org/10.2307/2076412
Gettler LT (2014) Applying socioendocrinology to evolutionary models: fatherhood and physiology. Evol Anthropol 23(4):146–160. https://doi.org/10.1002/evan.21412
Ghashghaei HT, Barbas H (2002) Pathways for emotion: interactions of prefrontal and anterior temporal pathways in the amygdala of the rhesus monkey. Neuroscience 115(4):1261–1279
Goetz SM, Tang L, Thomason ME, Diamond MP, Hariri AR, Carre JM (2014) Testosterone rapidly increases neural reactivity to threat in healthy men: a novel two-step pharmacological challenge paradigm. Biol Psychiatry 76(4):324–331. https://doi.org/10.1016/j.biopsych.2014.01.016
van Goozen SH, Fairchild G, Snoek H, Harold GT (2007) The evidence for a neurobiological model of childhood antisocial behavior. Psychol Bull 133(1):149–182. https://doi.org/10.1037/0033-2909.133.1.149
Gorrindo T, Blair RJ, Budhani S, Dickstein DP, Pine DS, Leibenluft E (2005) Deficits on a probabilistic response-reversal task in patients with pediatric bipolar disorder. Am J Psychiatry 162(10):1975–1977. https://doi.org/10.1176/appi.ajp.162.10.1975
Gregg TR, Siegel A (2001) Brain structures and neurotansmitters regulating aggression in cats: implications for human aggression. Prog Neuro-Psychopharmacol Biol Psychiatry 25(1):91–140. https://doi.org/10.1016/s0278-5846(00)00150-0
Guyon A, Conductier G, Rovere C, Enfissi A, Nahon JL (2009) Melanin-concentrating hormone producing neurons: activities and modulations. Peptides 30(11):2031–2039. https://doi.org/10.1016/j.peptides.2009.05.028
Haller J, van de Schraaf J, Kruk MR (2001) Deviant forms of aggression in glucocorticoid hyporeactive rats: a model for 'pathological' aggression? J Neuroendocrinol 13(1):102–107
Hamson DK, Jones BA, Watson NV (2004) Distribution of androgen receptor immunoreactivity in the brainstem of male rats. Neuroscience 127(4):797–803. https://doi.org/10.1016/j.neuroscience.2004.06.006
Hariri AR, Mattay VS, Tessitore A, Kolachana B, Fera F, Goldman D, Egan MF, Weinberger DR (2002) Serotonin transporter genetic variation and the response of the human amygdala. Science 297(5580):400–403. https://doi.org/10.1126/science.1071829
Hariri AR, Drabant EM, Munoz KE, Kolachana BS, Mattay VS, Egan MF, Weinberger DR (2005) A susceptibility gene for affective disorders and the response of the human amygdala. Arch Gen Psychiatry 62(2):146–152. https://doi.org/10.1001/archpsyc.62.2.146
Harris GC, Aston-Jones G (2006) Arousal and reward: a dichotomy in orexin function. Trends Neurosci 29(10):571–577. https://doi.org/10.1016/j.tins.2006.08.002
von Hausswolff-Juhlin Y, Brooks SJ, Larsson M (2015) The neurobiology of eating disorders—a clinical perspective. Acta Psychiatr Scand 131(4):244–255. https://doi.org/10.1111/acps.12335
Hazell PL, Stuart JE (2003) A randomized controlled trial of clonidine added to psychostimulant medication for hyperactive and aggressive children. J Am Acad Child Adolesc Psychiatry 42(8):886–894. https://doi.org/10.1097/01.CHI.0000046908.27264.00
Heils A, Teufel A, Petri S, Stober G, Riederer P, Bengel D, Lesch KP (1996) Allelic variation of human serotonin transporter gene expression. J Neurochem 66(6):2621–2624
Heinrichs M, Domes G (2008) Neuropeptides and social behaviour: effects of oxytocin and vasopressin in humans. Prog Brain Res 170:337–350
Heinrichs M, von Dawans B, Domes G (2009) Oxytocin, vasopressin, and human social behavior. Front Neuroendocrinol 30(4):548–557. https://doi.org/10.1016/j.yfrne.2009.05.005
Heisler LK, Cowley MA, Tecott LH, Fan W, Low MJ, Smart JL, Rubinstein M, Tatro JB, Marcus JN, Holstege H, Lee CE, Cone RD, Elmquist JK (2002) Activation of central melanocortin pathways by fenfluramine. Science 297(5581):609–611. https://doi.org/10.1126/science.1072327

Hermans EJ, Ramsey NF, van Honk J (2008) Exogenous testosterone enhances responsiveness to social threat in the neural circuitry of social aggression in humans. Biol Psychiatry 63(3):263–270. https://doi.org/10.1016/j.biopsych.2007.05.013

Hewitt JK (2012) Editorial policy on candidate gene association and candidate gene-by-environment interaction studies of complex traits. Behav Genet 42(1):1–2

Hillebrand JJ, Kas MJ, Adan RA (2006) To eat or not to eat; regulation by the melanocortin system. Physiol Behav 89(1):97–102. https://doi.org/10.1016/j.physbeh.2006.01.034

Holmes A, Murphy DL, Crawley JN (2002) Reduced aggression in mice lacking the serotonin transporter. Psychopharmacology 161(2):160–167. https://doi.org/10.1007/s00213-002-1024-3

Hostinar CE, Sullivan RM, Gunnar MR (2014) Psychobiological mechanisms underlying the social buffering of the hypothalamic-pituitary-adrenocortical axis: a review of animal models and human studies across development. Psychol Bull 140(1):256–282. https://doi.org/10.1037/a0032671

Huber D, Veinante P, Stoop R (2005) Vasopressin and oxytocin excite distinct neuronal populations in the central amygdala. Science 308(5719):245–248. https://doi.org/10.1126/science.1105636

Insel TR (2010) The challenge of translation in social neuroscience: a review of oxytocin, vasopressin, and affiliative behavior. Neuron 65(6):768–779

Ito M, Okazaki M, Takahashi S, Muramatsu R, Kato M, Onuma T (2007) Subacute postictal aggression in patients with epilepsy. Epilepsy Behav 10(4):611–614. https://doi.org/10.1016/j.yebeh.2007.02.016

Jokinen J, Chatzittofis A, Hellstrom C, Nordstrom P, Uvnas-Moberg K, Asberg M (2012) Low CSF oxytocin reflects high intent in suicide attempters. Psychoneuroendocrinology 37(4):482–490. https://doi.org/10.1016/j.psyneuen.2011.07.016

Kageyama H, Takenoya F, Shiba K, Shioda S (2010) Neuronal circuits involving ghrelin in the hypothalamus-mediated regulation of feeding. Neuropeptides 44(2):133–138. https://doi.org/10.1016/j.npep.2009.11.010

Kaitz M, Shalev I, Sapir N, Devor N, Samet Y, Mankuta D, Ebstein RP (2010) Mothers' dopamine receptor polymorphism modulates the relation between infant fussiness and sensitive parenting. Dev Psychobiol 52(2):149–157

Kampe J, Tschop MH, Hollis JH, Oldfield BJ (2009) An anatomic basis for the communication of hypothalamic, cortical and mesolimbic circuitry in the regulation of energy balance. Eur J Neurosci 30(3):415–430. https://doi.org/10.1111/j.1460-9568.2009.06818.x

Kelley AE, Baldo BA, Pratt WE, Will MJ (2005) Corticostriatal-hypothalamic circuitry and food motivation: integration of energy, action and reward. Physiol Behav 86(5):773–795. https://doi.org/10.1016/j.physbeh.2005.08.066

Keverne EB, Curley JP (2004) Vasopressin, oxytocin and social behaviour. Curr Opin Neurobiol 14(6):777–783

Kiehl KA, Smith AM, Hare RD, Mendrek A, Forster BB, Brink J, Liddle PF (2001) Limbic abnormalities in affective processing by criminal psychopaths as revealed by functional magnetic resonance imaging. Biol Psychiatry 50(9):677–684

Kim-Cohen J, Caspi A, Taylor A, Williams B, Newcombe R, Craig IW, Moffitt TE (2006) MAOA, maltreatment, and gene-environment interaction predicting children's mental health: new evidence and a meta-analysis. Mol Psychiatry 11(10):903–913. https://doi.org/10.1038/sj.mp.4001851

King BM (2006) The rise, fall, and resurrection of the ventromedial hypothalamus in the regulation of feeding behavior and body weight. Physiol Behav 87(2):221–244. https://doi.org/10.1016/j.physbeh.2005.10.007

Kis A, Kemerle K, Hernadi A, Topal J (2013) Oxytocin and social pretreatment have similar effects on processing of negative emotional faces in healthy adult males. Front Psychol 4:532. https://doi.org/10.3389/fpsyg.2013.00532

Knutson B, Wolkowitz OM, Cole SW, Chan T, Moore EA, Johnson RC, Terpstra J, Turner RA, Reus VI (1998) Selective alteration of personality and social behavior by serotonergic intervention. Am J Psychiatry 155(3):373–379. https://doi.org/10.1176/ajp.155.3.373

Kokras N, Sotiropoulos I, Pitychoutis PM, Almeida OFX, Papadopoulou-Daifoti Z (2011) Citalopram-mediated anxiolysis and differing neurobiological responses in both sexes of a genetic model of depression. Neuroscience 194:62–71. https://doi.org/10.1016/j.neuroscience.2011.07.077

Konner AC, Klockener T, Bruning JC (2009) Control of energy homeostasis by insulin and leptin: targeting the arcuate nucleus and beyond. Physiol Behav 97(5):632–638. https://doi.org/10.1016/j.physbeh.2009.03.027

Koshland DE Jr (2002) Special essay. The seven pillars of life. Science 295(5563):2215–2216. https://doi.org/10.1126/science.1068489

Krakowski MI, Czobor P, Citrome L, Bark N, Cooper TB (2006) Atypical antipsychotic agents in the treatment of violent patients with schizophrenia and schizoaffective disorder. Arch Gen Psychiatry 63(6):622–629. https://doi.org/10.1001/archpsyc.63.6.622

Lass-Hennemann J, Kuehl LK, Schulz A, Oitzl MS, Schachinger H (2011) Stress strengthens memory of first impressions of others' positive personality traits. PLoS One 6(1):e16389. https://doi.org/10.1371/journal.pone.0016389

Lee R, Ferris C, Van de Kar LD, Coccaro EF (2009) Cerebrospinal fluid oxytocin, life history of aggression, and personality disorder. Psychoneuroendocrinology 34(10):1567–1573. https://doi.org/10.1016/j.psyneuen.2009.06.002

Lesch KP, Bengel D, Heils A, Sabol SZ, Greenberg BD, Petri S, Benjamin J, Muller CR, Hamer DH, Murphy DL (1996) Association of anxiety-related traits with a polymorphism in the serotonin transporter gene regulatory region. Science 274(5292):1527–1531

Lieving LM, Cherek DR, Lane SD, Tcheremissine OV, Nouvion SO (2008) Effects of acute tiagabine administration on aggressive responses of adult male parolees. J Psychopharmacol 22(2):144–152. https://doi.org/10.1177/0269881107078489

Lin D, Boyle MP, Dollar P, Lee H, Lein ES, Perona P, Anderson DJ (2011) Functional identification of an aggression locus in the mouse hypothalamus. Nature 470(7333):221–226. https://doi.org/10.1038/nature09736

Lopez-Duran NL, Olson SL, Hajal NJ, Felt BT, Vazquez DM (2009) Hypothalamic pituitary adrenal axis functioning in reactive and proactive aggression in children. J Abnorm Child Psychol 37(2):169–182. https://doi.org/10.1007/s10802-008-9263-3

Lorenz KZ (1950) The comparative method in studying innate behaviour patterns. Symp Soc Exp Biol 4:221–268

Lorenz K (1952) King Solomon's ring. Crowell, New York, NY

Lubin DA, Elliott JC, Black MC, Johns JM (2003) An oxytocin antagonist infused into the central nucleus of the amygdala increases maternal aggressive behavior. Behav Neurosci 117(2):195–201

Lutter M, Nestler EJ (2009) Homeostatic and hedonic signals interact in the regulation of food intake. J Nutr 139(3):629–632. https://doi.org/10.3945/jn.108.097618

Machado CJ, Bachevalier J (2008) Behavioral and hormonal reactivity to threat: effects of selective amygdala, hippocampal or orbital frontal lesions in monkeys. Psychoneuroendocrinology 33(7):926–941. https://doi.org/10.1016/j.psyneuen.2008.04.012

Marino MD, Bourdelat-Parks BN, Cameron Liles L, Weinshenker D (2005) Genetic reduction of noradrenergic function alters social memory and reduces aggression in mice. Behav Brain Res 161(2):197–203. https://doi.org/10.1016/j.bbr.2005.02.005

Marsh AA, Blair RJ (2008) Deficits in facial affect recognition among antisocial populations: a meta-analysis. Neurosci Biobehav Rev 32(3):454–465. https://doi.org/10.1016/j.neubiorev.2007.08.003

Marsh DM, Dougherty DM, Moeller FG, Swann AC, Spiga R (2002) Laboratory-measured aggressive behavior of women: acute tryptophan depletion and augmentation. Neuropsychopharmacology 26(5):660–671. https://doi.org/10.1016/S0893-133X(01)00369-4

Marsh AA, Finger EC, Mitchell DG, Reid ME, Sims C, Kosson DS, Towbin KE, Leibenluft E, Pine DS, Blair RJ (2008) Reduced amygdala response to fearful expressions in children and adolescents with callous-unemotional traits and disruptive behavior disorders. Am J Psychiatry 165(6):712–720. https://doi.org/10.1176/appi.ajp.2007.07071145

Marsh AA, Finger EC, Fowler KA, Jurkowitz ITN, Schechter JC, Yu HH, Pine DS, Blair RJR (2011) Reduced amygdala-orbitofrontal connectivity during moral judgments in youths with disruptive behavior disorders and psychopathic traits. Psychiatry Res 194(3):279–286. https://doi.org/10.1016/j.pscychresns.2011.07.008
McBurnett K, Lahey BB, Rathouz PJ, Loeber R (2000) Low salivary cortisol and persistent aggression in boys referred for disruptive behavior. Arch Gen Psychiatry 57(1):38–43
Meister B (2007) Neurotransmitters in key neurons of the hypothalamus that regulate feeding behavior and body weight. Physiol Behav 92(1–2):263–271. https://doi.org/10.1016/j.physbeh.2007.05.021
Meyer-Lindenberg A, Buckholtz JW, Kolachana B, RH A, Pezawas L, Blasi G, Wabnitz A, Honea R, Verchinski B, Callicott JH, Egan M, Mattay V, Weinberger DR (2006) Neural mechanisms of genetic risk for impulsivity and violence in humans. Proc Natl Acad Sci U S A 103(16):6269–6274. https://doi.org/10.1073/pnas.0511311103
Meyer-Lindenberg A, Domes G, Kirsch P, Heinrichs M (2011) Oxytocin and vasopressin in the human brain: social neuropeptides for translational medicine. Nat Rev Neurosci 12(9):524–538
Miczek KA, Maxson SC, Fish EW, Faccidomo S (2001) Aggressive behavioral phenotypes in mice. Behav Brain Res 125(1–2):167–181
Miczek KA, Fish EW, De Bold JF, De Almeida RM (2002) Social and neural determinants of aggressive behavior: pharmacotherapeutic targets at serotonin, dopamine and gamma-aminobutyric acid systems. Psychopharmacology 163(3–4):434–458. https://doi.org/10.1007/s00213-002-1139-6
Miles DR, Carey G (1997) Genetic and environmental architecture of human aggression. J Pers Soc Psychol 72(1):207–217
Mileva-Seitz V, Fleming AS, Meaney MJ, Mastroianni A, Sinnwell JP, Steiner M, Atkinson L, Levitan RD, Matthews SG, Kennedy JL, Sokolowski MB (2012) Dopamine receptors D1 and D2 are related to observed maternal behavior. Genes Brain Behav 11(6):684–694. https://doi.org/10.1111/j.1601-183X.2012.00804.x
Murray S, Tulloch A, Gold MS, Avena NM (2014) Hormonal and neural mechanisms of food reward, eating behaviour and obesity. Nat Rev Endocrinol 10(9):540–552. https://doi.org/10.1038/nrendo.2014.91
Nelson RJ, Trainor BC (2007) Neural mechanisms of aggression. Nat Rev Neurosci 8(7):536–546. https://doi.org/10.1038/nrn2174
Neumann ID, Landgraf R (2012) Balance of brain oxytocin and vasopressin: implications for anxiety, depression, and social behaviors. Trends Neurosci 35(11):649–659
Olazabal DE, Pereira M, Agrati D, Ferreira A, Fleming AS, Gonzalez-Mariscal G, Levy F, Lucion AB, Morrell JI, Numan M, Uriarte N (2013) Flexibility and adaptation of the neural substrate that supports maternal behavior in mammals. Neurosci Biobehav Rev 37(8):1875–1892. https://doi.org/10.1016/j.neubiorev.2013.04.004
Ordonana JR, Bartels M, Boomsma DI, Cella D, Mosing M, Oliveira JR, Patrick DL, Veenhoven R, Wagner GG, Sprangers MA, Consortium G (2013) Biological pathways and genetic mechanisms involved in social functioning. Qual Life Res 22(6):1189–1200. https://doi.org/10.1007/s11136-012-0277-5
Panksepp J (1998) Affective neuroscience: the foundations of human and animal emotions. Oxford University Press, New York, NY
Panksepp J (2016) The cross-mammalian neurophenomenology of primal emotional affects: from animal feelings to human therapeutics. J Comp Neurol 524(8):1624–1635. https://doi.org/10.1002/cne.23969
Passamonti L, Fera F, Magariello A, Cerasa A, Gioia MC, Muglia M, Nicoletti G, Gallo O, Provinciali L, Quattrone A (2006) Monoamine oxidase-A genetic variations influence brain activity associated with inhibitory control: new insight into the neural correlates of impulsivity. Biol Psychiatry 59(4):334–340. https://doi.org/10.1016/j.biopsych.2005.07.027
Pezawas L, Meyer-Lindenberg A, Drabant EM, Verchinski BA, Munoz KE, Kolachana BS, Egan MF, Mattay VS, Hariri AR, Weinberger DR (2005) 5-HTTLPR polymorphism impacts human

cingulate-amygdala interactions: a genetic susceptibility mechanism for depression. Nat Neurosci 8(6):828–834. https://doi.org/10.1038/nn1463
Pope HG Jr, Kouri EM, Hudson JI (2000) Effects of supraphysiologic doses of testosterone on mood and aggression in normal men: a randomized controlled trial. Arch Gen Psychiatry 57(2):133–140. discussion 155–156
Popma A, Vermeiren R, Geluk CA, Rinne T, van den Brink W, Knol DL, Jansen LM, van Engeland H, Doreleijers TA (2007) Cortisol moderates the relationship between testosterone and aggression in delinquent male adolescents. Biol Psychiatry 61(3):405–411. https://doi.org/10.1016/j.biopsych.2006.06.006
Poulin F, Boivin M (2000) Reactive and proactive aggression: evidence of a two-factor model. Psychol Assess 12(2):115–122
Ragnauth AK, Devidze N, Moy V, Finley K, Goodwillie A, Kow LM, Muglia LJ, Pfaff DW (2005) Female oxytocin gene-knockout mice, in a semi-natural environment, display exaggerated aggressive behavior. Genes Brain Behav 4(4):229–239. https://doi.org/10.1111/j.1601-183X.2005.00118.x
Reif A, Rosler M, Freitag CM, Schneider M, Eujen A, Kissling C, Wenzler D, Jacob CP, Retz-Junginger P, Thome J, Lesch KP, Retz W (2007) Nature and nurture predispose to violent behavior: serotonergic genes and adverse childhood environment. Neuropsychopharmacology 32(11):2375–2383. https://doi.org/10.1038/sj.npp.1301359
Rilling JK, Young LJ (2014) The biology of mammalian parenting and its effect on offspring social development. Science 345(6198):771–776
Rodriguiz RM, Chu R, Caron MG, Wetsel WC (2004) Aberrant responses in social interaction of dopamine transporter knockout mice. Behav Brain Res 148(1–2):185–198
Rogge G, Jones D, Hubert GW, Lin Y, Kuhar MJ (2008) CART peptides: regulators of body weight, reward and other functions. Nat Rev Neurosci 9(10):747–758. https://doi.org/10.1038/nrn2493
Sabol SZ, Hu S, Hamer D (1998) A functional polymorphism in the monoamine oxidase A gene promoter. Hum Genet 103(3):273–279
Salvador A, Costa R (2009) Coping with competition: neuroendocrine responses and cognitive variables. Neurosci Biobehav Rev 33(2):160–170. https://doi.org/10.1016/j.neubiorev.2008.09.005
Saper CB, Chou TC, Elmquist JK (2002) The need to feed: homeostatic and hedonic control of eating. Neuron 36(2):199–211
Saudou F, Amara DA, Dierich A, LeMeur M, Ramboz S, Segu L, Buhot MC, Hen R (1994) Enhanced aggressive behavior in mice lacking 5-HT1B receptor. Science 265(5180):1875–1878
Seo D, Patrick CJ, Kennealy PJ (2008) Role of serotonin and dopamine system interactions in the neurobiology of impulsive aggression and its comorbidity with other clinical disorders. Aggress Violent Behav 13(5):383–395. https://doi.org/10.1016/j.avb.2008.06.003
Shin AC, Zheng H, Berthoud HR (2009) An expanded view of energy homeostasis: neural integration of metabolic, cognitive, and emotional drives to eat. Physiol Behav 97(5):572–580. https://doi.org/10.1016/j.physbeh.2009.02.010
Siegel A, Roeling TA, Gregg TR, Kruk MR (1999) Neuropharmacology of brain-stimulation-evoked aggression. Neurosci Biobehav Rev 23(3):359–389
Siever LJ (2008) Neurobiology of aggression and violence. Am J Psychiatry 165(4):429–442. https://doi.org/10.1176/appi.ajp.2008.07111774
Siever LJ, Buchsbaum MS, New AS, Spiegel-Cohen J, Wei T, Hazlett EA, Sevin E, Nunn M, Mitropoulou V (1999) d,l-fenfluramine response in impulsive personality disorder assessed with [18F]fluorodeoxyglucose positron emission tomography. Neuropsychopharmacology 20(5):413–423. https://doi.org/10.1016/S0893-133X(98)00111-0
Silver JM, Yudofsky SC, Slater JA, Gold RK, Stryer BL, Williams DT, Wolland H, Endicott J (1999) Propranolol treatment of chronically hospitalized aggressive patients. J Neuropsychiatry Clin Neurosci 11(3):328–335. https://doi.org/10.1176/jnp.11.3.328
Simon GE, Von Korff M, Saunders K, Miglioretti DL, Crane PK, van Belle G, Kessler RC (2006) Association between obesity and psychiatric disorders in the US adult population. Arch Gen Psychiatry 63(7):824–830. https://doi.org/10.1001/archpsyc.63.7.824

Skuse DH, Gallagher L (2009) Dopaminergic-neuropeptide interactions in the social brain. Trends Cogn Sci 13(1):27–35. https://doi.org/10.1016/j.tics.2008.09.007
Sokolowski K, Corbin JG (2012) Wired for behaviors: from development to function of innate limbic system circuitry. Front Mol Neurosci 5:55. https://doi.org/10.3389/fnmol.2012.00055
Soloff PH, Meltzer CC, Greer PJ, Constantine D, Kelly TM (2000) A fenfluramine-activated FDG-PET study of borderline personality disorder. Biol Psychiatry 47(6):540–547
Soloff PH, Price JC, Meltzer CC, Fabio A, Frank GK, Kaye WH (2007) 5HT2A receptor binding is increased in borderline personality disorder. Biol Psychiatry 62(6):580–587. https://doi.org/10.1016/j.biopsych.2006.10.022
Somerville LH, Fani N, McClure-Tone EB (2011) Behavioral and neural representation of emotional facial expressions across the lifespan. Dev Neuropsychol 36(4):408–428. https://doi.org/10.1080/87565641.2010.549865
Stein DJ (2015) Social anxiety disorder and the psychobiology of self-consciousness. Front Hum Neurosci 9:489. https://doi.org/10.3389/fnhum.2015.00489
Stein DJ, Vythilingum B (2009) Love and attachment: the psychobiology of social bonding. CNS Spectr 14(5):239–242
Storey AE, Walsh CJ, Quinton RL, Wynne-Edwards KE (2000) Hormonal correlates of paternal responsiveness in new and expectant fathers. Evol Hum Behav 21(2):79–95
Taylor SE (2006) Tend and befriend biobehavioral bases of affiliation under stress. Curr Dir Psychol Sci 15(6):273–277
Taylor SE, Klein LC, Lewis BP, Gruenewald TL, Gurung RA, Updegraff JA (2000) Biobehavioral responses to stress in females: tend-and-befriend, not fight-or-flight. Psychol Rev 107(3):411
Taylor SE, Saphire-Bernstein S, Seeman TE (2010) Are plasma oxytocin in women and plasma vasopressin in men biomarkers of distressed pair-bond relationships? Psychol Sci 21(1):3–7
Terburg D, Morgan B, van Honk J (2009) The testosterone-cortisol ratio: a hormonal marker for proneness to social aggression. Int J Law Psychiatry 32(4):216–223. https://doi.org/10.1016/j.ijlp.2009.04.008
Thompson RR, George K, Walton JC, Orr SP, Benson J (2006) Sex-specific influences of vasopressin on human social communication. Proc Natl Acad Sci U S A 103(20):7889–7894. https://doi.org/10.1073/pnas.0600406103
Tinbergen N (1951) The study of instinct. Clarendon Press, Oxford
Tsujino N, Sakurai T (2009) Orexin/hypocretin: a neuropeptide at the interface of sleep, energy homeostasis, and reward system. Pharmacol Rev 61(2):162–176. https://doi.org/10.1124/pr.109.001321
Van Anders SM, Goldey KL, Kuo PX (2011) The steroid/peptide theory of social bonds: integrating testosterone and peptide responses for classifying social behavioral contexts. Psychoneuroendocrinology 36(9):1265–1275
Vassos E, Collier DA, Fazel S (2014) Systematic meta-analyses and field synopsis of genetic association studies of violence and aggression. Mol Psychiatry 19(4):471–477. https://doi.org/10.1038/mp.2013.31
Verona E, Joiner TE, Johnson F, Bender TW (2006) Gender specific gene-environment interactions on laboratory-assessed aggression. Biol Psychol 71(1):33–41. https://doi.org/10.1016/j.biopsycho.2005.02.001
Viding E, Blair RJ, Moffitt TE, Plomin R (2005) Evidence for substantial genetic risk for psychopathy in 7-year-olds. J Child Psychol Psychiatry 46(6):592–597. https://doi.org/10.1111/j.1469-7610.2004.00393.x
Vukhac KL, Sankoorikal EB, Wang Y (2001) Dopamine D2L receptor- and age-related reduction in offensive aggression. Neuroreport 12(5):1035–1038
Wersinger SR, Caldwell HK, Christiansen M, Young WS 3rd (2007) Disruption of the vasopressin 1b receptor gene impairs the attack component of aggressive behavior in mice. Genes Brain Behav 6(7):653–660. https://doi.org/10.1111/j.1601-183X.2006.00294.x
WHO (2010) Injuries and violence: the facts. World Health Organization, Geneva

Williams KW, Elmquist JK (2012) From neuroanatomy to behavior: central integration of peripheral signals regulating feeding behavior. Nat Neurosci 15(10):1350–1355. https://doi.org/10.1038/nn.3217

Williams G, Bing C, Cai XJ, Harrold JA, King PJ, Liu XH (2001) The hypothalamus and the control of energy homeostasis: different circuits, different purposes. Physiol Behav 74(4–5):683–701

Wood RM, Rilling JK, Sanfey AG, Bhagwagar Z, Rogers RD (2006) Effects of tryptophan depletion on the performance of an iterated Prisoner's Dilemma game in healthy adults. Neuropsychopharmacology 31(5):1075–1084. https://doi.org/10.1038/sj.npp.1300932

Wundt W (1874) Principles of physiological psychology. Engelman, Leipzig

Wynne-Edwards KE (2001) Hormonal changes in mammalian fathers. Horm Behav 40(2):139–145. https://doi.org/10.1006/hbeh.2001.1699

Zink CF, Meyer-Lindenberg A (2012) Human neuroimaging of oxytocin and vasopressin in social cognition. Horm Behav 61(3):400–409. https://doi.org/10.1016/j.yhbeh.2012.01.016

4 Temperament-Personality-Character and Evolutionary Biology

Xenia Gonda and Kostas N. Fountoulakis

4.1 Introduction

One peculiarity of humans is the marked differences observable in personality and behaviour. While most characteristics of living organisms converge from differences to similarity and a single optimum which is the most efficient and adequate in the given environmental circumstances leading to maximum adaptation and fitness, human personality traits and temperaments show several remarkably distinct manifestation levels considering each trait, temperament or characteristic leading to a multimodal distribution along these dimensions. Research has also revealed that such differences in personality traits are heritable, based to a large extent on genetic variability, and are also associated and interact with environmental influences. Thus the question emerges: unlike other genetically and biologically based characteristics of living organisms which through evolution aim towards manifesting in a single optimal phenotype with maximal fitness underpinned by a universal, invariant and species-typical genome, why are there such significant heritable differences observable in human personality (Penke and Jokela 2016)? If evolution generally aims at eliminating variation of less adaptive forms and thus at zero variance around the most fitness-increasing optimal level of a given characteristic, then why does genetic and phenotypic variation persist in such adaptive systems as personality and temperament? The answer to this question is hidden under understanding the genetic architecture and interactions shaping the manifestation of personality traits and temperaments.

Personality and temperamental traits are the most basic systems which underlie and govern how humans interact with their environment; therefore such systems are

X. Gonda (✉)
Department of Psychiatry and Psychotherapy, Semmelweis University, Budapest, Hungary

K. N. Fountoulakis
3rd Department of Psychiatry, Faculty of Medicine, School of Health Sciences, Aristotle University of Thessaloniki, Thessaloniki, Greece

K. N. Fountoulakis, I. Nimatoudis (eds.), *Psychobiology of Behaviour*,
https://doi.org/10.1007/978-3-030-18323-3_4

designed to be aimed at survival and reproduction. Evolutionary psychology is built on the adaptations theory, guiding the identification of such particular ancestral problems which favoured psychological adaptations through natural selection (Penke and Jokela 2016), and the evolutionary perspective on personality focuses on consequential life-related outcomes in the variation of personality, encompassing its effect on health, relationships and reproductive efficiency and life expectancy which are the most basic components and indicators of biological fitness (Nettle 2011). Thus personality evolution and psychological adaptation proceeds by mechanisms which have been designed by natural selection through evolution to solve specific adaptive problems (MacDonald 2012).

Very highly specific emotional and motivational structural neural systems underlie human behaviour, with important variation among people in the function of these systems (MacDonald 2012). In our current understanding, personality traits and temperaments have biological bases and are therefore heritable to a substantial, possibly differing degree (Penke and Jokela 2016), yet we still do not sufficiently understand their genetic background. Temperament definitions build on the centrality of biology (Buss and Plomin 1984; Rothbart 2016) in determining emotional reactions to stimuli. In contrast, personality, in addition to biologically based characteristics, also includes such components as beliefs, skills, morals, social cognition, which are more determined by social influences (MacDonald 2012).

Individual differences in personality and temperament which manifest in the normal range are variations in the evolved systems (MacDonald 2012), but what is the reason for the existence of a persisting variation in personality traits temperaments, which develop based on substantial polymorphic genetic components contributing to differences in neurobiology (Nettle 2005)? Differences in health, mating, social and sexual behaviour influence reproductive success, and as the marked personality variation contributing to interindividual differences significantly influences related processes, personality variation may be subject to natural selection (Nettle 2005).

There are several theoretical frameworks aiming to understand how the genetic variation underpinning personality differences is maintained. Each personality trait and temperament is polygenic and multifactorial, which means that a given trait develops based on a large number of genes, interaction between these genes and interaction between the genes and the environment. Most of the heritable individual differences are not adaptations themselves, but rather reflect dimensions which tolerate some degree of genetic variability (Penke and Jokela 2016). Adaptive individual differences are in fact conditional strategies which are implemented in universal adaptations, and which are elicited by specific cues in the environment (Penke and Jokela 2016).

One possible answer to the question of personality differences is the presence of substantial environmental heterogeneity in space and time which must be confronted by individuals (MacDonald 1995). Phenotypic variation in the population is likely the effect of a regime of fluctuating selective pressures arising from these divergent environments, each of which favours specific phenotypic values under certain conditions while disfavours them under others (Nettle 2011). This approach argues for the role and importance of environmental heterogeneity in maintaining variation in personality and temperament, postulating that a personality profile which may be adaptive when an environment is rich in resources may be highly inadaptive in environments where resources are scarce and thus different personality traits and temperaments and constellations will be adaptive in different environments creating the basis for sustained personality and temperamental differences in humans (MacDonald 2012). Furthermore, as there is a correlation and covariation between facets and components of personality traits, the pattern of covariation may also provide clues about the synergistic relationship between behaviours in common situations (Nettle 2011).

4.2 Turning Points in the History of Evolutionary Psychology

Along with fashions in science, there has been a long history of increasing and decreasing interest evolutionary psychology, which saw a pike in attention in recent years. As evolutionary psychology aims at delineating how natural selection eventually locks the optimum of complex psychological adaptation in humans (Penke and Jokela 2016), it has been challenging to adapt evolutionary mechanisms to explain the differences observable in human personality traits and temperaments. Thus evolutionary psychology for a long time focused on the 'psychic unity of mankind' (Tooby and Cosmides 1990) rather than its individual differences, that is, on universally shared psychological mechanisms which contribute to limited phenotypic plasticity due to varying environmental input, but without much attention to genetic variability underlying observable personality differences or heritable traits (Figueredo et al. 2009).

This initial perspective meant that in the beginning evolutionary psychology aimed at uncovering species-typical mechanisms by focusing on situation-typical behaviours (Nettle 2005) and postulating that variability in behaviour is determined by the situation rather than underlied by genetic differences. Cosmides and Tooby (1995) stated that human genetic variation is largely reduced into functionally superficial biochemical differences, contributing to a universal and species-typical complex functional design in humans and observable individual differences arise mainly from noise (Tooby and Cosmides 1990). However, highly prevalent genetic polymorphisms described in the background of neurochemical and neuroanatomical systems giving rise to human behaviour indicate that between-human variation is far from superficial (Nettle 2005), but is rather strongly rooted in the genome and is the result of evolutionary processes and natural selection rather than accidental noise. This contributed to a shift of focus from species-typical mechanisms to inter-individual variations through the works of Buss (1991, 1999) and MacDonald (1995) as series of accumulating observations demanded attention and explanation for individual differences from an evolutionary psychology aspect. One such observation was that most early environment-induced phenotypic polymorphism showing lifelong stability observable in wild populations could have evolved only if there is not one universally optimal phenotype independent of context, but rather a series of contexts acting throughout evolution each selecting for a different optimal phenotype (Nettle 2011). Another observation reasoning for genetically based interindividual variation was that heritable variation in crucial characteristics is ubiquitous and significantly exceeds what can be expected from and predicted by population genetic models, which also demanded explanation (Nettle 2011). Thus evolutionary psychology refocused from conceptualising human nature as a unity built from universally shared psychological mechanisms to understand interindividual differences in such universal characteristics. While universally shared mechanisms may indeed lead to some degree of phenotypic plasticity due to varying environmental effects and influences, it was also increasingly accepted that the between-individual genetic variability of traits should no longer be neglected (Figueredo et al. 2009). However, there is still very little research targeting this important question.

While Tooby and Cosmides (1990) claimed that personality traits which are variable and heritable cannot result from adaptation, most evolutionary psychologists suggest that individual differences in the manifestation level of personality traits and temperaments, though differing in their effectiveness in adapting to different strategies in complex social groups, are indeed adaptive in their nature (Figueredo et al. 2005; Buss and Greiling 1999), and besides personality differences leading to increased adaptation in social interactions, personality variation allows different individuals to be better suited to particular niches in both social ecological and physical environments (MacDonald 1995). However, the adaptationist view of evolutionary personality psychology would predict that eventually there would be no heritable genetic variation, which is contradicted by research supporting the heritable variance of human personality (Penke and Jokela 2016).

4.3 Human Personality and Personality Genetics from an Evolutionary Aspect

Personality traits and temperaments are theoretical constructs which exist functionally and predict typical behavioural reactions and tendencies evoked in given situations or settings but giving space for diverse adaptive reactions. Appearing randomly, behavioural tendencies and underlying psychological mechanisms can be fixated and become prevalent if they enhance adapting and fitness, thus leading to little variation, at least according to classical evolutionary theory, since in case of varying manifestation levels of a trait, one would contribute to higher fitness than the other; therefore natural selection would remove the variant with the lower fitness. From this aspect any psychological trait with a history of evolution and selection would not show a significant variation but rather be universal and species-typical (Nettle 2006).

It seems, however, that human nature cannot reach a global optimum (Workman and Reader 2014). Personality traits and temperaments being multigenic, there are a large number of genes encoding for complex adaptation mechanisms, and variation in any of those genes would disrupt the complex pattern of function (Nettle 2006). To overcome this problem, it was postulated that if there is variation in a trait, it indicates that the given trait is not under natural selection, or, in other words, if there is heritable variation in a trait, then it indicates that it has no significance from an adaptive aspect (Nettle 2006).

From the 1930s, evolutionary genetics has been studying the origins and maintenance as well as the implications of genetic variations underlying personality traits, focused both on between-species and between-individual differences and on understanding the role of such evolutionary processes as mutation, selection, migration and drift in shaping personality besides providing information on its genetic background (Penke and Jokela 2016). While due to universal genomic identity there is a 99.9% overlap in the DNA base sequence in humans, with this shared genome giving rise to species-specific phenotypes in the focus of adaptationist evolutionary study, there is also significant phenotypic variation in one part due to phenotypic plasticity and in another part due to genotypic variation which is in the focus of evolutionary personality genetics (Penke and Jokela 2016). While inherited, this universal genome is never inert with a range of processes causing its physical alterations. Cell replication cycles inherently give rise to multiple copying errors and mutations leading to variation in the base sequence and contributing to the emergence of single nucleotide polymorphisms (SNPs), copy number variations (CNVs) or inversion and translocation of larger regions. Currently the 1000 Genomes Project reported more than 88 million variants including 84.7 million SNPs, 3.6 million insertion/deletions and 60,000 structural variants (1000 Genomes Project Consortium et al. 2015). The effect of mutation and polymorphism may or may not leave their marks in the expressed phenotype, with many not contributing to structural differences or not even impacting gene expression, while others severely disrupt either gene expression or regulation contributing to deleterious effects, and in rare cases improve function of the organism and its adaptation to the environment. While larger genomic rearrangements more severely disrupt expression and function contributing to prenatal death or birth defects, SNPs and CNVs, which are the most common sources of between-individual genetically based variations, can have a range of magnitude of effects on the resulting phenotype (Penke and Jokela 2016).

Personality differences are not governed by one or a few genetic variations but are built from quantitative traits with a multigenic or even omnigenic background, based on quantitative trait loci (Plomin et al. 1994) with possible additive effects or interactions between different alleles as well as with the environment contributing to different phenotypic manifestations which can be studied by comparing them across individuals in a given populations (Penke and Jokela 2016). Twin and adoption studies suggest that 30–60% of personality trait variation is accounted for by genetic factors, with the parents' level of a given personality trait being the strongest predictor of the level of that trait in the offspring (Polderman et al. 2015; Keller and Miller 2006). However, effect of individual genes and variants appears to be minute at most, and identified candidate gene effects can generally neither be replicated nor confirmed in genome-wide association studies (Genetics of Personality Consortium et al. 2015). In the background of personality traits thus a very large number of genes each with only a very small individual effect play a role. According to genome-wide complex trait analyses, common genetic variants explain 0–21% of variance in personality (Power and Pluess 2015). The heritable psychological differences are underpinned by gene sets encoding neurophysiological functions which can be studied as more discrete and better characterisable endophenotypes. Several of these endophenotypes contribute to reaction types and behavioural tendencies, which together and in correlation give rise to observable traits influencing and determining behavioural output in given situations, and through the adaptiveness of a given reaction in a given situation influence fitness (Penke and Jokela 2016).

What is the reason behind the persistence of mutation and variation introduced by mistakes in DNA copying during the cell cycle? The primary source of variation is the appearance of mutations, and the level of variation in a population reflects a balance between the introduction of variance by mutations and selection aimed at eliminating them. As mutations are infrequent, in case of any trait based on a single gene, no strong selection would be needed to keep the variation near zero; however, if a number of genes are involved in encoding for a given trait, which is the case for personality and temperamental traits, then there is a large number of possible mutations, and selection would not be able to remove these as efficiently. Therefore in case of polygenic traits, there would be a significant amount of variation in the genome giving rise to observable variations in reactions, behaviours, personality traits and temperaments (Nettle 2006; Houle 1998).

4.4 Evolutionary Processes in the Background of Personality Variation

Evolution is an adaptive process underlied by constant interaction between multiple genetic and environmental factors directed towards increased reproductive fitness resulting in a great number of gene × gene and gene × environment interactions in the background of personality. Thus the best way to conceive evolutional processes in the background of personality is by understanding the process of evolution in response to stressors, events, influences and complex ecological surroundings (Cloninger 2009).

4.4.1 The Role of the Environment in the Appearance of Personality Variation

The aforementioned 30–60% genetic determination and heritability of personality traits leave 50–70% to non-genetic factors provided by environmental influences (Workman and Reader 2014). One asset of humans is selecting such an environment which suits both their capabilities and their needs and preferences which is covered under the term gene-environment correlation or niche picking. This also means that individuals with a given personality are most likely to put themselves in such environments where their particular personality profiles would yield the best possible results. There are varied environments, so each individual is able to pick a niche corresponding best to their personality (Penke and Jokela 2016).

Evolutionary theory is environmentalistic in that it is about the extent an organism adaptively fits its environment. Animal studies suggest that environmental heterogeneity imposes varying selection pressures on traits (Dingemanse and Kalkman 2008). Thus observable characteristic are the results of gene × environment interactions (G × E). However, while selection acts at the level of the phenotype, G × E interactions take place at molecular genomic levels contributing to a double role of the environment in evolutionary genetics by interacting with the genotype in producing the phenotype and then as a selection force determining its fitness and also its destiny (Penke and Jokela 2016).

Environment has a profound role in determining adaptiveness of a given phenotype and behaviour. In case of humans, the social environment with its social niches exerts heterogeneous effects, contributing to the diversification of personality traits in order to fit different social niches (Figueredo et al. 2009). Environments show high variation both in time and in space, and during the 3.5 million years of human evolution, there have been significant fluctuations in the environment. Adapting to such heterogeneous environments has three possible solutions, including developmental plasticity, genetic diversity and spatial migration. If a variable occurs over time, then developmental phenotypic plasticity is adaptive as long as there are reliable cues available to signal the optimal alternative phenotype under the particular conditions. In case of spatial environmental heterogeneity, the emergence of genetically distinct individuals showing variation along locally optimal trait levels will be advantageous. Those individuals who cannot adapt to the stress of ecological variability can migrate to environments which are more ecologically supportive. A mixture of these three distinct processes play a role in the background of human personality trait variability, and they likely contribute to the partial heritability and partial environmentality observable in case of personality traits (Figueredo et al. 2009). The currently prevailing theory postulates that humans originated in Africa and from there progressed towards the current global distribution of the human species which means significant climate differences which, coupled with development, contributed to different self-sustaining practices (Figueredo et al. 2009). Altogether, early ancestral humans had to face three basic types of selective environmental pressures including climatic, ecological and social (Geary 2005), with most major ones being social, but also the other types of selective pressures may have enhanced social competition over the limited resources. Thus sociality is likely the most important cause for personality variations in humans (Figueredo et al. 2009).

4.4.2 Putative Processes Governing Evolution of Individual Differences

Adaptive evolution proceeds by the accidental appearance of rare but fitness-increasing mutations and elimination of the less adaptive ones by natural selection, towards one most optimal and most adaptive design without any variation in the population (Nettle 2005). Human personality traits, however, exhibit both a highly complex design and a high degree of continuous interindividual variation which indicate that they are underlied by a significant number of genetic polymorphisms at a large number of genetic loci (Penke and Jokela 2016). There are three possible approaches to explain the persisting genetic variation in the background of human personality traits, including selective neutrality, mutation-selection balance and balancing selection, which are not mutually exclusive but may rather operate in some combination (Verweij et al. 2012).

4.4.2.1 Selective Neutrality

Selective neutrality postulates that variation in personality traits does not contribute to differences in adaptation or fitness; therefore these would drift randomly in frequency unaffected by selection, and while some variants are lost to genetic drift, variation will be maintained by the appearance of new mutations through a process called mutation-drift balance (Verweij et al. 2012). Tooby and Cosmides (1990) hypothesised that in the background of personality differences, accumulating fitness-neutral and consequentially selectively neutral mutations play a role, which have no impact on survival or reproductive success. Other authors, however, find the role of selective neutrality in the background of variation of personality highly implausible (Penke and Jokela 2016), as a given variation could be selectively neutral only if there are no costs and benefits associated with it in any of the relevant environments, that is, no G × E interaction can be observed, meaning that a given level of a trait in question cannot be more adaptive in one environment compared to another. Personality traits are well-reported to influence performance and outcome in various domains of life (Penke and Jokela 2016) and are related to various well-established components or proxy variables of fitness including mortality (Mosing et al. 2012), physical and mental health (Lahey 2009), physical attractiveness (Lukaszewski and Roney 2011), mating behaviour (Zietsch et al. 2010) and number of offspring (Verweij et al. 2012; Jokela et al. 2009). However, a zero net effect may also result from positive correlations and negative correlations with different fitness components counterbalancing one another (Nettle 2005), or it is also possible that as personality trait dimensions are in fact a spectrum of fitness-maximising alternative strategies, in the normal personality range, average fitness will be uniform (MacDonald 1995; Verweij et al. 2012). Also, genetic polymorphisms which have an impact on several traits could be neutral to selection when multivariate genetic constraints yield little variation in fitness effects, such as when individual personality traits have a significant genetic polymorphism and correlate with fitness (Verweij et al. 2012).

4.4.2.2 Mutation-Selection Balance

The basic idea of mutation-selection balance is that genetic variance in personality traits would be maintained in a mechanism where any deviation from the optimal personality trait level is corrected through selection by the elimination of those alleles which predispose to trait levels outside the range of this optimum. This process also reduces genetic variation in the background of the trait but is complemented by the emergence of new mutations (Verweij et al. 2012). As human personality traits are multigenic, each of them is determined by a large number of genes, all of them are subjected to random, although rare and mildly harmful mutation effects, continuously counterbalanced by selection maintaining fitness-increasing ones. Mutation-selection balance is likely to influence components of cognitive fitness (Penke and Jokela 2016) or traits underpinned by a large number of rare variants each contributing a very small effect (Verweij et al. 2010).

From the aspect of mutation-selection balance, explanations postulate that fitness-maximising selection would contribute to the emergence of one optimal adaptive design. However, random mutations disrupting this design and therefore decreasing fitness should also arise. Most of such nonneutral dominant mutations with a strong effect will randomly disrupt the function of sophisticated systems and are thus harmful and are therefore easy selection targets shortly removed; so recessive mutations or ones with a weak effect are the ones which are likely to persist and spread to become common (Penke and Jokela 2016; Eyre-Walker 2010). With the constant appearance of new mutations, all individuals carry an individual pattern of the accumulated mutation load consisting of rare, partly recessive and mildly deleterious alleles (Verweij et al. 2012). In addition, everyone carries about 500 harmful mutations which is also likely to contribute greatly to the variance in fitness-related traits (Penke and Jokela 2016). More than half of the human genome is expressed in the brain (Sandberg et al. 2000), and personality traits arising from a large number of loci have a large mutational target size so have the potential to be more exposed to the effect of mutations (Verweij et al. 2012).

There are several characteristics of personality traits which cannot be explained by mutation-selection balance, including the various candidate genes associated with personality traits, the intermediate prevalence rates, and the fact that nonadditive variance is in many cases as high as additive variance which makes it unlikely for mutation-selection balance to account for variation in personality traits. Mutation-selection balance is rather a plausible mechanism to account for the genetic variation in the background of characteristics which affect overall functionality, such as intellectual function, than an explanation for the variation in personality traits and temperaments (Penke and Jokela 2016).

4.4.2.3 Balancing Selection

While in the previous cases mutations are either invisible to selection, or selection depletes but cannot eliminate all variation, in the process of balancing selection, selection actually maintains variation by fluctuating selection pressures. Environmental conditions show a significant spatiotemporal variation thus contributing to a spatiotemporal fluctuation of selection pressures for different alleles (Penke and Jokela 2016). Balancing selection by environmental heterogeneity thus means that the same personality trait in different places or at different times may be influenced by selection pressures acting in different directions; thus in general and on average, no genetic variant could consistently be favoured over the other variants, contributing to the persistence of variation in personality traits with environment-contingent fitness consequences (Penke and Jokela 2016). Personality traits are likely to be influenced by balancing selection underlied by a limited number of medium-effect common genetic variants as opposed to mental disorders which may arise as a consequence of a large number of mildly deleterious rare variants and mutations (Penke et al. 2007). As human beings show a tendency to actively search for and construct their environments and adapt to these, personality traits are plausible to be under balancing selection rather than either being neutral to selection or being under stabilising selection (Penke and Jokela 2016). While selective neutrality and mutation-selection balance maintained genetic variation because it couldn't be depleted by selection, if selective forces are balanced, it is selection itself which would maintain variation, for example, in cases where, due to both extremes of a personality trait being adaptive in the same magnitude but under different conditions, both extremes are favoured by selection (Penke and Jokela 2016).

Balancing selection has various types including environmental heterogeneity or negative frequency-dependent selection which latter favours traits occurring at a low frequency (Penke and Jokela 2016). In case of balancing selection, there should be several varying selection pressures which under different conditions favour different phenotypes and should be stronger than that unidirectional selection pressure which favours a single optimal trait level in a given environment. The consequence would be either a whole continuum of phenotypes or the appearance of multiple distinct phenotypes with identical average fitness averaged across all environments which cannot be optimised further (Penke and Jokela 2016).

As during evolutionary history humans were exposed to significantly varied and changing social and physical environments, the genetic variation underpinning personality traits is likely to have been maintained by balancing selection by environmental heterogeneity mediated by negative frequency-dependent selection on life history strategies (Penke et al. 2007; Penke and Jokela 2016).

4.5 Individual Differences from an Evolutionary Aspect

There are consistent differences in the behaviour between individuals belonging to the same species even under the same ecological conditions, contributing to a previously ignored but in fact substantial amount of within-species interindividual differences in reactions in a given environment, commonly referred to as personality differences (Nettle and Penke 2010). While personality differences are likely adaptive, and impact and determine suitability to different ecological and social niches, they also put a constraint on the behavioural repertoire and flexibility which is a function of the constellation of personality characteristics arising from an interaction between genetic and environmental factors during early development (MacDonald 2012). This constraint may appear maladaptive since as a consequence individuals would not be able to exploit the full range of possibilities within different situational contexts. However, the biological preparedness for certain behaviours and the developmental plasticity of these behaviours may vary independently (Figueredo et al. 2009). Individuals are prepared to pick environments where their prepared behaviours would be suitable, and individuals manifesting different personality and temperamental traits orient towards different niches best suiting their personality, and benefits of selecting suitable social niches allowing for the best performance overcome the constraints imposed by personality differences (Figueredo et al. 2005).

As investment in one component of fitness usually takes place at the cost of other components, it is hard to establish an obvious and clear advantage or disadvantage. However, at a given point in space and time, there could generally be an optimum value for a trait from the aspect of fitness, and this optimum shows variation across space and time. This means that there would be a complex interaction between spatiotemporal variation in selective optima and tradeoffs in different fitness components, also applying to personality traits. However, as there is a gene flow between populations, if all populations are taken together, a normal distribution of genetic polymorphism-based traits can be observed, while in different subpopulations at different times and in different spaces, the optimal levels of that given trait may differ (Nettle 2006). Due to the polygenic or even multigenic nature of personality traits and behavioural phenotypes coupled with the fluctuating nature of selection, variation is both normal and ubiquitous. Behavioural alternatives can be considered as tradeoffs with any level of a trait giving rise to a mixture of costs and benefits rather than an obvious advantage, so the optimal value of a trait would be a function of the given circumstances (Nettle 2006), contributing to several different optimal values in the manifestation of a given trait.

4.6 Evolutionary Aspects of Personality Trait and Temperament Dimensions

Personality as a constellation of behaviours and reactions has an obvious impact on fitness. In humans most of the between-individual behavioural variation can be accounted for by 3–7 basic personality dimensions or traits, and about 30–60% of between-individual variation of personality traits is due to genetic variation according to twin family and adoption studies (Verweij et al. 2012; Johnson et al. 2008). In spite of this high heritability, it is not fully understood how heritable variation in genetic background is maintained. There have been no common genetic variants consistently and replicably identified associated with personality factors and differences. This missing heritability observable in case of the majority of complex traits has been partially explained by the fact that the majority of variants with a small effect have not been identified as yet, or that available genotyping methods have a low capacity to detect rare variants, or that structural variants including CNVs cannot be sufficiently captured by current methods, or have been attributed to the low power of current methods to detect gene-gene (G × G) interactions. Lack of common genetic variants associated with personality traits even in GWAS-s suggest that other mechanisms should account for personality variation (Verweij et al. 2010). During development personality shows a genotype-determined and early environment-influenced canalization of the manifested behaviours, followed by stability after the developmental phase in spite of changes in environment and circumstances (Verweij et al. 2012). While personality is highly stable, between individuals it clearly varies along quantitative dimensions with continuity (Nettle and Penke 2010).

The most comprehensive and acceptable explanation of evolution of personality differences as discussed above postulates that selection effects on personality show variation in space and time and by condition; thus personality variation is adaptive. Payoffs may vary based on the frequency of personalities in the population leading to frequency dependence influencing fitness. Adaptive value of personality variation considering costs and benefits of specific dispositions and how it maintains multiple phenotypic equilibria can be conceptualised in case of individual personality and trait dimensions.

Personality traits and temperaments are defined as relatively stable dimensions of describing individual differences related to behaviour, affect and cognition, conceived as major reflections of causal agency embedded in humans by evolutionary processes (Bouchard Jr. and Loehlin 2001). While personality traits have distal causes, they are influenced, triggered and moderated by proximal internal and external stimuli (Bouchard Jr. and Loehlin 2001). Several personality models exist describing personality at the intersection of a varying number of traits starting from three in the Eysenck model and going up to 16 in Cattell's model just to name a few better known ones. The Big Five or five-factor personality models describing personality variation along five trait dimensions are the most prevalent in studies with several lines of evidence supporting their validity. Thus several researchers hypothesised selection regimes for personality traits at this hierarchical level, although some Big Five personality trait dimensions share common mechanisms and therefore are not entirely orthogonal. Furthermore each of the traits in the five-factor personality model consists of a number of facets expressing motivational and behavioural tendencies which often belong to different life domains, and it is not straightforward why they cluster together in their trait which is purely a result of factor analytic studies with little information on why the given facets are related (Nettle 2011).

4.6.1 Evolutionary Aspects of the Five-Factor Model of Personality

The axes of the five-factor model of personality capture the major dimensions in human dispositional variation by describing stable individual differences in people's reactions to circumscribed classes of environmental events, defining characteristic ways of thinking, feeling and behaving clustered in five independent traits. Descriptive work related to the five-factor models in personality research captures about 50% of dispositional variation reflected in a variety of languages concerning the most robust dimensions of personality. Twin studies suggested a 40–50% role of genetic variation in determining each of the five traits (South et al. 2018), while genome-wide complex trait analyses estimated that contribution of common genetic variants explains 0–21% of variance of the individual traits, with the highest heritability reported for neuroticism and openness and the smallest for conscientiousness and agreeableness (Power and Pluess 2015; Genetics of Personality Consortium et al. 2015). These latter figures are much lower than those reported in previous quantitative genetic studies reflecting a significant missing heritability possible explained either by rare mutations or substantial and widespread epistatic or gene × environment interaction effects (Penke and Jokela 2016; Johnson et al. 2008).

In case of the five-factor model traits, several lines of evidence were previously presented for balancing selection maintaining variance in the individual traits, specifying fitness costs and benefits for each dimensions (Nettle 2005, 2006, 2011). As discussed above, given the temporospatial fluctuation of selection pressures, differing environmental contexts in a given time favour different optimum trait manifestations with distinct cost-benefit curves, contributing to the maintenance of a range of distinct personality trait levels with separate costs and benefits. The large amount of evidence available for the effect of personality traits in the five-factor model on such fitness-relevant outcomes as life expectancy, relationships, mating success or health (Roberts et al. 2007) clearly indicates that personality traits are not neutral from an evolutionary and fitness-related perspective; thus there are both possible fitness benefits and costs of being high or low on each personality trait dimension. The key is to identify in case of each personality trait the costs and benefits corresponding to different phenotypes in different environmental contexts (Nettle 2006, 2011). Putting it very simply, the five higher-order traits in the five-factor personality model represent basic dimensions of social adaptation: who is good company (extraversion), who is kind and supportive (agreeableness), who puts in sustained effort (conscientiousness), who is emotionally undependable (neuroticism) and who has ideas that pan out (openness) (Bouchard Jr. and Loehlin 2001).

4.6.1.1 Evolutionary Aspects in Differences in Extraversion

Extraversion, part of both Eysenck's three-factor and all five-factor models of personality is one of the most researched and best described personality traits from various aspects. There are three types of conceptualisations of extraversion. The first type includes models associating extraversion with activation vs. inhibition of impulses, contributing to differences in behavioural approach or the sensitivity of the reward system. The second cluster of conceptualisations contains models which focus on the involvement of extraversion in hierarchical and leadership potential disposition to wield power as well as dominance and submission, while the third type of models conceptualises extraversion in terms of a motivational predisposition to experience social interactions as rewarding, expressed also in terms of assertiveness vs. passivity in initiating social contacts (Denissen and Penke 2008a). These various aspects of the single trait of extraversion also reflect that costs and benefits of distinct and varying manifestations along the extraversion dimension will also be manifold and complex.

From the aspect of activation and inhibition of impulses, extraverted behaviour may be a consequence of the strength of response to such naturally rewarding stimuli as sex, food, physical excitement and joy; thus these cues will be more salient for extraverted persons who consequentially invest more energy in them (Denissen and Penke 2008a). On a psychological level, extraversion is related to positive emotionality, exploration, and reward, and is thought to be underlied by the reward circuitry with dopamine as the chief neurochemical determinant. Extraversion shows a strong positive relationship with several evolutionally relevant behaviours such as the number of sexual partners which has an obviously positive impact on fitness at least in men (Nettle 2006), and is also related to the chance of partnering with higher quality partners. Beyond mating, high extraversion also correlates with higher sensation seeking, through more social behaviour initiation consequentially yielding more social support, higher physical activity and more exploration. The evolutionarily positive aspect of these behaviours is easy to grab. This all, however, is also associated with increased risk of exposure to dangerous and harmful situations, as evidenced by such contemporary consequences as higher hospitalisation risk for traumatic injury, accidents, and illnesses, higher chances of migrating, committing criminal acts or antisocial behaviour and being arrested, which in the ancestral environment probably meant ostracism or death (Nettle 2005, 2011). Contemporary consequences of behaviours related to higher extraversion also include an increased risk that children will be exposed to step parents which in turn is a risk for their well-being. So while the high side and benefits of extraversion include increased mating opportunities and exploration of novelties in the environment and thus increased access to possible novel resources, it also has risks for personal survival and welfare of offspring; thus this tradeoff curve probably has no universal optimum but is rather determined by such local conditions which cause a constant fluctuation of optimal value due to changes in density and behavioural strategies which contribute to the retainment of its genetic polymorphism (Nettle 2006).

Considering the second aspect of extraversion related to hierarchy and leadership, a cost-benefit analysis of tradeoffs along the continuum suggests that extraversion represents a variability in phenotypic strategies related to social exchange and hierarchy negotiation. High extroversion means more superficial participation in social exchange with a lot of people, while low extraversion means a smaller number of deep engagements and solitary activities. Thus the benefit of high extroversion will include large cooperative networks increasing potential gains in trade through dyadic exchange and collective action. Costs on the other hand arise from socialising with non-relatives which consumes time, energy and other resources which could be invested in close relationships or other fitness-relevant tasks, risks potential exploitation or increases exposure to pathogens. Optimal level of extroversion among ancestors varied with circumstances buffering either against potential costs or increased potential benefits of participating in social exchange with a large network of cooperative partners. So for anyone who is unlikely to be exploited, able to easily attract quality associates and unlikely to contract communicable illnesses, there would be a higher optimal level of extraversion (Lukaszewski and von Rueden 2015). Another facet of extraversion is motivation to attract social attention and compete for high social status and leadership positions. Extraversion aspects pertaining to individual differences in status motivation such as assertiveness, social boldness or desire for attention can be conceptualised as a leadership-followership gradient. Cost related to leadership orientation mostly arise from the costs of acquiring and maintaining influence, while benefits arise from the advantages related to high status and respect. On the other hand, followers unlike leaders benefit from participation in collective actions and without paying the costs of competing for and implementing leadership. Optimal level of status motivation for ancestors varied with circumstances altering potential benefit of leadership and followership, such as possession of characteristics determining leadership ability or the ratio of leadership-oriented individuals to followers in the local social world (Lukaszewski and von Rueden 2015). Phenotypic strategies related to cooperation and hierarchy are also relevant for mating (Lukaszewski and von Rueden 2015) as far as extraversion means proactively approaching and getting attention of others which facilitates mating, while females prefer resourceful and high status mates. Competing for access, however, also increases risk for conflicts with rivals, while mating with multiple partners increases risk of STD exposure and the burden of multiple offspring (Lukaszewski and von Rueden 2015).

Taken together, from the aspect of cost-benefit tradeoffs of the extraversion continuum, high extraversion means several benefits including a larger cooperative network, more success on the mating market, increased likelihood to attain positions of higher social status and leadership, a higher likelihood of increased social alliance formation, increased exploration, as well as access to resources, but it is also more likely that as costs those high on extraversion experience antagonistic conflict, spend limited time and energy in socialising, are exposed to multiple increased physical risks, are more likely to contract illnesses and sustain injuries and in general have a lower life expectancy (Lukaszewski and Roney 2011; Lukaszewski and von Rueden 2015; Friedman et al. 1995; Nettle 2005; Samuels et al. 2004). Weighing costs and benefits, high extraversion still positively predicts reproductive success, especially in males. A complete theory of the origins of extraversion would explain whether and how distinct personality strategies manifested by individuals are adaptively patterned in relation to variable circumstances and why natural selection maintained differences in extraversion despite the consistent positive association of its high levels with reproductive success (Lukaszewski and von Rueden 2015).

Considering also the environmental context, high extraversion carries a net fitness cost in adverse environments, while in safe environments a net benefit, environmental variability and fluctuation thus maintain genetic variation in extraversion (Nettle 2011). The end results of net increase or decrease in fitness arising from extraversion varies in two ways. First, there may be individual characteristics determining the optimal level of extraversion. Individuals with increased physical strength, attractiveness and an immune function are better suited to face the risks associated with extraversion; thus their optimal levels of extraversion are higher. Second, considering beyond-individual levels, there are ecological contexts, such as fluid social structures or novel habitats which specifically favour risk-taking attitudes, where the optimal level of extraversion will also be higher. On the other hand, in social structures with already saturated habitats and stable hierarchies increased cautiousness would be more optimal (Nettle 2011). In general, nomadism and new environments favour sociability, assertiveness, sexual motivation and high physical activity, whereas in ecologies with established social structures, none of these are favoured (Nettle 2011). Altogether, the optimal level of extraversion will vary across individuals sharing a given habitat, within one habitat over time, and will also vary across habitats which leads to the maintenance of the full range of phenotypic variation (Nettle 2011).

There is only little evidence supporting common alleles explaining personality traits or other aspects of personality including extraversion, as generally candidate gene studies either provide contradictory results or fail replication, and there are generally no SNPs with genome-wide significance identified for personality traits. Common variants collectively explain only about 10% of variance in extraversion which is a small percentage of heritable variance, leaving the remaining variance to be explained by very rare alleles due to recent mutations (Verweij et al. 2012). The distinction between adaptiveness of different levels of extraversion in different circumstances, however, is evident also at the genetic level. The long alleles of the *DRD4* gene which encodes the D4 dopamine receptor and have been associated with extraversion-related traits and behaviours are significantly more common in nomadic populations or in populations having completed long migrations historically, as compared to sedentary populations (Chen et al. 1999; Ebstein 2006), which we will discuss under novelty seeking, a construct related to one aspect of extraversion. This association, however, only informs about the association between nomadism and extraversion on a genetic level, but does not indicate whether nomadism selects for increasing extraversion or extraverted populations have an increased propensity for migrating lifestyles (Chen et al. 1999).

4.6.1.2 Evolutionary Aspects of the Neuroticism Trait

Neuroticism or emotional stability/lability is related to differences in affect regulation and intensity, handling of stress and facilitation of performance under pressure, and in general reflects sensitivity of a domain-general system to respond to environmental threats, and vigilance to environmental hazards as well as increased risk aversion (Denissen and Penke 2008a). Neuroticism reflects variability of the levels of negative emotions including fear, anxiety, guilt or sadness and is a strong predictor of psychiatric morbidity, especially anxious and depressive manifestations, besides showing an overall negative impact on physical health due to the overactivation of stress-related processes (Gurven et al. 2014; Nettle 2006).

While these associations of high neuroticism generally grab the negative and maladaptive side of the trait, there also have to be benefits as indicated by the globally normal distribution of this trait. In ancestral environments the adaptive aspect of neuroticism may have been the avoidance of acute dangers by enhanced detection of potentially threatening stimuli as reflected in the constant apprehension and increased startle response, as well as anxiety and also increasing reaction and its decreasing latency to these stimuli. From this aspect negatively interpreting ambiguous stimuli and fixing attention on such stimuli is also adaptive (Nettle 2006). In environments where threats are more prevalent, more sensitive threat detection mechanisms are needed and are advantageous even at the expense of false positives and the negative physiological consequences of these (Haselton and Nettle 2006). On a neuropsychological level, neuroticism is associated with increased performance on psychomotor tasks related to detecting predators. Animal studies also support the advantage of vigilance and wariness to avoid predators and dangers. However, in the absence of predation risk those benefits also disappear. Environments with high actual threats or where individuals are poorly able to deal with undetected threats favour high neuroticism, while selection reduces it in more benign environments. Furthermore, in modern life actual physical threats are less, while the negative psychophysiological and health consequences of increased neuroticism are more apparent. Nevertheless, those seeking highly risky environments such as alpinists show low neuroticism, and as these occupations show an increased lethality, avoiding them through neuroticism is indeed adaptive.

Also, there is a positive correlation between neuroticism and competitiveness and thus academic success and working on to bettering one's own position. Low neuroticism has further disadvantages like lack of striving, while high neuroticism is an achievement and competition motivator which is adaptive if you can succeed (Nettle 2006).

High neuroticism, as briefly touched upon above, is associated with serious drawbacks including increased risk of stress-related physical and mental illness and, due to increased negative affect and anxiety, difficulties in relationships (Neeleman et al. 2002). Neuroticism-related diseases are more common in women (Costa et al. 2001) even though the environment they experience is the same; thus it appears this is because the impact of undetected threats differ in genders, as succumbing to physical hazard or social ostracism has more serious effects on reproductive success in women because of greater effect on offspring survival (Nettle 2011). Neuroticism also predicts social isolation and failure (Gurven et al. 2014; Nettle 2006).

In summary, neuroticism is composed of vigilance to physical and social threats, perceived vulnerability to disease, and angry hostility, and has benefits related to increased vigilance to threats, while its costs include related mental and physical illnesses as well as its consequences on relationships (Nettle 2011; Denissen and Penke 2008b). Optimal level of neuroticism depends on actual conditions and other characteristics and traits of the subject (Nettle 2006).

4.6.1.3 Evolutionary Aspects of the Openness Trait

There are several approaches to conceptualise the trait of openness, but all of them involve a high level of cognitive activity and a broad, deep and permeable consciousness, a high capacity for innovation, advanced problem-solving skills, engagement in the intellectual and creative domain, capacity to process incomplete information as well as intrinsically motivated curiosity enhancing cognitive competence (Denissen and Penke 2008a).

Openness is related to artistic creativity which attracts mates as it is associated with increased sexual attractiveness. However, openness, as a cognitive style associated with seeking novelty and complexity, and making associations between apparently unconnected domains, correlates with unusual experiences; increases inclination to paranormal beliefs, schizotypy, a break with reality, and eventually delusions and psychosis; and is also elevated in schizophrenia patients. Thus individuals on the high end of the openness dimension can be both socially successful through creativity or become socioculturally marginalised through bizarre beliefs and related behaviour (Nettle and Clegg 2006). The emergence of disorganised psychotic or psychotic-like beliefs is a clear cost of high openness.

Schizotypy and similar conditions also significantly diminish mating chances that are otherwise associated with increased openness. It is hard to determine when openness leads to positive and when to negative consequences, and whether this outcome is determined by the level of the trait or rather by an interaction with life events; thus the outcome is highly context-dependent (Nettle 2006). In case of the openness dimension, not so much different ecologies select for different levels, but different levels of openness are optimal depending on other characteristics (Nettle 2011).

4.6.1.4 Evolutionary Aspects of the Conscientiousness Trait

Conscientiousness is involved in task-related behaviours including intensity of engagement, monitoring nonattainment of goals, executive regulation in the performance domain, capacity for reliable work and enduring commitment, as well as trustworthiness and dependability (Buss 1991). Conscientiousness involves self-control in pursuit of goals and orderliness, delaying immediate gratification for a long-term good manifested in and associated with goal-orientedness, being hard-working and cautious about health with such components as industriousness, and orderliness which lead to such benefits as planfulness and care in premediated tasks (Nettle 2011).

Both ends of the conscientiousness dimension have positive aspects related to promotion of immediate vs. distant goal striving (Denissen and Penke 2008a). While conscientiousness is associated with longer life expectancy through engaging in healthy behaviours and avoiding unhygienic risks, in extremely high levels conscientiousness-related traits such as perfectionism and self-control have been related to obsessive-compulsive personality disorder and eating disorders. In ancestral and also contemporary environments, being obsessional may be adaptive through increased safety as well as its relation to high achievement, but rigidly sticking to sometimes pathological routines and missing spontaneous opportunities which would enhance reproductive fitness are damaging. This latter is exemplified by the association of conscientiousness and missing short-term mating episodes, which means that in case of increased conscientiousness, by emphasising long-term payoffs, the opportunistic grabbing of middle payoffs would be decreased (Nettle 2006).

Higher levels of conscientiousness are associated with making plans and sticking to them which leads to high office and school attainment in the modern world, and being able to stick to internally generated plans and goals in the face of distraction may have been advantageous in ancestral contexts which involved repeating tasks when outcomes and optimal schedules were predictable (Nettle 2011). Other situations, however, like those including a sudden attack or opportunistic hunting can't be planned and thus require spontaneous reaction without the chance for extensive reflection. Costs associated with conscientiousness thus include rigidity, inflexibility, difficulty in adapting to changing circumstances as well as the likelihood to miss short-term mating and resource opportunities (Nettle 2011; Schmitt 2004). Those towards the high end of the conscientiousness dimension also tend to perform worse when spontaneous response is needed to changes in the environment because they stick to previously defined goals. Thus while both of the two major aspects of conscientiousness including industriousness and orderliness (DeYoung et al. 2007) can be useful in scenarios which involve recurrent maintainment, management and exploitation of the same resources in a predictable manner, neither of them is advantageous in case of opportunistic or unpredictable exploitation of resources which explains why variation in the level of conscientiousness with varying environmental contexts is adaptive (Nettle 2011). Optimal balance between planfulness and spontaneity depends on local ecology and personal role and the nature of resource extraction tasks (Nettle 2011).

4.6.1.5 Evolutionary Aspects of the Agreeableness Trait

Agreeableness is related to differences in motivation to cooperate versus acting selfishly in resource conflicts, and also reflects individual differences in the tenacity of goal pursuit under distracting circumstances (Denissen and Penke 2008a). Agreeableness can be interpreted in two major ways: on the one hand, from the aspect of fostering intimate family relationships and parental investment, and from a dispositional aspect, with a role in human cooperative behaviour, willingness to cooperate, acting cooperatively vs. acting competitively, trust vs. self-interest, and coordination vs. opposition of joint interests, with high scorers more likely to avoid interpersonal conflicts but also to be stuck in prisoners' dilemma games (Denissen and Penke 2008a). It shows correlation with empathy, which is in turn correlated with theory of mind and awareness of the mental states of others, and trust, while its absence is associated with antisocial traits and antisocial personality disorder. In highly social species ability to understand others' mental states is highly advantageous, as it facilitates interactions and cooperation, thus leading to the avoidance of violence and interpersonal hostility. However, excessive or unconditional trust also isn't adaptive. Furthermore, high agreeableness may lead to excessive attention to needs and interests of others in expense of one's own well-being and needs. Agreeableness is also negatively related to creativity and achieving higher status. Thus investing positively in others does not maximise fitness, and while empathic cognitive style is adaptive in social species, its costs are exploitation or inattention to personal material or fitness needs and gains. Also, a population where cooperativeness and agreeableness are widespread, offers large advantages for less agreeable cheaters or sociopaths for exploitation, so frequency-dependent advantages to agreeableness would maintain a mixture of phenotypes at an equilibrium (Nettle 2006, 2011).

Thus while benefits of high agreeableness include increased likelihood of cooperative ventures and engaging in harmonious alliances, its risks include failure to maximise personal returns and falling victim to cheaters (Nettle 2011). While high agreeableness is associated with investment in cooperative joint ventures and harmony in interpersonal relationships which is beneficial in human societies, and selection favours coordination with others, in certain contexts this can be counterprofitable with increased benefit for individuals fending themselves off. Whether increased or decreased level of agreeableness is adaptive depends on the available resources as well on the local social structure (Nettle 2007).

4.6.2 Evolutional Aspects of the Traits and Temperaments of the Psychobiological Model of Personality

Unlike the five-factor personality models which were developed exclusively based on a lexical approach, the Cloninger taxonomy of personality was developed to reflect the psychobiological aetiology of personality components, incorporating results of genetic, longitudinal developmental as well as neuroanatomical and pharmacological studies on behaviour and behaviour conditioning, and human and animal learning studies with psychometric research. The original four temperamental dimensions, novelty seeking, harm avoidance, reward dependence and persistence were theoretically associated with independent neurobiological systems developing on nonoverlapping genetic backgrounds, and each temperamental dimension has initially been associated with activity of specific neurotransmitter circuits (Verweij et al. 2010; Cloninger et al. 1993). The revised biopsychosocial model includes four temperamental and three character domains where temperaments are conceptualised as heritable biases in memory processes involving presemantic perceptual processing and encoding of visuospatial structural information and affective valence, processes functionally organised as independently varying brain systems aligned to monoaminergic systems for autonomic responses involved in activation, maintenance and inhibition of behaviour (Gillespie et al. 2003). Character traits on the other hand are conceptual memory biases involved in processing or conversion of sensory input into abstract symbols which translate into concepts of personal, social and universal identity (Gillespie et al. 2003). So while temperaments are involved in behavioural conditioning, character traits are in conceptual learning. Both temperaments and character traits are heritable, and they are likely to involve different brain systems evolving in different stages of evolution of learning in animals (Gillespie et al. 2003).

Heritability estimates for the temperamental traits measured by the Tridimensional Character Inventory (TCI) and the Tridimensional Personality Questionnaire (TPQ) are estimated between 30 and 60% (Gillespie et al. 2001, 2003; Keller et al. 2005) which magnitude corresponds to heritability of other personality traits. In spite of this, no specific genetic variants surviving replications have been identified, although meta-analyses indicated an association between *DRD4* encoding the D4 dopamine receptor and novelty seeking explaining a 3% variance (Munafo et al. 2008), while there are conflicting results concerning the association between harm avoidance and *5-HTTLPR* (Munafo et al. 2009; Minelli et al. 2011).

Temperamental dimensions in the biopsychosocial model carry a high likelihood of being related to fitness. More recently a genome-wide association study with SNP data from 8000 subjects indicated little, on average 7.2% variation in temperamental dimensions due to the combined effect of common, additive genetic variants across the genome, and concluded that common genetic variants do not contribute significantly to temperamental variation suggesting other mechanisms in its background, and that most heritable personality variation is due to rare variants and/or a combination of dominance and epistasis (Verweij et al. 2010). These findings are consistent with genetic variation in personality traits having been maintained by mutation-selection balance (Verweij et al. 2012).

While the current psychobiological model contains four temperamental and three character dimensions, only the temperament dimension of novelty seeking has been extensively examined from an evolutionary point of view.

4.6.2.1 Evolutionary Aspects of Novelty Seeking

Novelty seeking, linked to dopaminergic activity in the original psychobiological model (Cloninger et al. 1993), is associated with behaviour activation and initiation as well as with a tendency to react to novelty, cues of reward and cues of relief from punishment, and shows overlap with extraversion to a certain extent. In several studies genetic variation in the *DRD4* gene encoding the D4 dopamine receptor have been associated with both extraversion and novelty seeking, although the majority of studies focused on this latter (Golimbet et al. 2007).

Studies clearly indicate evolutionary drivers in the background of between-population variations observable in case of the *DRD4* exon III locus associated with novelty seeking. *DRD4* exon III allele frequency has been shown to exhibit significant geographical variation with striking differences described in studies comparing several populations (Chen et al. 1999; Gören 2016; Matthews and Butler 2011), and an association has been reported between this global distribution of variants and variation in novelty-seeking levels, which also provides insight on the evolutionary advantages of differing levels of novelty seeking and adaptiveness of between-individual personality variation as opposed to a single optimal level in general.

Adaptive nature of *DRD4* variants in migratory societies appears important in geographic areas and environmental conditions unsuitable for sedentary practice. Hyperactivity, inattention, impulsiveness, response-ready traits and explorative behaviour associated with novelty seeking may be adaptive in rapidly changing, hostile and resource-depleted environments, and in societies and populations constantly on the move through unfamiliar environments (Chen et al. 1999). Greater distance from East Africa was associated with higher *DRD4* exon III heterozygosity, explaining 21% of variance, while the remaining nonexplained variation in heterozygosity is due to other factors including genetic drift, population-specific histories, and environmental conditions (Gören 2016). There is a clear geographic pattern of *DRD4* exon III allele frequencies, with a clear shift from the 4-repeat to the 7-repeat allele since the exodus from Africa (Gören 2016). The 4-repeat allele is observed in every population suggesting that it is the ancestral variant (Wang et al. 2004). The 7-repeat allele is the second most observed predominantly in American and also European populations, and nearly absent in Asian populations, while the 2-repeat variant is frequent in Asian and Oceanic populations and uncommon in Africa, suggesting that the observable polymorphism in the *DRD4* gene may be the result of a recent mutational event of positive selection after the exodus from East Africa 40–50,000 years ago (Matthews and Butler 2011). That is, migratory distance increases the frequency of long allele variants with 6- and 7-repeats; however, frequency of an 8-repeat variant is not correlated with distance from Africa but rather with local biogeographic conditions. Altogether, research on the worldwide distribution of *DRD4* exon III allele frequencies in a large sample of indigenous populations clearly indicated that migratory distance from the east of Africa, considered as the origin of the development of humankind, and population-specific biogeographic factors including latitude, suitability of land for agriculture, pasture land and terrain ruggedness all contribute to between-population variance in *DRD4* polymorphisms and are most likely to be the key selective forces defining the appearance and distributions of its variants (Gören 2016). In harsh environments or where resources are scarce which are characteristic of hunter-gatherer societies, cooperation, strong pair bonds and family ties, and investment from both parents are necessary for survival and to secure effective reproduction which contributes to the retainment and maintenance of more ancestral and risk-averse *DRD4* variants (Harpending and Cochran 2002). On the other hand, the relatively evolutionarily newer 7-repeat allele of *DRD4* which has been associated with response-ready and extraverted novelty seeking shows a higher frequency in the western hemisphere including Europe and America compared to Asia (Chang et al. 1996). Specific circumstances, where this type of novelty-seeking behaviour is advantageous, such as migrating to new environments versus settling down in less harsh and more luxurious and resource-rich conditions as in agricultural and more modern societies, appear to have favoured selection of this novel and more risk-seeking allele, which also contributes to increased sexual promiscuity as well as intrasexual competition (Penke and Jokela 2016). Also in present day nutritionally stressed Ariaal population living in Kenya, DRD4 long alleles related to increased novelty seeking have been associated with increased body mass in those leading a nomadic lifestyle as this trait is advantageous under such conditions; however, in the settled Ariaal populations, where such traits are disadvantageous, long DRD4 alleles showed a significant association with lower body mass (Eisenberg et al. 2008). Human behaviour including inattention, impulsivity and risk-taking advantageous in hostile or unfamiliar environments during constant migration is less suitable in modern societies which require a minimum of hierarchical organisation and predictability in individual social outcomes (Williams and Taylor 2006), so focus on *DRD4* exon III variation also gained importance as possibly playing a role in ADHD and impulsiveness. It must also be noted that the hypothesis on natural selection operating on DRD4 is not supported by population genetic evidence yet.

4.6.2.2 Evolutionary Aspects of Other Temperamental Dimensions in the Psychobiological Model of Temperament and Character: Harm Avoidance, Reward Dependence and Persistence

Harm avoidance, linked to serotonergic activity and associated with acquired behavioural inhibition, expresses a tendency to react to aversive stimuli (Cloninger et al. 1993). There appears to be a straightforward association between harm avoidance and survival and reproduction, with extremely low levels leading to high risk of injury and death, whereas extremely high levels leading to extreme shyness and withdrawal severely impairing chances of survival and reproduction through diminished exploration and exploitation of resources and decreased success in mate acquisition, suggesting that different environmental conditions, depending on their inherent risk-advantage constellations, favour different manifestations along the harm avoidance spectrum.

Reward dependence, linked to noradrenergic activity, plays a role in behavioural maintenance and is associated with continuation of behaviour previously associated either with reward or with relief from punishment. Those showing high reward dependence are ambitious, sociable, and sympathetic, and are also more likely to recognise socially relevant cues, and are therefore effective communicators, show genuine care for others, and have warm social relations which increases fitness in social species. However, those with high reward dependence, as a cost are also more socially dependent and also have an increased capacity for dependence towards behaviours, which involve rewarding cues, and are characterised by behaviours that are easily influenced by such cues independently of their advantage with respect to survival and reproduction. Those who show low reward dependence, on the other hand, are more nonconformist and independent and reserved with respect to revealing their feelings and are consequentially socially detached and more insensitive to social cues and unmotivated to please others and respond to cues for immediate gratification, leading to social withdrawal and dissocial or antisocial behaviour, detachment and coldness in interpersonal relationships. Persistence reflects the nature and strength of responses to frustration and partial reinforcement in abandoning on maintaining behaviours in spite of these. Frustration and consequential cessation of behaviour may be adaptive and advantageous as a reduction in effort that is disproportionate for the current reward and thus is likely to be beneficial in rapidly changing environments making it possible to alternate behaviours more rapidly until finding the one most fruitful in case of the given environmental conditions (Cloninger 1994).

4.6.2.3 Character Traits in the Psychobiological Model of Personality

Even less attention was paid from an evolutionary or genetic aspect to the character traits of the biopsychosocial model of personality, in part because it is postulated that learning and social influences play a relatively larger part in their development compared to genetic or biological ones, and there is evidence of common environmental influence from recent research in their development. The self-directedness dimension spans between being responsible, purposeful, resourceful, self-accepting and disciplined at the high end of the trait vs. being blaming, aimless, inept, vain and undisciplined. In case of cooperativeness, high levels are manifested in being empathic, tender-hearted, compassionate, helpful, whereas those with low levels are intolerant, insensitive, hostile, revengeful and opportunistic. In case of self-transcendence, high levels are associated with self-forgetfulness, transpersonality, spirituality, enlightedness and being idealistic vs. low manifestations including being unimaginative, materialistic, possessive, controlling and practical (Cloninger 1994). A previous large sample study in Australian twins indicated that familial aggregation for the character dimensions could be explained by additive genetic action alone in spite of the hypotheses that shared environmental effects would account for a large proportion of character trait variance attributed at least partly to social learning. Heritability explained 26%, 37% and 10% of additive genetic variance in self-directedness, cooperativeness and self-transcendence, respectively, and almost all non-shared environmental variance was unique to each character dimension (Gillespie et al. 2003). A twin study involving monozygotic and dizygotic pairs

(Lester et al. 2016) found varying proportion of a shared environmental component in case of self-directedness and cooperativeness subscale components, while the shared environmental component was relatively stable in case of self-transcendence. In a subsequent twin study, however, support for environmental effects was reported in the character variability in case of adolescents reflecting initial concepts and subsequent reports on the effect of cultural and rearing influences in character development (Lester et al. 2016). The proportion of shared environmental components showed variation in case of the subscales within self-directedness and cooperativeness higher-order traits, while it was stable in case of self-transcendence (Lester et al. 2016).

4.7 Conclusion

There has recently been increased interest in the evolutionary aspects of personality and especially interindividual personality differences. It appears that in spite of the obvious disadvantages of the extreme ends of personality and temperamental trait dimensions encountered in clinical practice, there is not one mid-range optimum manifestation but several optimal levels carrying advantages especially in interaction with specific environmental settings. The appearance of such multiple optima is likely to be the result of fluctuating selection pressures.

There are several advantages of considering evolutionary aspects of personality. Integration with evolutionary theory may help to enhance and advance personality models by fostering prediction concerning the mechanisms which govern personality, and widen our understanding of the function of personality traits, and how these traits provide means of adaptation and how they lead to the development of pathology. When it comes to personality, we understand how it works but not why it works, why the architecture of personality is structured in the way it is, and what mechanisms and forces gave rise to the basic sets of personality traits and temperaments that operate in humans today (Figueredo et al. 2009). Evolutionary theory could also help to disentangle the effects of environmental influences both in shaping the evolution of personality and in determining its reactions today, and why and how they are responsive to the environment, and how epigenetic mechanisms may have an influence on them. In summary, evolutionary psychology may bring us closer to delineating the adaptive function of human personality.

References

Bouchard TJ Jr, Loehlin JC (2001) Genes, evolution, and personality. Behav Genet 31(3):243–273

Buss DM (1991) Evolutionary personality psychology. Annu Rev Psychol 42:459–491. https://doi.org/10.1146/annurev.ps.42.020191.002331

Buss DM (1999) Adaptive individual differences revisited. J Pers 67(2):259–264. https://doi.org/10.1111/1467-6494.00055

Buss DM, Greiling H (1999) Adaptive individual differences. J Pers 67(2):209–243. https://doi.org/10.1111/1467-6494.00053

Buss AH, Plomin R (1984) Temperament: early developing personality traits. Erlbaum, Hillsdale, NJ

Chang F-M, Kidd JR, Livak KJ, Pakstis AJ, Kidd KK (1996) The world-wide distribution of allele frequencies at the human dopamine D4 receptor locus. Hum Genet 98(1):91–101. https://doi.org/10.1007/s004390050166

Chen C, Burton M, Greenberger E, Dmitrieva J (1999) Population migration and the variation of dopamine D4 receptor (DRD4) allele frequencies around the globe. Evol Hum Behav 20(5):309–324. https://doi.org/10.1016/s1090-5138(99)00015-x

Cloninger CR (1994) The genetic structure of personality and learning: a phylogenetic model. Clin Genet 46(1 Spec):124–137

Cloninger CR (2009) Evolution of human brain functions: the functional structure of human consciousness. Aust N Z J Psychiatry 43(11):994–1006. https://doi.org/10.3109/00048670903270506

Cloninger CR, Svrakic DM, Przybeck TR (1993) A psychobiological model of temperament and character. Arch Gen Psychiatry 50(12):975–990

Cosmides L, Tooby J (1995) From evolution to adaptations to behavior. toward an integrated evolutionary psychology. In: Wong R (ed) Biological perspectives on motivated activities. Ablex Pub., Norwood NJ, pp 11–74

Costa PT, Terracciano A, McCrae RR (2001) Gender differences in personality traits across cultures: robust and surprising findings. J Pers Soc Psychol 81(2):322–331. https://doi.org/10.1037/0022-3514.81.2.322

Denissen JJA, Penke L (2008a) Motivational individual reaction norms underlying the five-factor model of personality: First steps towards a theory-based conceptual framework. J Res Pers 42(5):1285–1302. https://doi.org/10.1016/j.jrp.2008.04.002

Denissen JJA, Penke L (2008b) Neuroticism predicts reactions to cues of social inclusion. Eur J Personal 22(6):497–517. https://doi.org/10.1002/per.682

DeYoung CG, Quilty LC, Peterson JB (2007) Between facets and domains: 10 aspects of the big five. J Pers Soc Psychol 93(5):880–896. https://doi.org/10.1037/0022-3514.93.5.880

Dingemanse NJ, Kalkman VJ (2008) Changing temperature regimes have advanced the phenology of Odonata in the Netherlands. Ecol Entomol 33(3):394–402. https://doi.org/10.1111/j.1365-2311.2007.00982.x

Ebstein RP (2006) The molecular genetic architecture of human personality: beyond self-report questionnaires. Mol Psychiatry 11(5):427–445. https://doi.org/10.1038/sj.mp.4001814

Eisenberg DT, Campbell B, Gray PB, Sorenson MD (2008) Dopamine receptor genetic polymorphisms and body composition in undernourished pastoralists: an exploration of nutrition indices among nomadic and recently settled Ariaal men of northern Kenya. BMC Evol Biol 8:173. https://doi.org/10.1186/1471-2148-8-173

Eyre-Walker A (2010) Evolution in health and medicine Sackler colloquium: genetic architecture of a complex trait and its implications for fitness and genome-wide association studies. Proc Natl Acad Sci U S A 107(Suppl 1):1752–1756. https://doi.org/10.1073/pnas.0906182107

Figueredo AJ, Sefcek JA, Vasquez G, Brumbach BH, King JE, Jacobs WJ (2005) Evolutionary personality psychology. In: Buss DM (ed) The handbook of evolutionary psychology. John Wiley & Sons, Hoboken, NJ, pp 851–877

Figueredo AJ, Gladden P, Vásquez G, Wolf PSA, Jones DN (2009) Evolutionary theories of personality. In: Corr PJ, Matthews G (eds) The Cambridge handbook of personality psychology. Cambridge University Press, Cambridge, pp 265–274

Friedman HS, Tucker JS, Schwartz JE, Tomlinson-Keasey C, Martin LR, Wingard DL, Criqui MH (1995) Psychosocial and behavioral predictors of longevity. The aging and death of the "termites". Am Psychol 50(2):69–78

Geary DC (2005) The origin of mind : evolution of brain, cognition, and general intelligence, 1st edn. American Psychological Association, Washington, DC

Genetics of Personality Consortium, de Moor MH, van den Berg SM, Verweij KJ, Krueger RF, Luciano M, Arias Vasquez A, Matteson LK, Derringer J, Esko T, Amin N, Gordon SD, Hansell NK, Hart AB, Seppala I, Huffman JE, Konte B, Lahti J, Lee M, Miller M, Nutile T, Tanaka

T, Teumer A, Viktorin A, Wedenoja J, Abecasis GR, Adkins DE, Agrawal A, Allik J, Appel K, Bigdeli TB, Busonero F, Campbell H, Costa PT, Davey Smith G, Davies G, de Wit H, Ding J, Engelhardt BE, Eriksson JG, Fedko IO, Ferrucci L, Franke B, Giegling I, Grucza R, Hartmann AM, Heath AC, Heinonen K, Henders AK, Homuth G, Hottenga JJ, Iacono WG, Janzing J, Jokela M, Karlsson R, Kemp JP, Kirkpatrick MG, Latvala A, Lehtimaki T, Liewald DC, Madden PA, Magri C, Magnusson PK, Marten J, Maschio A, Medland SE, Mihailov E, Milaneschi Y, Montgomery GW, Nauck M, Ouwens KG, Palotie A, Pettersson E, Polasek O, Qian Y, Pulkki-Raback L, Raitakari OT, Realo A, Rose RJ, Ruggiero D, Schmidt CO, Slutske WS, Sorice R, Starr JM, St Pourcain B, Sutin AR, Timpson NJ, Trochet H, Vermeulen S, Vuoksimaa E, Widen E, Wouda J, Wright MJ, Zgaga L, Porteous D, Minelli A, Palmer AA, Rujescu D, Ciullo M, Hayward C, Rudan I, Metspalu A, Kaprio J, Deary IJ, Raikkonen K, Wilson JF, Keltikangas-Jarvinen L, Bierut LJ, Hettema JM, Grabe HJ, van Duijn CM, Evans DM, Schlessinger D, Pedersen NL, Terracciano A, McGue M, Penninx BW, Martin NG, Boomsma DI (2015) Meta-analysis of genome-wide association studies for neuroticism, and the polygenic association with major depressive disorder. JAMA Psychiat 72(7):642–650. https://doi.org/10.1001/jamapsychiatry.2015.0554

1000 Genomes Project Consortium, Auton A, Brooks LD, Durbin RM, Garrison EP, Kang HM, Korbel JO, Marchini JL, McCarthy S, McVean GA, Abecasis GR (2015) A global reference for human genetic variation. Nature 526(7571):68–74. https://doi.org/10.1038/nature15393

Gillespie NA, Johnstone SJ, Boyce P, Heath AC, Martin NG (2001) The genetic and environmental relationship between the interpersonal sensitivity measure (IPSM) and the personality dimensions of Eysenck and Cloninger. Personal Individ Differ 31(7):1039–1051. https://doi.org/10.1016/s0191-8869(00)00200-2

Gillespie NA, Cloninger CR, Heath AC, Martin NG (2003) The genetic and environmental relationship between Cloninger's dimensions of temperament and character. Pers Indiv Differ 35(8):1931–1946. https://doi.org/10.1016/s0191-8869(03)00042-4

Golimbet VE, Alfimova MV, Gritsenko IK, Ebstein RP (2007) Relationship between dopamine system genes and extraversion and novelty seeking. Neurosci Behav Physiol 37(6):601–606. https://doi.org/10.1007/s11055-007-0058-8

Gören E (2016) The biogeographic origins of novelty-seeking traits. Evol Hum Behav 37(6):456–469. https://doi.org/10.1016/j.evolhumbehav.2016.04.005

Gurven M, von Rueden C, Stieglitz J, Kaplan H, Rodriguez DE (2014) The evolutionary fitness of personality traits in a small-scale subsistence society. Evol Hum Behav 35(1). https://doi.org/10.1016/j.evolhumbehav.2013.09.002

Harpending H, Cochran G (2002) In our genes. Proc Natl Acad Sci U S A 99(1):10–12. https://doi.org/10.1073/pnas.012612799

Haselton MG, Nettle D (2006) The paranoid optimist: an integrative evolutionary model of cognitive biases. Personal Soc Psychol Rev 10(1):47–66. https://doi.org/10.1207/s15327957pspr1001_3

Houle D (1998) How should we explain variation in the genetic variance of traits? Genetica 102/103:241–253. https://doi.org/10.1023/a:1017034925212

Johnson AM, Vernon PA, Feiler AR (2008) Behavioral genetic studies of personality: an introduction and review of the results of 50+ years of research. In: Boyle GJ, Matthews G, Saklofske DH (eds) The SAGE handbook of personality theory and assessment, 1st edn. SAGE, Los Angeles, CA, pp 145–173

Jokela M, Kivimaki M, Elovainio M, Keltikangas-Jarvinen L (2009) Personality and having children: a two-way relationship. J Pers Soc Psychol 96(1):218–230. https://doi.org/10.1037/a0014058

Keller MC, Miller G (2006) Resolving the paradox of common, harmful, heritable mental disorders: which evolutionary genetic models work best? Behav Brain Sci 29(4):385–404.; discussion 405–352. https://doi.org/10.1017/S0140525X06009095

Keller MC, Coventry WL, Heath AC, Martin NG (2005) Widespread evidence for non-additive genetic variation in Cloninger's and Eysenck's personality dimensions using a twin plus sibling design. Behav Genet 35(6):707–721. https://doi.org/10.1007/s10519-005-6041-7

Lahey BB (2009) Public health significance of neuroticism. Am Psychol 64(4):241–256. https://doi.org/10.1037/a0015309
Lester N, Garcia D, Lundstrom S, Brandstrom S, Rastam M, Kerekes N, Nilsson T, Cloninger CR, Anckarsater H (2016) The genetic and environmental structure of the character sub-scales of the temperament and character inventory in adolescence. Ann General Psychiatry 15:10. https://doi.org/10.1186/s12991-016-0094-2
Lukaszewski AW, Roney JR (2011) The origins of extraversion: joint effects of facultative calibration and genetic polymorphism. Personal Soc Psychol Bull 37(3):409–421. https://doi.org/10.1177/0146167210397209
Lukaszewski AW, von Rueden CR (2015) The extraversion continuum in evolutionary perspective: a review of recent theory and evidence. Personal Individ Differ 77:186–192. https://doi.org/10.1016/j.paid.2015.01.005
MacDonald K (1995) Evolution, the five-factor model, and levels of personality. J Pers 63(3):525–567. https://doi.org/10.1111/j.1467-6494.1995.tb00505.x
MacDonald K (2012) Temperament and Evolution. In: Zentner M, Shiner RL (eds) Handbook of temperament. Guilford Press, New York, NY, pp 273–296
Matthews LJ, Butler PM (2011) Novelty-seeking DRD4 polymorphisms are associated with human migration distance out-of-Africa after controlling for neutral population gene structure. Am J Phys Anthropol 145(3):382–389. https://doi.org/10.1002/ajpa.21507
Minelli A, Bonvicini C, Scassellati C, Sartori R, Gennarelli M (2011) The influence of psychiatric screening in healthy populations selection: a new study and meta-analysis of functional 5-HTTLPR and rs25531 polymorphisms and anxiety-related personality traits. BMC Psychiatry 11:50. https://doi.org/10.1186/1471-244X-11-50
Mosing MA, Medland SE, McRae A, Landers JG, Wright MJ, Martin NG (2012) Genetic influences on life span and its relationship to personality: a 16-year follow-up study of a sample of aging twins. Psychosom Med 74(1):16–22. https://doi.org/10.1097/PSY.0b013e3182385784
Munafo MR, Yalcin B, Willis-Owen SA, Flint J (2008) Association of the dopamine D4 receptor (DRD4) gene and approach-related personality traits: meta-analysis and new data. Biol Psychiatry 63(2):197–206. https://doi.org/10.1016/j.biopsych.2007.04.006
Munafo MR, Freimer NB, Ng W, Ophoff R, Veijola J, Miettunen J, Jarvelin MR, Taanila A, Flint J (2009) 5-HTTLPR genotype and anxiety-related personality traits: a meta-analysis and new data. Am J Med Genet B 150B(2):271–281. https://doi.org/10.1002/ajmg.b.30808
Neeleman J, Sytema S, Wadsworth M (2002) Propensity to psychiatric and somatic ill-health: evidence from a birth cohort. Psychol Med 32(5):793–803. https://doi.org/10.1017/S0033291702005901
Nettle D (2005) An evolutionary approach to the extraversion continuum. Evol Hum Behav 26(4):363–373. https://doi.org/10.1016/j.evolhumbehav.2004.12.004
Nettle D (2006) The evolution of personality variation in humans and other animals. Am Psychol 61(6):622–631. https://doi.org/10.1037/0003-066X.61.6.622
Nettle D (2007) Traits and trade-offs are an important tier. Am Psychol 62(9):1074–1075. https://doi.org/10.1037/0003-066X.62.9.1074
Nettle D (2011) Evolutionary persepctives on the five factor model of personallity. In: Buss DM, Hawley PH (eds) The evolution of personality and individual differences. Oxford University Press, Oxford, p xix, 498
Nettle D, Clegg H (2006) Schizotypy, creativity and mating success in humans. Proc Biol Sci 273(1586):611–615. https://doi.org/10.1098/rspb.2005.3349
Nettle D, Penke L (2010) Personality: bridging the literatures from human psychology and behavioural ecology. Philos Trans R Soc B 365(1560):4043–4050. https://doi.org/10.1098/rstb.2010.0061
Penke L, Jokela M (2016) The evolutionary genetics of personality revisited. Curr Opin Psychol 7:104–109. https://doi.org/10.1016/j.copsyc.2015.08.021
Penke L, Denissen JJA, Miller GF (2007) The evolutionary genetics of personality. Eur J Personal 21(5):549–587. https://doi.org/10.1002/per.629

Plomin R, McClearn GE, Smith DL, Vignetti S, Chorney MJ, Chorney K, Venditti CP, Kasarda S, Thompson LA, Detterman DK et al (1994) DNA markers associated with high versus low IQ: the IQ Quantitative Trait Loci (QTL) Project. Behav Genet 24(2):107–118

Polderman TJ, Benyamin B, de Leeuw CA, Sullivan PF, van Bochoven A, Visscher PM, Posthuma D (2015) Meta-analysis of the heritability of human traits based on fifty years of twin studies. Nat Genet 47(7):702–709. https://doi.org/10.1038/ng.3285

Power RA, Pluess M (2015) Heritability estimates of the Big Five personality traits based on common genetic variants. Transl Psychiatry 5:e604. https://doi.org/10.1038/tp.2015.96

Roberts BW, Kuncel NR, Shiner R, Caspi A, Goldberg LR (2007) The power of personality: the comparative validity of personality traits, socioeconomic status, and cognitive ability for predicting important life outcomes. Perspect Psychol Sci 2(4):313–345. https://doi.org/10.1111/j.1745-6916.2007.00047.x

Rothbart MK (2016) Temperament, development, and personality. Curr Dir Psychol Sci 16(4):207–212. https://doi.org/10.1111/j.1467-8721.2007.00505.x

Samuels J, Bienvenu OJ, Cullen B, Costa PT Jr, Eaton WW, Nestadt G (2004) Personality dimensions and criminal arrest. Compr Psychiatry 45(4):275–280. https://doi.org/10.1016/j.comppsych.2004.03.013

Sandberg R, Yasuda R, Pankratz DG, Carter TA, Del Rio JA, Wodicka L, Mayford M, Lockhart DJ, Barlow C (2000) Regional and strain-specific gene expression mapping in the adult mouse brain. Proc Natl Acad Sci U S A 97(20):11038–11043

Schmitt DP (2004) The Big Five related to risky sexual behaviour across 10 world regions: Differential personality associations of sexual promiscuity and relationship infidelity. Eur J Personal 18(4):301–319. https://doi.org/10.1002/per.520

South SC, Jarnecke AM, Vize CE (2018) Sex differences in the Big Five model personality traits: a behavior genetics exploration. J Res Pers 74:158–165. https://doi.org/10.1016/j.jrp.2018.03.002

Tooby J, Cosmides L (1990) On the universality of human nature and the uniqueness of the individual: the role of genetics and adaptation. J Pers 58(1):17–67

Verweij KJ, Zietsch BP, Medland SE, Gordon SD, Benyamin B, Nyholt DR, McEvoy BP, Sullivan PF, Heath AC, Madden PA, Henders AK, Montgomery GW, Martin NG, Wray NR (2010) A genome-wide association study of Cloninger's temperament scales: implications for the evolutionary genetics of personality. Biol Psychol 85(2):306–317. https://doi.org/10.1016/j.biopsycho.2010.07.018

Verweij KJ, Yang J, Lahti J, Veijola J, Hintsanen M, Pulkki-Raback L, Heinonen K, Pouta A, Pesonen AK, Widen E, Taanila A, Isohanni M, Miettunen J, Palotie A, Penke L, Service SK, Heath AC, Montgomery GW, Raitakari O, Kahonen M, Viikari J, Raikkonen K, Eriksson JG, Keltikangas-Jarvinen L, Lehtimaki T, Martin NG, Jarvelin MR, Visscher PM, Keller MC, Zietsch BP (2012) Maintenance of genetic variation in human personality: testing evolutionary models by estimating heritability due to common causal variants and investigating the effect of distant inbreeding. Evolution 66(10):3238–3251. https://doi.org/10.1111/j.1558-5646.2012.01679.x

Wang E, Ding YC, Flodman P, Kidd JR, Kidd KK, Grady DL, Ryder OA, Spence MA, Swanson JM, Moyzis RK (2004) The genetic architecture of selection at the human dopamine receptor D4 (DRD4) gene locus. Am J Hum Genet 74(5):931–944. https://doi.org/10.1086/420854

Williams J, Taylor E (2006) The evolution of hyperactivity, impulsivity and cognitive diversity. J R Soc Interface 3(8):399–413. https://doi.org/10.1098/rsif.2005.0102

Workman L, Reader W (2014) Evolutionary psychology: an introduction, 3rd edn. Cambridge University Press, Cambridge

Zietsch BP, Verweij KJ, Bailey JM, Wright MJ, Martin NG (2010) Genetic and environmental influences on risky sexual behaviour and its relationship with personality. Behav Genet 40(1):12–21. https://doi.org/10.1007/s10519-009-9300-1

Psychobiology of Sexuality

5

Fotini Ferenidou, Loucas Athanasiadis,
and Kostas N. Fountoulakis

5.1 Introduction

Sexuality is widely seen as an intrinsic component of human experience and psychological function and a major aspect of human behavior. It includes biological, psychological-interpersonal, and sociocultural aspects and serves reproduction (biological target) but also other aims (including pleasure, communication, etc.) and develops through the various stages of human psychosexual development. Sexual attitudes and the approach to sexual issues such as female sexuality and sexual orientation (classified as a psychiatric disorder in the past) are dependent on religious and sociocultural factors; they change through centuries; however, this does not always happen toward a more tolerant direction. Gender identity, gender role, and sexual orientation are important aspects of one's identity. Sex refers to the biological characteristics.

There are still significant gaps in our knowledge, and since research on sexuality has been mainly male-oriented, these gaps exist especially regarding female sexuality. Sexuality is shaped in a multifactorial way; however, the psychological and biological components seem to play a major role. Therefore the study of the psychobiology of sexuality can be a very interesting and productive approach to the field.

F. Ferenidou (✉)
Department of Psychiatry, Aeginiteion Hospital, Athens, Greece

L. Athanasiadis
Faculty of Medicine, School of Health Sciences, Aristotle University of Thessaloniki, Thessaloniki, Greece

K. N. Fountoulakis
3rd Department of Psychiatry, Faculty of Medicine, School of Health Sciences, Aristotle University of Thessaloniki, Thessaloniki, Greece

K. N. Fountoulakis, I. Nimatoudis (eds.), *Psychobiology of Behaviour*,
https://doi.org/10.1007/978-3-030-18323-3_5

5.2 Sexual Differentiation of the Brain

There are both similarities and different aspects when human sexuality is compared to the sexual behavior of various nonhuman primates. In humans sexual activity may occur at any time during the menstrual cycle and regardless of age. Central neural control of sexual behavior involves brain regions, which are linked to motivation, consummation, and reward and are similar to those brain regions involved in the majority of primates. Many observational and experimental studies on the basic cellular mechanisms involved in sexuality come from animals. However, when such comparisons are made, the different epigenetic, cognitive, and sociocultural factors should be taken into account since they may also play an important role in humans (Martin 2007; Salonia et al. 2010).

The presence—or absence—of the Sry gene on the Y chromosome codifies the testis-determining factor (TDF) and initiates gene expression and protein products toward the development of testis, or accordingly the ovaries, from the bipotential gonads (Salonia et al. 2010). During the "sensitive period"—a restricted developmental period in male and female animals—the brain is sexually differentiated. The critical hormone for masculinization of the brain is estradiol (E2) which derives from testicular testosterone and leads to the male differentiation of the rodent brain, while male differentiation of the body tissues occurs due to dihydrotestosterone (DHT), through the effect of 5a-reductase on testosterone. In primates, including humans, this procedure begins during the second trimester of gestation and is largely prenatal, while the critical masculinizing hormone is testosterone (T). The "organizational/activational hypothesis" supports the early action of gonadal steroids which organize the main neural architecture which regulates sexual behavior (Phoenix et al. 1959). This behavior will be activated in adulthood after the secretion of steroids in response to appropriate stimuli. Therefore, shortly after birth and the formulation of brain differentiation during the sensitive period, the testis ceases androgen production. Only during puberty will these circuits be activated and hormones will permit sex-specific sexual behavior. It is hypothesized, but not proven, that male and female neuronal circuits are largely the same, but critical nodes are weighted differently (Salonia et al. 2010).

There seems to exist greater connectivity between the cerebral hemispheres in females, and this has been indicated since the anterior commissure, which connects anteriorly two-thirds of the right and left temporal neocortices and the posterior part of the orbital aspect of the frontal lobes (Demeter et al. 1988), was found to be 12% larger in women than in men (Swaab 2002).

In all animals, including humans, the hypothalamus is the main brain structure involved in sexual behavior, exerting an effect on the anterior lobe of the pituitary gland (hypophysis). The hormones that are secreted from the pituitary gland act upon the genitalia and specifically on the ovaries and on the testes, secreting estrogen and testosterone, respectively. Finally, these hormones affect the hypothalamus controlling sexual behavior.

The sexually dimorphic nucleus of the preoptic area (SDN-POA) has first been described in the rat brain (Gorski et al. 1978) and is located between the dorsolateral supraoptic nucleus (SON) and the mediorostral pole of the paraventricular nucleus (PVN) in the hypothalamus. Interestingly enough, it seems to be three to eight times larger in the male than in the female rat, due to different perinatal steroid levels. Sexual differentiation of the SDN-POA in humans occurs after 4 years of age according to sex, due to a decrease in both cell number and volume in women, while in men both cell number and volume remain the same until the fifth decade, where the number of cells also decreases. The SDN-POA in the young adult human brain is twice as large in men than in women on one side and contains twice as many cells (Swaab 2002). Although unclear, there are indications that the SDN-POA area is involved in sexual behavior since lesion experiments in rats in this area demonstrate its involvement in aspects of male sexual behavior, i.e., recognition of sensory stimulus as appropriate sexual targets, mounting, intromission, and ejaculation (Swaab 2002).

The median preoptic area (mPOA) is a critical hormonally sensitive region, sharing connections with the ventromedial nucleus (VMN) and other brain regions relevant to mating, and seems to be important regarding motive and completion of sexual encounter, based on lesions of the area in animal studies (Vaidakis 2005). The POA is considered the critical area for male and female sexual behavior, mainly through inhibitory mechanisms in its expression (Pfaff et al. 2009; Salonia et al. 2010). Interestingly enough, sexual intercourse seems to increase intrinsic opioids, and especially beta-endorphin, in the mPOA of the male rat. There are indications that beta-endorphin in the mPOA inhibits sexual activity, probably through the feeling of saturation (Vaidakis 2005).

The VMN of the hypothalamus is quite an essential brain region for receptive sexual behavior in female rodents, since its estrogen receptor-expressing neurons project to the estrogen-sensitive neurons of the midbrain and then to the spinal cord activating important back muscle groups. The sensory information is then transferred back to the brain-activating regions that are linked to reward, such as the ventral tegmental area and the nucleus accumbens. Additionally, the amygdala, receiving important olfactory information, also sends signals to the VMN (Pfaff et al. 2009; Salonia et al. 2010). Sexual dimorphism exists also in the VMN, since males have two to three more dendritic spine synapses that are also longer and branch more frequently than in the female brain, maybe due to higher steroid levels in male rodents during the critical period of brain development (Salonia et al. 2010; Schwarz et al. 2008).

The interstitial nucleus of the anterior hypothalamus-2 and interstitial nucleus of the anterior hypothalamus-3 (*INAH-2* and *INAH-3*) seem to have differentiations in males and females regarding its size and its secretions. More specifically they were found larger in the male than in the female brain, and they seem to have a continuous secretory pattern in men and a more "cyclic" one in women (Swaab 2002). Moreover, regarding hypothalamic peptides, oxytocin seems to be more representative of the female circuitry and vasopressin of the male one (Cuzin 2015).

The bed nucleus of the stria terminalis (*BNST*), situated at the junction of the hypothalamus, septum, and amygdala, contains nuclear androgen receptors and seems to play an important role in rodent sexual behavior (Fernandez-Guasti et al.

2000; Swaab 2002). Sexual dimorphism of this nucleus has not been detected so far, except for the volume of a region of the BNST called "the darkly staining posteromedial BST" (BNST-dspm) which is 2.5 times larger in men than in women (Allen and Gorski 1990).

The sexual differentiation of the hypothalamus (especially of the SDN-POA and the BNST) becomes apparent between 2 years of age and puberty, following earlier developmental periods (mid-pregnancy or neonatal period). Testosterone—deriving from estrogens through aromatization—is a major mechanism for androgenization, but the development of human male gender identity and male heterosexuality needs also the direct action of androgens on the brain, while other factors other than hormones (e.g., compounds that influence hormone or neurotransmitter metabolism such as alcohol and some drugs, kinases, enzymes, receptors) may affect sexual differentiation. Finally, continuous expression of critical genes (such as the Sry and the ZFY) may be necessary in order to maintain sex-specific structural or functional properties of differentiated male neurons (Swaab 2002).

Central nervous system dimorphisms are a biological basis for the differences and disparities in sexual desire, expression, content, and intensity of sexual thoughts and fantasies, erotic dreams, and daydreams, as well as differences in the perception and expression of arousal and orgasm (Cuzin 2015). Perhaps indicative of the human brain might be a spectrum of various degrees of masculinization and feminization and the potential for any individual to be divergent, following the variety of the steroid hormone action toward cellular mechanisms in the developing brain (Salonia et al. 2010).

5.3 Sexual Orientation

Sexual orientation may be described as a person's pattern of romantic, emotional, and sexual attraction as well as the gender/genders of the people the person is attracted to. Evidence suggests a genetic influence on sexual orientation, although environmental factors seem also to explain variation in orientation (Dean 2012b). There seems to be a linkage between DNA markers on the X chromosome and male sexual orientation, while sex hormones during development also seem to have an influence on sexual orientation, as shown on the increased percentage of bisexual or homosexual orientation in girls with congenital adrenal hyperplasia (Hamer et al. 1993). Maternal stress, as well as other gestational and perinatal factors, has been connected to increased occurrence of homosexuality in boys and girls. Finally, some authors suggested that postnatal social or environmental factors may play a role in the development of sexual orientation, but to date there are no solid evidence in support of this (Swaab 2002).

Regarding brain morphology studies, minor differences have been found between heterosexual and homosexual men, specifically in the interstitial nuclei of the anterior hypothalamus (INAH), which is larger in heterosexual men, and in the anterior commissure, which is larger in homosexual men, although it is not clear if these findings are causal or consequential. More likely, the determinants of both sexual orientation and identity are multifactorial and not the sole consequence of nature or nurture, while most authors seem to agree that they are probably not a matter of individual choice (Dean 2012b).

Other differences which were found most notably in the amygdala, the hippocampus, and the prefrontal cortex (PFC), as well as in parts of the hypothalamus (Allen et al. 2003; Salonia et al. 2010), might explain some of the differences between male and female sexual behavior and orientation (Salonia et al. 2010). Interestingly enough, there are studies showing similarities between homosexual women and heterosexual men regarding the activation of certain nuclei (the dorsomedial and the paraventricular nucleus of the hypothalamus) when smelling an estrogen-like odor (Savic et al. 2001; Berglund et al. 2006). Similar congruent activity is shown between homosexual men and heterosexual women when smelling androgen-like odor (the ventromedial hypothalamus and POA) (Salonia et al. 2010; Savic et al. 2001; Savic et al. 2005). Moreover, female sexual orientation has been connected to symmetrical cerebral volumes, while male sexual orientation has been connected to rightward cerebral asymmetry. Additionally, the amygdala has differences according to sexual orientation, showing specifically more connections from the left than from the right in heterosexual women and homosexual men (female sexual orientation) and the opposite for individuals who have a male sexual orientation (Savic and Lindstrom 2008; Salonia et al. 2010).

Regarding animals, the SDN-POA is presumed to be involved in sexual orientation, as lesions in this area seem to cause a significant shift from male-typical pattern of sexual behavior to a more female-typical pattern (Swaab 2002). Its possible involvement in human sexual orientation has been suggested (Swaab 2002), as well as theories that homosexual men have a "female hypothalamus," although studies did not reveal any differences in the size or the cell numbers of this nucleus in relation to sexual orientation. A study based on a small sample of homosexual men found that the INAH-3 nucleus in these subjects is closer to female size, while on the contrary it is two times bigger in heterosexual men (Levay 1991).

Regarding the suprachiasmatic nucleus (SCN), which seems to be involved in sexual behavior, orientation, and reproduction both in animals and in humans, morphometric analysis in a small study sample revealed that its volume was 1.7 times larger and it contained 2.1 times as many cells in homosexual than heterosexual men. Additionally, the number of vasopressin-expressing neurons was increased in homosexual men, reflecting perhaps a difference that is formed during the early stages of brain development. Regarding the BST-dspm which, as mentioned before, is found larger in men than in women, no relationship was observed between its volume or somatostatin cell number and sexual orientation (Swaab 2002).

5.4 Sexual Identity

Sexual or gender identity is the individual's personal sense, and subjective experience, of their own gender (Dean 2012a). Some individuals may be also self-referred as non-binary. Research on this filed is limited and often presents controversial findings. Some evidence in the literature supports the involvement of multiple factors in its development, which may be biological (genes and their regulation, pre- and postnatal hormone levels), sociocultural (family upbringing, norms and expectations, sociocultural or religious requirements), as well as social learning factors (which assumes that gender identity is also developed through observing and copying gender-specific behaviors, which may then be rewarded or punished) (Dean 2012a). Interestingly enough, it has been recently proposed that gender identity disorders may be considered disorders of sexual differentiation limited to the central nervous system, without involvement of the reproductive tract (Meyer-Bahlburg 2010).

Regarding biological/genetic variables, only a few reports have found chromosomal abnormalities in transsexual people (Turan et al. 2000), while some indications exist that it is a disorder of the hypothalamic-pituitary-gonadal axis, considering the high frequency of polycystic ovary syndrome, oligomenorrhea, and amenorrhea in female-to-male transsexuals (Sadeghi and Fakhrai 2000; Swaab 2002). Congenital adrenal hyperplasia, characterized by high androgen levels during prenatal development, seems to be a risk factor for the development of gender identity disorders, as well as maternal use of some medicine (e.g., phenobarbital and diphantoin) during pregnancy which are assumed to alter steroid levels (Swaab 2002). These observations support the theory that intrauterine or perinatal factors, which influence sex hormone levels, may permanently affect gender identity, in interaction to the developing brain (Swaab 2002).

Interestingly enough, some evidence exist regarding brain morphology of transsexuals. Both male-to-female (MtF) and female-to-male (FtM) transsexuals have been reported to have sex reversal in terms of volume and cell number of sex-dimorphic nuclei, such as the central portion of the BNST (BNSTc) (Kruijver et al. 2000), the gray matter in the right putamen (Luders et al. 2009), and the interstitial nuclei 3 and 4 of the anterior hypothalamus (INAH 3 and INAH 4) (Garcia-Falgueras and Swaab 2008). More specifically, the BNSTc was found to be smaller (40% of the male volume and neuron number) in a small sample of MtF transsexual persons, suggesting that this size, formed during psychosexual development, may be part of a network that is involved in gender (the feeling of being male or female) (Kruijver et al. 2000; Swaab 2002). Moreover, gender-atypical brain activation patterns regarding the process of steroid-based odors and erotic stimuli have been demonstrated (Berglund et al. 2008; Gizewski et al. 2009). Finally, genetically based systemic sex hormone abnormalities that do not cause abnormalities of the reproductive tract, but influence the brain and its functions have been proposed, e.g., androgen receptor gene polymorphisms (Hare et al. 2009; Meyer-Bahlburg 2010).

5.5 Normal Sexual Function and the Sexual Response Cycle

Biological, psychological, and sociocultural factors influence human sexual function, while sexual behavior involves several behaviors such as motivation to seek a partner, evaluation of the appropriate cues, execution of the appropriate behavior, as well as processes which may include reward that derives from the experience and may reinforce the behavior to be repeated (Salonia et al. 2010). Many brain regions and neural nodes are activated in order to collect and process relevant information and organize the corresponding behavior. In general, high levels of steroid receptors are expressed in these critical brain regions, while a hormonally sensitive network is created (Salonia et al. 2010).

"Sexual response models" represent the interaction of the several factors that seem to be important during sexual response, explaining the emotional and physical changes that take place when a person participates in a sexually stimulating activity (Tripodi and Silvaggi 2013).

Masters and Johnson were the first to describe a human sexual response model which was analogous for men and women, involving excitement (arousal), plateau, orgasm, and resolution (known also as EPOR model, taken from the initials of the words) (Masters and Johnson 1966). In 1979, Kaplan added the concept of "desire" as a prerequisite and separate phase of the human sexual response cycle which now consisted of sexual desire, excitement (arousal), orgasm, and resolution (known also as DEOR model, taken from the initials of the words) (Kaplan 1979). Until recently this has been the main model in clinical practice, for the diagnosis of sexual dysfunctions and for research activities. Nevertheless, the presence of exact sequential sexual response phases in women (desire-arousal-orgasm-resolution) has been criticized and questioned (Basson et al. 2004). Women's conscious sexual motivation may be more related to a desire for emotional connection or the avoidance of negative consequences rather than by internal feelings of genital tension, or by sexual thoughts or fantasies, which seem to describe the meaning of sexual desire. On the other hand, sexual desire in men is more readily and more reliably accessed, is usually palpable, and has a direct association with the experience or sight of an erection and the feeling of subjective sexual arousal, while in women there isn't always an association between subjective arousal and genital vasocongestion (Leiblum 2007). Additionally, in females, the phases of sexual response (especially desire and arousal) are not always discreet since they may overlap (Kirana 2013a, b).

Research suggests that there are many differences in the arousal response between men and women. In many females, the sexual arousal precedes the conscious feelings of sexual desire, whereas in most men desire precedes arousal and is a prerequisite in order to achieve or maintain an erection (Leiblum 2007). Based on the above, a circular/dynamic sexual response model that depicts a different conceptualization of women's sexual function than the previous models was suggested (Basson et al. 2004). Spontaneous/intrinsic desire is not necessarily a prerequisite since their sexual response may begin for a number of other reasons, which may even not be sexual such as her willingness to be receptive (this has been termed responsive sexual desire). Motivation to be sexually active again is increased if the experienced sexual activity is physically or emotionally positive (Kirana 2013a, b).

Genital arousal in females does not necessarily reflect their stated sexual interest in the same way that it does for men and therefore is considered nonspecific. This is supported by studies that show that non-preferred sexual partners, or even nonhuman stimuli, may provoke physiological changes of the genital system (e.g., vasocongestion) in women, but not in men (Leiblum 2007; Salonia et al. 2010). An evolutionally perspective may explain this finding, since females are vulnerable to several forms of aggressive sexuality, and therefore a rapid vasocongestive response and concurrent lubrication protect their genitals (Chivers 2005). Physiological genital response in females has also been proposed to be an automatic reflex that precedes conscious processing of sexual stimuli and cues and can occur in the absence of subjective experience of sexual arousal (Chivers 2005). Concluding, solely genital response may be inaccurate to reflect women's sexual desire and sexual attractions (Salonia et al. 2010). Female sexuality has evolved, highly influenced by cultural, social, and interpersonal variables, which also play an important role in the nature of female sexual problems. Nowadays, it seems quite differentiated from male sexuality, while among other things, the influence of biological, anatomical, and hormonal deficits is not the same (Leiblum 2007).

5.5.1 Sexual Desire

Mammals and other species respond to hormonal and neurochemical changes that signal sexual desire and arousal (Goldstein et al. 2004). In humans this appropriate "internal state" of arousability is often referred to as "sex drive." Besides that, instinct, learning, and feedback involve several neural structures in order for the interaction with the external environment to take place. Additionally, neural mechanisms that assess the feedback of a sexual act as positive and rewarding may reinforce or not a similar behavior in the future (Goldstein et al. 2004).

Sexual desire is part of the reproduction instinct, including both biological and cognitive parameters. More specifically, it may be divided into three separate components: sexual drive, sexual motives, and sexual expectations. *Sexual drive* includes all the neuroendocrinological pathways that are involved and are mainly controlled by androgens. *Sexual motives* describe the individual reasons for which somebody wishes to engage in a sexual activity. This parameter is quite complex, since sexual motives are multiple, may be sexual or nonsexual (as mentioned before, especially regarding women's sexual behavior), and seem to differ between genders and individuals or throughout the lifespan. Sexual motive, as a component of sexual desire, is actually describing the reasons that partners approach each other for a sexual encounter. Finally, *sexual expectations* are supposed to be the cognitive dimension of sexual desire and refer to the needs that an individual is expecting to fulfill through sexual function.

The biological basis of human sexual desire seems to include, as mentioned before, cognitive (such as thoughts or fantasies) and affective parameters, as well as central neurophysiological mechanisms of arousal (Corona and Maggi 2012). It has been postulated that sexual excitation and sexual inhibition brain systems seem to be responsible for an individual's behavior related to sexual desire and arousal (two separate functions, which in the brain are often difficult to differentiate).

Hypothalamic and mesolimbic dopamine (DA) transmission is involved in sexual excitation, including three major systems (Corona and Maggi 2012):

- Diencephalic incertohypothalamic DA systems terminating in the medial preoptic area (mPOA) of the anterior hypothalamus.
- The mesolimbic and mesocortical DA system, which terminates in the nucleus accumbens and medial prefrontal cortex, respectively.
- The nigrostriatal system, which terminates in the striatum (caudate and putamen).

Initiation of sexual desire activates the limbic system both in males and in females, including sexual fantasies, sexual dreams, mental sexual arousal, as well as the neurovascular events responsible for body and genital reactions. Additionally, a problem at any level of the limbic system may cause sexual dysfunction in both sexes, affecting desire, central arousal, and socially appropriate sexual behavior (Cuzin 2015). The tuberoinfundibular DA system controlling hormone release from the anterior pituitary gland is also involved in the control of sexual desire. The mPOA is the central core of this pathway, projecting to the ventral tegmental area of the midbrain which contains DA cell bodies. These bodies project to various limbic and cortical regions, such as the prefrontal cortex, olfactory tubercle, nucleus accumbens, anterior cingulate cortex, lateral septum, and corticomedial amygdala (Corona and Maggi 2012).

Regarding the inhibition of desire and arousal, the prefrontal lobe seems to play an important role over the basic instinctual drives, while opioid, endocannabinoid, and serotonin systems are activated and block the normal excitatory pathway (Cuzin 2015). Hormones, particularly testosterone, are involved in the modulation of the proposed excitatory and inhibitory sexual desire systems. Testosterone (T) acts on androgen receptors (AR) which are present in several areas of the brain, such as the temporal and the preoptic areas, the hypothalamus, the amygdala, the midbrain, the frontal and prefrontal areas, and the cingulated gyrus (Corona and Maggi 2012).

Specifically regarding the male sexual desire, several cerebral areas that have been related to sexual drive are androgen related. For example, Brodmann area 24 (BA24) is a sexual dimorphic AR area, related to emotional behavior and arousal reaction, since it is activated during the demonstration of erotic films. Interestingly enough, when T is administered to hypogonadal men, it seems to increase blood perfusion in this area and in other areas (the midbrain, superior frontal gyrus, middle occipital gyrus) which are involved in the processing of visual stimuli (Corona and Maggi 2012). T is also modulating the strength of activation in the brain areas that are related to female sexual response (Levin et al. 2016). Regarding the role of other hormones on male sexual desire, data is unclear. For example, reduction of dihydrotestosterone (DHT) may or may not reduce sexual desire, while estrogens—although evidence is scarce—seem not to play an important role for normal male sexual behavior. Elevation of prolactin has a negative effect on sexual desire, since it induces a decrease in the secretion of luteinizing hormone (LH) and therefore a decrease in testosterone. Finally, elevation or reduction of cortisol levels may negatively affect sexual desire, through the modification of the hypothalamic-pituitary-gonadal axis and testosterone production (Corona and Maggi 2012).

Neuroimaging studies using visual sexual stimulation (VSS), with erotic or sexually explicit photos or film excerpts, reveal interesting results regarding brain regions that are activated during sexual response, especially in women (Levin et al. 2016). It has been reported that women activate the following areas of the brain when they watch erotic material: occipitotemporal cortex, superior parietal lobule, orbitofrontal cortex, inferior frontal gyrus, anterior cingulate cortex, anterior insula, ventral striatum, amygdala, thalamus, and hypothalamus. Interestingly enough, similar brain areas are activated during the first phases of sexual response (detection of sexual stimuli, sexual interest, attraction, and expectancy), which has been demonstrated through pheromone stimulation (Levin et al. 2016). The same brain areas are also activated when watching more vivid type of VSS (video instead of photos), as well as during the experience of genital sexual arousal and consequent body responses or during objective sexual arousal without genital response (Levin et al. 2016). This is not the case in males, where different brain areas seem to be activated when exposure to VSS is longer or more vivid, and this exposure leads to different affective or behavioral consequences with the analogous body response (Levin et al. 2016). VSS in males produce greater activations than in females in the amygdala and the hypothalamus, and visual stimuli seem to have fast access to primordial systems. This may reflect the presence of a higher "sensitivity" to sexual cues and stimuli in males, through a system which includes mainly the hypothalamus and the amygdala, and is more phylogenetically tuned toward this direction. This supports the theory that females take a more thoughtful approach regarding sexual encounters, while males follow a more instinctive mechanism through a different brain system (Salonia et al. 2010). Finally low gray matter density in these areas or enhanced activity in the prefrontal cortex has been related to hypoactive sexual desire disorder in women (Levin et al. 2016).

5.5.2 Male Sexual Arousal and Ejaculation

5.5.2.1 Arousal

Male sexual arousal resulting in penile erection takes place after central processing of several stimuli (tactile, visual, olfactory, and imaginative) (Gratzke et al. 2010). Neurotransmitters and specific CNS areas that are involved are not completely understood yet. Some brain anatomical areas that have a role in sexual function have been identified, including the medial amygdala, medial preoptic area (MPOA), paraventricular nucleus, periaqueductal gray, and ventral tegmentum (Gratzke et al. 2010). The spinal network consists of primary signals from the genitals, spinal interneurons, as well as sympathetic, parasympathetic, and somatic nuclei. Information from the periphery is gathered, eliciting erections, while it also receives supraspinal information (Gratzke et al. 2010; Giuliano and Rampin 2000).

The first level of reflex organization in sexual function is located in the lumbar spinal cord. Spinal reflexes can be regulated by descending signals from the midbrain and higher centers. A cascade of parasympathetic activity, reduced sympathetic activity, and somatic muscular support is activated by both local sensors in the penis and descending spinal signals (Saenz de Tejada et al. 2004a, b). The thalamus seems to receive significant sensory inputs from the male genitalia relevant to sexual response. In particular, the nucleus paragigantocellularis (nPG) receiving sensory input has neurons that innervate the penis and appears to have a role in orgasm. Although maybe not crucial for erection, the paraventricular nucleus (PVN) sends neurons to the penis via the nPG. Additionally, the PVN has direct projections to pelvic and autonomic neurons and is connected to the medial preoptic area (mPOA). The mPOA seems to be important for the control of sexual behavior (especially regarding mate selection) and is connected to the nPG and the periaqueductal gray (PAG), an area connected to the hypothalamic site involved in sexual response (Saenz de Tejada et al. 2004a, b).

Regarding higher brain centers, studies in rats show that the medial amygdala has a role in recognizing the appropriate partner for sexual arousal, while mainly peptide neurotransmitters are associated with these mechanisms. Functional MRI (fMRI) studies have identified the activation of the dominant (L > R) brain hemisphere after imagery of sexual stimuli, while VSS seems to activate the occipital cortex, inferior frontal lobe, cingulate gyrus, insula gyrus, corpus callosum, thalamus, caudate nucleus, globus pallidus, and inferior temporal lobe. Penile erection has also shown activation in the right sub-insular region including the claustrum, left caudate and putamen, right middle occipital and middle temporal gyri, bilateral cingulate gyrus, right sensorimotor and premotor regions, and right hypothalamus (Saenz de Tejada et al. 2004a, b). Activation of the somatosensory thalamus and of the nucleus of the solitary tract—which is the brainstem nucleus to which the vagus nerves project—is shown in PET and MRI studies (Saenz de Tejada et al. 2004a, b).

Multiple neurotransmitters seem to have a role in sexual function, while for a few of them, evidence is more concrete. Dopamine (DA), for example, clearly has a role in the PVN and the mPOA, since central dopaminergic neurons project there (Saenz de Tejada et al. 2004a, b). In general terms, DA is supportive of copulation and is increased in the mPOA at the time of ejaculation. Changes in DA and 5-HT (serotonin) in different areas of the brain may promote copulation and sexual satiety, respectively (Saenz de Tejada et al. 2004a, b). Both the two major types of dopamine receptors (D1-like and D2-like) have been associated with central erectile functions, while D2-like receptors seem to have a predominating effect (Gratzke et al. 2010). Testosterone enhances DA release in the mPOA at rest and with sexual challenge possibly by upregulating NOS, which increases nitric oxide (NO), thereby increasing DA release (Saenz de Tejada et al. 2004a, b).

Serotonin (5-HT) and thyrotropin-releasing hormone (TRH), both seem to inhibit penile erection through common or parallel sets of neurons (Saenz de Tejada et al. 2004a, b). Also in animals, 5-HT seems to generally inhibit male sexual behavior. Nevertheless, its action may be inhibitory or facilitatory depending upon the different sites and receptors within the central nervous system, explaining conflicting reports regarding enhancing or depressing effects of 5-HT agonists in sexual function (Gratzke et al. 2010).

Oxytocin is supposed to induce erection when injected into the lateral cerebral ventricle, the paraventricular nucleus, or the hippocampus of laboratory animals, while intracavernous oxytocin is not effective (Gratzke et al. 2010; Saenz de Tejada et al. 2004a, b). Nitric oxide (NO) influences sexual behavior and penile erection, especially at the level of the paraventricular nucleus of the hypothalamus and at other levels of the neural pathway, supporting sexual response (Gratzke et al. 2010; Saenz de Tejada et al. 2004a, b). Gamma aminobutyric acid (GABA) activity in the PVN seems to balance (inhibit) pro-erectile signaling, since it is supposed to modulate autonomic and somatic reflex pathways that inhibit penile erection (Gratzke et al. 2010; Saenz de Tejada et al. 2004a, b). Prolactin is associated with changes in striatal dopaminergic activity, since it is supposed to inhibit the dopaminergic incertohypothalamic pathway to the mPOA. Therefore hyperprolactinemia in the long term may inhibit sexual behavior, diminish sexual potency in men, and suppress genital reflexes in rats. Nevertheless, it is still unclear if this effect is mediated centrally by reduction in sexual interest and drive or peripherally through the effect of prolactin on corpus cavernosum smooth muscle (Gratzke et al. 2010). Endogenous and exogenous cannabinoids have been associated with changes in erectile function and sexual activity, possibly by modulation of PVN oxytocinergic neurons, while opioid peptides, acting centrally, prevent penile erection affecting oxytocinergic neurotransmission (Gratzke et al. 2010). ACTH and α-melanocyte-stimulating hormones (α-MSH) are able to induce penile erection, probably via stimulation of melanocortin receptors (MC). Particularly MC4 receptor subtype seems to contribute to the pro-erectile effects observed with MC pan-receptor agonists (Gratzke et al. 2010).

Finally androgens, and testosterone in particular, are necessary—but not sufficient—for sexual function in men. They are essential for the maintenance of libido and for the regulation of erectile capacity. Circulating testosterone levels within normal limits do not correlate however with sexual interest, activity, or erectile function in men with normal gonadal function. Literature findings seem controversial, since castration in humans may result in complete loss of libido along with erectile and ejaculatory dysfunctions and may be restored with testosterone administration, although in other cases sexual function following castration may be normal. Thus, the role of androgens in erectile function is complex, and androgen deprivation may not always cause erectile dysfunction, either in man or in rats (Gratzke et al. 2010).

5.5.2.2 Ejaculation

Orgasm seems to be a spinal reflex, since it may persist even after spinal cord injury, while reward deriving from orgasm is represented in higher brain centers (Saenz de Tejada et al. 2004a, b). Sensory receptors and areas, cerebral sensory and motor regions, spinal motor centers, and efferent and afferent pathways are also involved in this function (Rowland et al. 2010). In general, cerebral blood flow seems to decrease during orgasm, except for the right prefrontal cortex, where it increases (Saenz de Tejada et al. 2004a, b). Additionally, other limbic cortical areas are activated, the cerebellum and the ventral tegmental area of the meso-diencephalic junction, where dopamine neural circuits in animal sexual arousal and psychostimulant reward are located (Deak and Panksepp 2004).

The ejaculatory process is typically divided into three phases, emission, ejection, and orgasm, and is a result of actions in the central as well as the peripheral nervous system. Sympathetic nerves (T10-L2) provoke emission, while ejection is mediated by somatic nerves (S2–S4) but also involves a sympathetic spinal reflex upon which little voluntary control exists (Rowland et al. 2010). Orgasm though is the result of CNS processing of pudendal nerve sensory stimuli resulting from sexual organs (e.g., pressure and muscle contraction) (Rowland et al. 2010).

The main CNS structures that are involved in ejaculation include the mPOA, the nucleus paragigantocellularis (nPGi), the posteromedial bed nucleus of the stria terminalis, the posterodorsal medial amygdala, and the magnocellular neurons of the PVN of the anterior hypothalamus (Rowland et al. 2010; McMahon et al. 2004). More specifically, the mPOA, situated rostral to the anterior hypothalamus, seems to elicit seminal emission or ejaculation in rats and monkeys, as well as the urethrogenital reflex in rats, similar to orgasm in humans (McMahon et al. 2004). MPOA seems to stimulate ejaculatory response, while the nPGi is inhibitory (Rowland et al. 2010). Descending serotoninergic signals from the nPGi to the lumbosacral motor nuclei inhibit ejaculation, while the mPOA inhibits the nPGi and provokes ejaculation (Rowland et al. 2010). Finally, the parvocellular neurons of the hypothalamic PVN mediates erectile function in rats, while oxytocinergic activation in the magnocellular neurons of the PVN mediate ejaculation (McMahon et al. 2004).

The neurotransmitters involved in ejaculation include dopamine, norepinephrine, serotonin, acetylcholine, oxytocin, GABA, and nitric oxide (NO) (Rowland et al. 2010). Dopamine and serotonin seem to play the major role. Specifically, dopamine via D2 receptors and mainly through the anterior hypothalamus seems to provoke ejaculation, since it is released in the mPOA of male rats in the presence of an estrous female and increases during copulation, triggering ejaculation (Rowland et al. 2010; Hull et al. 1995). Serotonin, on the other hand, seems to inhibit ejaculation, based mainly on the hypothesis of premature ejaculation that supports the hyposensitivity of the 5-HT2C and/or the hypersensitivity of the 5-HT1A receptor. More specifically, stimulation of the 5-HT2C receptor with 5-HT2C agonists results in delay of ejaculation in rats, while stimulation of postsynaptic 5-HT1A receptors results in shortening of ejaculation latency time (Rowland et al. 2010).

5.5.3 Female Sexual Arousal and Orgasm

5.5.3.1 Arousal

Sexual arousal in females can be described by *subjective sexual arousal* and by *genital sexual arousal*.

Subjective Sexual Arousal

It includes the cognitive characteristics of arousal, such as the ease with which sexual stimuli arouse a woman. Most women express difficulties differentiating sexual desire from subjective sexual arousal; thus, DSM-5 is describing sexual interest/arousal disorder as one clinical entity (American Psychiatric Association 2013; Kirana 2013a, b).

Subjective sexual arousal would probably be considered as a slower response involving brain circuits which consciously and subconsciously recognize and process the stimulus. Genital sexual arousal therefore may precede, since it includes an automated, reflexive response of the autonomic nervous system (Basson 2000). Emotional stimuli are probably processed in the sensory thalamus and the amygdala, while supraspinal centers organize genital autonomic nerve activity. Subjective sexual arousal which seems to pursue genital sexual arousal, apart from the sensory thalamus and the amygdala (for further information processing), also involves the hippocampus (important in relating present experience to memories of past experiences), as well as cortical centers (Basson 2000).

Genital Sexual Arousal

It involves vaginal lubrication, swelling, and tingling and does not necessarily relate to subjective sexual arousal, which involves the sexual excitement component (Kirana 2013a, b).

Functional MRI (fMRI) studies have identified areas of the brain which are activated during sexual arousal, including areas subserving the genital vasocongestive response, such as the posterior hypothalamus, which correlate with subjective arousal response in men but not in women (Basson 2000; Karama et al. 2002). Interestingly enough, the location of the genitalia on the somatosensory brain area has not yet reached a consensus. Nevertheless, the primary somatosensory cortex, the operculum (secondary somatosensory cortex), and the posterior part of the insula have been proposed (Levin et al. 2016). These areas have emerged based mainly on studies of women with sexual disorders (vulvar vestibular pain and persistent genital arousal disorder) and seem to be important for processing the emotional meaning of the genital stimulation (Levin et al. 2016).

Visual sexual stimulation (VSS), as mentioned above, has been used to document, through PET and fMRI, the areas that are activated during sexual arousal (Salonia et al. 2010). Multiple cortical and subcortical areas are activated during VSS. More specifically, extrastriate visual areas in the occipitotemporal cortex, the lateral prefrontal cortex (PFC), the inferior parietal lobule, the orbitofrontal cortex (OFC), the ventral striatum, the amygdala, the insula, the cingulate cortex, the inferior temporal lobe, the thalamus, and basal ganglia seem to be activated both in women and in men, during the arousal phase (Salonia et al. 2010; Park et al. 2001). Interestingly enough, although an increase in sexual arousal presented in women while they were watching an erotic video, no correlations were found between perception of sexual arousal and brain activity (Park et al. 2001).

Other studies have demonstrated the activation of the amygdala and the hypothalamus, as well as decrease in the activation of bilateral temporal lobe. This is quite interesting, since sites in the latter area have been associated with moral (Salonia et al. 2010). Therefore, decreased activity of these sites would not inhibit the response to a sexual stimulus. Greater activation (especially of the entorhinal cortex) has been reported in women with reduced sexual desire disorder (Salonia et al. 2010; Maravilla and Yang 2008; Arnow et al. 2009). Similarly to these findings, studies with partnered clitoral stimulation resulted in increased activity in mainly cortical somatosensory areas and decreased activity in prefrontal areas and medial temporal areas (including the amygdala) (Levin et al. 2016). Decreased brain activity of these areas could mediate the disinhibition in order for a woman to engage in intense sexual activity and feelings (Levin et al. 2016). On the other hand, arousal resulting from vaginal self-stimulation in women with spinal cord injury did not result in decreased activity of the aforementioned areas (although it resulted in activation of the similar cortical and subcortical areas) (Levin et al. 2016; Komisaruk et al. 2004). Activation of the amygdala and the hypothalamus in heterosexual women seems to need longer VSS periods with erotic videos, than the rest of the brain areas involved (e.g., extrastriate visual areas, the inferior parietal lobule, the anterior cingulated cortex, and the ventral striatum, all of which are also activated in men) (Salonia et al. 2010). Heterosexual and lesbian women's brain responses showed specificity in brain areas activated, with regard to woman's sexual orientation (Salonia et al. 2010). Hypothalamic response however did not show category-specific response for either group (heterosexual and lesbian women), while specificity in amygdala function was found for lesbian women only and was non-specific in heterosexual women (Salonia et al. 2010; Chivers 2005).

Although useful information emerges through the interpretation of VVS-induced brain responses, it is important to point out the epiphenomenal, "contaminating" effect of the general emotional arousal that is present in any neuroimaging experiment. If controlled for the general emotional effect, only a restricted amount of brain areas seem to be activated due only to sexual content processing (Salonia et al. 2010; Walter et al. 2008).

5.5.3.2 Orgasm

Studying orgasm in women includes some difficulties, such as the lack of an objective measure to document that orgasm is occurring, as well as the lack of an agreed-upon definition of orgasm (Heiman 2007). The brain is undoubtedly an important source of sexual arousal and orgasm, since there is evidence of individuals who reach orgasm without direct genital stimulation. Phantom orgasm of paraplegics, orgasms induced through hypnosis, or orgasms induced from fantasy alone are only some paradigms suggesting that the brain is part of the anatomical requirements for orgasmic experience (Heiman 2007). On the other hand, persistent genital arousal and orgasm can occur not only in the absence of any genital or mental stimulation but even when both systems provide with negative feedback concerning orgasm (Fountoulakis et al. 2017).

Many brain areas, cortical and subcortical, show activation during orgasm. Orgasm resulting from vaginal self-stimulation in women with spinal cord injury reveals that the orgasm-related brain activity seems to be an expansion of that found in genital stimulation alone (Levin et al. 2016). In general, women's brain responses (brain areas that are activated) during orgasms seem to be similar to men's, suggesting that the main differentiations exist regarding the use of different cerebral pathways in order to reach orgasms (Salonia et al. 2010). Clitoral stimulation and clitoral orgasm, compared to passive nonsexual resting state, seem to increase regional cerebral blood flow (rCBF) in the bilateral primary somatosensory cortex (SI), left secondary somatosensory cortex (SII), and left supplementary motor area (Georgiadis et al. 2006). Probably, SII activation is a reflection of the context of the sensory stimulus, "weighting" the salience of somatosensory stimuli before they become conscious (Salonia et al. 2010).

Interestingly enough, rCBF decreased during orgasm in some brain regions. The inferomedial temporal lobe—including the amygdala—showed decreased rCBF (Salonia et al. 2010). Temporal lobe probably inhibits sexual arousal, and release of this inhibition may be imperative for sexual opportunity (Salonia et al. 2010). The female amygdala, although activated during long periods of VSS, shows deactivation in response to sexual stimulation crucial for sexual encounters to take place (Georgiadis et al. 2006). Similarly, CBF seems to decrease in the prefrontal cortex (PFC) and the orbitofrontal cortex (OFC), with the largest blood flow decrease being measured between clitoral stimulation and orgasm in the OFC, and between the nonsexual resting state and orgasm for the ventromedial area of the PFC (vmPFC) (Georgiadis et al. 2006). The PFC is a crucial part of the brain network that has to do with social behavior and executive function, while the vmPFC is an important brain area for self-monitoring and self-referential thought (Salonia et al. 2010). The OFC consists of lateral, middle, and medial parts (Salonia et al. 2010). The middle OFC is activated with increasing satisfaction and subjective pleasantness and deactivated with feelings of satiety, playing a role in hedonic experience, while the lateral OFC is linked to urge suppression (Salonia et al. 2010; Beauregard et al. 2001; Rolls 2000). It is suggested that the activity level of OFC reflects conscious control over the sexual urge, with high rCBF during clitoral stimulation and low rCBF during orgasm, indicating the reduced conscious control that is necessary (Georgiadis et al. 2006). Failed orgasm attempts are also associated with increased rCBF in the OFC (since it is associated to sexual arousal), revealing that possibly the release of tension and the loss of conscious control that is experienced with orgasm are strongly related to decreased OFC activity (Salonia et al. 2010; Georgiadis et al. 2006).

Increased brain activity during orgasm seems to take place in the cerebellum, although its contribution is quite unclear. More specifically, the left anterior lobe of the cerebellar vermis and adjacent deep cerebellar nuclei activate during orgasm in both sexes (Salonia et al. 2010; Georgiadis et al. 2006). Additionally, a strong positive association was found between activity of the left vermis of the cerebellum and rectal pressure (van Netten et al. 2008). The vermis takes part in axial motor control, but also in autonomic regulation and effect, explaining its activation during orgasm, since substantial cardiovascular and respiratory arousals take place (Salonia et al. 2010).

Cholinergic and adrenergic nerves, as well as a number of peptides, have been found to play a crucial role in the arousal of the female genitalia (Heiman 2007). Orgasm is correlated with increased secretion of prolactin and oxytocin—which is reflected by the elevated activation of the pituitary gland (Huynh et al. 2013)—and vasopressin and

Table 5.1 Major brain sites activated and deactivated during orgasm and clitoral stimulation in women

Brain region	Orgasm vs. rest	Orgasm vs. clitoris stimulation
Frontal lobe		
Primary motor cortex (pelvic floor)	Activation	Activation
Lateral orbitofrontal cortex	Activation	Deactivation
Ventromedial prefrontal cortex	Deactivation	Deactivation
Dorsomedial prefrontal cortex	Deactivation	Deactivation
Temporal lobe		
Inferior temporal gyrus	Deactivation	Deactivation
Middle temporal gyrus	Deactivation	Deactivation
Fusiform gyrus	Deactivation	Deactivation
Temporal pole	Deactivation	Deactivation
Parietal lobe		
Primary somatosensory cortex (SI) (genital)	Activation	Deactivation
Secondary somatosensory cortex (SII)	Activation	Deactivation
Superior parietal lobule	Activation	Deactivation
Precuneus	Activation	Deactivation
Occipital lobe		
Lingual gyrus	Deactivation	Activation
Limbic system		
Posterior insula	Activation	Activation
Amygdala	Deactivation	Activation
Cerebellum		
Deep cerebellar nuclei	Activation	Activation
Anterior vermis	Activation	Activation
Cerebellar hemisphere	Activation	Activation

Courtesy of Georgiadis JR. Reprinted from Journal of Sexual Medicine, volume 7, Salonia A, Giraldi A, Chivers ML, Georgiadis JR, Levin R, Maravilla KR, McCarthy MM, Physiology of Women's Sexual Function: Basic Knowledge and New Findings, 2637–2660, Copyright 2010, with the permission of Elsevier)

vasoactive intestinal peptide (VIP), while serotonergic and dopaminergic factors play an important role (Meston et al. 2004). More specifically, increased serotonin levels and decrease in dopamine seem to have a negative impact on sexual response, although the complete mechanism of this action is not clear (Heiman 2007). A list of brain regions activated and deactivated during orgasm in females is shown in Table 5.1.

5.6 Conclusions

Human sexuality and therefore human sexual behavior are influenced by factors that have to do with one's genitalia but also far from them, such as the central nervous system. Therefore, one's *sexual identity* and *normal sexual function* rely upon central and peripheral mechanisms and anatomic structures in the brain but are also strongly influenced by social morals, personality traits, religious values, moral beliefs, and family upbringing, which all affect the way sexual issues are approached (Vaidakis 2013). The critical masculinization hormone of the brain is estradiol which derives from testicular testosterone. The main neural structure that regulates sexual behavior seems to be influenced mainly by gonadal steroids that act during sensitive periods of development. The hypothalamus and specific areas, such as the SDN-POA, the mPOA, the VMN, the INAH-2 and INAH-3, the BNST, and the SCN, seem to play a major role in sexual behavior including sexual orientation and sexual identity. Sexual responses of men and women seem to have similarities as well as differences and are linked to regions of the CNS that modulate sexual function. Hypothalamic structures (e.g., the mPOA) and limbic and cortical sites (e.g., the prefrontal lobe) activate and deactivate during sexual desire, arousal, orgasm, and ejaculation in both genders (Fig. 5.1). Visual sexual stimulation (VSS) using neuroimaging techniques have given interesting results, regarding specific brain regions that are involved during sexual response. Hormones (especially testosterone) and numerous neurotransmitters (especially dopamine and serotonin) with their receptors are all important contributors to the male and female neuronal network, mediating particular sexual functions. More research is needed in order to achieve a better understanding of the various aspects of the psychobiology of sexuality.

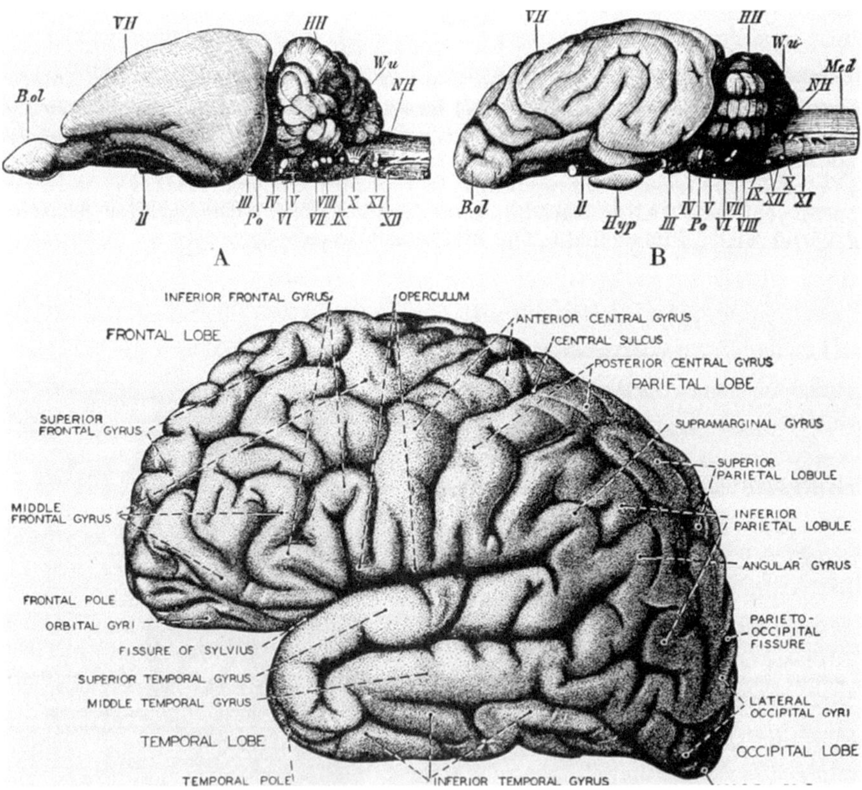

Fig. 5.1 Morphology of the cerebral cortex. Various areas are involved in human sexual response

References

Allen LS, Gorski RA (1990) Sex difference in the bed nucleus of the stria terminalis of the human brain. J Comp Neurol 302(4):697–706. https://doi.org/10.1002/cne.903020402
Allen JS, Damasio H, Grabowski TJ, Bruss J, Zhang W (2003) Sexual dimorphism and asymmetries in the gray-white composition of the human cerebrum. NeuroImage 18(4):880–894
American Psychiatric Association (2013) Diagnostic and statistical manual of mental disorders, fifth edition. DSM-5. American Psychiatric Publishing, Arlington, VA
Arnow BA, Millheiser L, Garrett A, Lake Polan M, Glover GH, Hill KR, Lightbody A, Watson C, Banner L, Smart T, Buchanan T, Desmond JE (2009) Women with hypoactive sexual desire disorder compared to normal females: a functional magnetic resonance imaging study. Neuroscience 158(2):484–502. https://doi.org/10.1016/j.neuroscience.2008.09.044
Basson R (2000) Sexual desire/arousal disorders in women. In: Leiblum S (ed) Principles and practice of sex therapy. Guilford, New York, nY, p 25
Basson R, Leiblum S, Brotto L, Derogatis L, Fourcroy J, Fugl-Meyer K, Graziottin A, Heiman JR, Laan E, Meston C, Schover L, van Lankveld J, Schultz WW (2004) Revised definitions of women's sexual dysfunction. J Sex Med 1(1):40–48. https://doi.org/10.1111/j.1743-6109.2004.10107.x
Beauregard M, Levesque J, Bourgouin P (2001) Neural correlates of conscious self-regulation of emotion. J Neurosci 21(18):RC165
Berglund H, Lindstrom P, Savic I (2006) Brain response to putative pheromones in lesbian women. Proc Natl Acad Sci U S A 103(21):8269–8274. https://doi.org/10.1073/pnas.0600331103
Berglund H, Lindstrom P, Dhejne-Helmy C, Savic I (2008) Male-to-female transsexuals show sex-atypical hypothalamus activation when smelling odorous steroids. Cereb Cortex 18(8):1900–1908. https://doi.org/10.1093/cercor/bhm216
Chivers ML (2005) A brief review and discussion of sex differences in the specificity of sexual arousal. Sex Relatsh Ther 20(4):377–390. https://doi.org/10.1080/14681990500238802
Corona G, Maggi M (2012) Hypoactive sexual desire (libido) disorder. In: Porst H, Reisman Y (eds) The ESSM syllabus of sexual medicine. MEDIX, Amsterdam, pp 401–431
Cuzin B (2015) Anatomy and physiology of female sexual organs. In: Reisman Y, Porst H, Lowenstein L, Tripodi F, Kirana P (eds) The ESSM syllabus of sexual medicine, 2nd edn, pp 40–68
Deak T, Panksepp J (2004) Stress, sleep, and sexuality in psychiatric disorders. In: Panksepp J (ed) Textbook of biological psychiatry. Wiley-Liss, Hoboken, NJ, pp 111–143
Dean J (2012a) Gender identity and related disorders. In: Kirana P, Tripodi F, Reisman Y, Porst H (eds) The EFS and ESSM syllabus of clinical sexology. MEDIX, Amsterdam, pp 1253–1278
Dean J (2012b) Sexual orientation. In: Porst H, Reisman Y (eds) The ESSM syllabus of sexual medicine. MEDIX, Amsterdam, pp 175–184
Demeter S, Ringo JL, Doty RW (1988) Morphometric analysis of the human corpus callosum and anterior commissure. Hum Neurobiol 6(4):219–226
Fernandez-Guasti A, Kruijver FP, Fodor M, Swaab DF (2000) Sex differences in the distribution of androgen receptors in the human hypothalamus. J Comp Neurol 425(3):422–435
Fountoulakis KN, Tegos T, Goulis DG, Nimatoudis I, Kimiskidis V (2017) Treatment of a female patient with persistent genital arousal and Parkinson's disease with paliperidone. Aust N Z J Psychiatry 51(1):98–99. https://doi.org/10.1177/0004867416660200
Garcia-Falgueras A, Swaab DF (2008) A sex difference in the hypothalamic uncinate nucleus: relationship to gender identity. Brain 131(Pt 12):3132–3146. https://doi.org/10.1093/brain/awn276
Georgiadis JR, Kortekaas R, Kuipers R, Nieuwenburg A, Pruim J, Reinders AA, Holstege G (2006) Regional cerebral blood flow changes associated with clitorally induced orgasm in healthy women. Eur J Neurosci 24(11):3305–3316. https://doi.org/10.1111/j.1460-9568.2006.05206.x
Giuliano F, Rampin O (2000) Central neural regulation of penile erection. Neurosci Biobehav Rev 24(5):517–533
Gizewski ER, Krause E, Schlamann M, Happich F, Ladd ME, Forsting M, Senf W (2009) Specific cerebral activation due to visual erotic stimuli in male-to-female transsexuals compared with male and female controls: an fMRI study. J Sex Med 6(2):440–448. https://doi.org/10.1111/j.1743-6109.2008.00981.x

Goldstein I, Giraldi A, Kodigliu A, Van Lunsen H, Marson L, Nappi R, Pfaus A, Salonia A, Traish A, Vardi Y (2004) Physiology of female sexual function and pathophysiology of female sexual dysfunction. In: Lue T, Basson R, Rosen R, Giuliano F, Khoury S, Montorsi F (eds) Sexual medicine: sexual dysfunctions in men and women. Health Publications, Paris, pp 683–748

Gorski RA, Gordon JH, Shryne JE, Southam AM (1978) Evidence for a morphological sex difference within the medial preoptic area of the rat brain. Brain Res 148(2):333–346

Gratzke C, Angulo J, Chitaley K, Dai YT, Kim NN, Paick JS, Simonsen U, Uckert S, Wespes E, Andersson KE, Lue TF, Stief CG (2010) Anatomy, physiology, and pathophysiology of erectile dysfunction. J Sex Med 7(1 Pt 2):445–475. https://doi.org/10.1111/j.1743-6109.2009.01624.x

Hamer DH, Hu S, Magnuson VL, Hu N, Pattatucci AML (1993) A linkage between DNA markers on the X-chromosome and male sexual orientation. Science 261(5119):321–327. https://doi.org/10.1126/science.8332896

Hare L, Bernard P, Sanchez FJ, Baird PN, Vilain E, Kennedy T, Harley VR (2009) Androgen receptor repeat length polymorphism associated with male-to-female transsexualism. Biol Psychiatry 65(1):93–96. https://doi.org/10.1016/j.biopsych.2008.08.033

Heiman J (2007) Orgasmic disorders in women. In: Leiblum S (ed) Principles and practice of sex therapy. Guilford Press, New York, NY, pp 84–124

Hull EM, Du J, Lorrain DS, Matuszewich L (1995) Extracellular dopamine in the medial preoptic area: implications for sexual motivation and hormonal control of copulation. J Neurosci 15(11):7465–7471

Huynh HK, Willemsen AT, Holstege G (2013) Female orgasm but not male ejaculation activates the pituitary. A PET-neuro-imaging study. NeuroImage 76:178–182. https://doi.org/10.1016/j.neuroimage.2013.03.012

Kaplan H (1979) Disorders of sexual desire and other new concepts and techniques in sex therapy. Brunner/Mazel, New York, NY

Karama S, Lecours AR, Leroux JM, Bourgouin P, Beaudoin G, Joubert S, Beauregard M (2002) Areas of brain activation in males and females during viewing of erotic film excerpts. Hum Brain Mapp 16(1):1–13

Kirana P (2013a) Female sexual desire disorders. In: Kirana P, Tripodi F, Reisman Y, Porst H (eds) The EFS and ESSM syllabus of clinical sexology. MEDIX, Amsterdam, pp 756–784

Kirana PS (2013b) Female sexual arousal disorders. In: Kirana PS, Tripodi F, Reisman Y, Porst H (eds) The EFS and ESSM syllabus of clinical sexology. MEDIX, Amsterdam, pp 786–801

Komisaruk BR, Whipple B, Crawford A, Liu WC, Kalnin A, Mosier K (2004) Brain activation during vaginocervical self-stimulation and orgasm in women with complete spinal cord injury: fMRI evidence of mediation by the vagus nerves. Brain Res 1024(1–2):77–88. https://doi.org/10.1016/j.brainres.2004.07.029

Kruijver FP, Zhou JN, Pool CW, Hofman MA, Gooren LJ, Swaab DF (2000) Male-to-female transsexuals have female neuron numbers in a limbic nucleus. J Clin Endocrinol Metab 85(5):2034–2041. https://doi.org/10.1210/jcem.85.5.6564

Leiblum S (2007) Sex therapy today current issues and future perpectives. In: Leiblum S (ed) Principles and practice of sex therapy. Guilford Press, New York, NY, pp 3–22

Levay S (1991) A difference in hypothalamic structure between heterosexual and homosexual men. Science 253(5023):1034–1037. https://doi.org/10.1126/science.1887219

Levin RJ, Both S, Georgiadis J, Kukkonen T, Park K, Yang CC (2016) The physiology of female sexual function and the pathophysiology of female sexual dysfunction (committee 13A). J Sex Med 13(5):733–759. https://doi.org/10.1016/j.jsxm.2016.02.172

Luders E, Sanchez FJ, Gaser C, Toga AW, Narr KL, Hamilton LS, Vilain E (2009) Regional gray matter variation in male-to-female transsexualism. NeuroImage 46(4):904–907. https://doi.org/10.1016/j.neuroimage.2009.03.048

Maravilla KR, Yang CC (2008) Magnetic resonance imaging and the female sexual response: overview of techniques, results, and future directions. J Sex Med 5(7):1559–1571. https://doi.org/10.1111/j.1743-6109.2008.00839.x

Martin RD (2007) The evolution of human reproduction: a primatological perspective. Am J Phys Anthropol Suppl 45:59–84. https://doi.org/10.1002/ajpa.20734

Masters W, Johnson V (1966) Human sexual response. Little Brown & Co, Boston, MA

McMahon CG, Abdo C, Incrocci L, Perelman M, Rowland D, Waldinger M, Xin ZC (2004) Disorders of orgasm and ejaculation in men. J Sex Med 1(1):58–65. https://doi.org/10.1111/j.1743-6109.2004.10109.x

Meston CM, Levin RJ, Sipski ML, Hull EM, Heiman JR (2004) Women's orgasm. Annu Rev Sex Res 15:173–257

Meyer-Bahlburg HF (2010) From mental disorder to iatrogenic hypogonadism: dilemmas in conceptualizing gender identity variants as psychiatric conditions. Arch Sex Behav 39(2):461–476. https://doi.org/10.1007/s10508-009-9532-4

van Netten JJ, Georgiadis JR, Nieuwenburg A, Kortekaas R (2008) 8-13 Hz fluctuations in rectal pressure are an objective marker of clitorally-induced orgasm in women. Arch Sex Behav 37(2):279–285. https://doi.org/10.1007/s10508-006-9112-9

Park K, Kang HK, Seo JJ, Kim HJ, Ryu SB, Jeong GW (2001) Blood-oxygenation-level-dependent functional magnetic resonance imaging for evaluating cerebral regions of female sexual arousal response. Urology 57(6):1189–1194

Pfaff D, McCarthy M, Schwartz-Giblin S, Kow L (2009) Cellular and molecular mechanisms of female reproductive behaviors. In: Knobil E, Neill JD (eds) Physiology of reproduction: Female reproductive behavior. Raven Press, New York, NY, pp 107–220

Phoenix CH, Goy RW, Gerall AA, Young WC (1959) Organizing action of prenatally administered testosterone propionate on the tissues mediating mating behavior in the female guinea pig. Endocrinology 65:369–382. https://doi.org/10.1210/endo-65-3-369

Rolls ET (2000) The orbitofrontal cortex and reward. Cereb Cortex 10(3):284–294

Rowland D, McMahon CG, Abdo C, Chen J, Jannini E, Waldinger MD, Ahn TY (2010) Disorders of orgasm and ejaculation in men. J Sex Med 7(4 Pt 2):1668–1686. https://doi.org/10.1111/j.1743-6109.2010.01782.x

Sadeghi M, Fakhrai A (2000) Transsexualism in female monozygotic twins: a case report. Aust N Z J Psychiatry 34(5):862–864. https://doi.org/10.1080/j.1440-1614.2000.00804.x

Saenz de Tejada I, Angulo J, Cellek S, Gonzalez-Cadavid N, Heaton J, Pickard R, Simonsen U (2004a) Physiology of erectile function and pathophysiology of erectile dysfunction. In: Lue T, Basson R, Rosen R, Giuliano F, Khoury S, Montorsi F (eds) Sexual medicine: sexual dysfunctions in men and women. 2nd International consultation in Sexual Medicine Health Publications, Paris

Saenz de Tejada I, Angulo J, Cellek S, Gonzalez-Cadavid N, Heaton J, Pickard R, Simonsen U (2004b) Physiology of erectile function. J Sex Med 1(3):254–265. https://doi.org/10.1111/j.1743-6109.04038.x

Salonia A, Giraldi A, Chivers ML, Georgiadis JR, Levin R, Maravilla KR, McCarthy MM (2010) Physiology of women's sexual function: basic knowledge and new findings. J Sex Med 7(8):2637–2660. https://doi.org/10.1111/j.1743-6109.2010.01810.x

Savic I, Lindstrom P (2008) PET and MRI show differences in cerebral asymmetry and functional connectivity between homo- and heterosexual subjects. Proc Natl Acad Sci U S A 105(27):9403–9408. https://doi.org/10.1073/pnas.0801566105

Savic I, Berglund H, Gulyas B, Roland P (2001) Smelling of odorous sex hormone-like compounds causes sex-differentiated hypothalamic activations in humans. Neuron 31(4):661–668

Savic I, Berglund H, Lindstrom P (2005) Brain response to putative pheromones in homosexual men. Proc Natl Acad Sci U S A 102(20):7356–7361. https://doi.org/10.1073/pnas.0407998102

Schwarz JM, Liang SL, Thompson SM, McCarthy MM (2008) Estradiol induces hypothalamic dendritic spines by enhancing glutamate release: a mechanism for organizational sex differences. Neuron 58(4):584–598. https://doi.org/10.1016/j.neuron.2008.03.008

Swaab DF (2002) Gender issues in brain structures and functions and their relevance for psychopathology. In: D'Haenen H, den Boer J, Willner P (eds) Biological psychiatry. John Wiley & Sons, Hoboken, NJ, pp 189–209. https://doi.org/10.1002/0470854871.chxiv

Tripodi F, Silvaggi C (2013) Sexual response: motivation and models. In: Kirana P, Tripodi F, Reisman Y, Porst H (eds) The EFS and ESSM syllabus of clinical sexology. MEDIX, Amsterdam, pp 230–278

Turan MT, Esel E, Dundar M, Candemir Z, Basturk M, Sofuoglu S, Ozkul Y (2000) Female-to-male transsexual with 47,XXX karyotype. Biol Psychiatry 48(11):1116–1117

Vaidakis N (2005) Human sexual behavior. BHTA, Athens

Vaidakis N (2013) Historical aspects of human sexuality and sex research. In: Kirana P, Tripodi F, Reisman Y, Porst H (eds) The EFS and ESSM syllabus of clinical sexology. MEDIX, Amsterdam, pp 22–34

Walter M, Bermpohl F, Mouras H, Schiltz K, Tempelmann C, Rotte M, Heinze HJ, Bogerts B, Northoff G (2008) Distinguishing specific sexual and general emotional effects in fMRI-subcortical and cortical arousal during erotic picture viewing. NeuroImage 40(4):1482–1494. https://doi.org/10.1016/j.neuroimage.2008.01.040

Neurobiology of Aging

6

Sebastien Tassy, Anastasia Konsta, and Jean Michel Azorin

6.1 Aging in Our Developed Society

We evolve in a consumer society in which, every passing minute, the sellers of pleasures solicit our mind. Responding to these calls, we are constantly looking to have more and to win more, and we are launched into a performance race. However, this performance is measured by the results of our competitors, and the spirit of competition infuses.To succeed, the required standard appears to be beauty and youth, whose images are sent back continuously to us through all the media channels. Unfortunately beauty and youth are ephemeral and very subjective. To resist the stress caused by the difference between what we are and these images, we exhaust ourselves in a continuous struggle, of which depression can be the pathological consequence. Another mechanism for responding to this stress is avoidance. Among the images we refuse to see, that of old age is probably one of those carrying the strongest negative emotions. This association between old age and negative emotions is not new. Both in philosophy and the arts, there exists a multitude of examples of this bias referring to disgust, sadness, or nostalgia.

The image of our aging body refers to the painful feeling of no longer arousing the desire of the other. But the desire of the other nourishes us and gives us the impression of existing. It flatters our ego. It empowers us by giving us the possibility to use the other and to access our own desires more easily. Accessing our desires by using the other is also given to us by physical force or cognitive abilities, by subjecting him to our will. Nevertheless beauty/youth, physical strength, and cognitive abilities decline with age. These losses, therefore, lead to the fear of dependence and

S. Tassy (✉) · J. M. Azorin
Département de Psychiatrie, Assistance Publique – Hôpitaux de Marseille, CHU Sainte Marguerite, Marseille Cedex, France

A. Konsta
1st Department of Psychiatry, Faculty of Medicine, Aristotle University of Thessaloniki, Thessaloniki, Greece

K. N. Fountoulakis, I. Nimatoudis (eds.), *Psychobiology of Behaviour*,
https://doi.org/10.1007/978-3-030-18323-3_6

the seizure of our own power and self by the other. At least, the satisfaction of our most basic needs will also depend to a great extent on others. Old age will then be felt as a disease in the sense that dependence will be directly threatening for the survival of the individual.

6.2 Physical and Cognitive Weakening

6.2.1 Brain Aging

The mechanisms of aging are multifactorial and concern the rest of the body like the brain. One of these main changes is the alteration of DNA by intrinsic and external factors. The resulting modifications will modify gene expression and protein synthesis. Oxidative stress participates in this process. It also leads to alteration of membrane lipids or cytoskeletal proteins. Because of its high lipid content, the brain is particularly vulnerable to this phenomenon. Oxidative stress is also accompanied by inflammatory phenomena with multiple repercussions including modifications in cytoskeletal proteins that lead to changes in the transport of neurotransmitters or neurotrophins. Microcirculation is also affected by this phenomenon, which consequently affects oxygenation and glucose supply resulting in a cascade of self-sustained microdysfunctions. This condition promotes the appearance of pathology including neurodegenerative diseases.

The different major neurotransmission systems are affected to varying degrees by this set of processes. The alteration of the dopaminergic system is the earliest, the most constant, and the most important. It leads to a relative frontal deafferentation and the alteration of sensitivo-motor performances. Cholinergic involvement is more moderate and would be responsible for changes in memory capacity outside pathological processes such as that of Alzheimer's disease. In a more marginal way, serotonergic and noradrenergic transmission is impaired.

From a macroscopic point of view, these microscopic modifications lead to a change in the appearance of the brain. The loss of white matter is relatively greater than that of the gray matter. They are the result of processes such as the neuronal loss, the dendritic arborization decline, and the demyelination of neurons. Frontal and associative regions are those whose volume is the most affected compared to primary regions.

6.2.2 Cognitive Aging

Cognitive psychology bases its classical distinction of the two dimensions of intelligence on the evolution, with age, of the cognitive capacities that underlie them (Belsky 1990). Fluid intelligence is considered a form of intelligence very sensitive to the effects of age. On the contrary, crystallized intelligence is very robust to passing time. The fluid intelligence relies heavily on visual-motor and/or spatial capabilities and is required for tasks requiring instantaneous use of information to solve a problem. The crystallized intelligence is mobilized during the resolution of verbal tasks and/or involving the knowledge and general culture that develops with experience. The nonverbal capacities would reach their maximum toward 30 years and decline afterward. This maximum is reached at about 60 years for the verbal capacities.

The executive functions are the elementary cognitive abilities that allow planning, organization, and synchronization of processes involved in controlling the execution of the action. Three of main cognitive processes that allow this task are inhibition of automatic responses, updating of relevant information in working memory, and flexibility which is the ability to alternate the multiple cognitive tasks required for execution of a planned global task. These three processes can be independently altered. However the disruption of one will tend to resonate on the other two. During aging, it is the inhibition and updating abilities that would be the most disturbed, the flexibility process being more spared. Higher-order executive functions such as planning and fluid intelligence require simultaneous use of such multiple basic executive functions. These higher-order processes are then also affected by aging.

One of the classic models of *memory* is that proposed by Tulving (1995) who organizes it into five major types of process. The first two, the *procedural memory* and *the system of perceptual representation*, are said to be implicit. They involve an unconscious treatment of information. Procedural memory is the process of learning perceptual motor skills or behavioral algorithms. The system of perceptual representations is involved in the maintenance and encoding of perceptual information allowing the perceptive identification of objects and words but without reference to their meaning. These two types of implicit memory would be little affected by age.

Working memory is a mnesic representation of a very recent, real or mental, event. It contains a limited number of information kept active for a short time. It allows the temporary storage and the mental manipulation of information relevant to the realization of a task. This memory is affected by aging, but for the more demanding tasks.

Semantic memory is the memory of general knowledge acquired on the world. This memory is an essential constitutive element of crystallized intelligence. It allows the connection between knowledge by ordering them in conceptual categories ("I know the distinction living/non-living; dog and cat are living objects"). This memory of knowledge is little affected by age; however certain categories of knowledge seem to be disturbed, such as proper names.

Episodic memory, which is the memory of personal events as a sensory event ("I remember seeing …"), is sensitive to the effects of age. The difficulty in rendering the information presented results from not only a weakening of the strategies allowing the encoding of this information for their storage but also difficulties growing in recovery strategies of already stored information.

As memory, *attention* is a cognitive function that is not unitary, and it can be defined as the ability to maintain cognitive processes directed toward a goal while filtering the continuous flow of the information. The changes that affect it during aging are therefore heterogeneous (Verhaeghen and Cerella 2002).

Selective attention is the ability to extract information relevant to the realization of the current cognitive task from the mass of information that constantly stimulates the nervous system. Aging alters this faculty incidentally by disrupting the ability to disengage from the cognitive processing of a first object when performing complex and multiple tasks.

Sustained attention is the ability to maintain a consistent behavioral response during continuous and repetitive activity. This corresponds to vigilance which is little changed by aging. However, elderly subjects are more vulnerable to alterations in alertness and especially to delirium. The physical or psychological stresses that trigger delirium in the elderly are much lower than those with the same consequences in younger subjects.

Alternative attention or mental flexibility is the ability to shift the focus of attention and move between tasks requiring different cognitive processes which decline with age.

The divided attention is the higher form of attention. It allows the distribution of attentional resources on several simultaneous stimuli according to the relative importance of each of them. All things being equal, older subjects are obliged to mobilize more cognitive resources than younger subjects have. When the task become particularly complex, they are therefore more penalized.

6.2.3 Affects in Aging: The Paradox of Aging and High Well-Being

Psychoanalysts have identified an existential crisis between the beginning of adulthood and the end of life which they called the "midlife crisis." This crisis would be part of the normal maturation of individuals. It will intervene as a transition in what Erickson (1963), the theoretician who proposed a development in eight stages, called the middle adulthood. During this period, individuals are particularly concerned about the meaning and purpose of their lives. This then allows them to achieve more stable personal satisfaction at the end of their lives, which he calls "late adulthood."

The idea that a lesser period exists around the age of 40 refers symmetrically to the idea that psychological well-being would be superior before and after this turning point in life. This is indeed what many studies show (Blanchflower and Oswald 2008; Stone et al. 2010).

The emotional well-being would follow a U-shaped curve with a low point around 45 years. The result would be that the well-being around 80 years would be equivalent to that of 20 years. Indeed, even with aging-related hardships, older adults are relatively satisfied with their emotional well-being, decreasing in negative effects (Grühn et al. 2005).

Some authors have argued that these results would mainly concern the populations of the most developed countries (Steptoe et al. 2015). However, there also seems to be a midlife crisis in apes (Weiss et al. 2012), suggesting this phenomenon could exist outside of any cultural context. A recent review of the literature found these results to be robust and the criticisms rather limited (Ulloa et al. 2013).

Much of our well-being and stress depends on our social relationships. Indeed, Almeida and Kessler (1998) demonstrated that interpersonal relationships are the main source of daily stress and can induce very strong emotional distress. However, older people are more satisfied with their social networks than younger (Carstensen 1992). Their relation with their family members is more a source of positive emotion than for younger adults(Charles and Piazza 2007), and they experience more positive than negative exchanges. The more positive emotions they generally report could thus reflect these social experiences.

The socioemotional selectivity theory (SST) (Carstensen et al. 1999) argues that two broad categories of life goals are evolving in relation to perceived time: the ones related to the acquisition of knowledge and the ones concerning the regulation of states of emotion. When the horizon of time to live appears as unlimited, the individuals prioritize their objectives around optimization of the collection of information, the exploration of the novelty, and the widening of the knowledge. When time seems limited, the priority objectives are those that can be achieved in the short term. From then on, emotional well-being becomes an objective in itself, depending on the affective experience and the regulation of emotional states. This emotion regulation aims to maximize positive experiences and minimize negative ones. In this view, older adults would thus be more focused on their emotional well-being and behave in accordance with this goal. They tend to reduce their exposure to negative situations and increase their exposure to positive events. They report fewer social conflicts in the social world they navigate in and are more effective in solving the interpersonal problems they encounter than younger adults (Birditt et al. 2005).

Happiness has long been neglected from the field of psychology and psychiatry. The main objectives of these disciplines were to understand the mechanisms of pathological states and to provide therapeutic solutions. But the researchers have found that positive psychosocial characteristics are associated with objectively better health outcomes and greater longevity. Moreover happiness predicts success in many areas and helps develop resilience to the difficult circumstances of existence (Cohn et al. 2009). In contradiction with the widespread idea that people become less happy as they age, in studies on well-being, happiness appears to be specifically higher when the age advances (Stone et al. 2010). If we easily imagine that the generation of baby boomers can say that they are happy because of the improvement of the quality of life and health for the 50- to 70-year-olds, it is difficult to imagine that the very old ones can keep the happiness. Diseases and griefs multiply. But above all the loss of autonomy can significantly alter the idea that one can have of one's dignity. However again contrary to the a priori, a study of German centenarians showed that they were as happy as their younger compatriots (Jopp and Rott 2006). Moreover in the context of highly controlled study, the happiness experience by ones seems to correlate with age (Gross et al. 1997).

The frequency of *anger* increases in the young adult but then decreases markedly with aging. In laboratory studies, elderly patients respond with less anger than younger subjects while maintaining an identical sensitivity to stimuli causing sadness (Charles and Carstensen 2008). Visual evoked potentials in response to angry faces are lower in elderly subjects, but this is not true for faces expressing sadness or happiness

(Mienaltowski et al. 2011). This tendency toward decreased reactivity to anger appears to be beneficial for health, as it is recognized that anger increases cardiovascular risk (Mostofsky et al. 2014). This mechanism would therefore be adaptive.

Moreover, as previously underline, the negative interactions with members of their social networks are scarcer than for youngers, and interpersonal tensions are less distressing for them (Birditt et al. 2005). Eventually, when these tensions exist, they lead to less anger, and older subjects regulate them by milder strategies (Kunzmann and Thomas 2014).

Sadness is triggered by loss and older adults are as sensitive as younger to sad stimuli (Charles and Carstensen 2008). Losses and deficits in personal abilities increase with age, and the contentment with performance should then decline with age; however it does not (Rothermund and Brandtstadter 2003).Older adults have relationships with older individuals, so they are confronted with more frequent mourning in their social relationships in general. Old age is also accompanied by more symbolic losses such as physical abilities or autonomy. Old age therefore refers to a greater probability of sadness. Yet when sadness is measured at a given point in time, elderly people are no more so than other parts of the population (Kunzmann and Thomas 2014). It seems even that the feeling of sadness shows a slight peak in the middle of life, to decline then (Stone et al. 2010).

Regret involves sadness or remorse over past acts. Having a longer life to look back on means that there are more things to regret and also potentially fewer opportunities to address the regret via new behaviors. But older adults are less likely than younger adults to report regrets (Timmer et al. 2005). There is a difference in how to look at situations that can generate regrets depending on age. In a game in which a choice is presented with several options that can generate different gains, older subjects do not use the same strategy to guide their decisions. Elderly subjects are more likely to focus on potential gains than avoid potential regrets in such situations (Mather et al. 2012). Regrets also influence the elderly subjects' choice behaviors much less than those of young or depressed elderly subjects (Brassen et al. 2012). It appears that the old age is associated with a greater emotion regulation, involving the medial prefrontal areas, during the choices inducing regrets. This regulatory strategy does not therefore seem to imply a cognitive strategy of the reappraisal type requiring significant executive resources but more probably a strategy of regulation by external attribution.

6.2.4 Positivity Effect

As we have already pointed out, aging is accompanied by physical decline and the likelihood of greater loss. However, aging is also accompanied by greater emotional stability and often a feeling of "successful aging," a term associated with both the physical and psychological dimensions. A possible explanation of this apparent paradox is that there is an age by valence interaction in cognitive processes. The older adults remember the smiling faces (Charles et al. 2003) or their happy stories but less vividly feel the financial losses (Samanez-Larkin et al.

2007). This bias is called "positivity effect" and would rely on a memory and attention bias (Mather and Carstensen 2005). The SST discussed above explains this psychological bias. The time available to live for the elders appears shorter which leads them to regulate emotional states in order to optimize psychological well-being.

6.2.5 Emotion Regulation

Emotion regulation involves all the processes aiming at modifying the nature and intensity, the duration, or the expression of emotions. It relies on various strategies and cognitive processes in order to be efficient. Both the paradox of aging and well-being and the positivity effect support the idea that emotion regulation is specifically efficient in older individuals. However in experimental conditions during which they are given emotion regulation tasks, older individuals don't have better emotion regulation skills than their younger counterparts (Mather 2012). How then could older subjects be more successful in eliminating negative emotions? One explanation would be that older adults are more focused on positive emotions and devote more daily resources to regulate their emotions. Younger subjects, who can change their future by modifying it, have less need to be positive, and their negative emotions can motivate them in this perspective. They have less need to eliminate negative emotions and therefore devote fewer cognitive resources (Mather and Johnson 2000). Another explanation is a difference in emotion regulation strategy. When SAs and young people are compared on specific tasks, SAs perform worst or not as good. Since SAs have a decreasing executive capacity relying on the lateral prefrontal areas, they must rely on strategies involving the more medial cerebral areas that favor distraction rather than reappraisal. In favor of this hypothesis, a study showed that in AS the distractibility by positive stimuli is parallel to the activity of the anterior cingulated cortex (ACC) and that this activity of the ACC is correlated with the emotional stability (Brassen et al. 2011).

6.2.6 Brain Connectivity and Aging

For a long time, the paradigm that allowed the study of the brain was a function/a brain area. It is now widely accepted that most cerebral functions are supported by the coordinated activity of a set of structures that make up a more or less complex network. Thus, one talks of cerebral connectivity. We talk about connectivity to translate the idea of links between brain areas. This link is reflected either by a direct connection by fibers or a synchronous activity over time. This method of connectivity has made it possible to identify networks of structures whose activity is intense when the brains are at rest. The DMN has been identified as a network of areas that activity typically decreases below baseline level during cognitive tasks requiring externally directed attention. It is a deactivation of this global network. In healthy older adults, this deactivation is however reduced (Damoiseaux et al. 2008). This reduced deactivation would result from a decrease ability to suppress task-irrelevant processes, compared to young adults (Grady et al. 2006). An alternative view is that when cognitive tasks require important resources, older individuals rely more on processes mediated by the DMN (Maillet and Schacter 2016), in which the medial prefrontal cortex is a key hub. Interestingly, elderly subjects rely more on strategies involving the medial prefrontal areas for emotion regulation as cited above. Elderly better emotion regulation and well-being would then be linked to this DMN overuse.

6.3 Conclusion

The very reason why man has become a social animal is because his survival depends very much on his ability to integrate among his fellows and maintain himself in the social hierarchy. However, this capacity seems to diminish with age.

Yet every strength that gives us our place in society does not necessarily disappear with age, so emotion management and temperance are qualities that are attributed to older individuals. These abilities are becoming more important as people are confined in the immediacy of emotions aroused by our digital world.

Moreover, our social position lies not only in our ability to dominate the other but also in our ability to arouse sympathy. Compassion and empathy are motivational forces as powerful as envy in initiating behaviors. Vulnerability can thus also lead to a strengthening of the social bond. This shows that there exist reasons to not be pessimistic.

Ultimately, the relationship of the individual to himself and his regrets changes with age as we have explained. The frequency of regrets decreases. The feeling of having accomplished one's life according to one's abilities allows the disappearance of the anxiety of success, and the fear of the future decreases as the future shrinks.

References

Almeida DM, Kessler RC (1998) Everyday stressors and gender differences in daily distress. J Pers Soc Psychol 75(3):670–680

Belsky J (1990) The psychology of aging: theory, research, and interventions. Brooks/Cole Publishing Company, Pacific Grove, CA

Birditt KS, Fingerman KL, Almeida DM (2005) Age differences in exposure and reactions to interpersonal tensions: a daily diary study. Psychol Aging 20(2):330–340. https://doi.org/10.1037/0882-7974.20.2.330

Blanchflower DG, Oswald AJ (2008) Is well-being U-shaped over the life cycle? Soc Sci Med 66(8):1733–1749

Brassen S, Gamer M, Büchel C (2011) Anterior cingulate activation is related to a positivity bias and emotional stability in successful aging. Biol Psychiatry 70(2):131–137

Brassen S, Gamer M, Peters J, Gluth S, Buchel C (2012) Don't look back in anger! Responsiveness to missed chances in successful and nonsuccessful aging. Science 336(6081):612–614. https://doi.org/10.1126/science.1217516

Carstensen LL (1992) Social and emotional patterns in adulthood: support for socioemotional selectivity theory. Psychol Aging 7(3):331–338
Carstensen LL, Isaacowitz DM, Charles ST (1999) Taking time seriously. A theory of socioemotional selectivity. Am Psychol 54(3):165–181
Charles ST, Carstensen LL (2008) Unpleasant situations elicit different emotional responses in younger and older adults. Psychol Aging 23(3):495–504. https://doi.org/10.1037/a0013284
Charles ST, Piazza JR (2007) Memories of social interactions: age differences in emotional intensity. Psychol Aging 22(2):300–309. https://doi.org/10.1037/0882-7974.22.2.300
Charles ST, Mather M, Carstensen LL (2003) Aging and emotional memory: the forgettable nature of negative images for older adults. J Exp Psychol Gen 132(2):310–324
Cohn MA, Fredrickson BL, Brown SL, Mikels JA, Conway AM (2009) Happiness unpacked: positive emotions increase life satisfaction by building resilience. Emotion 9(3):361–368. https://doi.org/10.1037/a0015952
Damoiseaux J, Beckmann C, Arigita ES, Barkhof F, Scheltens P, Stam C, Smith S, Rombouts S (2008) Reduced resting-state brain activity in the "default network" in normal aging. Cereb Cortex 18(8):1856–1864
Erikson EH (1963) Youth: change and challenge. Basic Books, New York, NY
Grady CL, Springer MV, Hongwanishkul D, McIntosh AR, Winocur G (2006) Age-related changes in brain activity across the adult lifespan. J Cogn Neurosci 18(2):227–241
Gross JJ, Carstensen LL, Pasupathi M, Tsai J, Götestam Skorpen C, Hsu AY (1997) Emotion and aging: experience, expression, and control. Psychol Aging 12(4):590
Grühn D, Smith J, Baltes PB (2005) No aging bias favoring memory for positive material: evidence from a heterogeneity-homogeneity list paradigm using emotionally toned words. Psychol Aging 20(4):579
Jopp D, Rott C (2006) Adaptation in very old age: exploring the role of resources, beliefs, and attitudes for centenarians' happiness. Psychol Aging 21(2):266
Kunzmann U, Thomas S (2014) Multidirectional age differences in anger and sadness. Psychol Aging 29(1):16
Maillet D, Schacter DL (2016) Default network and aging: beyond the task-negative perspective. Trends Cogn Sci 20(9):646–648
Mather M (2012) The emotion paradox in the aging brain. Ann N Y Acad Sci 1251(1):33–49
Mather M, Carstensen LL (2005) Aging and motivated cognition: the positivity effect in attention and memory. Trends Cogn Sci 9(10):496–502
Mather M, Johnson MK (2000) Choice-supportive source monitoring: do our decisions seem better to us as we age? Psychol Aging 15(4):596
Mather M, Mazar N, Gorlick MA, Lighthall NR, Burgeno J, Schoeke A, Ariely D (2012) Risk preferences and aging: the "certainty effect" in older adults' decision making. Psychol Aging 27(4):801
Mienaltowski A, Corballis PM, Blanchard-Fields F, Parks NA, Hilimire MR (2011) Anger management: age differences in emotional modulation of visual processing. Psychol Aging 26(1):224
Mostofsky E, Penner EA, Mittleman MA (2014) Outbursts of anger as a trigger of acute cardiovascular events: a systematic review and meta-analysis. Eur Heart J 35(21):1404–1410
Rothermund K, Brandtstadter J (2003) Coping with deficits and losses in later life: from compensatory action to accommodation. Psychol Aging 18(4):896–905
Samanez-Larkin GR, Gibbs SE, Khanna K, Nielsen L, Carstensen LL, Knutson B (2007) Anticipation of monetary gain but not loss in healthy older adults. Nat Neurosci 10(6):787–791. https://doi.org/10.1038/nn1894
Steptoe A, Deaton A, Stone AA (2015) Subjective wellbeing, health, and ageing. Lancet 385(9968):640–648
Stone AA, Schwartz JE, Broderick JE, Deaton A (2010) A snapshot of the age distribution of psychological well-being in the United States. Proc Natl Acad Sci 107(22):9985–9990
Timmer E, Westerhof GJ, Dittmann-Kohli F (2005) "When looking back on my past life I regret…": retrospective regret in the second half of life. Death Stud 29(7):625–644

Tulving E (1995) Organization of memory: quo vadis. In: Gazzaniga MS (ed) The cognitive neurosciences. The MIT Press, Cambridge, MA, pp 839–847
Ulloa BFL, Møller V, Sousa-Poza A (2013) How does subjective well-being evolve with age? A literature review. J Popul Ageing 6(3):227–246
Verhaeghen P, Cerella J (2002) Aging, executive control, and attention: a review of meta-analyses. Neurosci Biobehav Rev 26(7):849–857
Weiss A, King JE, Inoue-Murayama M, Matsuzawa T, Oswald AJ (2012) Evidence for a midlife crisis in great apes consistent with the U-shape in human well-being. Proc Natl Acad Sci 109(49):19949–19952

Psychobiology of Addictions

7

Icro Maremmani, Matteo Pacini,
and Kostas N. Fountoulakis

7.1 Overview

Investigation has been increasing into which function is closer to the core of the addictive neurobiology that we witness as a globally subverted ensemble of psychological parameters (Table 7.1).

First of all, one should take the pre-relapse status as the apparent normality following temporary detachment from the addictive cycle and compare it with the renewal dynamics of already initiated relapse.

Abnormality does not loom as an altered sensitivity to a substance, even if it also includes that; it mainly involves an altered appetitive drive, followed by a paradoxical project of pleasure and balance.

Along with the comparison that can be drawn with diabetes (the 'thrifty gene' hypothesis), we may say that human self-stimulation is a system built to be engaged in or else disengaged from external cues. Its evolutionary interpretation may be that of anticipating events that may otherwise take place in a state of need or of happening fortuitously. For instance, appetite should be seen as the dynamic that leads to eating food before the same behaviour is evoked as hunger, that is, in extremis, so complying with the need to avoid the running out of supplies. On the other hand, the instinctual drive towards reproduction leads to systematic intercourse by linking arousal to acting out, instead of waiting for the environment to put together a

I. Maremmani (✉)
V.P. Dole Dual Disorder Unit, Santa Chiara University Hospital, University of Pisa, Pisa, Italy

G. De Lisio Institute of Behavioural Sciences, Pisa, Italy
e-mail: icro.maremmani@med.unipi.it

M. Pacini
G. De Lisio Institute of Behavioural Sciences, Pisa, Italy

K. N. Fountoulakis
3rd Department of Psychiatry, Faculty of Medicine, School of Health Sciences, Aristotle University of Thessaloniki, Thessaloniki, Greece

K. N. Fountoulakis, I. Nimatoudis (eds.), *Psychobiology of Behaviour*,
https://doi.org/10.1007/978-3-030-18323-3_7

Table 7.1 Structure of addiction psychobiology

Components	Clinical aspects	Neurobiological substrate
Psychological and psychiatric precursors	Temperamental disposition, symptoms related to mood, anxiety, impulse-control dimensions	• Dysfunction in reward and stress system • Hypo-activity of prefrontal brain region • Other molecular changes (CREB, etc.)
Acute substance-induced effect	Symptoms related to mood, anxiety, impulse-control dimensions	• Activation/inhibition of specific receptors and neuronal circuits • Activation of dopamine release in nucleus accumbens
Addictive processes	Repeated brain-substance interaction resulting in: • Craving for and dyscontrol over substances • Symptoms of mood, anxiety, impulse-control dimensions	• Dysfunction of limbic dopamine circuits and reward system • Hypo-activity of prefrontal brain regions • Changes in stress system • Other molecular changes (CREB, etc.)
Interaction of addictive processes and psychological-psychiatric precursors	Worsening of craving, loss of control, symptoms of mood, anxiety, impulse-control dimensions	• Positive feedback between assets preceding substance use and those following addictive processes (reward system, impulsive-reflective system, stress system)

Longitudinal perspective (from Pani et al. 2010, modified)

favourable environment. Appetite and sexual drive are structures, which anticipate and amplify the behavioural template, and answer to the environment by enabling the achievement of certain objectives.

These systems have a high threshold of overstimulation, with respect to an average baseline, so that an individual may handle a wide range of stimuli and increase stimulation in order to maintain high levels of performance, in a reverberating circuit. Once that threshold is exceeded, the system fails, but its failure does not take place as a loss of functionality but rather as a decoupling between stimulation seeking and performance. In other words, the systems suddenly go 'out of order' in conditions of irreversible overdrive, which may be intermittent or continuous.

7.2 Theories of Addiction

Theories about addiction should usually try to illustrate the mutual relationships between the three factors involved: the individual, the substance and the social or environmental function of consumption. Some of them actually fail to focus on the dynamics of basic factors, so providing only a possible explanation of what addiction may mean to others, rather than what addiction means intrinsically. Those portrayals do not deserve the status of theories of addiction.

Apart from descriptions of the intrinsic dynamics of addiction, other explanations actually present consumption and addiction as variants of a unique phenomenon, addiction being its extreme and excessive form, while seeming to recognize no qualitative threshold between habitual and heavy consumption and addiction itself. Such theories, although commonly accounted for as theories of addiction, should, more precisely, be referred to as theories of consumption.

We now come to real theories of addiction. In an attempt to summarize the theoretical views on addiction, we may distinguish between the genetically oriented theory, on one hand, and the environmentally oriented theory, on the other, where the substance itself functions as the environmental agent.

7.2.1 Genetic Theories of Addiction

Obviously, the development of addiction requires the individual to interact with an environmental element, which is the source of stimulation, so that no direct genetic determinism of the phenotype is possible in a stimulus-free environment. In this connection it might be argued that, given the presence of some addictive stimulus, a genetic array can influence those who are prone to engaging in self-stimulation (toxicophilia). Recent research has proven the high frequency of certain affective temperamental profiles among drug addicts, regardless of drug type (alcohol, heroin, cocaine polyabuse). Such a genetically based, early-onset disposition to addiction is quite a common phenotype, so that it is better classified as a risk factor rather than a specific substrate for the development of addictive ties.

In other words, a large number of individuals are theoretically predisposed to engage in addictive drug use. Curiously, the temperamental profile seems to vary according to the stage and type of drug: hyperthymia creates a risk of early-onset alcohol consumption, whereas cyclothymia is a correlate of addiction (to alcohol or other substances). Since addiction also looms as a later stage of consumption, this difference may be induced by chronic intoxication rather than by an originally different substrate (Maremmani et al. 2009; Pacini et al. 2009; Rovai et al. 2013).

In individuals who come into contact with psychoactive substances, some genetic variables are concentrated among addicts, though not to the point of allowing a definition of addiction as a genetic result of a 'per se' variable environmental factor (substance use) (Ouzir and Errami 2016). In other words, genetic factors may favour but are not necessary to the development of addiction. It is likely that a number of individuals lacking genetic proneness to addiction will never engage in substance use, so that we cannot state that the generic array is enough by itself to bring about a switch from consumption to addiction, though such a hypothesis remains plausible. All things considered, the psychopathological phenomenon known as addiction comes about either in the presence or in the absence of genetic determinants.

7.2.2 Addictogenic Potential

Exposure to the consumed substance is the only constant of drug addiction, as a premorbid common ground (Pani et al. 2010). On the other hand, cases share the same clinical features and have been described as having common neurobiological substrates. Reviewing these neurobiological findings, some may just mirror the effects chronic exposure has on the human brain, without there being any difference between heavy consumers and addicts. Others may be rather specific to addiction, such as those displaying the different receptorial arrays of detoxified alcoholics and the sensitivity to cueing in preconditioned human brains with respect to craving.

Also, some neuroendocrine features have been described as predictors of relapse, such as prolactin for cocaine users or the hypothalamus-hypophysis-surrenal axis for detoxified heroin addicts (Teoh et al. 1990).

According to the substance-centred hypothesis, exposure to certain addictogenic substances is enough by itself, given a variable period of time, to induce a stable change in the human brain, to be located as overlapping with the reward-motivational circuit. The stability of such a new array is enough to constitute a bridge across the symptom-free periods in between and reveal itself as long-term relapsing behaviour. Cueing, stressful life events and other factors may shorten the intervals of short- and medium-term latency before each relapse but would make no difference in a longer-term observation. On the whole, the diagnostic criteria show a good fit with this view of addiction, as long as the development of certain behaviours makes it automatic to formulate a prognosis of automatic relapsing (chronic-relapsing course).

The definition of addictogenic potential seems to correspond to a common pharmacological feature, shared by all addictive stimuli, that is, the time-to-peak (the shorter it is, the higher the non-medical use potential), regardless of other factors. Changes in time-to-peak that correspond to a different route of administration (i.e. oral instead of inhalation) do have an impact on the non-medical use potential of the same molecule, as also happens with any change in the kinetics of release.

Among benzodiazepines, for instance, the more lipophilic (sharper alpha phase) they are, the more they are welcomed by drug users and the more implicated they are in cases of sedative use, regardless of duration, chemical structure and potency (Lalive et al. 2011). The same goes for free base products, which are sometimes so highly addictive that they provide a perfect fit for the dead-end subpopulation of street addicts (from a pusher's point of view).

An addictogenic profile must, therefore, be conceived as a combination of a molecular target (receptor) and a pharmacokinetic profile characterized by a short time-to-peak, together with a high ratio of brain/blood distribution (partial distribution volume).

Highly addictive 'preparations' include snorted, inhaled and injected heroin, buprenorphine, morphine, cocaine and methamphetamine, while their slow-acting or oral variants show little or no problematic use liability. Such a radical difference makes it clear why certain substances may be employed for the treatment of an addiction that was induced by other substances belonging to the same class (e.g. methadone treatment for heroin addiction). In some cases, the same molecule in a different preparation becomes the treatment agent for addiction to its fast-acting form (as happens with slow-release morphine and slow-release heroin for heroin addiction).

7.2.3 The Self-Medication and Self-Selection Hypothesis

This hypothesis bears the name of Khantzian, who originally described it as a possible explanation for two case reports concerning cocaine and heroin use (Khantzian 1980, 1985, 1997).

The theory assumes that consumers become addicts in an attempt to treat symptoms of discomfort, which is equivalent to granting them an average balance, from a homeostatic perspective. Once they learn that certain substances fit their discomfort, they are likely to use them on a regular basis, only to discover later on that the effect is transient and the risk/benefit ratio becomes inverted with respect to the initial phase. Unfortunately, they cannot detach themselves from their acquired use, due to the exacerbation of the original discomfort while tapering, let alone interrupting that use altogether, so that they find themselves hooked: in other words, they came to depend on a substance in order to avoid suffering, but at a later stage they no longer gain any benefit from it.

Although this theory is easy to understand and is intrinsically consistent, it actually fails to describe addiction as a relapsing phenomenon. Despite the spread of psychopharmacological treatment, reaching out to a variety of psychopathological areas (including depression, psychosis and anxiety), drug addiction has continued to spread. Moreover, substance-using trends among psychiatric patients fail to show a choice, which is consistent with a homeostatic link: depressed patients use sedatives and hypnotics, while overexcited patients use anything, with a special liking for stimulants (Maremmani et al. 2012). The majority of psychotic patients endure in psychotomimetic use, despite having experienced hospitalization (Maremmani et al. 2004).

Epidemiological findings following Khantzian's work have failed to confirm his hypothesis, either on grounds of personality profiles or of DSM axis I psychiatric disorders.

On conceptual grounds, detachment from the substance is far from being unlikely in the course of an addict's career. The feasibility of detoxification and the opportunity to have it tailored to one's personal needs make it implausible that addicts stay hooked on substances due to their incapability to handle withdrawal. Also, evidence of being tied to self-medication can no longer be found in the addictive phase, where all parameters of psychic well-being worsen. Lastly, the experience of the failure of addictive drugs as a valid means of long-term self-medication should immunize former users from relapse and make them likely to resort to psychiatric facilities for the management of their psychic discomfort. Neither of these happen to be true (Lovrecic et al. 2004).

7.2.4 Environmental Induction

This theoretical view assumes that the external objective conditions in which the individual meets the substance may play a crucial role in the development of addiction. The environment could make the difference in raising the utility of certain substance-induced effects or encouraging a substance-based lifestyle (Enoch 2012; Meyers and Dick 2010; Mayfield et al. 2008). Yet neither type of explanation is consistent with the diagnostic evidence of the disadaptive condition of drug addicts, which does not allow them to fit any environment, no matter how substance-based it may be. In other words, addicts are unhappy, disorganized and clumsy consumers, and they certainly have fewer resources at their disposal in handling environmental stress.

More recently, this theory has been revisited in epigenetic terms: environmental factors may change the brain's sensitivity to substances, thus playing an important role in the issue of how much addictive power substances express into the brain (Kalda and Zharkovsky 2015; Enoch 2011). The extent to which exposure to substances leads to addiction may be conditioned by an epigenetic effect that makes individuals prone to addiction. However, any such common epigenetic ground does not have yet to be defined, nor has any precise correspondence between epigenetic molecular changes and life events been ascertained so far.

7.2.5 Behavioural Addictions

Theories of behavioural addiction do not differ substantially. Gambling addiction has been investigated in some depth as far as addictogenic features are concerned. The short time gaps between game opening and betting and between betting and its outcome are powerful variables, as well as the time elapsing between repetitions of the betting cycle within the same prepaid game session.

Although expectations about winning chances and the overrating of the rational strategies for successful betting may play a role in getting involved in habitual gambling, the addictive phase is intrinsically independent of such factors. At this stage evidence of negative balance is there to prove how strongly addiction is distinguished by loss of sensitivity to losing and winning. In other words, no reward or negative reinforcement is enough to stop the overwhelming automacy of behavioural relapse (Wiehler and Peters 2014; Gaher et al. 2015).

7.3 The Reward and the Decision-Making System

Addiction is usually described as a disease of the reward system or the incentivizing, decision-making system, which is reinforced by rewarding stimuli and held back by drawbacks and negative reinforcement (Bjork et al. 2009).

Yet, it may not be clear whether we are referring to the impairment of liking, an abnormal intensity of reward seeking, a twist in the attribution of reward or a failure by regulatory areas in the brain to filter the development of craving.

7.3.1 Habit

Habit differs from addiction in the intentionality of use or the lack of any effort to refrain from a certain behaviour that is performed frequently or regularly. Although 'the habit' is a term often used as a synonym for the drug problem, or addiction, the issue of control is not involved: it may happen that people do not give up habits or stick to them simply because they actually feel better when exposed to some stimulus and suffer from its absence. It is, in fact, accurate to say that the revealing moment of addiction is the loss of control when the stimulus is widely available and exposure to it becomes frequent.

Habitual exposure to certain stimuli may also imply a homeostatic change known as 'tolerance', which implies a susceptibility to transient withdrawal discomfort as a consequence of abrupt interruption of such exposure. So too, habitual stimulation may imply toxic consequences that individuals may only become aware of much later—consequences that may lead them to quit or reduce their exposure.

As far as giving up the habit is concerned, addicts are probably the human category that is best at doing so, before any relapse takes place. It is relapsing, not the inability to detach, that is the crucial feature of addiction.

7.3.2 Reward

The central role of reward is sometimes made equal to craving, which is quite a serious conceptual mistake. Reward should never be thought of as 'pathological', not even in addiction. What ceases to work properly in the system of reward, once addiction has developed, is the capability to optimize reward. Although drug addicts worsen their chances of getting rewards from sources other than the drug, it is also true that their first failure is that of handling reward by the drug itself. Addiction develops in the place of physiological reward, so that the best way to grasp its meaning is to compare addictive use with use that is pre-addictive or is still under physiological control.

The presence of other sources of pleasure, interest into them and endurance in searching for pleasure from the same source may vary from individual to individual. On the other hand, all addicts share the same incapability to pursue reward by the substance they are dependent on.

Habitual use, far from meaning how frequently addicts need to be stimulated, is, rather, a proof of their inability to learn how to prevent the development of a state (tolerance) that is unfavourable to internal development. In fact, they find themselves striving again and again to overcome withdrawal, if possible very rapidly, in order to be able to sense the drug again. Addicts probably end up spending more time far from reward than occasional users are, regardless of the other harmful correlates of their condition.

7.3.3 Reinforcement

Addicts become insensitive to either kind of reinforcement. On one hand, individuals usually go through a refractory period after being exposed to rewarding stimulation, so that some time passes before they engage in stimulation again. Although sessions of repetitive stimulation may be featured in physiological self-stimulation, spontaneous intervals take place, which are distinguished by enduring well-being. It also happens that repetitive stimulation ends up in saturation of reward, together with boredom and loss of interest in a specific source of stimulation (Rovai et al. 2013).

On physiological grounds, positive reinforcement is expected to work both to make the individual closer to, and more efficient in, self-stimulation and also to promote the completion of the stimulation cycle and the onset of boredom.

Negative reinforcement is usually rooted in the experience of negative consequences after drug-taking. Real-time negative reinforcement (e.g. vomiting after heroin use) may be a weaker kind of conditioning than positive reinforcement, but it is resorted to as a strategy of aversion from drug use, as in disulfiram treatment for alcohol and cocaine addiction. Later reinforcement consists of the negative consequences of addictive behaviours and a drug-related lifestyle. Such reinforcement, however, is more likely to be experienced once addiction has developed. Early negative reinforcements are likely to be effective in preventing drug use in a pre-addictive phase, although in that stage, they are powerfully counteracted by the increasing positive reward ('honeymoon') phase (Heinz et al. 2004).

A major source of negative reinforcement is the withdrawal syndrome (Heinz et al. 2004). Some users may be held back from enduring in drug use just because of the transition from a positive balance (prevailing positive reward) to a negative balance (emerging withdrawal, decreasing reward). As long as such a transition takes place in the not-yet-addicted individual, it may be perceived as a reason strong enough for breaking the habit. It may be hypothesized that such a balance soon becomes unsatisfactory for those drug users who start using drugs for environmental reasons or self-medication.

Negative reinforcement shows a different degree of influence on drug use according to the drug use stage. In nonaddicted individuals, negative reinforcement may lead to quitting, cutting down or temporarily detaching. Conversely, negative reinforcement during addiction may produce the opposite effect, especially when it is linked to drug deprivation. The experience of withdrawal, far from being a 'rock-bottom' experience that strengthens the individual's motivation to quit, may leave a memory similar to that of a panic attack. In the subsequent addictive cycle, the individual becomes more and more phobic about being on withdrawal, so that his/her phobia becomes a stronger and stronger path to craving, so bringing the individual closer to the essential appetitive stance. In other words, addicts are led to come to grips with withdrawal, or the fear of it, by enduring drug use rather than trying to detach from it. Eventually, they may experience spontaneous withdrawal or cue-elicited withdrawal starting in a drug-free condition and be led to relapse by the abrupt emergence of craving (Grace 2000; Maremmani and Pacini 2003)

On the whole, positive and negative reinforcement become neutral for the addicted individual and may even be subverted. A smaller reward becomes enough to maintain a heavier habit, whereas a worse trail of consequences becomes ineffective in encouraging detachment from the drug, possibly ending up by working in the opposite direction, as auxiliary to craving.

The output behavioural array that appears in response to a rewarding stimulus (more simply indicated as 'wanting') is a variable dependent on other psychobiological variables and may be divided into two main subtypes: appetitive and avoidant (Weiss et al. 2001). Appetitive behaviour aims to achieve exposure to a certain stimulus, whether passively or actively, in order to reach the highest peak stimulation possible. Avoidant behaviour aims to prevent suffering, pain or discomfort. When the individual's array has been conditioned by a rewarding dynamic, the balance between appetite and avoidance shifts towards the former, as a result. Despite this, to a certain extent, the behavioural output stays sensitive to pro-avoidant inputs. When the lack of stimulation is followed by the onset of acute withdrawal, a special form of avoidance is strengthened, and it converges with appetite in aiming for the same objective. At this stage, the behavioural array may be redefined in terms of detachment from exposure to the addictive stimulus, which is the source of a vicious circle between reward impairment, appetite endurance and harm increase. However, appetitive drive may already have developed so far that it overwhelms any attempt by the individual to become detached or stay abstinent.

It should also be kept in mind that the appetitive subsystem mostly grows through a fast-acting subcortical mechanism, which is reward, whereas the avoidant subsystem mostly grows by a slow-acting, cortical mechanism (Milton et al. 2008). Once appetite has taken over in a stable mode, the behavioural array becomes rigidly oriented towards a certain output, which is hardly influenced by any other harm-avoidant drive, let alone intention. In other words, the array is not influenced by recurrent harm, nor is it discouraged by the lack of reward. In fact, appetite itself is set at an overdrive level, where no reward can further reinforce it and no frustration is able to decrease it.

Awareness of such a condition may vary across the different stages. Addicted people tend to lose their awareness of (insight into) the change they have gone through. The hierarchical position of the cortical 'inhibitory' areas seems to be subordinate to the subcortical, instinctual ones, so that the individual's attitude may change along with increasing appetite, so that he/she becomes self-unaware when addiction has set in. The expectancy that inhibitory areas will come to grips with increasing appetite and put a limit to it is true below a certain threshold. Over that threshold, the array is subverted: not only are inhibitory areas weaker and slower than antagonist instinctual drives, but they also tend to be obscured by them. As a result, the psychobiological state of the addicted individual becomes wholly consistent with the appetitive purpose, with no room left for a different outcome.

The reinforced appetite drives towards self-stimulation by the reinforcing objectives. Stressful experience leads to the instinctual activation of withdrawal-related harm avoidance and thus eventually to self-stimulation, though theoretically in a withdrawal-relieving perspective. Inhibitory areas fail to hold back appetite or alarm, while the thought about this dynamic is focused on the pursuit of stimulus optimization and control rather than detachment.

7.4 Factors Affecting Non-medical Use Potential

The psychobiology of risk dispositions includes genetic, epigenetic and environmental factors and is to be understood as a focus on the psychobiological dynamics corresponding to the different correlates of problematic use and addiction (Pani et al. 2010).

In other words, the question arises: at what level may different factors influence the circuit of drug-related decision-making, regardless of whether those factors are genetic, epigenetic or environmental?

Relying on these premises, we may now try to describe how addiction may be facilitated, or enhanced, at different levels, by pre-existing dispositions affecting drug-liking, impulsivity and behavioural inhibition vs. elation, reward and reinforcement.

Some of those factors may be psychologically 'silent', that is, have no cognitive translation, only a behavioural ground of expression. Others may contribute to psychological variants of addictive phenotypes. Furthermore, some factors seem to affect the severity of expression of core addictive features (expressivity) and the threshold of exposure to the substance at which addiction sets in (penetrance).

To date, several studies have indicated the concentration of genetic variants among populations of addicts, though without clarifying whether they indicate a genetic disposition to risk behaviours, substance use or a specific vulnerability to developing addiction with respect to nonaddicted users. In fact, control groups are often made up of non-users, and they often fail to produce evidence about the role of genetic features in determining the addictive potential of substances (Blum et al. 2012).

Clinical research into the dynamics of addiction has revealed how the feature of cyclothymia is maintained across different substance use groups (cocaine, heroin, alcohol) and stands out as a discriminant trait of addicts regardless of dual disorder, even of dimension-related psychiatric pictures (major mood disorders) (Maremmani et al. 2009; Pacini et al. 2009; Rovai et al. 2013). Heroin addicts, cocaine users/addicts and alcoholics are more cyclothymic than control subjects. Dual Disorder addicts are as cyclothymic as single-disorder ones, and both groups are more cyclothymic than controls. Looking deeper into the typology of dual disorder, some interesting data emerge. In a group of alcoholics, cyclothymia is higher in lifetime polyabusers (of heroin and alcohol), but the stronger link does not apply to the more deviant substance (heroin) but to the more common one (alcohol). In fact, when heroin addicts are compared with a control group composed of non-users and pure alcoholics, the level of cyclothymia shows no difference, whereas a difference is measurable when alcoholics are compared with non-users grouped together with heroin users.

It should also be noted that the same temperamental disposition may not play the same role at all stages of the pathophysiological process, leading from first exposure to addiction. In fact, adolescents with early-onset recreational alcohol use show a higher raw score for hyperthymia and are more likely to have a dominant hyperthymic profile (Placidi et al. 1998). Hyperthymia, however, is not the most likely profile among addicts, as mentioned above. Otherwise, in the comparison between heroin addicts and controls, hyperthymia appears to be discriminant for nonaddicts, whereas cyclothymia, variably mixed up with other 'polar' features, is typical of addicts (Maremmani and Gerra 2010).

We may, therefore, hypothesize that hyperthymic traits play a role in favouring the first contact with either legal or illegal drugs, whereas it is the cyclothymic profile as a dominant structure that favours the transition from physiological to pathological use and addiction.

On psychiatric grounds, major psychiatric disorders seem to be neutral, although some may be underrated as to the presence of 'para-physiological phases' such as hypomania for bipolar disorder II and also due to the interpretation of substance-related mood disturbances as intoxication rather than as clinical variants of the psychiatric disorder (bipolar phenotype III) (Akiskal et al. 2003).

In order to avoid the classic overlap between an ambiguous diagnosis of depression (as an episode, syndrome or longitudinal unipolar disease) and minor bipolar disorders, we have compared the diagnoses received by alcoholics according to their substance use status, dedicating special attention to the recognition of hypomania. Our comparative analyses reveal that depression and bipolar disorders (which also include depressive phases) play opposite roles with respect to alcohol problematic use and addiction. First of all, the anamnesis of a major depressive episode is related to the absence of polyabuse (heroin-alcohol), whereas the presence of polyabuse, as anticipated above, was predicted by cyclothymia. At a later stage, we divided the sample into two subgroups (dual addiction or alcoholism coupled with a history of nonaddictive heroin use). The negative correlation was still evident between a history of depression and dual addiction, but not in cases of single drug use (Pacini et al. 2009).

On the whole, the pathway to addiction seems to start on the wings of hyperthymic traits, proceed along cyclothymia and be interrupted by the occurrence of major depression. Prominent hyperthymia, on one hand, and depressive phases loom as obstacles to the progression of substance use in moving towards the addictive stage.

The addictive potential of substances, in other words, may be enhanced by cyclothymia and minor bipolar syndromes ('less than manic') and diminished by a reward-refractory state, such as depression.

The course of polyabuse is also related to the dual disorder of bipolar disorders, so that bipolar heroin addicts or alcoholics are those who mostly tend to engage in use of cocaine as a secondary substance (Maremmani et al. 2008; Pacini et al. 2010).

A reduced latency to addiction is reported for aggressive patients. The meaning of such a link is not known, although we can hypothesize that it has to do with the opioid system, which is involved both in the modulation of aggressiveness and in the reception of opiates. The malignancy of addiction appears to run parallel with the loss of control over aggressiveness, which is weaker from the beginning, and is later hampered by opiate intoxication. The fragility of the opiate system therefore seems to independently underlie aggressiveness and addiction proneness (Bacciardi et al. 2013; Maremmani et al. 2014; Bacciardi et al. 2012).

The self-medicating pathway may shorten the period of latency prior to regular use. It should, however, be remembered that self-medicating would-be addicts do not engage in regular use just as a consequence of the substance's medicating potential. A study on a group of social phobic patients clarified the differences between alcohol-abusing and non-alcohol-abusing ones (Perugi et al. 2002). Although a majority of socially phobic patients agreed on the medicating, anti-phobic potential of alcohol on the basis of personal experience, only one subgroup had chosen to use it wilfully to antagonize social anxiety on a regular basis. This latter subgroup was characterized by a higher prevalence both of individual and family diagnosis of bipolar disorder.

As regards alcohol use in depressed patients, available data for a group of atypically depressed patients show that dominant cyclothymic temperament is far more concentrated among alcohol-abusing than non-abusing depressed peers. Comorbidity with bulimia displays the same correlation with cyclothymia, whereas other anxious or somatic forms of comorbidity do not (Perugi et al. 2002).

It is difficult to establish whether cyclothymic patients feel the effects of substances as sharper, in terms of reward intensity. The self-reports of reward intensity do not allow any easy comparisons between different diagnostic groups, due to the different thresholds of satisfaction they may refer to. In any case, the professional literature is lacking in studies about the influence of psychiatric diagnosis on the salience and the reward correlates of non-medically used drugs, which should become a future field of clinical research.

Epigenetic research has shown that exposure to stress triggers biochemical cascades, which end up stably influencing DNA transcription rates (McGowan and Roth 2015). Persistent changes lead to a second genetic array, resulting from a modulation of the basic array, and may justify a stable change in function at a local level, involving single cells, circuitry or cell types (Bale 2015). It is known that chronic intoxication is related to abnormalities in stress-related endocrine circuits, such as the cerebral-surrenal axis, which accounts for the acquired difficulty of heroin addicts in handling normal or exceptional stress (Kreek et al. 2004). Endocrine abnormalities do persist after the accomplishment of detoxification and are linked with a chronic-relapsing course. Stressful events, although to a lower degree than drug cues, are effective in provoking relapses, by a sharp increase in craving levels. Beyond the subjective meaning that can be attributed to these events, the behavioural output either of negative emotions or a positive tickling of desire for the chosen drug ends up leading the individual to relapse.

Nevertheless, data are unable to clarify whether stressful stimulation has an influence and, if so, of what kind, on the liking for drugs, and the intensity of reinforcement after occasional or habitual drug use, before addiction has settled in. On the other hand, indirect or concurrent rewards may play a crucial role in enhancing the reinforcing power of drugs and accelerating the process of addictive binding, starting from equal levels of exposure.

7.5 Human Biological Correlates of Drug Using Behaviour: From Physiology to Pathophysiology

Non-medically used substances share a common acute effect, which connects up the subjective effects of euphoria, 'high' pleasure, with the increase occurring in dopamine release in certain subcortical areas, notably the ventral striatum, in particular in a smaller area known as the nucleus accumbens.

While both slow-onset and rapid-onset dopamine increases do correspond to a neuroimaging change, only fast-onset dopamine increases are real-time correlates of a subjective pleasurable change (Grace 2000; Volkow and Swanson 2003).

In comparable conditions of metabolic hyperstimulation, the reinforcing effect is thus due to a fast-acting dopamine high tide (peak) (Volkow et al. 1996). Non-medically used substances may differ as to their primary receptor, the quality of their effects and the duration of otherwise similar effects, but their use liability is rooted in their rapidity of dopamine-mediated action. In order to produce a rapid action, the route of administration is crucial, since oral administration is unlikely to produce a fast blood peak, as happens with intranasal absorption, inhalation or intravenous injection (Volkow et al. 2000; Chait 1994; Volkow et al. 2001).

An expectation of pleasant effects, which is automatically conditioned by previous experiences, corresponds to an amplification of the drug's effects (Kalivas and Volkow 2005). As a result, people who try a drug which matches the effects they expect experience a higher level of reward and are more strongly reinforced. Also, at an addictive stage, a heightened expectation running in an automatic way, regardless of any recent negative experiences, does increase the reinforcing power of even low doses or allows some effect to break through in a condition of theoretical tolerance.

During chronic intoxication, the common ground is represented by a downregulation of D2 receptors in the striatum, which is attributed to a competition with endogenous dopamine. Once dopamine release dwindles, due to the discontinuation of substance use, the D2 downregulation endures long after the end of acute withdrawal and corresponds to a subjective feeling of dysphoria, dullness, reduced sensitivity to and interest in previously pleasant stimulation. On the other hand, sensitivity to drug-related cues is amplified (Heinz et al. 2004; Volkow et al. 1993, 2001, 2007).

Consequently, symptoms of chronic intoxication, residual 'late-withdrawal' symptoms, as well as acute withdrawal increase the discrepancy between the indifference shown towards general stimulation and the exacerbation of drug-selective craving (Childress et al. 1988; Martinez et al. 2004; McClernon et al. 2009; Volkow et al. 1997).

The activity level of functionally antagonist systems seems to be enhanced in chronic drug users. For instance, acute cocaine withdrawal is distinguished by supersensitivity to GABAergic medications, as well as to opioidergic medication. From a chronic perspective, cocaine exposure may become less and less pleasant, and its side-effects more and more likely, so that people may resort to polyabuse in order to buffer cocaine toxicity or switch to substances they have become more sensitive to, such as benzodiazepines, alcohol or opiates (Volkow et al. 1998; Zubieta et al. 1996)

All of these phenomena that belong to a reversible pattern feature the sequence: hyperstimulation-desensitization, followed by functional impairment and hypersensitivity. Moreover, features of chronic intoxication emerge over time as signs of chronic, persistent (sometimes permanent) brain damage.

As to addiction, we have to map out a different pattern that underlies the features that are peculiar to addictive diseases, namely:

1. Chronic relapse, despite the intention to stay detached
2. Enduring use, despite the absence of expected effects and the presence of individual knowledge about the dynamics of tolerance
3. Loss of insight into the right perspective for handling substance use and becoming detached from it

Chronic cocaine users show an attenuated response to stimulants, which indicates that their capability to self-stimulate has been hampered. The hypothesis that repeated exposure results in sensitization is probably true in the medium term, but further changes take place later on, which account for the chronic-relapsing course, extending beyond each single cycle of non-medical use.

Repeated exposure to drugs induces a chain of reaction that starts from the ventral tegmental area, proceeds throughout the nucleus accumbens and is projected onto the dorsal striatum. The dorsal striatum shows dopamine increases which are rather specific to addiction, and correlate with subjective craving, as long as they correspond to spikes of dopamine (phasic), rather than a slow-onset tonic increase. Studies on cueing have also suggested that craving runs parallel with activation of the dorsal striatum and does not depend on exposure to substances: any such chain of reactivation would, in fact, be unlikely, due to the downregulation or desensitization of D2 receptors after chronic exposure to drugs.

It has been hypothesized that addicted individuals differ from nonaddicted ones due to the presence of an inner source of dorsal striatum dopamine increases, which is no longer related to reward from the outside but can be triggered by cueing (McClernon et al. 2009). This source may correspond to the anatomic area of the cingulate gyrus, orbitofrontal cortex and extended amygdala and its glutamatergic connection with the striatum (Franklin et al. 2007; Volkow et al. 2004, 2005, 2008).

The global brain change originating from outreach of the areas involved in the affective encoding of experience, and in long-term memory, is likely to correspond to the development of a strong incentivization and motivation 'pole', which hijacks the pre-existing moral array of the individual and prevents any other array from replacing the one that stays focused on drug availability and consumption.

7.6 Clinical Issues

On psychobiological grounds, we feel the need to overcome the clinical limitations and uncertainty resulting from the evolution of DSM layouts. Addiction had, in fact, been defined in a qualitative way in the DSM-IV era, with DSM-IV TR presenting a set of criteria that first opened the way for a clinically based solution (APA 2000). Going now into greater detail, the patient's history was accounted for as diagnostic criteria with prognostic weight, and intoxication-related aspects were isolated so as not to weigh as core criteria. Later on, DSM-5 provided a completely different layout, with addiction and use aligned along a quantitative continuum (APA 2013). The failure to justify the distinction separating the putative 'non-medically used' category from vice, by reference to physiology, on one hand, and addiction, on the other (by considering it as a disease), was eventually overcome by dissolving all firm qualitative barriers between substance use disorders. The difficulty encountered in delineating the borders between pathology and physiology can be called something of a constant in this sequence of classification attempts.

We propose a three-stage clinical classification leading from physiological use to addiction, possibly through an intermediate problematic use stage (Table 7.2).

At stage I, the picture features a variable pattern of use, displaying global post hoc satisfaction after episodes of fulfilling acute consumption. People may develop a physiological habit and dislike the absence of the stimulus that rewards them. At this stage they do show interest and arousal when the stimulus is available.

At stage I-a the individual is engaged in an intensive way with self-stimulation and may experience negative consequences or implications arising from this behaviour, whether it is habitual or not. Nevertheless, this behaviour cannot be said to be out of control, as long as the reason for increasing commitment can be found in an increasing level of reward. This kind of picture corresponds to what is usually meant by 'problematic use' and can be assigned to the area of pathology, as far as intoxication symptoms are concerned, but it could also be assigned to the area of physiology, as far as control is concerned.

At stage II, a leap has taken place, and what is shown is a different kind of behavioural disarray. The person is dissatisfied, although he/she may not identify the problem as being due to the loss of control. The acute reward is still present but is now weaker, spoiled by side-effects or the deterioration of other sources of pleasure. Withdrawal may be exacerbated, but what counts most is the urgency of substance-bound behaviour that is enhanced to an overdrive level: in other words, the consistency between behavioural engagement and reward has been lost (Maremmani and Pacini 2003). The frequency of consumption has risen to a ceiling level, which by itself is not enough to constitute an isolated criterion but is a logical consequence of the increase of craving.

Table 7.2 Stadiation of addictive processes

	Stage I	Stage I-a	Stage II
Consumption mode	Consumption	Problematic use	Addiction
Post hoc satisfaction	+	+/–	–
'High'	+	++	(+)/–
Withdrawal-related discomfort	–	+/–	(+)
Frequency of behaviour	(+)	++	+++
Substance-elicited desire	+	+	+++

On the whole, we propose a control-over-stimulus area, comprising both use and problematic use, which can be indicated as a reversible state still rooted in the dynamics of reward. That leaves the 'out of control area' corresponding to addiction, which is rooted in an acquired, no longer reversible, chronic-relapsing change in the brain network relationship between reward and motivational drive (Bacciardi et al. 2013).

Treatments for addiction have a variety of psychobiological targets, according to which they may be classified (Maremmani and Pacini 2003; Pacini and Maremmani 2007a, b).

Direct anticraving treatments: these aim to restore the balance between desire and behavioural control. In addition to the obvious improvement to be experienced, patients are also relieved and amazed by recovering a substance-free perspective, which they had often given up after years of failures. Therapeutic engagement and motivation may be enhanced after realizing that one's brain has not gone wrong, or been born wrong, but that just a single piece of it is responsible for what is seen from the outside and experienced internally as moral disarray. Methadone, buprenorphine and other emerging treatments exert this kind of effect (Maremmani and Gerra 2010).

Anti-reward treatment: the psychological effect of such treatment continues to be incisive as long as the clinical picture shows it is still rooted in the dynamics of rewarding use (I and I-a stages) and may even help the gradual reversal of stage II pictures, so allowing a return to a state of behavioural control. Even so, it must be pointed out that interference with reward alone is unlikely to stop the course of addiction. Most addicts have already gone through the various phases of reward deterioration and have relapsed into it after detoxifying – a development that indicates that reward-centred logic is no longer dominant within that addiction. In fact, the best acting anti-reward regimen is provided gradually by a process of 'induction' into methadone or buprenorphine treatment, which is coupled with a state of narcotic blockade (due to high tolerance or stronger blockade at lower levels of tolerance), so producing a direct anticraving effect.

Aversive treatments: these are designed to abolish the chance to experience acute reward, as an anti-reward treatment, while also making this approach overlap with an aversive effect. Both the previously cited treatment modalities require external control over the treatment intake, since the psychobiological orientation of the patient is not favourable to self-administration. In fact, one reaction to be expected in response to reward impairment will be for the patient to act in such a way as to restore potential reward by discontinuing treatment.

Treatment for nonaddictive problematic use varies and may spring from different roots, either psychopathological or psychological, which underlie a problematic relationship with useful or rewarding stimuli.

Lastly, addicts differ from problematic users because they are expected to react to any treatment effort by presenting an ambivalent behaviour. Addicts, in fact, will ask for help because of the collateral damage done to their drug-related lifestyle or the need to change current states of intoxication but tend to keep away from treatment as an instinctive reaction, as if engaging in any treatment meant being kept in a state of separation from the substance. In the absence of an inner substance-free perspective, such a reaction should be expected. Moreover, the reasons for requests for treatment should not be mistaken for what a nonaddicted person would look for in response to the same interventions: for instance, what addicts requiring detoxification are basically struggling with is not withdrawal but the absence of reward produced by tolerance. It should also be borne in mind that recovery may mean restoring the patient's global resources, only for him/her to engage in a new bout of addiction, instead of building up a barrier against addiction. While addicts may sincerely intend to stay free from addiction, they are often subject to a stronger drive, which can be defined as getting control over the substance, given that they can no longer do without it. This also explains why they often sign on for long-term programmes from which they drop out during the early phases and why they show enthusiasm for rapid and uncertain interventions but avoid long-term commitments.

Appendix: Multiple-Choice Questions

1. Craving corresponds to
   - The perception of the stimulating agent
   - The positive quality of stimulation (liking)
   - The increase of liking for experienced drugs through time
   - *The abnormally urgent and overwhelming desire to stimulate oneself*
2. The liking of a drug, which induced addiction
   - Is always stronger than that of other drugs
   - *Is usually stronger and sharper*
   - Becomes highest than ever once addiction has developed
   - Depends on a general reward-deficient personality
3. Nonaddictive craving
   - Does not exist
   - *Does not actually make sense, although it is used to indicate amplified appetite for self-stimulation*
   - Is the craving for nonaddictive substances
   - Is the craving for legal substances
4. Addiction corresponds to a brain state
   - Forerunning the first contact with the drug
   - Starting at the time of first drug use
   - *Developing during drug use*
   - Developing after several relapses
5. Addictions to different substances
   - *Share a similar brain-imaging profile*
   - Concern different brain areas
   - Concern different cortical areas
   - Share a similar withdrawal-related brain state

6. The shared ground between addictions from different substances includes
   - *All of the following*
   - Late withdrawal
   - Cross-reactive craving
   - Reduction of cortical inhibiting activity
7. The release of dopamine in the nucleus accumbens
   - Is a key effect of dopaminergic drugs
   - Is a key effect of heavy drugs
   - *Is shared by all addictive stimuli*
   - Is mediated by a unique specific receptor
8. Once addiction has established
   - The effect of the substance is no longer possible
   - The effect of the substance is amplified
   - *The expectation of the effect is amplified*
   - The relevance of general reward is amplified
9. The addictive potential of stimuli depends on
   - The mechanism of action
   - The legal status
   - The potency
   - *The short latency of action*
10. Late withdrawal
    - *Is not specific of addiction*
    - Is specific of addiction
    - Is predictive of relapse
    - Is a milder trail of early withdrawal
11. Cortical inhibitory areas
    - Are underdeveloped in subject who then become addicted
    - Are atrophic in addicted persons
    - *Are inhibited by craving-related areas*
    - Are not recruited effectively by addicted persons
12. Cueing is
    - *An intense recall of appetite by drug-related memories*
    - The subconscious recall of drug-related memories
    - The persistence of drug-related memories
    - The recall of affectively relevant drug-related memories
13. The objective of anticraving treatment is
    - To suppress the desire for the drug
    - *To normalize craving to normal desire*
    - To suppress the urgency to get the drug
    - To suppress the fear of withdrawal
14. Craving during withdrawal
    - Is basically due to the fear of withdrawal
    - *Is basically due to the increase of appetition*
    - Is the irresistible component of craving
    - Is the core component of craving

15. Non-chemical addictions
    - Are not true addictions
    - *Share similar biological features with chemical addictions*
    - Are not chemically mediated, but psychologically
    - Are a side aspect of certain psychiatric disorders
16. The psychiatric syndrome closely resembling addiction is
    - *Mania*
    - Depression
    - Panic
    - Personality disorders
17. The best fitting model of addictive diseases can be identified in
    - A general neurotransmitter deficiency
    - Subcortical plastic changes in some input areas, affecting substance effects
    - *Subcortical plastic changes in some feedback areas, affecting substance-seeking behaviour*
    - A cortical area
18. Opiate withdrawal has a relevant biological overlap with
    - Depression
    - *Panic attacks*
    - Mania
    - Obsessive-compulsive disorder
19. Relapse differs from a simple new-use episode
    - *The brain reacts differently to the substance and starts back from the addictive stage*
    - The substance is the same
    - Doses are higher
    - The brain is still under withdrawal when relapsing
20. The most effective mechanisms against craving are
    - Motivational treatment
    - Blockade of the substance's effects
    - Direct receptor-mediated reduction of baseline craving
    - *The coupling between the previous two*

References

Akiskal HS, Hantouche EG, Allilaire J-F, Sechter D, Bourgeois M, Azorin JM, Chatenêt-Duchêne L, Lancrenon S (2003) Validating antidepressant-associated hypomania (bipolar III): a systematic comparison with spontaneous hypomania (bipolar II). J Affect Disord 73:65–74

APA (2000) DSM-IV-TR. Diagnostic and statistical manual of mental disorders. American Psychiatric Association, Washington, DC

APA (2013) DSM-5. Diagnostic and statistical manual of mental disorders. American Psychiatric Association, Washington, DC

Bacciardi S, Maremmani AGI, Rugani F, Pacini M, Dell'Osso L, Maremmani I (2012) Aggressive behaviour and heroin addiction. Heroin Addict Relat Clin Probl 14(4):81–84

Bacciardi S, Maremmani AGI, Rovai L, Rugani F, Pani PP, Pacini M, Dell'Osso L, Akiskal HS, Maremmani I (2013) Drug (heroin) addiction, bipolar spectrum and impulse control disorders. Heroin Addict Relat Clin Probl 15(2):29–36
Bale TL (2015) Epigenetic and transgenerational reprogramming of brain development. Nat Rev Neurosci 16(6):332–344. https://doi.org/10.1038/nrn3818
Bjork JM, Momenan R, Hommer DW (2009) Delay discounting correlates with proportional lateral frontal cortex volumes. Biol Psychiatry 65(8):710–713. https://doi.org/10.1016/j.biopsych.2008.11.023
Blum K, Werner T, Carnes S, Carnes P, Bowirrat A, Giordano J, Oscar-Berman M, Gold M (2012) Sex, drugs, and rock 'n' roll: hypothesizing common mesolimbic activation as a function of reward gene polymorphisms. J Psychoactive Drugs 44(1):38–55
Chait LD (1994) Reinforcing and subjective effects of methylphenidate in humans. Behav Pharmacol 5(3):281–288
Childress AR, McLellan AT, Ehrman R, O'Brien CP (1988) Classically conditioned responses in opioid and cocaine dependence: a role in relapse. NIDA Res Monogr 84:25–43
Enoch MA (2011) The role of early life stress as a predictor for alcohol and drug dependence. Psychopharmacology (Berl) 214(1):17–31. https://doi.org/10.1007/s00213-010-1916-6
Enoch MA (2012) The influence of gene-environment interactions on the development of alcoholism and drug dependence. Curr Psychiatry Rep 14(2):150–158. https://doi.org/10.1007/s11920-011-0252-9
Franklin TR, Wang Z, Wang J, Sciortino N, Harper D, Li Y, Ehrman R, Kampman K, O'Brien CP, Detre JA, Childress AR (2007) Limbic activation to cigarette smoking cues independent of nicotine withdrawal: a perfusion fMRI study. Neuropsychopharmacology 32(11):2301–2309. https://doi.org/10.1038/sj.npp.1301371
Gaher RM, Hahn AM, Shishido H, Simons JS, Gaster S (2015) Associations between sensitivity to punishment, sensitivity to reward, and gambling. Addict Behav 42:180–184. https://doi.org/10.1016/j.addbeh.2014.11.014
Grace AA (2000) The tonic/phasic model of dopamine system regulation and its implications for understanding alcohol and psychostimulant craving. Addiction 95(Suppl 2):S119–S128
Heinz A, Siessmeier T, Wrase J, Hermann D, Klein S, Grusser SM, Flor H, Braus DF, Buchholz HG, Grunder G, Schreckenberger M, Smolka MN, Rosch F, Mann K, Bartenstein P (2004) Correlation between dopamine D(2) receptors in the ventral striatum and central processing of alcohol cues and craving. Am J Psychiatry 161(10):1783–1789. https://doi.org/10.1176/appi.ajp.161.10.1783
Kalda A, Zharkovsky A (2015) Epigenetic mechanisms of psychostimulant-induced addiction. Int Rev Neurobiol 120:85–105. https://doi.org/10.1016/bs.irn.2015.02.010
Kalivas PW, Volkow ND (2005) The neural basis of addiction: a pathology of motivation and choice. Am J Psychiatry 162(8):1403–1413. https://doi.org/10.1176/appi.ajp.162.8.1403
Khantzian EJ (1980) An ego/self theory of substance dependence: a contemporary psychoanalitic perspective. NIDA Res Monogr 30:184–191
Khantzian EJ (1985) The self-medication hypothesis of addictive disorders: focus on heroin and cocaine dependence. Am J Psychiatry 142:1259–1264
Khantzian EJ (1997) The self-medication hypothesis of substance use disorders: a reconsideration and recent applications. Harv Rev Psychiatry 4(5):231–244. https://doi.org/10.3109/10673229709030550
Kreek MJ, Zhon Y, Schussman S (2004) Craving in opiate, cocaine and alcohol addiction. Heroin Addict Relat Clin Probl 6(2):5–52
Lalive AL, Rudolph U, Luscher C, Tan KR (2011) Is there a way to curb benzodiazepine addiction. Swiss Med Wkly 141:w13277. https://doi.org/10.4414/smw.2011.13277
Lovrecic M, Lovrecic B, Dernovsek MZ, Tavcar R, Maremmani I (2004) Unreported double frequency of heroin addicts visiting psychiatric services and addiction treatment services. Heroin Addict Relat Clin Probl 6(3):27–32
Maremmani I, Gerra G (2010) Buprenorphine-based regimens and methadone for the medical management of opioid dependence: Selecting the appropriate drug for treatment. Am J Addict 19(6):557–568

Maremmani I, Pacini M (2003) Understanding the pathogenesis of drug addiction in order to implement a correct pharmacological intervention. Heroin Addict Relat Clin Probl 5(3):5–12

Maremmani I, Lazzeri A, Pacini M, Lovrecic M, Placidi GF, Perugi G (2004) Diagnostic and symptomatological features in chronic psychotic patients according to cannabis use status. J Psychoactive Drugs 36(2):235–241

Maremmani I, Pacini M, Perugi G, Deltito J, Akiskal HS (2008) Cocaine abuse and the bipolar spectrum in 1090 heroin addicts: clinical observations and a proposal of a pathophysiologic model. J Affect Disord 106:55–61

Maremmani I, Pacini M, Popovic D, Romano A, Maremmani AG, Perugi G, Deltito J, Akiskal K, Akiskal H (2009) Affective temperaments in heroin addiction. J Affect Disord 117(3):186–192. https://doi.org/10.1016/j.jad.2009.01.007

Maremmani I, Maremmani AGI, Rugani F, Rovai L, Pacini M, Bacciardi S, Deltito J, Dell'Osso L, Akiskal HS (2012) Clinical presentations of substance abuse in bipolar heroin addicts at time of treatment entry. Ann Gen Psychiatry 11:23

Maremmani AGI, Rugani F, Bacciardi S, Rovai L, Pacini M, Dell'Osso L, Maremmani I (2014) Does dual diagnosis affect violence and self-harm in heroin addicts at treatment entry? J Addict Med 8(2):116–122

Martinez D, Broft A, Foltin RW, Slifstein M, Hwang DR, Huang Y, Perez A, Frankle WG, Cooper T, Kleber HD, Fischman MW, Laruelle M (2004) Cocaine dependence and d2 receptor availability in the functional subdivisions of the striatum: relationship with cocaine-seeking behavior. Neuropsychopharmacology 29(6):1190–1202. https://doi.org/10.1038/sj.npp.1300420

Mayfield RD, Harris RA, Schuckit MA (2008) Genetic factors influencing alcohol dependence. Br J Pharmacol 154(2):275–287. https://doi.org/10.1038/bjp.2008.88

McClernon FJ, Kozink RV, Lutz AM, Rose JE (2009) 24-h Smoking abstinence potentiates fMRI-BOLD activation to smoking cues in cerebral cortex and dorsal striatum. Psychopharmacology (Berl) 204(1):25–35. https://doi.org/10.1007/s00213-008-1436-9

McGowan PO, Roth TL (2015) Epigenetic pathways through which experiences become linked with biology. Dev Psychopathol 27(2):637–648. https://doi.org/10.1017/S0954579415000206

Meyers JL, Dick DM (2010) Genetic and environmental risk factors for adolescent-onset substance use disorders. Child Adolesc Psychiatr Clin N Am 19(3):465–477. https://doi.org/10.1016/j.chc.2010.03.013

Milton AL, Lee JL, Everitt BJ (2008) Reconsolidation of appetitive memories for both natural and drug reinforcement is dependent on {beta}-adrenergic receptors. Learn Mem 15(2):88–92. https://doi.org/10.1101/lm.825008

Ouzir M, Errami M (2016) Etiological theories of addiction: a comprehensive update on neurobiological, genetic and behavioural vulnerability. Pharmacol Biochem Behav 148:59–68. https://doi.org/10.1016/j.pbb.2016.06.005

Pacini M, Maremmani I (2007a) Addiction treatment: when will medical principles matter? Heroin Addict Relat Clin Probl 9(3):5–8

Pacini M, Maremmani I (2007b) The need for scientifically based ethical principles in dealing with drug-addicted persons. Heroin Addict Relat Clin Probl 9(4):5–8

Pacini M, Maremmani I, Vitali M, Santini P, Romeo M, Ceccanti M (2009) Affective temperaments in alcoholic patients. Alcohol 43(5):397–404. https://doi.org/10.1016/j.alcohol.2009.05.002

Pacini M, Maremmani I, Vitali M, Romeo M, Santini P, Vermeil V, Ceccanti M (2010) Cocaine abuse in 448 alcoholics: evidence for a bipolar connection. Addict Disord Treat 9(4):164–171

Pani PP, Maremmani I, Trogu E, Gessa GL, Ruiz P, Akiskal HS (2010) Delineating the psychic structure of substance abuse and addictions: should it include anxiety, mood and impulse-control dysregulation. J Affect Disord 122:185–197

Perugi G, Frare F, Madaro D, Maremmani I, Akiskal HS (2002) Alcohol abuse in social phobic patients: is there a bipolar connection? J Affect Disord 68(1):33–39

Placidi GF, Signoretta S, Liguori A, Gervasi R, Maremmani I, Akiskal HS (1998) The semi-structured affective temperament interview (TEMPS-I). Reliability and psychometric properties in 1010 14–26-year-old students. J Affect Disord 47(1-3):1–10

Rovai L, Maremmani AGI, Bacciardi S, Rugani F, Pacini M, Perugi G, Dell'Osso L, Akiskal H, Maremmani I (2013) Drug addiction: affective temperaments as risk factor. Heroin Addict Relat Clin Probl 15(2):19–28

Teoh SK, Mendelson JH, Mello NK, Weiss R, McElroy S, McAfee B (1990) Hyperprolactinemia and risk for relapse of cocaine abuse. Biol Psychiatry 28(9):824–828

Volkow ND, Swanson JM (2003) Variables that affect the clinical use and abuse of methylphenidate in the treatment of ADHD. Am J Psychiatry 160(11):1909–1918. https://doi.org/10.1176/appi.ajp.160.11.1909

Volkow ND, Fowler JS, Wang GJ, Hitzemann R, Logan J, Schlyer DJ, Dewey SL, Wolf AP (1993) Decreased dopamine D2 receptor availability is associated with reduced frontal metabolism in cocaine abusers. Synapse 14(2):169–177. https://doi.org/10.1002/syn.890140210

Volkow ND, Wang GJ, Fowler JS, Gatley SJ, Ding YS, Logan J, Dewey SL, Hitzemann R, Lieberman J (1996) Relationship between psychostimulant-induced "high" and dopamine transporter occupancy. Proc Natl Acad Sci U S A 93(19):10388–10392

Volkow ND, Wang GJ, Fowler JS, Logan J, Gatley SJ, Hitzemann R, Chen AD, Dewey SL, Pappas N (1997) Decreased striatal dopaminergic responsiveness in detoxified cocaine-dependent subjects. Nature 386(6627):830–833. https://doi.org/10.1038/386830a0

Volkow ND, Wang GJ, Fowler JS, Hitzemann R, Gatley SJ, Dewey SS, Pappas N (1998) Enhanced sensitivity to benzodiazepines in active cocaine-abusing subjects: a PET study. Am J Psychiatry 155(2):200–206. https://doi.org/10.1176/ajp.155.2.200

Volkow ND, Wang GJ, Fowler JS, Franceschi D, Thanos PK, Wong C, Gatley SJ, Ding YS, Molina P, Schlyer D, Alexoff D, Hitzemann R, Pappas N (2000) Cocaine abusers show a blunted response to alcohol intoxication in limbic brain regions. Life Sci 66(12):PL161–PL167

Volkow ND, Wang G, Fowler JS, Logan J, Gerasimov M, Maynard L, Ding Y, Gatley SJ, Gifford A, Franceschi D (2001) Therapeutic doses of oral methylphenidate significantly increase extracellular dopamine in the human brain. J Neurosci 21(2):RC121

Volkow ND, Fowler JS, Wang GJ, Swanson JM (2004) Dopamine in drug abuse and addiction: results from imaging studies and treatment implications. Mol Psychiatry 9(6):557–569. https://doi.org/10.1038/sj.mp.4001507

Volkow ND, Wang GJ, Ma Y, Fowler JS, Wong C, Ding YS, Hitzemann R, Swanson JM, Kalivas P (2005) Activation of orbital and medial prefrontal cortex by methylphenidate in cocaine-addicted subjects but not in controls: relevance to addiction. J Neurosci 25(15):3932–3939. https://doi.org/10.1523/JNEUROSCI.0433-05.2005

Volkow ND, Wang GJ, Telang F, Fowler JS, Logan J, Jayne M, Ma Y, Pradhan K, Wong C (2007) Profound decreases in dopamine release in striatum in detoxified alcoholics: possible orbitofrontal involvement. J Neurosci 27(46):12700–12706. https://doi.org/10.1523/JNEUROSCI.3371-07.2007

Volkow ND, Wang GJ, Telang F, Fowler JS, Logan J, Childress AR, Jayne M, Ma Y, Wong C (2008) Dopamine increases in striatum do not elicit craving in cocaine abusers unless they are coupled with cocaine cues. Neuroimage 39(3):1266–1273. https://doi.org/10.1016/j.neuroimage.2007.09.059

Weiss F, Ciccocioppo R, Parsons LH, Katner S, Liu X, Zorrilla EP, Valdez GR, Ben Shahar O, Angeletti S, Richter RR (2001) Compulsive drug-seeking behavior and relapse. Neuroadaptation, stress, and conditioning factors. Ann N Y Acad Sci 937:1–26

Wiehler A, Peters J (2014) Reward-based decision making in pathological gambling: the roles of risk and delay. Neurosci Res. https://doi.org/10.1016/j.neures.2014.09.008

Zubieta JK, Gorelick DA, Stauffer R, Ravert HT, Dannals RF, Frost JJ (1996) Increased mu opioid receptor binding detected by PET in cocaine-dependent men is associated with cocaine craving. Nat Med 2(11):1225–1229

Sleep and Dreams

8

Theocharis Kyziridis and Ioannis Nimatoudis

8.1 Brief Historical Overview

Descriptions about sleep and dreaming can be found in texts of ancient civilizations such as those of Egypt and Greece. In Greek mythology, the state of sleep (Hypnos) was considered similar to that of death (Thanatos) (Norman et al. 2011). Both Aristotle and Hippocrates wrote about sleep; the former dedicated a book to sleep entitled *On Sleep and Sleeplessness* (Kirsch 2013). The Bible also contains mentions about dreaming through which prophecies occur, and Maimonides, a Hebrew scholar, gave instructions concerning sleep habits (Rosner 1965).

Major contributions toward understanding sleep were made though only during the twentieth century, and empirical studies were undertaken for the first time around 1950 (Hobson 1990) (Table 8.1). Current knowledge about sleep and dreaming is the result of research efforts made since only the previous century (Hobson 1990), and this was made possible after the discovery of electroencephalography (EEG) by Hans Berger in 1928. Using EEG to measure and record the brain electrical activity (Kirsch 2013), Berger showed that brain's rhythms were different between awake or asleep individuals (Berger 1930).

The anatomical localization of sleep was made possible by Constantin von Economo. His careful observations of patients during the epidemic of lethargic encephalitis led him to an important conclusion concerning the role of hypothalamus in sleep-wake cycle regulation: the anterior hypothalamus (AH) contained sleep-promoting neurons, whereas the posterior hypothalamus (PH) contained wakefulness-promoting neurons (Economo 1930). Nathaniel Kleitman's

T. Kyziridis (✉) · I. Nimatoudis
3rd Department of Psychiatry, Faculty of Medicine, School of Health Sciences, Aristotle University of Thessaloniki, Thessaloniki, Greece

University Hospital AHEPA, Thessaloniki, Greece

K. N. Fountoulakis, I. Nimatoudis (eds.), *Psychobiology of Behaviour*,
https://doi.org/10.1007/978-3-030-18323-3_8

Table 8.1 Important hallmarks in the history of sleep medicine (Lee-Chiong 2008)

1818	John Cheyne (1777–1836) and William Stokes (1804–1878) described the respiration that bears their name
1880	Jean-Baptiste-Edouard Gelineau (1828–1906) described narcolepsy
1907	Rene Legendre and Henri Pieron (1881–1964) induced sleep in dogs by administering serum from sleepy dogs
1920	Nathaniel Kleitman (1895–1999) reported that increased sleepiness was a result of sleep deprivation
1928–1930	Hans Berger (1873–1941) discovered EEG and described beta waves for the first time
1936	Edmund Newton Harvey (1887–1959) and Alfred Lee Loomis (1887–1975) proposed sleep EEG classification (stages A, B, C, D, and E)
1939	Kleitman published his book *Sleep and Wakefulness*
1944	William Grey Walter (1910–1977) and F.J. Dovey first described delta and theta waves
1945	Karl-Axel Ekbom (1907–1977) described restless legs syndrome
1949	Giuseppe Moruzzi (1910–1986) and Horace Winchell Magoun (1907–1991) discovered reticular formation
1953	Eugene Aserinski (1921–1998) and Kleitman discovered REM sleep
1960	Gerald Vogel (1928–2012) described sleep onset REM periods in narcolepsy
1965	Clinical description of obstructive sleep apnea
1986	Carlos H. Schenck (1951–) reported on features of REM sleep behavior disorder
1999	Description of link of hypocretin and narcolepsy

contribution is also important: he conducted sleep deprivation studies (Dement 2005) and, together with his student Eugene Aserinsky, described rapid eye movement (REM) sleep; this was followed by the discovery of atonia during it (Aserinsky and Kleitman 1953; Berger 1961). The different EEG stages of sleep had already been described (Loomis et al. 1937) as well as the promotion of arousal by the electrical stimulation of the reticular formation (Lindsley et al. 1949; Moruzzi and Magoun 1949). Finally, Jouvet with his colleagues established the role of relevant circuitry in the pons and discovered the characteristic elements of the "paradoxical" REM sleep (Jouvet and Michel 1959; Jouvet et al. 1959). Jouvet, who passed away in 2017, is considered a great pioneer in sleep research (Jones 2018).

8.2 Sleep in Humans and Animals

8.2.1 Definition of Sleep

Despite the enormous scientific advances, an exact definition of sleep has not been yet found, and its fundamental action remains largely unclear (Norman et al. 2011). Sleep is not a passive state or, to put it simply, just the absence of wakefulness and the reduction of motility and responsiveness to external stimuli. On the contrary, it is a reversible, active state which is essential for mental and physical well-being (Stenberg 2007; Weber 2017) and has behavioral and physiologic characteristics that differentiate it from wakefulness (Tables 8.2, 8.3, 8.4, and 8.5).

Table 8.2 Behavioral characteristics of wakefulness and sleep (Chokroverty 2013)

Characteristics	Wakefulness	Sleep
Posture	Erect, sitting, or recumbent	Recumbent
Mobility	Normal	Mildly reduced to absent with postural shifts (NREM) Myoclonic jerks (REM)
Response to stimulation	Normal	Mildly to moderately reduced (NREM) Moderately reduced to absent (REM)
Level of alertness	Alert	Unconscious but reversible
Eye position	Open	Closed
Eye movements	Waking eye	Slow eye movements (NREM) Rapid eye movements (REM)

Table 8.3 Physiologic characteristics of wakefulness and sleep (Chokroverty 2013)

	Wakefulness	Sleep
Electroencephalography	Parieto-occipital α-waves Fronto-central β-rhythms	θ and δ-waves, sleep spindles, vertex waves and K-complexes (NREM) θ waves and β-rhythms (REM)
Electromyography	Normal	Mildly reduced (NREM) Markedly reduced to absent (REM)
Electrooculography	Waking eye movements	Slow eye movements (NREM) Rapid eye movements (REM)

Table 8.4 Characteristics of wakefulness and sleep (Hobson and Pace-Schott 2002)

	Wakefulness	NREM sleep	REM sleep
Behavior	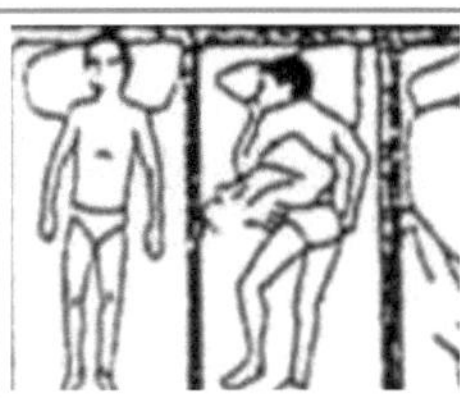	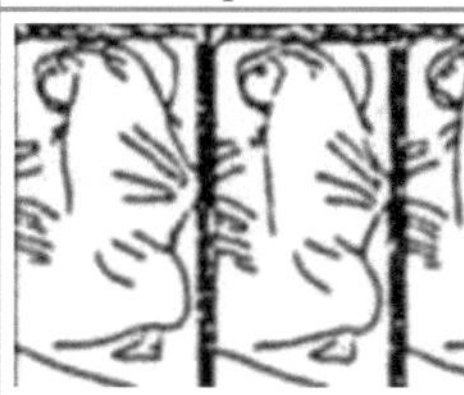	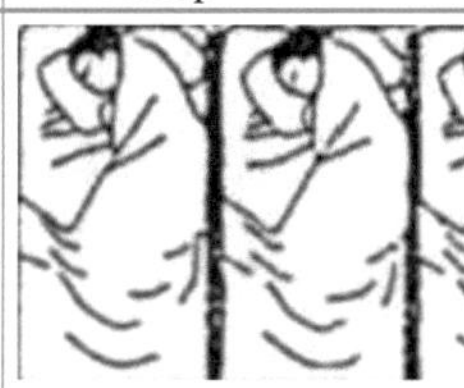
Awake Polygraph	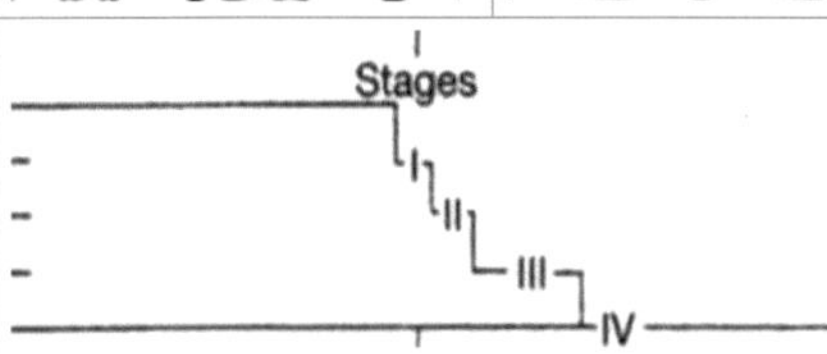		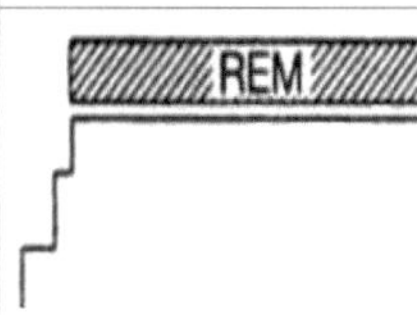
EMG EEG EOG	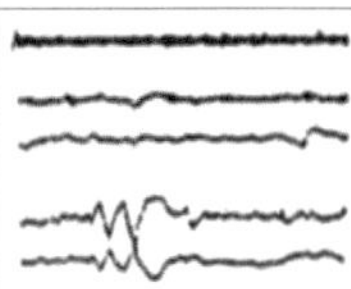	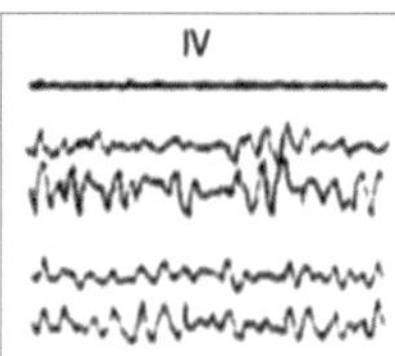	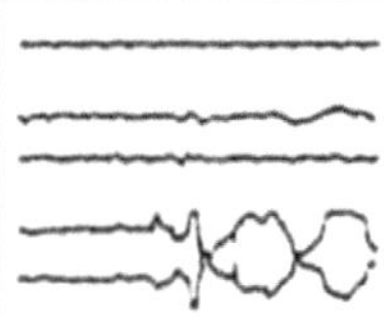
Sensation/perception	Vivid and externally generated	Dull or absent	Vivid and internally generated
Thought	Logical Progressive	Logical Perseverative	Illogical Bizarre
Movement	Voluntary Continuous	Involuntary Episodic	Inhibited Commanded

Table 8.5 EEG wave frequencies (Silber et al. 2010)

Beta (β)	>13 Hz
Alpha (α)	8–13 Hz
Theta (θ)	4–7.9 Hz
Delta (δ)	<4 Hz

8.2.2 Evolutionary Perspective

Sleep is universally present in the animal kingdom (Greenspan et al. 2001); all mammals and most of the animals sleep, and its deprivation usually leads to the need for recovery sleep (Cirelli and Tononi 2008). Sleep has a vital function at cellular level; thus all animals have an absolute need for it; from an evolutionary point of view, it seems to be important for the survival of animals (Reading 2013). Even unicellular organisms have a sleeplike behavior manifesting as sequential phases of activity and rest (Hobson 1990). Rest-activity cycle is a basic function of living organisms. Rest is a rather passive process that, in homeothermic vertebrates, evolved as sleep, providing both rest and a biological advantage through memory formation (Hobson and Pace-Schott 2002). Sleep is characterized by reduction of responsiveness to stimuli from the environment, reduction of mobility, and stereotyped postures (Tononi and Cirelli 2011), while wakefulness is a behavioral state characterized by response to stimuli and voluntary motor activation of animals (Scammell et al. 2017). Reptiles and birds have diminished responsiveness and EEG synchrony but no REM phase (with the exception of birds), and amphibians have none of the characteristic features of sleep in mammals (Hobson 1990). In order to speak about sleep in animals, they must present a minimum number of the following characteristics:

- A reversible state with motor repose and elevated sensory threshold.
- Protected sleeping site.
- Circadian organization.
- "Hunger"
- "Satiety".

Animals with lower number of these characteristics do not sleep in fact but have periods of rest and activity. It could be said that sleep in animals is a state of behavior alternative to wakefulness the regulation of which is based on circadian and homeostatic systems (Esteban et al. 2005).

8.2.3 Sleep During the Human Life Cycle

Approximately one third of the human life involves sleeping—90% of adults need at least 7 h of good sleep every day (Reading 2013). During life-span, sleep patterns and sleep requirement change (Table 8.6) following an order that depends upon maturation of the central nervous system (CNS) (Lavie et al. 2002; Roffwarg et al. 1966). Duration

Table 8.6 Changes in content and length of sleep with age (Lavie et al. 2002)

	Sleep time (h)	NREM Stages 1–2 (%)	NREM Stages 3–4 (%)	REM (%)
Infants	13–16	10–30	30–40	40–50
Children	8–12	40–60	20–30	20–30
Adults	6–9	45–60	15–25	15–25
Elderly	5–8	50–80	5–15	15–25

of the sleep cycles shows correlation with brain size; rodents have a sleep cycle of a few minutes, while the duration is estimated to be some hours in large mammals (Weber 2017). The sleep of normal adults is fairly consistent as far as it concerns both the time spent in each stage and the pattern of stages across the night (Tononi and Cirelli 2011).

Newborns sleep approximately 16 h per day, in the form of short naps. These naps begin to disappear at about the age of 6 months (Parmelee 1961). Total sleep time and REM sleep proportion are reduced during early childhood and napping stops (Iglowstein et al. 2003). Slow-wave sleep (SWS) (a phase of NREM sleep) increases during the first year of life and reaches a peak. In adolescents and adults, it is reduced and by age 60 may even disappear (Tononi and Cirelli 2011). Elderly usually sleep less, have frequent awakenings, and wake earlier in the morning (Ehlers and Kupfer 1989).

8.3 Importance and Functions of Sleep

8.3.1 Importance of Sleep

Sleep serves important biological (immunity, hormonal regulation, thermoregulation) (Morrissey et al. 2004; Opp 2009; Van Cauter et al. 2008) and neurocognitive functions concerning emotion, memory, and reward (Lansink et al. 2008; Lena et al. 2005). Information processing and energy homeostasis are also among the important functions of sleep (Hobson 1990) (Table 8.7).

Its role in memory function is achieved through stabilization and enhancement of certain memory processes (Maquet 2001; Walker and Stickgold 2006) via memory consolidation; by this process, reorganization of new memories takes place first. After that, these memories are transferred to preexisting long-term memory networks (Wang and Morris 2010). This process may take place during sleep in humans (Diekelmann and Born 2010; Maquet 2001), and declarative (hippocampus-dependent) memory likely benefits from NREM sleep and particularly SWS (Born et al. 2006). Sleep seems to enhance changes taking place in neural networks after perceptual, motor, or emotional learning tasks (Maquet et al. 2003; Payne and Kensinger 2011; Schwartz et al. 2002; Sterpenich et al. 2007). The mechanism of such enhancement involves activation of the cells of two different brain regions: hippocampus proper (HC) and ventral striatum (VS). This coordinated activation of both HC and VS during NREM sleep may facilitate:

Table 8.7 Sleep functions (Chokroverty 2013; Silber et al. 2010)

Restoration • Body • Brain tissue
Energy conservation
Adaptation to environmental conditions
Learning and unlearning
Memory consolidation
Synaptic and neural network integrity
Synaptic homeostasis
Thermoregulation
Immunocompetence

- The replay of memories with contextual, emotional, and motivational components.
- The consolidation of memory-reward information.
- The selection of memories with a high storage priority (Lansink et al. 2008, 2009).

Apart from declarative memory, NREM sleep also consolidates motor skills; both are influenced by emotional relevance and motivational biases. This consolidation may be enhanced when linked with an anticipation of reward; thus memories could serve as salient stimuli for the neurons of ventral tegmental area (VTA), a region known as the reward center of the brain which is influenced by needs and memories (Fischer and Born 2009; Schultz 2010; Sterpenich et al. 2009; Wilhelm et al. 2011).

There seems to be a link between sleep and emotion regulation processes (Gujar et al. 2011; Lara-Carrasco et al. 2009; Pace-Schott et al. 2012; Talamini et al. 2013). REM sleep could possibly affect the emotional well-being in a negative way. This idea is based on studies showing that awakenings from an REM episode are associated with high scores in the Hamilton Depression Rating Scale (McNamara et al. 2010). Furthermore, depression is usually preceded by increases in REM density and decreases in REM latency. This might indicate that REM sleep, which is under circadian and homeostatic control (Wurts and Edgar 2000), may constitute a biomarker for depressive disease (Gottesmann and Gottesman 2007).

REM sleep may play a role in creative thinking and problem-solving by integrating information that is initially unassociated (Cai et al. 2009; Walker et al. 2002). Sleep may also facilitate insight since it can reconstruct new memory representations (Hobson and Pace-Schott 2002; Wagner et al. 2004). Furthermore, it may have a role of re-establishing metabolism of the cell and of cleansing brain from neurotoxins (Brown et al. 2012; Xie et al. 2013) benefiting as well the integrity and reorganization of neuronal synapses (Kavanau 1997; Krueger et al. 1995).

8.3.2 Effects of Sleep Deprivation

Sleep deprivation, which can be the result of various factors, such as physical disease or mental health problems, lifestyle factors, vocational demands, and others (Banks et al. 2017), may have detrimental effects (Table 8.8). It can lead rats to thermoregulation impairment, to metabolic dyscontrol (Bach et al. 2011; Hobson 1990), and, in extreme cases, even to death sooner than food deprivation, while humans, suffering from the rare genetic prion disease of familial fatal insomnia, usually die within 2 years (Higgins and George 2007).

The underlying pathophysiologic mechanism is rather the excess of wakefulness than the lack of sleep per se. Indeed, the excess of wakefulness seems to activate pathways that are harmful for the cell. Furthermore, there seems to be temporary dysfunction of the frontal lobe; the brain of sleep-deprived young adults has been shown to function like the brain of elderly people (Reading 2013).

Sleep deprivation may also increase blood pressure and the risk for diabetes. It can lead to obesity, growth hormone decrease (Van Cauter et al. 2008), and, along with poor sleep quality, deficits in both synaptic plasticity and memory processes probably leading this way to mood disorders (Casement et al. 2006; Prince et al. 2014). The most prominent effect of sleep deprivation in humans is neurocognitive impairment which may have serious consequences (Dinges 2006). If humans stay awake much longer than the usual 16 h a day, they are soon overcome by sleepiness and become cognitively impaired (Tononi and Cirelli 2011).

Even a few hours of sleep loss or just one night of sleep deprivation may cause impairment of tasks requiring sustained attention or the use of higher cognitive functions; lack of sleep makes the performance of these tasks inconsistent and unreliable without completely eliminating the capacity to perform them (Tononi and Cirelli 2011).

Table 8.8 Consequences of sleep deprivation (Lee-Chiong 2008; Reading 2013)

Sleepiness
Increased morbidity and mortality • Increased risk for motor vehicle accidents • Cardiovascular morbidity • Endocrine dysfunction • Immune dysfunction • Neurological dysfunction
Negative impact on mood
Cognitive dysfunction and performance difficulties • Increased reaction times • Perseveration and reduced flexibility • Impaired sense of humor • Increased risk taking • Impaired moral judgment • Reduced emotional intelligence • Increased "negativity" with enhanced memory for adverse events • Increased distractibility

8.3.3 Conclusions

Memory and learning processes seem to be specific to distinct NREM or REM sleep and dreaming states, and this makes them important for performance during wakefulness. There may also exist a difference among these states concerning their contribution to off-line reprocessing of emotional and reward information; REM sleep is probably responsible for emotional memory and synaptic consolidation, and NREM sleep may be more specialized in linking memory traces with motivational values (Diekelmann and Born 2010; Lansink et al. 2009; Payne and Kensinger 2011; Pennartz et al. 2004; Popa et al. 2010; Sterpenich et al. 2009; Wagner et al. 2001).

8.4 Sleep Architecture

8.4.1 Introduction

Sleep comprises two stages (NREM and REM), which take place approximately 4–6 times in one night (period of nearly 90 min) (Higgins and George 2007; Silber et al. 2010) (Fig. 8.1; Table 8.9). These two stages have distinct physiological parameters. NREM is the stage through which initiation of sleep normally takes place. It comprises four phases (light phases: 1, 2 and deep phases 3, 4, collectively known as SWS) based on EEG (Carskadon and Dement 2017).

SWS dominates the first third of sleep in adult humans (Chokroverty 2013) and usually occurs within an hour of sleep onset (Reading 2010). EEG during NREM

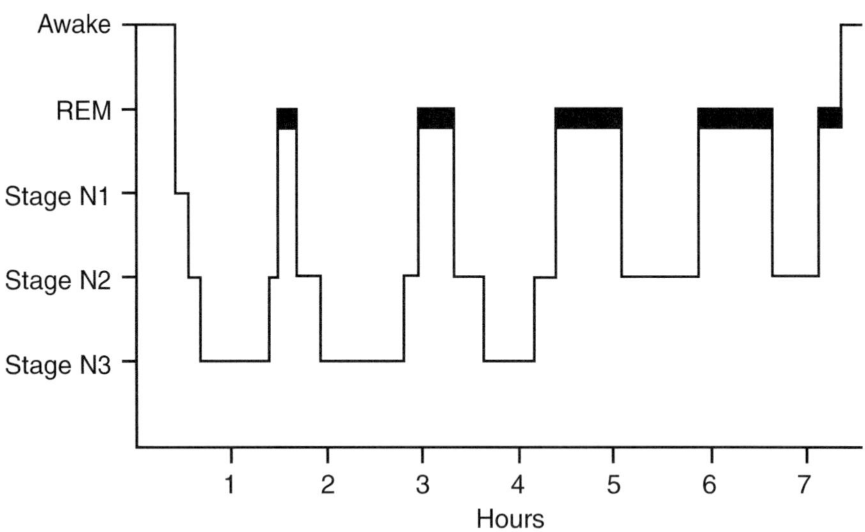

Fig. 8.1 A typical night's sleep of a young adult (Silber et al. 2010)

Table 8.9 Summary of various sleep states (Chokroverty 2013)

NREM sleep	*75–80% of sleep time*
N1 stage	3–8% of sleep time
N2 stage	45–55% of sleep time
N3 stage (SWS)	15–23% of sleep time
REM sleep	20–25% of sleep time

sleep is characterized as synchronized and has characteristic waveforms: sleep spindles, K-complexes, and high-voltage slow waves. Phase 1 of NREM sleep has the lowest arousal threshold; on the contrary, phase 4 has the highest (Carskadon and Dement 2017).

REM sleep, which follows one or more NREM periods (Weber 2017), accounts for around 20% of sleep and is characterized by EEG activation and wake-like activity, loss of muscle tone, episodic bursts of rapid eye movements (hence the name rapid eye movement sleep), and intense activation of the cortex and the limbic system (Carskadon and Dement 2017; Hobson and Pace-Schott 2002; Reading 2010). Loss of muscle tone and areflexia during this stage are the effects of inhibitory impulses from the brainstem. REM sleep, which predominates the last third of nocturnal sleep (Chokroverty 2013; Reading 2010), is also known as paradoxical, desynchronized, or wake-like. The reason for that is that the level of metabolic activity and the cortical EEG during this stage are similar to that observed during wakefulness usually corresponding to the vivid dreaming that takes place during this stage (Reading 2010) (Fig. 8.2).

Various mechanisms are responsible for the regulation of these two sleep stages while mammals are asleep (Datta and Maclean 2007). Under normal circumstances, humans first enter the sleep state via NREM sleep. Initially, the eyes drift slowly and the sleep position changes frequently. Individuals gradually lose awareness of the outside world and may even experience hypnagogic hallucinations and illusions of movements of the body in space. During NREM sleep, physiological changes take place. These changes include decreases in body temperature, blood pressure, and heart and respiratory rate and increases in the pulsatile release of growth and sex hormones from the pituitary gland and in the production of antibodies. This might imply that NREM sleep may be functionally associated with anabolic processes that benefit the somatic tissues (Pace-Schott and Hobson 2002).

8.4.2 Stages of Sleep

8.4.2.1 NREM Sleep: Stage 1

Sleep is usually entered through stage 1, and falling asleep takes place gradually accompanied by progressive disconnection from the environment. This transitory stage is characterized by loss of alpha activity; low-voltage, mixed-frequency EEG pattern; and prominent theta activity. Eye movements become slow and rolling, and

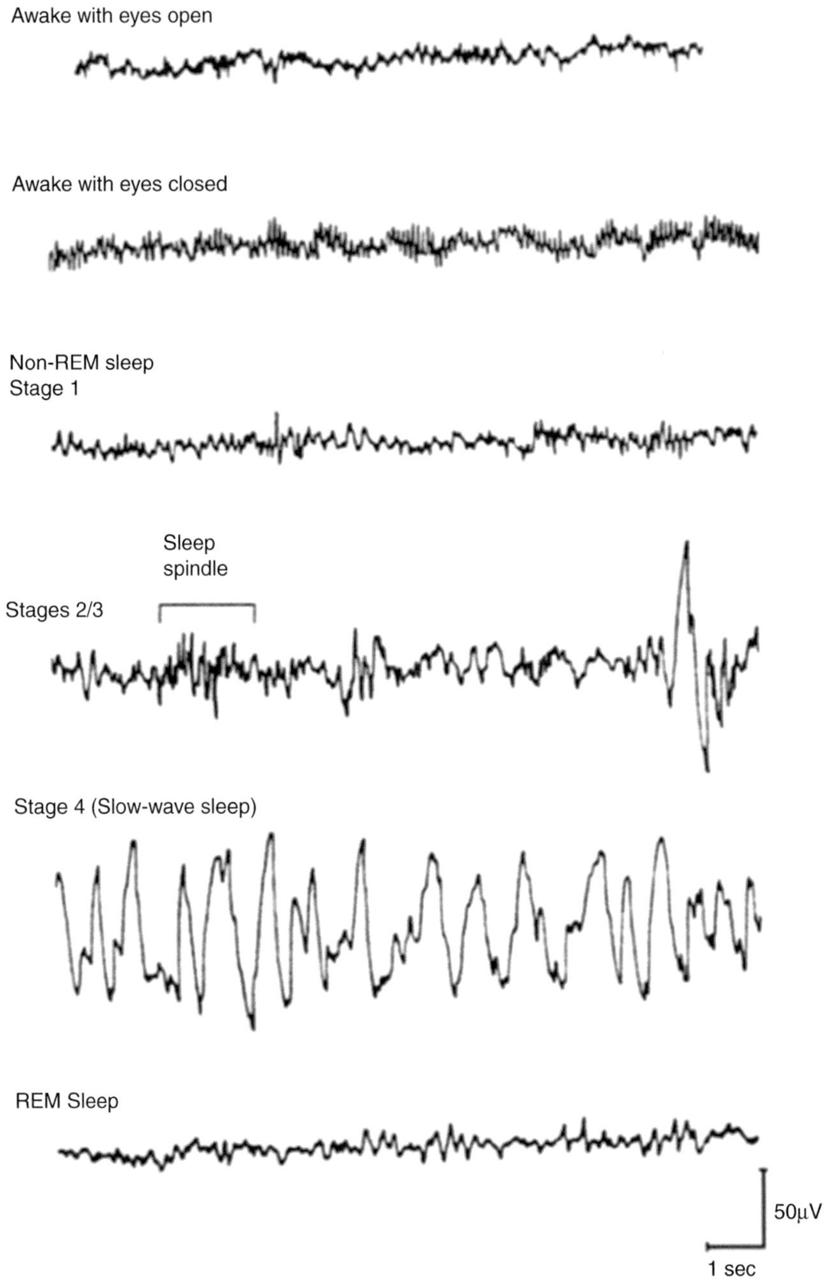

Fig. 8.2 EEG recordings during wakefulness and sleep (Siegel 2002)

muscle tone relaxes even though sudden muscle contractions may take place occasionally; these may sometimes be accompanied by a sense of falling and dreamlike imagery. Awareness of sensory stimuli is reduced, and motor activity may persist for a few seconds (Reading 2013).

8.4.2.2 NREM Sleep: Stage 2

Stage 2 follows after a few minutes. It is characterized by K-complexes and sleep spindles in EEG. Eye movements and muscle tone are further reduced. There is partial disconnection from the environment, and the arousal threshold is increased (Reading 2013).

8.4.2.3 NREM Sleep: Stages 3 and 4 (SWS)

Stage 2 is followed, especially at the beginning of the night, by a period of SWS, which is characterized by slow waves in the delta range. Eye movements cease and electromyographic (EMG) activity decreases further. After deepening through stages 2–4, NREM sleep lightens and returns to stage 2, followed by REM sleep (Reading 2013).

8.4.2.4 REM Sleep

EEG during REM sleep is similar to that during wakefulness or stage 1. Contrary to NREM, REM sleep is not subdivided into stages but comprises tonic and phasic components (Table 8.10). The first include penile erections, the activated EEG and a generalized loss of muscle tone, except for the extraocular muscles that drive the REMs and the diaphragm that keeps breathing. Phasic components include muscle twitches and irregular bursts of REMs. Behaviorally, REM sleep is deep sleep, with an arousal threshold that is as high as in SWS (Reading 2013).

Table 8.10 Phenomena of REM sleep (Silber et al. 2010)

Tonic phenomena
Desynchronized EEG
Theta activity in hippocampus
Loss of voluntary muscle tone
Increased cerebral blood flow
Thermal regulation impairment
Dreaming
Erections of penis
Engorgement of clitoris
Phasic phenomena
Rapid eye movements
Transient muscle activity
Increased respiratory and heart rate
Ponto-geniculo-occipital waves
Sawtooth waves

8.5 Neuroanatomy of Wakefulness and Sleep

Anatomically, brainstem, diencephalon, and telencephalon contain regions that regulate sleep and wakefulness (McGinty and Szymusiak 2017). Three functional systems regulate sleep:

- The homeostatic system, which is responsible for the regulation of intensity, amount, and duration of sleep, is mainly seated in the preoptic area.
- The ultradian system, regulating REM and NREM sleep alternation, is seated in mesencephalic structures and pons.
- The circadian system, responsible for the regulation of timing of wakefulness and sleep, is mainly seated in the AH (Borb and Achermann 2016; Economo 1930; Pace-Schott and Hobson 2002).

Multiple wake-promoting systems exist in mammals (Datta 1995). Each of these systems may be individually activated thus contributing to the maintenance of the general state of wakefulness. Wake-promoting neuronal groups are found in the basal forebrain (BF), the PH, and in the upper brain stem, whereas NREM sleep-promoting neurons are located in the AH and the BF (Jones 2003, 2005; McGinty and Szymusiak 2003). REM sleep-promoting areas include the meso-pontine reticular formation, thalamic nuclei, and the forebrain (Steriade and McCarley 1990) (Figs. 8.3, 8.4, and 8.5; Table 8.11).

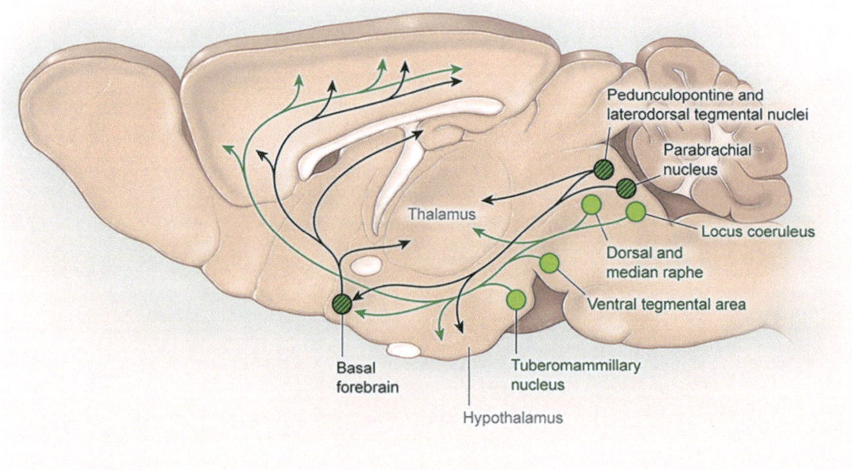

Fig. 8.3 Wake-promoting pathways (Scammell et al. 2017)

The neuronal groups of REM sleep generation are cholinergic cell groups located in the dorsal part of the pons and in the medulla (laterodorsal tegmentum (LDT) and pedunculopontine tegmentum (PPT)) as well as in the medial pontine reticular formation (Tononi and Cirelli 2011). The critical neurons are localized mainly in the pontine tegmentum, as Jouvet showed, and these neurons have a wide distribution in various nuclei (Hobson 1990).

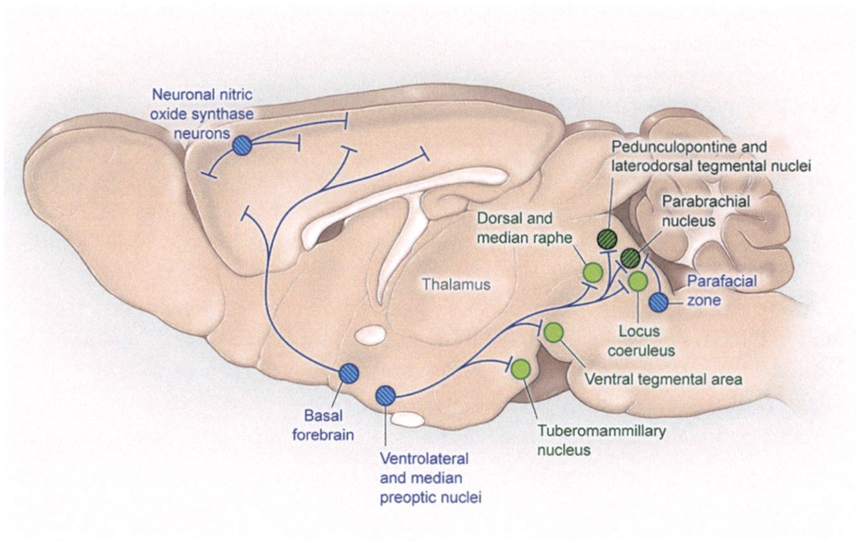

Fig. 8.4 NREM sleep-promoting pathways (Scammell et al. 2017)

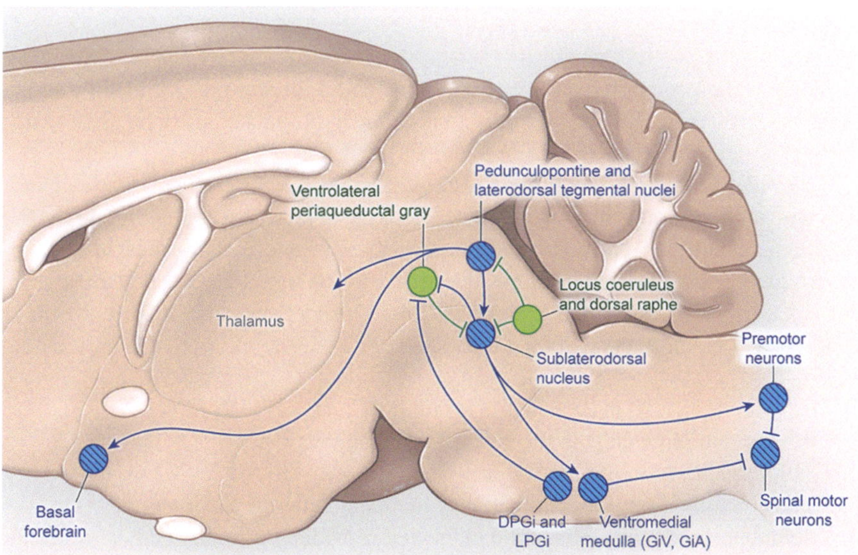

Fig. 8.5 REM sleep-promoting pathways (Scammell et al. 2017)

Table 8.11 Activity of wake and sleep centers (Lu and Zee 2010; Norman et al. 2011)

	Wake	NREM sleep	REM sleep
Laterodorsal tegmental (LDT)/pedunculopontine tegmental (PPT) (acetylcholine)	↑↑	–	↑↑
Locus coeruleus (norepinephrine) (brain stem) Dorsal raphe (serotonin) Tuberomammillary nucleus (histamine) (forebrain) Substantia nigra/ventral tegmental area (dopamine)	↑↑	↑	–
Lateral hypothalamus (orexin/hypocretin)	↑↑	–	–
Ventrolateral preoptic—cluster (galanin and GABA)	–	↑↑	–
Ventrolateral preoptic—extended (galanin and GABA)	–	–	↑↑

Many of the aforementioned areas are under the control of the circadian clock which is located on the suprachiasmatic nucleus of the hypothalamus (SCN). This mechanism ensures that sleep occurs at the appropriate time of the 24-h light-dark cycle (Zee and Manthena 2007).

During REM sleep, increases in glucose metabolism or regional blood flow are found in the pontine tegmentum, thalamic nuclei, limbic, paralimbic, and temporo-occipital areas as shown by functional neuroimaging (Braun et al. 1997, 1998; Maquet et al. 1996; Nofzinger et al. 1997).

Neurochemically specific wake-promoting cell groups are found within the ascending reticular activating system (ARAS):

1. Noradrenergic cells in the locus coeruleus (LC).
2. Serotonergic cells in the raphe nuclei (RN).
3. Cholinergic cells in the PPT and LDT.
4. Glutamatergic cells in the midbrain.
5. Dopaminergic cells in the substantia nigra compacta (SNc) and ventral tegmental area (VTA).

These wake-promoting cell groups of pons and midbrain project dorsally to the thalamo-cortical system and ventrally to both the hypothalamo-cortical and basalo-cortical systems (Garcia-Rill 2002; Sakai and Crochet 2003).

The wake-promoting cell groups in brainstem ARAS are in functional cooperation with similar groups in the forebrain. There are at least five cellular groups in the forebrain, which can promote wakefulness either in coordination with the wake-promoting cells of the brainstem or independently (Alam et al. 2002; Deboer et al. 2003; Easton et al. 2004; Koyama et al. 2002; Mendelson et al. 2003; Parmentier et al. 2002; Takahashi et al. 2006; Vertes 2006). These forebrain cell groups include:

1. Histaminergic cells in the tuberomammillary nuclei (TMN) of the PH.
2. Orexin-containing cells in the lateral hypothalamus (LH).
3. Cholinergic cells in the BF.
4. Cells containing neuropeptide-Y; they are located on the SCN.
5. Glutamate-producing cells found in the ventromedial prefrontal cortex (vmPFC).

When these systems are activated, either by experimental procedures or spontaneously, they result in activation patterns of the cortex which is necessary to maintain wakefulness (Gerashchenko and Shiromani 2004).

The ARAS was initially described by Moruzzi and Magoun, who made the hypothesis that it was implicated in the regulation of wakefulness. This system contains two branches:

- (a) Cholinergic neurons from LDT and PPT of the dorsal midbrain and pons, thalamic relay nuclei-thalamic reticular nuclei-thalamocortical tract-cortex (b) Cholinergic neurons from ventral projection of LDT and PPT-BF-substantia innominata-medial septum-diagonal band of Broca-cerebral cortex
- Monoaminergic neurons to LH-BF-cerebral cortex (Lu and Zee 2010).

Each of the arousal systems may promote wakefulness independently. Nevertheless, these systems usually cooperate to generate behavioral arousal. Anatomically, they are characterized by the existence of many interconnections between them, which gives them a functional advantage; that is, even if one of these systems is injured, wakefulness will still occur. In fact, there are only a few brain regions in which lesions produce lasting reductions in arousal. One of them is the rostral reticular formation in the midbrain and PH. Lesions from strokes or tumors in this formation can produce severe hypersomnolence or even coma, probably from damage to many of the ascending monoaminergic and cholinergic pathways (Higgins and George 2007).

Sleep-active neurons have been found in the ventrolateral preoptic area (VLPO) and the median preoptic area (MNPO) in animal models (Gong et al. 2004). Many neurons in these nuclei fire most frequently during NREM sleep and to a lesser degree during REM sleep. During wakefulness, they are virtually silent (Takahashi et al. 2009).

Lesions of these VLPO neurons reduce sleep to a great degree. Furthermore, the sleep that does occur is light and fragmented. These observations suggest that MNPO neurons may help initiate sleep, whereas VLPO neurons may be necessary for the maintenance of sleep. Anatomically, the VLPO and MNPO are well positioned to promote sleep. The neurons in these nuclei contain the inhibitory neurotransmitter GABA and the inhibitory neuropeptide galanin (Gaus et al. 2002), and they innervate all the arousal-promoting regions, including the LDT/PPT, LC, DR, TMN, and also the orexin neurons. Thus, the VLPO and MNPO are hypothesized to promote sleep by coordinating the inhibition of arousal regions during NREM and REM sleep (Saper et al. 2005).

8.6 Neurochemistry of Wakefulness and Sleep

8.6.1 Introduction

Various neurotransmitters have been linked to the sleep-wake cycle regulation; the most recent discovery has been the finding of the link between wakefulness

Table 8.12 Neurotransmitters in sleep and wakefulness (Lee-Chiong 2008)

REM sleep	Acetylcholine GABA Glycine
NREM sleep	Adenosine GABA Norepinephrine Peptides Serotonin
Wakefulness	Acetylcholine Dopamine Glutamate Histamine Hypocretin Norepinephrine Peptides Serotonin

and orexin (hypocretin) which is produced in the LH (Norman et al. 2011). Neurotransmitter systems that promote wakefulness include:

- The cholinergic systems of the basal nuclei and the reticular formation in pons.
- The noradrenergic system in LC.
- The dopaminergic system in SN and VTA.
- The histaminergic system in PH (Pandi-Perumal and Kramer 2011) (Table 8.12).

Cholinergic neurons stimulate REM sleep, while the opposite is true for aminergic neurotransmitter systems (Hobson 1990).

8.6.2 Neurotransmitters of Wakefulness and Sleep

8.6.2.1 Acetylcholine (ACh)

Cholinergic agonists augment vigilance and promote fast activity in the brain cortex, while muscarinic receptor antagonists, such as atropine, reduce vigilance and produce dissociation between EEG activity and behavior (Norman et al. 2011). Ach has an important role in the control of vigilance, the function of neuromuscular and parasympathetic junctions, and cortical EEG activation (McCarley 2007; Steriade and McCarley 2005). Cholinergic neurons are important for REM sleep, wakefulness, cognition, learning, and memory. They are found in BF and brainstem. BF neurons project to the cortex and hippocampus promoting fast EEG rhythms. Brainstem neurons are found within the LDT/PPT nuclei projecting primarily to subcortical regions, such as the thalamus, hypothalamus, and BF (Gritti et al. 1997). These neurons are mainly active during wakefulness and REM sleep, and release of Ach is increased. During NREM sleep release of Ach is decreased (Lee-Chiong 2008).

REM sleep is believed by many researchers to be controlled by cholinergic neurons in the LDT/PPT. These are the same nuclei that contain wake-promoting cells, but a subpopulation of these cholinergic neurons are active in both wakefulness and REM sleep or are selectively active in REM sleep (el Mansari et al. 1989; McCarley 2007; Steriade 2004; Steriade and McCarley 1990).

Drugs that enhance ACh signaling elicit intense and long-lasting REM sleep when they are injected into the lateral pontine tegmentum (Baghdoyan 1997). Large lesions that include the LDT/PPT produce significant reductions in REM sleep (Steriade and McCarley 2005), suggesting that the LDT/PPT is necessary for REM sleep.

8.6.2.2 Adenosine (ADE)

The stimulant action of caffeine is achieved through the blockade of ADE receptors, and this finding led to the hypothesis that there exists a link between ADE release and sleep (Norman et al. 2011). Administration of ADE has been shown to induce sleep-like behavior in cats and to increase NREM sleep in rodents (Stenberg 2007). ADE neurons are located in BF (Lee-Chiong 2008) and may promote sleep in response to high metabolic activity observed during wakefulness (Basheer et al. 2004). When cells have high levels of energy, ADE levels are low because it is phosphorylated to ATP. On the contrary, when cells are fatigued, ADE levels are high, and it acts as an inhibitory neuromodulator (Porkka-Heiskanen et al. 1997).

8.6.2.3 Cytokines

Among cytokines, interleukin-1β (IL-1β) and tumor necrosis factor-α (TNF-α) have been shown to promote sleep (Imeri and Opp 2009). Administration of both cytokines into the preoptic area reduces firing rates of wake-active neurons and promotes NREM sleep (Alam et al. 2004; Kubota et al. 2002).

8.6.2.4 Dopamine (DA)

DA mediates processes such as motor function, motivation, reward, and learning and has potent wake-promoting effects. Neuronal cells that produce DA are mainly found in the SN and VTA, but they have not been found to alter their rates of firing during wakefulness and sleep (Gaykema and Zaborszky 1996). Nevertheless, wakefulness has been found to be characterized by high extracellular levels of DA, while levels are lower during NREM sleep. This might suggest that some DA neurons are wake-active. Such neurons are found in the ventral periaqueductal area of the pons, and their lesions have been shown to lead to moderate reductions in wakefulness (Lu et al. 2006; Mignot and Nishino 2005; Nishino and Mignot 1997).

8.6.2.5 Gamma-Aminobutyric Acid (GABA)

The SLD nucleus produces GABA or glutamate, and many of its neurons are active during REM sleep (Lu et al. 2006; Verret et al. 2005; Xi et al. 2004). These neurons give projections to both the ventromedial medulla and the ventral horn of the spinal cord, through which they may inhibit motor neurons. When SLD nucleus is activated, the results are loss of muscle tone and REM sleep-like EEG activity. Inhibition of the same region promotes wakefulness and reduces REM sleep. Lesions of the SLD nucleus disrupt REM sleep atonia and reduce REM sleep (Hendricks et al. 1982; Sakai et al. 1979).

8.6.2.6 Glutamate (Glu)

Glutamate is an excitatory neurotransmitter and the ARAS is probably glutamatergic (Stenberg 2007). Even though the firing patterns of cells in both the ARAS and the BF are not sufficiently characterized, it is known that they release glutamate and play an important role in wakefulness (Jones 2003, 2005).

8.6.2.7 Histamine (His)

His is essential in wakefulness promotion, and older antihistamines that cross the blood-brain barrier are known for their sedating effects (Brown et al. 2001). His is found in the forebrain and the brainstem, regions which are innervated by the TMN. TMN fires mainly during wakefulness and less during sleep; thus His levels are higher when awake and lower when asleep (John et al. 2004; Monti 1993; Panula et al. 1984; Strecker et al. 2002; Watanabe et al. 1984). The crucial role of His for the initiation of arousal has also been shown in experimental models (John et al. 2004; Lin et al. 1989; Monti 1993).

8.6.2.8 Melanin-Concentrating Hormone (MCH)

MCH is produced by REM sleep-active neurons found, mixed in with orexin neurons, in the LH (Koyama et al. 2003; Verret et al. 2005). Both neuronal cell groups innervate nearly the same target regions, including the DR and LC (Bittencourt et al. 1992). MCH neurons are inhibitory; they fire at a high rate during REM sleep, while, during NREM sleep, they fire at a much lower rate being completely inactive during wakefulness (Ahnaou et al. 2009).

8.6.2.9 Monoamines

Monoamines, such as NE and 5-HT, cause excitation of motor neurons thus increasing muscle tone (Fedirchuk and Dai 2004; Lai et al. 2001). They inhibit REM sleep, and, during wakefulness and to some degree NREM sleep, the REM-active cholinergic neurons are inhibited by 5-HT, NE, and HA (Leonard and Llinas 1994). The interaction between cholinergic and monoaminergic neuronal groups forms the basis of the model explaining the NREM and REM sleep alternations during night (McCarley and Hobson 1975).

8.6.2.10 Norepinephrine (NE)

The major site of NE is the LC. This neurotransmitter is helpful for arousal under conditions that require high levels of attention and task performance. Firing patterns of LC neurons are higher during wakefulness and lower during NREM sleep; during REM sleep, these neurons do not fire at all (Aston-Jones and Bloom 1981; Hobson et al. 1975; Rasmussen and Jacobs 1986). Furthermore, NE neurons in the ventral medulla have an active role during stress (Saper et al. 2005). NE projections to arousal systems are excitatory, while those to sleep-promoting systems are inhibitory (Aghajanian and VanderMaelen 1982; Brown et al. 2002; Williams and Reiner 1993).

8.6.2.11 Orexin (Orx)

Orexins (Orx) or hypocretins are excitatory neuropeptides discovered in 1998 (Norman et al. 2011). They are produced in the LH and PH, and their role in the sleep-wake cycle regulation is important (de Lecea et al. 1998; Sakurai and Sasaki 1998). Furthermore, the Orx neurons in the LH are related to emotional processing and motivated behaviors (Harris et al. 2005; Ponz et al. 2010; Schwartz et al. 2008; Thompson and Borgland 2011). Their absence may lead to problems such as transition to sleep at inopportune moments, which is the case in narcolepsy (Norman et al. 2011). Experimental animals lacking Orx or the receptor for this neuropeptide have been shown to have narcolepsy with cataplexy (Chemelli et al. 1999; Lin et al. 1999). Orx neurons innervate the LC and TMN (Peyron et al. 1998). Their firing patterns are high during wakefulness, especially during active exploration, and very low (they are practically silent) during NREM and REM sleep (Mileykovskiy et al. 2005), even though they may have occasional burst discharges during REM sleep of rodents (Mileykovskiy et al. 2005; Takahashi et al. 2008). Orx levels are highest during wakefulness (Zeitzer et al. 2003), and, when injected into the brain, they increase arousal and behavioral activity while suppressing NREM and REM sleep (Espana et al. 2001).

8.6.2.12 Prolactin (PRL)

PRL neurons are found in LH and innervate hypothalamic areas, LC and DR. PRL is secreted during the second half of sleep. Experiments have shown increase of REM sleep after PRL injection and decrease of REM sleep in genetically PRL-deficient mice (Obal and Krueger 2004; Sassin et al. 1972).

8.6.2.13 Prostaglandin D2 (PGD2)

Prostaglandin, which is probably synthesized in the basal meninges (Mizoguchi et al. 2001), has sleep-promoting properties. Injection of PGD2 in the preoptic area of experimental animals has been shown to lead to neuronal activation in the VLPO, increasing NREM and REM sleep (Scammell et al. 1998; Ueno et al. 1982). Furthermore, its levels in the cerebrospinal fluid are highest during the sleep period (Pandey et al. 1995) and increase with sleep deprivation (Ram et al. 1997).

8.6.2.14 Serotonin (5-HT)

The relationship between 5-HT and sleep-wake cycle has been noticed since the 1950s when it was observed that chemicals inhibiting 5-HT catabolism enhanced SWS while those inhibiting 5-HT production led to insomnia (Jones 1972). 5-HT-producing neurons are found in the RN and innervate many brain regions that can influence sleep-wake cycle, such as the preoptic area, the BF, the hypothalamus, and the thalamus. Their firing patterns are highest during wakefulness, much lower during NREM sleep, and lowest during REM sleep. Thus, during wakefulness, extracellular 5-HT levels are higher indicating that 5-HT promotes wakefulness (Jacobs and Azmitia 1992; McGinty and Harper 1976; Portas and McCarley 1994; Trulson and Jacobs 1979).

8.6.2.15 Somnogens

They include substances that may function as natural sleep-inducing molecules, such as adenosine, cytokines, and prostaglandins (Obal and Krueger 2004).

8.6.2.16 Vasoactive Intestinal Peptide (VIP)

VIP is a PRL-inducing factor (Stenberg 2007) that has been shown to increase REM sleep in experimental animals (Riou et al. 1982). In humans, VIP administration seems to stimulate REM sleep when given in doses high enough to stimulate PRL (Murck et al. 1996).

8.7 Neurophysiology of Wakefulness and Sleep

8.7.1 Cortical and Thalamic Activity During Wakefulness and Sleep

Even though BF, hypothalamus, or brainstem (all of them are subcortical systems) are essential in generating wakefulness and sleep and in regulating the transitions between them, both activity of the EEG and consciousness itself arise from interactions between these subcortical systems, the thalamus, and the cortex (Higgins and George 2007).

The thalamus and the cortex share information; cortical neurons relay information to the neurons of thalamus which, in turn, relay information to the cortex. These thalamic neurons have intrinsic electrical characteristics, and some of the cortical rhythms in NREM sleep are due to them (McCormick and Bal 1997). The thalamus contains:

1. Glutamatergic neurons that project to the cortex and relay sensory, motor, and limbic information to it.
2. GABAergic neurons in its reticular nucleus. They are innervated by the projection neurons and cortex and in turn inhibit the projection neurons.

These connections are believed to be responsible for some cortical rhythms, such as sleep spindles (Steriade et al. 1987).

NREM sleep is characterized by hyperpolarization of the thalamic neurons leading to burst firing and reduction of their responsiveness to sensory stimuli from the environment. On the contrary, REM sleep and wakefulness are characterized by depolarization of thalamic neurons. This is caused by ACh and results to spindle and slow-wave suppression. Furthermore, it promotes the transmission of single spikes that efficiently transmit information to the cortex and drive desynchronized cortical activity. Despite the fact that thalamus has an important role, even extensive damage to it does not change the general patterns of wakefulness, REM, and NREM sleep but impairs consciousness and the ability to interact with the environment. Thus, it is believed that thalamus is not required for the basic generation of sleep states (Fuller et al. 2011).

Apart from the thalamus, GABAergic neurons are found within the cortex. They have wide projections, and they are active during NREM sleep, implying that they may broadly inhibit other cortical neurons, thus helping generate slow waves during NREM sleep (Gerashchenko et al. 2008). Furthermore, the intensity of cortical slow waves may reflect prior local activity and changes in synaptic strength, as slow waves during NREM sleep are increased over supplementary motor cortex after learning a motor task but decreased with arm immobilization (Vyazovskiy et al. 2008).

8.7.2 Process C and Process S: The Two-Process Model

This is a model that emerged in the mid-1980s and provided a useful macroscopic perspective on the dynamic control of wakefulness and sleep (Borbely et al. 2016; Daan et al. 1984). According to this model, process S refers to a homeostatic factor that accumulates during wakefulness and declines during sleep. It is defined as the drive to sleep depending on the duration individual is awake. It interacts with process C, a circadian process responsible for the regulation of the timing of wakefulness and REM sleep (Achermann 2004).

Since process C is a circadian process, it is under the control of both cues from the environment and the SCN, which regulates the circadian cycle of most physiologic rhythms, including sleep and wakefulness (Dibner et al. 2010). The neurons of SCN have a rhythmic function both individually and, especially, when coupled with other SCN neurons (Welsh et al. 2010). The synchronization of the activity of these neurons with the environmental light-dark cycle is achieved with the help of melanopsin. Melanopsin is a photosensitive molecule using luminance information and is found in retinal ganglion cells (Hattar et al. 2002).

The information arriving to SCN neurons is then transmitted to the adjacent subparaventricular zone (Kramer et al. 2001) passing through the dorsomedial nucleus of the hypothalamus and to brain regions that regulate sleep and wakefulness such as the LC, VLPO, and LH (Chou et al. 2003). Besides these, SCN neurons regulate the daily rhythm of body temperature. This is important since regulation of this rhythm entrains circadian activity throughout the body (Buhr et al. 2010).

Circadian rhythms and metabolism are closely linked. Disordered coordination of central and peripheral rhythms may give rise to problems, such as obesity and glucose intolerance which are common in people with shift work sleep disorder or insufficient sleep (Bass and Takahashi 2010).

8.7.3 Mechanisms That Regulate the Transitions Between Sleep and Wakefulness

A dynamic interaction among the systems that promote sleep and wakefulness takes place in various ways. The aim is to ensure rapid and complete transitions between sleep and wake states. The VLPO and other sleep-promoting preoptic neurons inhibit monoaminergic and cholinergic wake-promoting neurons, and the preoptic neurons themselves are inhibited by NE, 5-HT, and Ach (Chou et al. 2003; Gallopin et al. 2004).

During wakefulness, regions containing arousal-promoting neurons are disinhibited. This is made possible through high monoaminergic and cholinergic tone which silences the VLPO. Conversely, during sleep, arousal regions are inhibited. This is achieved through the activation of preoptic neurons that disinhibit their own firing. These mutually inhibitory systems are probably reinforced by orexins. Orexins may stabilize wakefulness by enhancing activity in the arousal systems ensuring alertness and long periods of wakefulness despite rising homeostatic pressure across the day. During wakefulness (and, perhaps to a lesser degree, in NREM sleep), orexins may excite a variety of neurons that inhibit REM sleep, such as monoaminergic neurons, the vlPAG/LPT, and GABAergic inputs to the SLD (Arrigoni et al. 2006; Brown et al. 2001; Eggermann et al. 2001).

8.7.4 Activation of Emotional and Reward Circuits During Sleep and Dreaming

Neuroimaging, neurophysiological, and behavioral studies in both humans and animals have shown that, during sleep and dreaming, emotional and reward-related processes and networks are activated. Furthermore, lesion and pharmacological studies of dreaming and dream content analysis in humans also support such a hypothesis (Gaillard and Moneme 1977; Lena et al. 2005; Merritt et al. 1994 Nielsen-Bohlman et al. 1991; Perogamvros and Schwartz 2012; Schenck and Mahowald 2002; Solms 2000).

8.7.5 Neural Mechanisms of REM Sleep

REM sleep is characterized by the following events:

1. Cortical EEG activity pattern similar to that during wakefulness.
2. Loss of tone of the postural muscles resulting from the activation of neurons in the locus coeruleus alpha (LCα).
3. Rapid eye movements caused by the activation of neurons in the peri-abducens reticular formation (PAb).
4. Theta rhythm within the hippocampus produced via the activation of neurons in the pontis oralis (PO).
5. Field potentials in the pons (P wave) and spikes in the lateral geniculate nucleus and the occipital cortex. These waves emerge from activation of neurons in the caudo-lateral peribrachial area (C-PBL) of predator mammals and in the dorsal part of the nucleus subcoeruleus (SubCD) of prey mammals.
6. Myoclonic twitches, especially in the muscles of the face and the distal limbs, due to the activation of neurons in the nucleus gigantocellularis.

7. Fluctuations in core body temperature, cardiac and respiratory rate which result from activation of neurons in the parabrachial nucleus (PBN).
8. Penile erection in men and clitoral tumescence in women (Domhoff 2003; Higgins and George 2007; Kirsch 2013; Lee-Chiong 2008).

Each of these events is likely to be mediated by distinct cell groups in the brainstem (Vertes 1984). During REM sleep, aminergic cell activities are markedly reduced or absent, and cholinergic cell activities are comparatively high, even though much less than that during wakefulness (Datta et al. 2009; Lydic et al. 1983; McGinty and Harper 1976). To date, almost all studies have consistently demonstrated that the activation of cholinergic cells in the PPT is one of the most critical steps for the generation and maintenance of REM sleep (Datta et al. 2009).

8.7.6 Neural Mechanisms of SWS

Electrophysiological and behavioral signs of SWS-I in animals are comparable to those of stage II sleep in humans. The defining EEG signs of stage II sleep are the sleep spindles and the K-complexes. The former are waxing and waning waves in group sequences every 2–5 s. Sleep spindles are caused by the GABAergic neurons of the thalamic reticular nucleus (Steriade et al. 1993). The thalamus is also a gateway for most sensory and internal signals in their way to the cerebral cortex (Jones et al. 1985).

The thalamus contains two functionally different types of neurons:

- Thalamocortical relay neurons, through which incoming sensory information is transmitted to the cerebral cortex where they cause rhythmical excitation of the cortical neurons that ultimately generate cortical sleep spindles (Llinas and Steriade 2006).
- Thalamic reticular neurons, the activation of which has the opposite results.

The sleep spindle mechanism by thalamocortical relay neurons combines both passive and active physiological processes. The SWS/SWS-II generating mechanisms are active processes. The active steps of the SWS generating mechanisms, compared to the passive steps of the initiation of sleep, operate on shorter timescales. Yet, these behavioral states remain conducive to transfer short-term memory information from the cortex to the hippocampus and/or amygdala for the sorting and deleting steps of long-term memory formation processes (Datta 2006).

Studies have shown that lesions of cell bodies in the POA of the AH effectively suppress SWS in mammals (John et al. 1994; John and Kumar 1998; Srividya et al. 2006). A large population of cells within the POA are more active during SWS than during wakefulness or REM sleep (Alam et al. 1995; Koyama and Hayaishi 1994; McGinty and Szymusiak 2000; Suntsova et al. 2002; Szymusiak et al. 1998).

A study using functional magnetic resonance imaging (fMRI) in the behaving rat has demonstrated that the POA is more active than other parts of the hypothalamus and BF during SWS (Khubchandani et al. 2005). Immunohistochemical analysis of sleep-active neurons in the POA has revealed that the majority contain the inhibitory neurotransmitters GABA and galanin (Gaus et al. 2002; Gong et al. 2004; Gvilia et al. 2006). These sleep-active neurons innervate many wake-promoting areas of the brain, including the TMN, LH, LC, DRN, and PPT/LDT (Gritti et al. 1994). Thus, it is possible that the increased activity of SWS-active GABAergic cells in the POA could release GABA to targets within the wake-promoting areas of the brain.

Released GABA may then suppress activity in these areas in two different ways:

1. GABA receptor activation-mediated inhibition of wake-promoting cells.
2. Inhibition of presynaptic neurotransmitter release that is necessary for the activation of wake-promoting cells (Gottesmann 2002).

Neuropharmacological studies have shown that SWS may be also induced by sedative and hypnotic drugs that involve potentiation of POA GABAergic neurotransmission (Mendelson and Ziegler 2001; Sallanon et al. 1989; Tung et al. 2001).

8.7.7 Physiological Changes During Sleep (Table 8.13)

8.7.7.1 Autonomic Nervous System

The sympathetic activity decreases during NREM and the tonic phase of REM sleep, while the parasympathetic increases and reaches a peak in SWS (Brandenberger et al. 2001; Trinder et al. 2001).

8.7.7.2 Cardiovascular and Respiratory System

During sleep, cardiac output, blood pressure, and heart and respiratory rate are reduced. Episodes of arrhythmia may take place during REM sleep thus explaining the higher incidence of myocardial infarctions early in the morning (Verrier et al. 1996). Air exchange and ventilation decrease, whereas muscle relaxation increases resistance in the upper airways (Penzel et al. 2007).

8.7.7.3 Endocrine System

Release of various hormones takes place during sleep. These include growth hormone, prolactin, cortisol, and adrenocorticotropic hormone (Freeman et al. 2000; Obal and Krueger 2004; Wagner and Born 2008) contrary to thyroid-stimulating hormone which is inhibited by sleep (Luboshitzky 2000). Generally, during sleep, increased secretion of anabolic hormones and synthesis of proteins and nucleic acid of the brain take place; the secretion of catabolic hormones is decreased (Nakanishi et al. 1997).

Table 8.13 Physiological changes during NREM and REM sleep (Colten and Altevogt 2006)

Physiological process	NREM	REM
Brain activity	Decreased	Increased in motor and sensory areas Other areas similar to NREM
Heart rate	Decreased	Increased Varies from NREM
Blood pressure	Decreased	Increased (up to 30 percent) Varies from NREM
Sympathetic nerve activity	Decreased	Increased significantly
Muscle tone	Similar to wakefulness	Absent (atonia)
Blood flow to brain	Decreased	Increased from NREM Depends on brain region
Respiration	Decreased	Increased and varies from NREM Coughing suppressed
Airway resistance	Increased	Increased and varies from wakefulness
Body temperature	Regulated at lower set point than wakefulness; shivering initiated at lower temperature than during wakefulness	Not regulated; no shivering or sweating; temperature drifts toward that of the local environment
Sexual arousal	Occurs infrequently	Greater than NREM

8.7.7.4 Reproductive System

During REM sleep, men may experience nocturnal penile tumescence, and women may have increased vaginal blood flow and engorgement of clitoris (Abel et al. 1979; Schmidt and Schmidt 2004).

8.7.7.5 Thermoregulation

Body temperature is reduced following a decrease in hypothalamic temperature set point. Sweating and shivering are also reduced (Krauchi 2007; Lack et al. 2008).

8.8 Neurobiology of Dreaming

Dreaming, a universal human phenomenon, takes place in all stages of sleep. Usually, if one is not awaken during sleep, there is no recall of the dreams. Recall of dreaming depends on both brain activation during sleep and awakening conditions (Hobson and Pace-Schott 2002). Compared to memories of everyday life, memory for dreams is rather poor. This is due to the specific brain states in which dreaming occurs and its reduced sensory and contextual anchoring (Schwartz et al. 2002). Nightmares are frightening dreams that seem to be related to intense and anxiety-provoking situations of everyday life. They are more common in children and decrease in old age (Table 8.14) (Chokroverty 2013; Colten and Altevogt 2006).

Table 8.14 Types of dreams (Schredl 2010)

Dream types	Definition
REM/NREM dreams	Recollection of mental activity during REM/NREM sleep
Sleep-onset dreams	Recollection of mental activity during stage 1 of NREM sleep
Nightmares	REM dreams with strong negative emotions which cause awakening
Night terrors	Sudden arousal with intense anxiety out of SWS Sometimes accompanied by short NREM dreams
Post-traumatic reenactments	REM or NREM dreams which replay the original traumatic experience in a less distorted way
Lucid dreams	REM dreams in which the dreamer is aware that she or he is dreaming

The dreams of REM sleep account for about 80% of dreams during sleep, while the rest are taking place during NREM sleep (Lee-Chiong 2008). Neurobiology of dreaming involves activation of cells and synapses in the brainstem and transmission of the signals to the hemispheres (Chokroverty 2013), activation of higher visual centers, activation of limbic system, and loss of working memory (Hobson and Pace-Schott 2002).

Their content differs. It is more bizarre, irrational, and complex in REM sleep and looks more like problem-solving, simple, and realistic in NREM sleep (Higgins and George 2007; Lee-Chiong 2008). Dreams during REM sleep contain formed hallucinatory perceptions in practically every sensory modality. They are delusional and lack stability in orientation. The imagery is often bizarre, and there may exist images and events that are common in everyday life (Hobson and Pace-Schott 2002).

They contain people, animals, and things with various colors and shapes. They may also contain sounds and experiences with a sensory character (Nir and Tononi 2010). Self-reflection and self-control are absent. Emotions, especially fear or anxiety, are intensified, and volitional control is attenuated (Hobson and Pace-Schott 2002). Emotional involvement levels are high even though negative feelings, such as sadness, are rare (Fosse et al. 2001; Foulkes et al. 1988; Hobson et al. 2000).

Dreams reflect our personality, interests, and anxieties of everyday life, our past, and our current waking concerns (Schredl and Hofmann 2003) and are quite stable during adulthood (Domhoff 1996, 2003). They are usually novel constructions and rarely reproductions of past events (Fosse et al. 2003; Schwartz et al. 2002). Contrary to NREM dreams, REM dreams are easier to remember. They are also easier to remember if subjects are awaken immediately after the onset of REM dreams (Chokroverty 2013).

Administration of dopaminergic agents in humans has been shown by pharmacological studies to elicit vivid dreams (Balon 1996; Thompson and Pierce 1999). On the other hand, administration of D_2 antagonists (a common mechanism for almost

Table 8.15 Theories of function of dreaming (Chokroverty 2013)

Activation of neural networks of brain
Restructuring and reinterpretation of data
Memory consolidation
Removal of useless information
Dreams as epiphenomenon
Iterative programming
Guardian of sleep
Compensation
Reverse learning hypothesis
Mastery hypothesis
Mood regulation
Systemic desensitization
Threat simulation theory

all antipsychotics) reduces vivid dreaming (Gaillard and Moneme 1977) and nightmares (Lambert 2006). This dopaminergic theory of dreams has been criticized though (Doricchi and Violani 2000), and effects of dopamine on dreaming may be a result of its interactions with other neuromodulatory systems, such as cholinergic cells of BF that activate cortical and limbic structures (Perry and Piggott 2000). Other theories of dreaming include the cognitive and the activation synthesis hypothesis (Lee-Chiong 2008).

Dreams are believed to be associated with memory-related processes and activation of neural networks in CNS (Lee-Chiong 2008). Their neurobiological significance is as yet unknown even though various theories have been postulated (Table 8.15). Nevertheless, there is no scientific evidence that dreams foretell the future (Hobson and Pace-Schott 2002).

8.9 Clinical Aspects

8.9.1 Introduction

Sleep disorders are considered an important aspect of public health. Depending on their etiology, they are categorized into two broad groups: those of primary etiology and those that have an underlying neurologic disease or mental health disorder (Chokroverty 2009; Mitler et al. 2000; Saper and Scammell 2013).

8.9.2 Circadian Rhythm Sleep Disorders

Patients with circadian rhythm sleep disorders (CRSL) have adequate sleep quantity and quality, but they cannot sleep when they want or they are expected to do so; this may lead both to insomnia and excessive daytime sleepiness. Jet lag and shift work are two common causes of CRSD (Silber et al. 2010). The exact pathophysiologic basis is not fully elucidated but likely involves genetic and molecular mechanisms (Iwase et al. 2002; Reid et al. 2001; Toh et al. 2001).

8.9.3 Fatal Familial Insomnia

This is a rare, prion disease defined by agrypnia excitata syndrome. Patients cannot sleep and concurrently have overactivation of the sympathetic nervous system. There is no treatment for the disease, and patients die from it (Guaraldi et al. 2011; Montagna et al. 2003).

8.9.4 Insomnia

It is estimated that insomnia is the most common sleep disorder (Ohayon 2002) and many of those complaining for it may have an underlying psychiatric disorder (Ohayon 1997). It can lead to many adverse effects, such as anxiety and mood disorders, fatigue, and problems in daily functioning. In clinical practice, insomnia is a term used to denote difficulties in falling or staying asleep. Insomnia is believed to be a disorder of increased physiological arousal since, contrary to what was generally believed, arousal and sleep are not mutually exclusive states (Nofzinger et al. 2004). Potential mechanisms include stress response, inhospitable environment (temperature, light, noises), homeostatic factors, and maladaptive coping mechanisms (Silber et al. 2010).

8.9.5 Narcolepsy

Narcolepsy, a disease of REM sleep mechanism, is characterized by excessive sleepiness due to activation or disfacilitation of sleep-active or wake-active neurons, respectively. Most patients with narcolepsy exhibit also a sudden loss of muscle tone, which is called cataplexy (Siegel 2009). Apart from cataplexy, there are two other REM-related symptoms that define the disease: hypnagogic hallucinations and sleep paralysis (Guilleminault and Anagnos 2000). The majority of narcolepsy cases are caused by the loss of hypocretin-containing hypothalamic cells (Thannickal et al. 2000), and this has been shown in brains of patients with the disease (Nishino et al. 2000; Thannickal et al. 2000).

8.9.6 Obstructive Sleep Apnea

This disorder is caused by dysfunction of the upper airway or the control mechanisms of the respiratory system (Silber et al. 2010). Factors predisposing to obstructive sleep apnea include small upper airway, such as in obesity and in lesions of the upper airway, reduction of dilator activity of the upper airway muscles (may be caused by alcohol and drugs, such as benzodiazepines, and neurological disorders), and increased resistance of chest wall in obesity (Reading 2013).

8.9.7 Parasomnias

Parasomnias are sleep disorders characterized by abnormal behaviors during sleep (Howell and Schenck 2012). REM sleep behavior disorders are parasomnias occurring during REM sleep; instead of the normal atonia in this stage, patients have increased muscle tone resulting in overt motor behavior during dreaming (Schenck and Mahowald 2002). Parasomnias during NREM sleep are characterized by the concurrent occurrence of wake and sleep states (Bassetti et al. 2000). The underlying pathophysiological mechanisms of parasomnias are not fully known. Clinically, they are characterized by instinctual and exploratory behaviors, such as oriented locomotion, aggression, sexual behaviors, feeding, chewing, or swallowing (Morrison et al. 1992; Vetrugno et al. 2006; Winkelman 2006).

8.9.8 Sleep-Related Movement Disorders

8.9.8.1 Restless Legs Syndrome (Willis-Ekbom's Disease)

Various movements have been described to occur when falling for sleep or during sleep. Restless legs syndrome (RLS) is the most common disease in this group. Its pathophysiology is not fully known yet, but it is likely due to dopamine dysregulation. Iron may also be implicated since it is cofactor for the production of dopamine (Raman 2013) and has been shown to be reduced in the SN of patients with RLS (Allen et al. 2001; Connor et al. 2003). CNS pathology might exist since it is known that centrally active dopamine antagonists exacerbate the disease. Studies also indicate a possible pathology of subcortical areas or of subcortical inhibition of flexor reflex (Bara-Jimenez et al. 2000; Tergau et al. 1999).

8.9.8.2 Periodic Limb Movement Disorder

This is a disease entity characterized by periodic, stereotyped limb movements during sleep, usually in N1 and N2 stages (Raman 2013).

8.9.9 Sleep and Psychiatric Disorders

8.9.9.1 Sleep and Alcohol

Alcohol is frequently used as hypnagogic substance since it results to decreased sleep latency and increased SWS during the first hours of the night. However, later in the night, alcohol ingestion leads to nightmares and sleep fragmentation (Pressman 2007). Chronic alcohol use leads to increased sleep latency, poor quality of sleep, and exacerbation of preexisting sleep disorders (Stein and Friedmann 2006).

8.9.9.2 Sleep and Anxiety Disorders

Many patients suffering from anxiety complain about sleep disorders, especially insomnia, which, in turn, may lead to further anxiety. This leads to a vicious cycle: insomnia deteriorates anxiety and mood symptoms suggesting that there may exist a common pathophysiologic mechanism (Mellman 2006, 2008; Ohayon et al. 1998; Ohayon and Roth 2003).

8.9.9.3 Sleep and Bipolar Disorder

Patients with hypomania and/or mania have a reduced need for sleep (Plante and Winkelman 2008; Wehr 1991). Furthermore, a decrease in sleep may lead to hypomania or mania in patients with bipolar disorder, while the opposite is true for bipolar depression (Bauer et al. 2006). Sleep disturbances may also be common prodromal signs of both mania and depression (Jackson et al. 2003). This connection between sleep disturbances and mood disorders has been proposed by research to form a closed loop (Harvey 2008) with circadian (Li et al. 2013; Roybal et al. 2007; Wirz-Justice et al. 2009) and neuroendocrine (Spiegel 2000) underlying mechanisms.

8.9.9.4 Sleep and Depression

Circadian dysfunction may be partially responsible for the initiation or maintenance of mood disorders. Depression has a diurnal variation and may be alleviated by sleep deprivation, which could possibly be explained by circadian or homeostatic processes (Boivin et al. 1997). Misalignment between circadian and sleep phase can deteriorate mood (Danilenko et al. 2003); this could likely explain mood-related problems in jet lag or shift work (Kolla and Auger 2011), as well as severity of unipolar depression (Hasler et al. 2010). Furthermore, prior wakefulness seems to deteriorate mood (Boivin et al. 1997). Sleep deprivation may alleviate depressive mood in patients immediately even though this effect is usually of short duration (Barbini et al. 1998; Giedke and Schwarzler 2002). Researchers have attempted to explain this effect by dopaminergic (Ebert and Berger 1998), noradrenergic (Payne et al. 2002), circadian (Wirz-Justice et al. 1981), and homeostatic mechanisms (Endo et al. 1997). On the other hand, insomnia is an independent risk factor of subsequent onset of major depression (Buysse et al. 2008; Ford and Kamerow 1989; Johnson et al. 2006; Riemann and Voderholzer 2003).

8.9.9.5 Sleep and Psychotic Disorders

Insomnia is common in patients with schizophrenia and may be a clinical prodromal sign of first-episode psychosis or relapse (Benca et al. 1992; Chouinard et al. 2004). These patients usually have short REM latency and SWS deficits (Zarcone et al. 1987) (Table 8.15).

References

Abel GG, Murphy WD, Becker JV, Bitar A (1979) Women's vaginal responses during REM sleep. J Sex Marital Ther 5(1):5–14. https://doi.org/10.1080/00926237908403713

Achermann P (2004) The two-process model of sleep regulation revisited. Aviat Space Environ Med 75(3 Suppl):A37–A43

Aghajanian GK, VanderMaelen CP (1982) Alpha 2-adrenoceptor-mediated hyperpolarization of locus coeruleus neurons: intracellular studies in vivo. Science 215(4538):1394–1396. https://doi.org/10.1126/science.6278591

Ahnaou A, Dautzenberg FM, Geys H, Imogai H, Gibelin A, Moechars D, Steckler T, Drinkenburg WH (2009) Modulation of group II metabotropic glutamate receptor (mGlu2) elicits common changes in rat and mice sleep-wake architecture. Eur J Pharmacol 603(1–3):62–72. https://doi.org/10.1016/j.ejphar.2008.11.018

Alam MN, McGinty D, Szymusiak R (1995) Neuronal discharge of preoptic/anterior hypothalamic thermosensitive neurons: relation to NREM sleep. Am J Phys 269(5 Pt 2):R1240–R1249. https://doi.org/10.1152/ajpregu.1995.269.5.R1240

Alam MN, Gong H, Alam T, Jaganath R, McGinty D, Szymusiak R (2002) Sleep-waking discharge patterns of neurons recorded in the rat perifornical lateral hypothalamic area. J Physiol 538(Pt 2):619–631. https://doi.org/10.1113/jphysiol.2001.012888

Alam MN, McGinty D, Bashir T, Kumar S, Imeri L, Opp MR, Szymusiak R (2004) Interleukin-1beta modulates state-dependent discharge activity of preoptic area and basal forebrain neurons: role in sleep regulation. Eur J Neurosci 20(1):207–216. https://doi.org/10.1111/j.1460-9568.2004.03469.x

Allen RP, Barker PB, Wehrl FW, Song HK, Earley CJ (2001) MRI measurement of brain iron in patients with restless legs syndrome. Neurology 56(2):263–265. https://doi.org/10.1212/wnl.56.2.263

Arrigoni E, Chamberlin NL, Saper CB, McCarley RW (2006) Adenosine inhibits basal forebrain cholinergic and noncholinergic neurons in vitro. Neuroscience 140(2):403–413. https://doi.org/10.1016/j.neuroscience.2006.02.010

Aserinsky E, Kleitman N (1953) Regularly occurring periods of eye motility, and concomitant phenomena, during sleep. Science 118(3062):273–274. https://doi.org/10.1126/science.118.3062.273

Aston-Jones G, Bloom FE (1981) Activity of norepinephrine-containing locus coeruleus neurons in behaving rats anticipates fluctuations in the sleep-waking cycle. J Neurosci 1(8):876–886. https://doi.org/10.1523/jneurosci.01-08-00876.1981

Bach V, Telliez F, Chardon K, Tourneux P, Cardot V, Libert J-P (2011) Thermoregulation in wakefulness and sleep in humans. Handb Clin Neurol 98:215–227. https://doi.org/10.1016/b978-0-444-52006-7.00014-9

Baghdoyan HA (1997) Location and quantification of muscarinic receptor subtypes in rat pons: implications for REM sleep generation. Am J Phys 273(3 Pt 2):R896–R904. https://doi.org/10.1152/ajpregu.1997.273.3.R896

Balon R (1996) Bupropion and nightmares. Am J Psychiatry 153(4):579–580

Banks S, Dorrian J, Basner M, Dinges DF (2017) Sleep deprivation. In: Kryger M, Roth T, Dement W (eds) Principles and practice of sleep medicine. Elsevier, Amsterdam, pp 49–55. https://doi.org/10.1016/b978-0-323-24288-2.00005-2

Bara-Jimenez W, Aksu M, Graham B, Sato S, Hallett M (2000) Periodic limb movements in sleep: state-dependent excitability of the spinal flexor reflex. Neurology 54(8):1609–1616. https://doi.org/10.1212/wnl.54.8.1609

Barbini B, Colombo C, Benedetti F, Campori E, Bellodi L, Smeraldi E (1998) The unipolar–bipolar dichotomy and the response to sleep deprivation. Psychiatry Res 79(1):43–50. https://doi.org/10.1016/s0165-1781(98)00020-1

Basheer R, Strecker RE, Thakkar MM, McCarley RW (2004) Adenosine and sleep-wake regulation. Prog Neurobiol 73(6):379–396. https://doi.org/10.1016/j.pneurobio.2004.06.004

Bass J, Takahashi JS (2010) Circadian integration of metabolism and energetics. Science 330(6009):1349–1354. https://doi.org/10.1126/science.1195027

Bassetti C, Vella S, Donati F, Wielepp P, Weder B (2000) SPECT during sleepwalking. Lancet 356(9228):484–485. https://doi.org/10.1016/S0140-6736(00)02561-7

Bauer M, Grof P, Rasgon N, Bschor T, Glenn T, Whybrow PC (2006) Temporal relation between sleep and mood in patients with bipolar disorder. Bipolar Disord 8(2):160–167. https://doi.org/10.1111/j.1399-5618.2006.00294.x

Benca RM, Obermeyer WH, Thisted RA, Gillin JC (1992) Sleep and psychiatric disorders. A meta-analysis. Arch Gen Psychiatry 49(8):651–668.; discussion 669–670. https://doi.org/10.1001/archpsyc.1992.01820080059010

Berger H (1930) On the electroencephalogram of man. J Psychol Neurol 40:160–179

Berger RJ (1961) Tonus of extrinsic laryngeal muscles during sleep and dreaming. Science 134(3482):840. https://doi.org/10.1126/science.134.3482.840

Bittencourt JC, Presse F, Arias C, Peto C, Vaughan J, Nahon JL, Vale W, Sawchenko PE (1992) The melanin-concentrating hormone system of the rat brain: an immuno- and hybridization histochemical characterization. J Comp Neurol 319(2):218–245. https://doi.org/10.1002/cne.903190204

Boivin DB, Czeisler CA, Dijk DJ, Duffy JF, Folkard S, Minors DS, Totterdell P, Waterhouse JM (1997) Complex interaction of the sleep-wake cycle and circadian phase modulates mood in healthy subjects. Arch Gen Psychiatry 54(2):145–152. https://doi.org/10.1001/archpsyc.1997.01830140055010

Borb AA, Achermann P (2016) Sleep homeostasis and models of sleep regulation. J Biol Rhythm 14(6):559–570. https://doi.org/10.1177/074873099129000894

Borbely AA, Daan S, Wirz-Justice A, Deboer T (2016) The two-process model of sleep regulation: a reappraisal. J Sleep Res 25(2):131–143. https://doi.org/10.1111/jsr.12371

Born J, Rasch B, Gais S (2006) Sleep to remember. Neuroscientist 12(5):410–424. https://doi.org/10.1177/1073858406292647

Brandenberger G, Ehrhart J, Piquard F, Simon C (2001) Inverse coupling between ultradian oscillations in delta wave activity and heart rate variability during sleep. Clin Neurophysiol 112(6):992–996. https://doi.org/10.1016/S1388-2457(01)00507-7

Braun AR, Balkin TJ, Wesenten NJ, Carson RE, Varga M, Baldwin P, Selbie S, Belenky G, Herscovitch P (1997) Regional cerebral blood flow throughout the sleep-wake cycle. An H2(15)O PET study. Brain 120(Pt 7):1173–1197. https://doi.org/10.1093/brain/120.7.1173

Braun AR, Balkin TJ, Wesensten NJ, Gwadry F, Carson RE, Varga M, Baldwin P, Belenky G, Herscovitch P (1998) Dissociated pattern of activity in visual cortices and their projections during human rapid eye movement sleep. Science 279(5347):91–95. https://doi.org/10.1126/science.279.5347.91

Brown RE, Sergeeva O, Eriksson KS, Haas HL (2001) Orexin A excites serotonergic neurons in the dorsal raphe nucleus of the rat. Neuropharmacology 40(3):457–459. https://doi.org/10.1016/s0028-3908(00)00178-7

Brown FC, Buboltz WC Jr, Soper B (2002) Relationship of sleep hygiene awareness, sleep hygiene practices, and sleep quality in university students. Behav Med 28(1):33–38. https://doi.org/10.1080/08964280209596396

Brown RE, Basheer R, McKenna JT, Strecker RE, McCarley RW (2012) Control of sleep and wakefulness. Physiol Rev 92(3):1087–1187. https://doi.org/10.1152/physrev.00032.2011

Buhr ED, Yoo SH, Takahashi JS (2010) Temperature as a universal resetting cue for mammalian circadian oscillators. Science 330(6002):379–385. https://doi.org/10.1126/science.1195262

Buysse DJ, Angst J, Gamma A, Ajdacic V, Eich D, Rossler W (2008) Prevalence, course, and comorbidity of insomnia and depression in young adults. Sleep 31(4):473–480. https://doi.org/10.1093/sleep/31.4.473

Cai DJ, Mednick SA, Harrison EM, Kanady JC, Mednick SC (2009) REM, not incubation, improves creativity by priming associative networks. Proc Natl Acad Sci U S A 106(25):10130–10134. https://doi.org/10.1073/pnas.0900271106

Carskadon MA, Dement WC (2017) Normal human sleep. In: Kryger M, Roth T, Dement W (eds) Principles and practice of sleep medicine. Elsevier, Amsterdam. https://doi.org/10.1016/b978-0-323-24288-2.00002-7

Casement MD, Broussard JL, Mullington JM, Press DZ (2006) The contribution of sleep to improvements in working memory scanning speed: a study of prolonged sleep restriction. Biol Psychol 72(2):208–212. https://doi.org/10.1016/j.biopsycho.2005.11.002

Chemelli RM, Willie JT, Sinton CM, Elmquist JK, Scammell T, Lee C, Richardson JA, Williams SC, Xiong Y, Kisanuki Y, Fitch TE, Nakazato M, Hammer RE, Saper CB, Yanagisawa M (1999) Narcolepsy in orexin knockout mice: molecular genetics of sleep regulation. Cell 98(4):437–451. https://doi.org/10.1016/s0092-8674(00)81973-x

Chokroverty S (2009) Sleep and neurodegenerative diseases. Semin Neurol 29(4):446–467. https://doi.org/10.1055/s-0029-1237124

Chokroverty S (2013) An overview of normal sleep. Sleep and movement disorders. Oxford University Press, Oxford. https://doi.org/10.1093/med/9780199795161.003.0003
Chou TC, Scammell TE, Gooley JJ, Gaus SE, Saper CB, Lu J (2003) Critical role of dorsomedial hypothalamic nucleus in a wide range of behavioral circadian rhythms. J Neurosci 23(33):10691–10702. https://doi.org/10.1523/jneurosci.23-33-10691.2003
Chouinard S, Poulin J, Stip E, Godbout R (2004) Sleep in untreated patients with schizophrenia: a meta-analysis. Schizophr Bull 30(4):957–967. https://doi.org/10.1093/oxfordjournals.schbul.a007145
Cirelli C, Tononi G (2008) Is sleep essential? PLoS Biol 6(8):e216. https://doi.org/10.1371/journal.pbio.0060216
Colten H, Altevogt B (2006) Committee on Sleep Medicine and Research, Institute of the National Academies. Sleep disorders and sleep deprivation: an unmet public health problem. National Academies Press, Washington, DC
Connor JR, Boyer PJ, Menzies SL, Dellinger B, Allen RP, Ondo WG, Earley CJ (2003) Neuropathological examination suggests impaired brain iron acquisition in restless legs syndrome. Neurology 61(3):304–309. https://doi.org/10.1212/01.wnl.0000078887.16593.12
Daan S, Beersma DG, Borbely AA (1984) Timing of human sleep: recovery process gated by a circadian pacemaker. Am J Phys 246(2 Pt 2):R161–R183. https://doi.org/10.1152/ajpregu.1984.246.2.R161
Danilenko KV, Cajochen C, Wirz-Justice A (2003) Is sleep per se a zeitgeber in humans? J Biol Rhythm 18(2):170–178. https://doi.org/10.1177/0748730403251732
Datta S (1995) Neuronal activity in the peribrachial area: relationship to behavioral state control. Neurosci Biobehav Rev 19(1):67–84. https://doi.org/10.1016/0149-7634(94)00043-z
Datta S (2006) Activation of phasic pontine-wave generator: a mechanism for sleep-dependent memory processing. Sleep Biol Rhythms 4(1):16–26. https://doi.org/10.1111/j.1479-8425.2006.00202.x
Datta S, Maclean RR (2007) Neurobiological mechanisms for the regulation of mammalian sleep-wake behavior: reinterpretation of historical evidence and inclusion of contemporary cellular and molecular evidence. Neurosci Biobehav Rev 31(5):775–824. https://doi.org/10.1016/j.neubiorev.2007.02.004
Datta S, Siwek DF, Stack EC (2009) Identification of cholinergic and non-cholinergic neurons in the pons expressing phosphorylated cyclic adenosine monophosphate response element-binding protein as a function of rapid eye movement sleep. Neuroscience 163(1):397–414. https://doi.org/10.1016/j.neuroscience.2009.06.035
Deboer T, Vansteensel MJ, Detari L, Meijer JH (2003) Sleep states alter activity of suprachiasmatic nucleus neurons. Nat Neurosci 6(10):1086–1090. https://doi.org/10.1038/nn1122
Dement WC (2005) History of sleep medicine. Neurol Clin 23(4):945–965., v. https://doi.org/10.1016/j.ncl.2005.07.001
Dibner C, Schibler U, Albrecht U (2010) The mammalian circadian timing system: organization and coordination of central and peripheral clocks. Annu Rev Physiol 72(1):517–549. https://doi.org/10.1146/annurev-physiol-021909-135821
Diekelmann S, Born J (2010) The memory function of sleep. Nat Rev Neurosci 11(2):114–126. https://doi.org/10.1038/nrn2762
Dinges DF (2006) The state of sleep deprivation: from functional biology to functional consequences. Sleep Med Rev 10(5):303–305. https://doi.org/10.1016/j.smrv.2006.07.001
Domhoff GW (1996) Finding meaning in dreams. Springer US, New York, NY. https://doi.org/10.1007/978-1-4899-0298-6
Domhoff GW (2003) The scientific study of dreams: neural networks, cognitive development, and content analysis. Am Psychol Assoc. https://doi.org/10.1037/10463-000
Doricchi F, Violani C (2000) Mesolimbic dopamine and the neuropsychology of dreaming: some caution and reconsiderations. Behav Brain Sci 23(6):930. https://doi.org/10.1017/S0140525x00374027

Easton A, Meerlo P, Bergmann B, Turek FW (2004) The suprachiasmatic nucleus regulates sleep timing and amount in mice. Sleep 27(7):1307–1318. https://doi.org/10.1093/sleep/27.7.1307

Ebert D, Berger M (1998) Neurobiological similarities in antidepressant sleep deprivation and psychostimulant use: a psychostimulant theory of antidepressant sleep deprivation. Psychopharmacology 140(1):1–10. https://doi.org/10.1007/s002130050732

Economo CV (1930) Sleep as a problem of localization. J Nerv Ment Dis 71(3):249–259. https://doi.org/10.1097/00005053-193003000-00001

Eggermann E, Serafin M, Bayer L, Machard D, Saint-Mleux B, Jones BE, Muhlethaler M (2001) Orexins/hypocretins excite basal forebrain cholinergic neurones. Neuroscience 108(2):177–181. https://doi.org/10.1016/s0306-4522(01)00512-7

Ehlers CL, Kupfer DJ (1989) Effects of age on delta and REM sleep parameters. Electroencephalogr Clin Neurophysiol 72(2):118–125. https://doi.org/10.1016/0013-4694(89)90172-7

Endo T, Schwierin B, Borbely AA, Tobler I (1997) Selective and total sleep deprivation: effect on the sleep EEG in the rat. Psychiatry Res 66(2–3):97–110

Espana RA, Baldo BA, Kelley AE, Berridge CW (2001) Wake-promoting and sleep-suppressing actions of hypocretin (orexin): basal forebrain sites of action. Neuroscience 106(4):699–715. https://doi.org/10.1016/S0306-4522(01)00319-0

Esteban S, Nicolau MC, Gamundi A, Akaârir M, Rial RV (2005) Animal sleep: phylogenetic correlations. The physiologic nature of sleep. Imperial College Press, London. https://doi.org/10.1142/9781860947186_0010

Fedirchuk B, Dai Y (2004) Monoamines increase the excitability of spinal neurones in the neonatal rat by hyperpolarizing the threshold for action potential production. J Physiol 557(Pt 2):355–361. https://doi.org/10.1113/jphysiol.2004.064022

Fischer S, Born J (2009) Anticipated reward enhances offline learning during sleep. J Exp Psychol Learn Mem Cogn 35(6):1586–1593. https://doi.org/10.1037/a0017256

Ford DE, Kamerow DB (1989) Epidemiologic study of sleep disturbances and psychiatric disorders. An opportunity for prevention? JAMA 262(11):1479–1484. https://doi.org/10.1001/jama.1989.03430110069030

Fosse R, Stickgold R, Hobson JA (2001) The mind in REM sleep: reports of emotional experience. Sleep 24(8):947–955. https://doi.org/10.1093/sleep/24.8.1

Fosse MJ, Fosse R, Hobson JA, Stickgold RJ (2003) Dreaming and episodic memory: a functional dissociation? J Cogn Neurosci 15(1):1–9. https://doi.org/10.1162/089892903321107774

Foulkes D, Sullivan B, Kerr NH, Brown L (1988) Appropriateness of dream feelings to dreamed situations. Cognit Emot 2(1):29–39. https://doi.org/10.1080/02699938808415227

Freeman ME, Kanyicska B, Lerant A, Nagy G (2000) Prolactin: structure, function, and regulation of secretion. Physiol Rev 80(4):1523–1631. https://doi.org/10.1152/physrev.2000.80.4.1523

Fuller PM, Sherman D, Pedersen NP, Saper CB, Lu J (2011) Reassessment of the structural basis of the ascending arousal system. J Comp Neurol 519(5):933–956. https://doi.org/10.1002/cne.22559

Gaillard JM, Moneme A (1977) Modification of dream content after preferential blockade of mesolimbic and mesocortical dopaminergic systems. J Psychiatr Res 13(4):247–256. https://doi.org/10.1016/s0022-3956(77)90020-6

Gallopin T, Fort P, Luppi P-H (2004) In vitro identification of the presumed sleep-promoting neurons of the ventrolateral preoptic nucleus (VLPO). Sleep. CRC Press, Boca Raton, FL. https://doi.org/10.1201/9780203496732.ch3

Garcia-Rill E (2002) The sleep state-dependent midlatency auditory evoked P50 potential in various disorders. Thalamus Relat Syst 2(1):9–19. https://doi.org/10.1016/s1472-9288(02)00032-8

Gaus SE, Strecker RE, Tate BA, Parker RA, Saper CB (2002) Ventrolateral preoptic nucleus contains sleep-active, galaninergic neurons in multiple mammalian species. Neuroscience 115(1):285–294. https://doi.org/10.1016/s0306-4522(02)00308-1

Gaykema RP, Zaborszky L (1996) Direct catecholaminergic-cholinergic interactions in the basal forebrain. II Substantia nigra-ventral tegmental area projections to cholinergic neurons. J Comp Neurol 374(4):555–577. https://doi.org/10.1002/(SICI)1096-9861(19961028)374:4<555::AID-CNE6>3.0.CO;2-0

Gerashchenko D, Shiromani PJ (2004) Different neuronal phenotypes in the lateral hypothalamus and their role in sleep and wakefulness. Mol Neurobiol 29(1):41–59. https://doi.org/10.1385/Mn:29:1:41

Gerashchenko D, Wisor JP, Burns D, Reh RK, Shiromani PJ, Sakurai T, de la Iglesia HO, Kilduff TS (2008) Identification of a population of sleep-active cerebral cortex neurons. Proc Natl Acad Sci U S A 105(29):10227–10232. https://doi.org/10.1073/pnas.0803125105

Giedke H, Schwarzler F (2002) Therapeutic use of sleep deprivation in depression. Sleep Med Rev 6(5):361–377. https://doi.org/10.1053/smrv.2002.0235

Gong H, McGinty D, Guzman-Marin R, Chew KT, Stewart D, Szymusiak R (2004) Activation of c-fos in GABAergic neurones in the preoptic area during sleep and in response to sleep deprivation. J Physiol 556(Pt 3):935–946. https://doi.org/10.1113/jphysiol.2003.056622

Gottesmann C (2002) GABA mechanisms and sleep. Neuroscience 111(2):231–239. https://doi.org/10.1016/s0306-4522(02)00034-9

Gottesmann C, Gottesman I (2007) The neurobiological characteristics of rapid eye movement (REM) sleep are candidate endophenotypes of depression, schizophrenia, mental retardation and dementia. Prog Neurobiol 81(4):237–250. https://doi.org/10.1016/j.pneurobio.2007.01.004

Greenspan RJ, Tononi G, Cirelli C, Shaw PJ (2001) Sleep and the fruit fly. Trends Neurosci 24(3):142–145. https://doi.org/10.1016/s0166-2236(00)01719-7

Gritti I, Mainville L, Jones BE (1994) Projections of GABAergic and cholinergic basal forebrain and GABAergic preoptic-anterior hypothalamic neurons to the posterior lateral hypothalamus of the rat. J Comp Neurol 339(2):251–268. https://doi.org/10.1002/cne.903390206

Gritti I, Mainville L, Mancia M, Jones BE (1997) GABAergic and other noncholinergic basal forebrain neurons, together with cholinergic neurons, project to the mesocortex and isocortex in the rat. J Comp Neurol 383(2):163–177. https://doi.org/10.1002/(sici)1096-9861(19970630)383:2<163::aid-cne4>3.3.co;2-t

Guaraldi P, Calandra-Buonaura G, Terlizzi R, Montagna P, Lugaresi E, Tinuper P, Cortelli P, Provini F (2011) Oneiric stupor: the peculiar behaviour of agrypnia excitata. Sleep Med 12(Suppl 2):S64–S67. https://doi.org/10.1016/j.sleep.2011.10.014

Guilleminault C, Anagnos A (2000) Narcolepsy. In: Kryger M, Roth T, Dement W (eds) Principles and practice of sleep medicine, 3rd edn. WB Saunders, Philadelphia, PA, pp 676–686

Gujar N, McDonald SA, Nishida M, Walker MP (2011) A role for REM sleep in recalibrating the sensitivity of the human brain to specific emotions. Cereb Cortex 21(1):115–123. https://doi.org/10.1093/cercor/bhq064

Gvilia I, Turner A, McGinty D, Szymusiak R (2006) Preoptic area neurons and the homeostatic regulation of rapid eye movement sleep. J Neurosci 26(11):3037–3044. https://doi.org/10.1523/JNEUROSCI.4827-05.2006

Harris GC, Wimmer M, Aston-Jones G (2005) A role for lateral hypothalamic orexin neurons in reward seeking. Nature 437(7058):556–559. https://doi.org/10.1038/nature04071

Harvey AG (2008) Sleep and circadian rhythms in bipolar disorder: seeking synchrony, harmony, and regulation. Am J Psychiatry 165(7):820–829. https://doi.org/10.1176/appi.ajp.2008.08010098

Hasler BP, Buysse DJ, Kupfer DJ, Germain A (2010) Phase relationships between core body temperature, melatonin, and sleep are associated with depression severity: further evidence for circadian misalignment in non-seasonal depression. Psychiatry Res 178(1):205–207. https://doi.org/10.1016/j.psychres.2010.04.027

Hattar S, Liao HW, Takao M, Berson DM, Yau KW (2002) Melanopsin-containing retinal ganglion cells: architecture, projections, and intrinsic photosensitivity. Science 295(5557):1065–1070. https://doi.org/10.1126/science.1069609

Hendricks JC, Morrison AR, Mann GL (1982) Different behaviors during paradoxical sleep without atonia depend on pontine lesion site. Brain Res 239(1):81–105. https://doi.org/10.1016/0006-8993(82)90835-6

Higgins E, George M (2007) Neuroscience of clinical psychiatry, the pathophysiology of behavior and mental illness. Lippincott Williams & Wilkins, Philadelphia, PA

Hobson JA (1990) Sleep and dreaming. J Neurosci 10(2):371–382. https://doi.org/10.1523/jneurosci.10-02-00371.1990
Hobson JA, Pace-Schott EF (2002) The cognitive neuroscience of sleep: neuronal systems, consciousness and learning. Nat Rev Neurosci 3(9):679–693. https://doi.org/10.1038/nrn915
Hobson JA, McCarley RW, Wyzinski PW (1975) Sleep cycle oscillation: reciprocal discharge by two brainstem neuronal groups. Science 189(4196):55–58. https://doi.org/10.1126/science.1094539
Hobson JA, Pace-Schott EF, Stickgold R (2000) Dreaming and the brain: toward a cognitive neuroscience of conscious states. Behav Brain Sci 23(6):793–842.; discussion 904–1121. https://doi.org/10.1017/s0140525x00003976
Howell MJ, Schenck CH (2012) Restless nocturnal eating: a common feature of Willis-Ekbom Syndrome (RLS). J Clin Sleep Med 8(4):413–419. https://doi.org/10.5664/jcsm.2036
Iglowstein I, Jenni OG, Molinari L, Largo RH (2003) Sleep duration from infancy to adolescence: reference values and generational trends. Pediatrics 111(2):302–307. https://doi.org/10.1542/peds.111.2.302
Imeri L, Opp MR (2009) How (and why) the immune system makes us sleep. Nat Rev Neurosci 10(3):199–210. https://doi.org/10.1038/nrn2576
Iwase T, Kajimura N, Uchiyama M, Ebisawa T, Yoshimura K, Kamei Y, Shibui K, Kim K, Kudo Y, Katoh M, Watanabe T, Nakajima T, Ozeki Y, Sugishita M, Hori T, Ikeda M, Toyoshima R, Inoue Y, Yamada N, Mishima K, Nomura M, Ozaki N, Okawa M, Takahashi K, Yamauchi T (2002) Mutation screening of the human Clock gene in circadian rhythm sleep disorders. Psychiatry Res 109(2):121–128. https://doi.org/10.1016/s0165-1781(02)00006-9
Jackson A, Cavanagh J, Scott J (2003) A systematic review of manic and depressive prodromes. J Affect Disord 74(3):209–217. https://doi.org/10.1016/s0165-0327(02)00266-5
Jacobs BL, Azmitia EC (1992) Structure and function of the brain serotonin system. Physiol Rev 72(1):165–229. https://doi.org/10.1152/physrev.1992.72.1.165
John J, Kumar VM (1998) Effect of NMDA lesion of the medial preoptic neurons on sleep and other functions. Sleep 21(6):587–598. https://doi.org/10.1093/sleep/21.6.587
John J, Kumar VM, Gopinath G, Ramesh V, Mallick H (1994) Changes in sleep-wakefulness after kainic acid lesion of the preoptic area in rats. Jpn J Physiol 44(3):231–242. https://doi.org/10.2170/jjphysiol.44.231
John J, Wu MF, Boehmer LN, Siegel JM (2004) Cataplexy-active neurons in the hypothalamus: implications for the role of histamine in sleep and waking behavior. Neuron 42(4):619–634. https://doi.org/10.1016/s0896-6273(04)00247-8
Johnson EO, Roth T, Schultz L, Breslau N (2006) Epidemiology of DSM-IV insomnia in adolescence: lifetime prevalence, chronicity, and an emergent gender difference. Pediatrics 117(2):e247–e256. https://doi.org/10.1542/peds.2004-2629
Jones B (1972) The respective involvement of noradrenaline and its deaminated metabolites in waking and paradoxical sleep: a neuropharmacological model. Brain Res 39(1):121–136. https://doi.org/10.1016/0006-8993(72)90789-5
Jones BE (2003) Arousal systems. Front Biosci 8(6):s438–s451. https://doi.org/10.2741/1074
Jones BE (2005) From waking to sleeping: neuronal and chemical substrates. Trends Pharmacol Sci 26(11):578–586. https://doi.org/10.1016/j.tips.2005.09.009
Jones BE (2018) In memoriam Michel Jouvet 1925–2017. Sleep Med 41:116–117. https://doi.org/10.1016/j.sleep.2017.12.002
Jones D, Kelwala S, Bell J, Dube S, Jackson E, Sitaram N (1985) Cholinergic REM sleep induction response correlation with endogenous major depressive subtype. Psychiatry Res 14(2):99–110. https://doi.org/10.1016/0165-1781(85)90054-x
Jouvet M, Michel F (1959) Correlations electromyographiques du sommeil chez le chat decortique et mesencephalique chronique. C R Seances Soc Biol Fil 153:422–425
Jouvet M, Michel F, Courjon J (1959) Sur un stade d'activite electrique cerebrale rapide au cours du sommeil physiologique. C R Seances Soc Biol Fil 153:1024–1028

Kavanau JL (1997) Memory, sleep and the evolution of mechanisms of synaptic efficacy maintenance. Neuroscience 79(1):7–44. https://doi.org/10.1016/s0306-4522(96)00610-0
Khubchandani M, Jagannathan NR, Mallick HN, Mohan Kumar V (2005) Functional MRI shows activation of the medial preoptic area during sleep. NeuroImage 26(1):29–35. https://doi.org/10.1016/j.neuroimage.2005.01.002
Kirsch DB (2013) Introduction to sleep medicine. Sleep medicine in neurology. John Wiley & Sons, Hoboken, NJ. https://doi.org/10.1002/9781118764152.ch1
Kolla BP, Auger RR (2011) Jet lag and shift work sleep disorders: how to help reset the internal clock. Cleve Clin J Med 78(10):675–684. https://doi.org/10.3949/ccjm.78a.10083
Koyama Y, Hayaishi O (1994) Firing of neurons in the preoptic/anterior hypothalamic areas in rat: its possible involvement in slow wave sleep and paradoxical sleep. Neurosci Res 19(1):31–38. https://doi.org/10.1016/0168-0102(94)90005-1
Koyama Y, Kodama T, Takahashi K, Okai K, Kayama Y (2002) Firing properties of neurones in the laterodorsal hypothalamic area during sleep and wakefulness. Psychiatry Clin Neurosci 56(3):339–340. https://doi.org/10.1046/j.1440-1819.2002.00975.x
Koyama Y, Takahashi K, Kodama T, Kayama Y (2003) State-dependent activity of neurons in the perifornical hypothalamic area during sleep and waking. Neuroscience 119(4):1209–1219. https://doi.org/10.1016/s0306-4522(03)00173-8
Kramer A, Yang FC, Snodgrass P, Li X, Scammell TE, Davis FC, Weitz CJ (2001) Regulation of daily locomotor activity and sleep by hypothalamic EGF receptor signaling. Science 294(5551):2511–2515. https://doi.org/10.1126/science.1067716
Krauchi K (2007) The human sleep-wake cycle reconsidered from a thermoregulatory point of view. Physiol Behav 90(2–3):236–245. https://doi.org/10.1016/j.physbeh.2006.09.005
Krueger JM, Obal F Jr, Kapas L, Fang J (1995) Brain organization and sleep function. Behav Brain Res 69(1–2):177–185. https://doi.org/10.1016/0166-4328(95)00015-l
Kubota T, Uchiyama M, Suzuki H, Shibui K, Kim K, Tan X, Tagaya H, Okawa M, Inoue S (2002) Effects of nocturnal bright light on saliva melatonin, core body temperature and sleep propensity rhythms in human subjects. Neurosci Res 42(2):115–122. https://doi.org/10.1016/s0168-0102(01)00310-8
Lack LC, Gradisar M, Van Someren EJ, Wright HR, Lushington K (2008) The relationship between insomnia and body temperatures. Sleep Med Rev 12(4):307–317. https://doi.org/10.1016/j.smrv.2008.02.003
Lai Y-Y, Kodama T, Siegel JM (2001) Changes in monoamine release in the ventral horn and hypoglossal nucleus linked to pontine inhibition of muscle tone: an in vivomicrodialysis study. J Neurosci 21(18):7384–7391. https://doi.org/10.1523/jneurosci.21-18-07384.2001
Lambert D (2006) Minimax Tests. In: Encyclopedia of statistical sciences. John Wiley & Sons, Inc., Hoboken, NJ. https://doi.org/10.1002/0471667196.ess1639.pub2
Lansink CS, Goltstein PM, Lankelma JV, Joosten RN, McNaughton BL, Pennartz CM (2008) Preferential reactivation of motivationally relevant information in the ventral striatum. J Neurosci 28(25):6372–6382. https://doi.org/10.1523/JNEUROSCI.1054-08.2008
Lansink CS, Goltstein PM, Lankelma JV, McNaughton BL, Pennartz CM (2009) Hippocampus leads ventral striatum in replay of place-reward information. PLoS Biol 7(8):e1000173. https://doi.org/10.1371/journal.pbio.1000173
Lara-Carrasco J, Nielsen TA, Solomonova E, Levrier K, Popova A (2009) Overnight emotional adaptation to negative stimuli is altered by REM sleep deprivation and is correlated with intervening dream emotions. J Sleep Res 18(2):178–187. https://doi.org/10.1111/j.1365-2869.2008.00709.x
Lavie P, Pillar G, Malhotra A (2002) Sleep disorders: diagnosis, management and treatment. a handbook for clinicians. Martin Dunitz, London
de Lecea L, Kilduff TS, Peyron C, Gao X, Foye PE, Danielson PE, Fukuhara C, Battenberg EL, Gautvik VT, Bartlett FS 2nd, Frankel WN, van den Pol AN, Bloom FE, Gautvik KM, Sutcliffe JG (1998) The hypocretins: hypothalamus-specific peptides with neuroexcitatory activity. Proc Natl Acad Sci U S A 95(1):322–327. https://doi.org/10.1073/pnas.95.1.322
Lee-Chiong T (2008) Sleep medicine: essentials and review. Oxford University Press, Oxford

Lena I, Parrot S, Deschaux O, Muffat-Joly S, Sauvinet V, Renaud B, Suaud-Chagny MF, Gottesmann C (2005) Variations in extracellular levels of dopamine, noradrenaline, glutamate, and aspartate across the sleep—wake cycle in the medial prefrontal cortex and nucleus accumbens of freely moving rats. J Neurosci Res 81(6):891–899. https://doi.org/10.1002/jnr.20602

Leonard CS, Llinas R (1994) Serotonergic and cholinergic inhibition of mesopontine cholinergic neurons controlling REM sleep: an in vitro electrophysiological study. Neuroscience 59(2):309–330. https://doi.org/10.1016/0306-4522(94)90599-1

Li JZ, Bunney BG, Meng F, Hagenauer MH, Walsh DM, Vawter MP, Evans SJ, Choudary PV, Cartagena P, Barchas JD, Schatzberg AF, Jones EG, Myers RM, Watson SJ Jr, Akil H, Bunney WE (2013) Circadian patterns of gene expression in the human brain and disruption in major depressive disorder. Proc Natl Acad Sci U S A 110(24):9950–9955. https://doi.org/10.1073/pnas.1305814110

Lin JS, Sakai K, Vanni-Mercier G, Jouvet M (1989) A critical role of the posterior hypothalamus in the mechanisms of wakefulness determined by microinjection of muscimol in freely moving cats. Brain Res 479(2):225–240. https://doi.org/10.1016/0006-8993(89)91623-5

Lin L, Faraco J, Li R, Kadotani H, Rogers W, Lin XY, Qiu XH, de Jong PJ, Nishino S, Mignot E (1999) The sleep disorder canine narcolepsy is caused by a mutation in the hypocretin (orexin) receptor 2 gene. Cell 98 (3):365–376. https://doi.org/10.1016/S0092-8674(00)81965-0

Lindsley DB, Bowden JW, Magoun HW (1949) Effect upon the eeg of acute injury to the brain stem activating system. Electroencephalogr Clin Neurophysiol 1(4):475–486. https://doi.org/10.1016/0013-4694(49)90221-7

Llinas RR, Steriade M (2006) Bursting of thalamic neurons and states of vigilance. J Neurophysiol 95(6):3297–3308. https://doi.org/10.1152/jn.00166.2006

Loomis AL, Harvey EN, Hobart GA (1937) Cerebral states during sleep, as studied by human brain potentials. J Exp Psychol 21(2):127–144. https://doi.org/10.1037/h0057431

Lu BS, Zee PC (2010) Neurobiology of sleep. Clin Chest Med 31(2):309–318. https://doi.org/10.1016/j.ccm.2010.02.004

Lu J, Sherman D, Devor M, Saper CB (2006) A putative flip-flop switch for control of REM sleep. Nature 441(7093):589–594. https://doi.org/10.1038/nature04767

Luboshitzky R (2000) Endocrine activity during sleep. J Pediatr Endocrinol Metab 13(1):13–20. https://doi.org/10.1515/jpem.2000.13.1.13

Lydic R, McCarley RW, Hobson JA (1983) The time-course of dorsal raphe discharge, PGO waves, and muscle tone averaged across multiple sleep cycles. Brain Res 274(2):365–370. https://doi.org/10.1016/0006-8993(83)90720-5

el Mansari M, Sakai K, Jouvet M (1989) Unitary characteristics of presumptive cholinergic tegmental neurons during the sleep-waking cycle in freely moving cats. Exp Brain Res 76(3):519–529. https://doi.org/10.1007/bf00248908

Maquet P (2001) The role of sleep in learning and memory. Science 294(5544):1048–1052. https://doi.org/10.1126/science.1062856

Maquet P, Peters J, Aerts J, Delfiore G, Degueldre C, Luxen A, Franck G (1996) Functional neuroanatomy of human rapid-eye-movement sleep and dreaming. Nature 383(6596):163–166. https://doi.org/10.1038/383163a0

Maquet P, Schwartz S, Passingham R, Frith C (2003) Sleep-related consolidation of a visuomotor skill: brain mechanisms as assessed by functional magnetic resonance imaging. J Neurosci 23(4):1432–1440. https://doi.org/10.1523/jneurosci.23-04-01432.2003

McCarley RW (2007) Neurobiology of REM and NREM sleep. Sleep Med 8(4):302–330. https://doi.org/10.1016/j.sleep.2007.03.005

McCarley RW, Hobson JA (1975) Neuronal excitability modulation over the sleep cycle: a structural and mathematical model. Science 189(4196):58–60. https://doi.org/10.1126/science.1135627

McCormick DA, Bal T (1997) Sleep and arousal: thalamocortical mechanisms. Annu Rev Neurosci 20(1):185–215. https://doi.org/10.1146/annurev.neuro.20.1.185

McGinty DJ, Harper RM (1976) Dorsal raphe neurons: depression of firing during sleep in cats. Brain Res 101(3):569–575. https://doi.org/10.1016/0006-8993(76)90480-7
McGinty D, Szymusiak R (2000) The sleep-wake switch: a neuronal alarm clock. Nat Med 6(5):510–511. https://doi.org/10.1038/74988
McGinty D, Szymusiak R (2003) Hypothalamic regulation of sleep and arousal. Front Biosci 8(6):s1074–s1083. https://doi.org/10.2741/1159
McGinty D, Szymusiak R (2017) Neural control of sleep in mammals. In: Kryger M, Roth T, Dement W (eds) Principles and practice of sleep medicine. Elsevier, Amsterdam. https://doi.org/10.1016/b978-0-323-24288-2.00007-6
McNamara P, Auerbach S, Johnson P, Harris E, Doros G (2010) Impact of REM sleep on distortions of self-concept, mood and memory in depressed/anxious participants. J Affect Disord 122(3):198–207. https://doi.org/10.1016/j.jad.2009.06.030
Mellman TA (2006) Sleep and anxiety disorders. Psychiatr Clin North Am 29(4):1047–1058. https://doi.org/10.1016/j.psc.2006.08.005
Mellman TA (2008) In psychiatric clinics of North America. Sleep Med Clin 3(2):261–268. https://doi.org/10.1016/j.jsmc.2008.01.010
Mendelson H, Ziegler J (2001) Organisations-intelligenz IQ. Gabler Verlag, Wiesbaden. https://doi.org/10.1007/978-3-322-82344-1
Mendelson WB, Bergmann BM, Tung A (2003) Baseline and post-deprivation recovery sleep in SCN-lesioned rats. Brain Res 980(2):185–190. https://doi.org/10.1016/s0006-8993(03)02896-8
Merritt JM, Stickgold R, Paceschott E, Williams J, Hobson JA (1994) Emotion profiles in the dreams of men and women. Conscious Cogn 3(1):46–60. https://doi.org/10.1006/ccog.1994.1004
Mignot E, Nishino S (2005) Emerging therapies in narcolepsy-cataplexy. Sleep 28(6):754–763. https://doi.org/10.1093/sleep/28.6.754
Mileykovskiy BY, Kiyashchenko LI, Siegel JM (2005) Behavioral correlates of activity in identified hypocretin/orexin neurons. Neuron 46(5):787–798. https://doi.org/10.1016/j.neuron.2005.04.035
Mitler MM, Harsh J, Hirshkowitz M, Guilleminault C (2000) Long-term efficacy and safety of modafinil (PROVIGIL®) for the treatment of excessive daytime sleepiness associated with narcolepsy. Sleep Med 1(3):231–243. https://doi.org/10.1016/s1389-9457(00)00031-9
Mizoguchi A, Eguchi N, Kimura K, Kiyohara Y, Qu WM, Huang ZL, Mochizuki T, Lazarus M, Kobayashi T, Kaneko T, Narumiya S, Urade Y, Hayaishi O (2001) Dominant localization of prostaglandin D receptors on arachnoid trabecular cells in mouse basal forebrain and their involvement in the regulation of non-rapid eye movement sleep. Proc Natl Acad Sci U S A 98(20):11674–11679. https://doi.org/10.1073/pnas.201398898
Montagna P, Gambetti P, Cortelli P, Lugaresi E (2003) Familial and sporadic fatal insomnia. Lancet Neurol 2(3):167–176. https://doi.org/10.1016/s1474-4422(03)00323-5
Monti R (1993) Preface. Acta Astronaut 30:ix. https://doi.org/10.1016/0094-5765(93)90093-c
Morrison DN, McGee R, Stanton WR (1992) Sleep problems in adolescence. J Am Acad Child Adolesc Psychiatry 31(1):94–99. https://doi.org/10.1097/00004583-199201000-00014
Morrissey MJ, Duntley SP, Anch AM, Nonneman R (2004) Active sleep and its role in the prevention of apoptosis in the developing brain. Med Hypotheses 62(6):876–879. https://doi.org/10.1016/j.mehy.2004.01.014
Moruzzi G, Magoun HW (1949) Brain stem reticular formation and activation of the EEG. Electroencephalogr Clin Neurophysiol 1(4):455–473. https://doi.org/10.1016/0013-4694(49)90219-9
Murck H, Guldner J, Colla-Muller M, Frieboes RM, Schier T, Wiedemann K, Holsboer F, Steiger A (1996) VIP decelerates non-REM-REM cycles and modulates hormone secretion during sleep in men. Am J Phys 271(4 Pt 2):R905–R911. https://doi.org/10.1152/ajpregu.1996.271.4.R905
Nakanishi H, Sun Y, Nakamura RK, Mori K, Ito M, Suda S, Namba H, Storch FI, Dang TP, Mendelson W, Mishkin M, Kennedy C, Gillin JC, Smith CB, Sokoloff L (1997) Positive cor-

relations between cerebral protein synthesis rates and deep sleep in *Macaca mulatta*. Eur J Neurosci 9(2):271–279. https://doi.org/10.1111/j.1460-9568.1997.tb01397.x
Nielsen-Bohlman L, Knight RT, Woods DL, Woodward K (1991) Differential auditory processing continues during sleep. Electroencephalogr Clin Neurophysiol 79(4):281–290. https://doi.org/10.1016/0013-4694(91)90124-m
Nir Y, Tononi G (2010) Dreaming and the brain: from phenomenology to neurophysiology. Trends Cogn Sci 14(2):88–100. https://doi.org/10.1016/j.tics.2009.12.001
Nishino S, Mignot E (1997) Pharmacological aspects of human and canine narcolepsy. Prog Neurobiol 52(1):27–78. https://doi.org/10.1016/s0301-0082(96)00070-6
Nishino S, Ripley B, Overeem S, Lammers GJ, Mignot E (2000) Hypocretin (orexin) deficiency in human narcolepsy. Lancet 355(9197):39–40. https://doi.org/10.1016/S0140-6736(99)05582-8
Nofzinger EA, Mintun MA, Wiseman M, Kupfer DJ, Moore RY (1997) Forebrain activation in REM sleep: an FDG PET study. Brain Res 770(1–2):192–201. https://doi.org/10.1016/s0006-8993(97)00807-x
Nofzinger EA, Buysse DJ, Germain A, Price JC, Miewald JM, Kupfer DJ (2004) Functional neuroimaging evidence for hyperarousal in insomnia. Am J Psychiatry 161(11):2126–2128. https://doi.org/10.1176/appi.ajp.161.11.2126
Norman W, Hayward L, Geyer J, Carney P (2011) The neurobiology of sleep. In: Carney P, Berry R, Geyer J (eds) Clinical sleep disorders, 2nd edn. Wolters Kluwer, Alphen aan den Rijn, pp 35–51
Obal F Jr, Krueger JM (2004) GHRH and sleep. Sleep Med Rev 8(5):367–377. https://doi.org/10.1016/j.smrv.2004.03.005
Ohayon MM (1997) Prevalence of DSM-IV diagnostic criteria of insomnia: distinguishing insomnia related to mental disorders from sleep disorders. J Psychiatr Res 31(3):333–346. https://doi.org/10.1016/s0022-3956(97)00002-2
Ohayon MM (2002) Epidemiology of insomnia: what we know and what we still need to learn. Sleep Med Rev 6(2):97–111. https://doi.org/10.1053/smrv.2002.0186
Ohayon MM, Roth T (2003) Place of chronic insomnia in the course of depressive and anxiety disorders. J Psychiatr Res 37(1):9–15. https://doi.org/10.1016/s0022-3956(02)00052-3
Ohayon MM, Caulet M, Lemoine P (1998) Comorbidity of mental and insomnia disorders in the general population. Compr Psychiatry 39(4):185–197. https://doi.org/10.1016/s0010-440x(98)90059-1
Opp MR (2009) Sleeping to fuel the immune system: mammalian sleep and resistance to parasites. BMC Evol Biol 9(1):8. https://doi.org/10.1186/1471-2148-9-8
Pace-Schott EF, Hobson JA (2002) The neurobiology of sleep: genetics, cellular physiology and subcortical networks. Nat Rev Neurosci 3(8):591–605. https://doi.org/10.1038/nrn895
Pace-Schott EF, Nave G, Morgan A, Spencer RM (2012) Sleep-dependent modulation of affectively guided decision-making. J Sleep Res 21(1):30–39. https://doi.org/10.1111/j.1365-2869.2011.00921.x
Pandey HP, Ram A, Matsumura H, Satoh S, Hayaishi O (1995) Circadian variations of prostaglandins D2, E2, and F2 alpha in the cerebrospinal fluid of anesthetized rats. Biochem Biophys Res Commun 213(2):625–629. https://doi.org/10.1006/bbrc.1995.2177
Pandi-Perumal SR, Kramer M (2011) Preface. Sleep and mental illness. Cambridge University Press, Cambridge. https://doi.org/10.1017/cbo9781139042734.002
Panula P, Yang HY, Costa E (1984) Histamine-containing neurons in the rat hypothalamus. Proc Natl Acad Sci U S A 81(8):2572–2576. https://doi.org/10.1073/pnas.81.8.2572
Parmelee AH (1961) Sleep patterns in infancy a study of one idant from birth to eight months of age. Acta Paediatr 50(2):160–170. https://doi.org/10.1111/j.1651-2227.1961.tb08035.x
Parmentier R, Ohtsu H, Djebbara-Hannas Z, Valatx J-L, Watanabe T, Lin J-S (2002) Anatomical, physiological, and pharmacological characteristics of histidine decarboxylase knock-out mice: evidence for the role of brain histamine in behavioral and sleep–wake control. J Neurosci 22(17):7695–7711. https://doi.org/10.1523/jneurosci.22-17-07695.2002
Payne JD, Kensinger EA (2011) Sleep leads to changes in the emotional memory trace: evidence from FMRI. J Cogn Neurosci 23(6):1285–1297. https://doi.org/10.1162/jocn.2010.21526

Payne JL, Quiroz JA, Zarate CA, Manji HK (2002) Timing is everything: does the robust upregulation of noradrenergically regulated plasticity genes underlie the rapid antidepressant effects of sleep deprivation? Biol Psychiatry 52(10):921–926. https://doi.org/10.1016/s0006-3223(02)01676-1

Pennartz CM, Lee E, Verheul J, Lipa P, Barnes CA, McNaughton BL (2004) The ventral striatum in off-line processing: ensemble reactivation during sleep and modulation by hippocampal ripples. J Neurosci 24(29):6446–6456. https://doi.org/10.1523/JNEUROSCI.0575-04.2004

Penzel T, Wessel N, Riedl M, Kantelhardt JW, Rostig S, Glos M, Suhrbier A, Malberg H, Fietze I (2007) Cardiovascular and respiratory dynamics during normal and pathological sleep. Chaos 17(1):015116. https://doi.org/10.1063/1.2711282

Perogamvros L, Schwartz S (2012) The roles of the reward system in sleep and dreaming. Neurosci Biobehav Rev 36(8):1934–1951. https://doi.org/10.1016/j.neubiorev.2012.05.010

Perry EK, Piggott MA (2000) Neurotransmitter mechanisms of dreaming: implication of modulatory systems based on dream intensity. Behav Brain Sci 23(6):990. https://doi.org/10.1017/S0140525x00774024

Peyron C, Tighe DK, van den Pol AN, de Lecea L, Heller HC, Sutcliffe JG, Kilduff TS (1998) Neurons containing hypocretin (orexin) project to multiple neuronal systems. J Neurosci 18(23):9996–10015. https://doi.org/10.1523/jneurosci.18-23-09996.1998

Plante DT, Winkelman JW (2008) Sleep disturbance in bipolar disorder: therapeutic implications. Am J Psychiatry 165(7):830–843. https://doi.org/10.1176/appi.ajp.2008.08010077

Ponz A, Khatami R, Poryazova R, Werth E, Boesiger P, Schwartz S, Bassetti CL (2010) Reduced amygdala activity during aversive conditioning in human narcolepsy. Ann Neurol 67(3):394–398. https://doi.org/10.1002/ana.21881

Popa D, Duvarci S, Popescu AT, Lena C, Pare D (2010) Coherent amygdalocortical theta promotes fear memory consolidation during paradoxical sleep. Proc Natl Acad Sci U S A 107(14):6516–6519. https://doi.org/10.1073/pnas.0913016107

Porkka-Heiskanen T, Strecker RE, Thakkar M, Bjorkum AA, Greene RW, McCarley RW (1997) Adenosine: a mediator of the sleep-inducing effects of prolonged wakefulness. Science 276(5316):1265–1268. https://doi.org/10.1126/science.276.5316.1265

Portas CM, McCarley RW (1994) Behavioral state-related changes of extracellular serotonin concentration in the dorsal raphe nucleus: a microdialysis study in the freely moving cat. Brain Res 648(2):306–312. https://doi.org/10.1016/0006-8993(94)91132-0

Pressman MR (2007) Factors that predispose, prime and precipitate NREM parasomnias in adults: clinical and forensic implications. Sleep Med Rev 11(1):5–30.; discussion 31–33. https://doi.org/10.1016/j.smrv.2006.06.003

Prince TM, Wimmer M, Choi J, Havekes R, Aton S, Abel T (2014) Sleep deprivation during a specific 3-hour time window post-training impairs hippocampal synaptic plasticity and memory. Neurobiol Learn Mem 109:122–130. https://doi.org/10.1016/j.nlm.2013.11.021

Ram A, Pandey HP, Matsumura H, Kasahara-Orita K, Nakajima T, Takahata R, Satoh S, Terao A, Hayaishi O (1997) CSF levels of prostaglandins, especially the level of prostaglandin D2, are correlated with increasing propensity towards sleep in rats. Brain Res 751(1):81–89. https://doi.org/10.1016/s0006-8993(96)01401-1

Raman M (2013) Restless legs syndrome, periodic limb movements, and other movement disorders in sleep. Sleep medicine in neurology. John Wiley & Sons, Hoboken, NJ. https://doi.org/10.1002/9781118764152.ch12

Rasmussen K, Jacobs BL (1986) Single unit activity of locus coeruleus neurons in the freely moving cat. Brain Res 371(2):335–344. https://doi.org/10.1016/0006-8993(86)90371-9

Reading PJ (2010) Sleep disorders in neurology. Pract Neurol 10(5):300–309. https://doi.org/10.1136/jnnp.2010.224097

Reading P (2013) ABC of Sleep medicine. Wiley-Blackwell, Chichester

Reid KJ, Chang AM, Dubocovich ML, Turek FW, Takahashi JS, Zee PC (2001) Familial advanced sleep phase syndrome. Arch Neurol 58(7):1089–1094. https://doi.org/10.1001/archneur.58.7.1089

Riemann D, Voderholzer U (2003) Primary insomnia: a risk factor to develop depression? J Affect Disord 76(1–3):255–259. https://doi.org/10.1016/s0165-0327(02)00072-1
Riou F, Cespuglio R, Jouvet M (1982) Endogenous peptides and sleep in the rat: III the hypnogenic properties of vasoactive intestinal polypeptide. Neuropeptides 2(5):265–277. https://doi.org/10.1016/0143-4179(82)90016-6
Roffwarg HP, Muzio JN, Dement WC (1966) Ontogenetic development of the human sleep-dream cycle. Science 152(3722):604–619. https://doi.org/10.1126/science.152.3722.604
Rosner F (1965) The hygienic principles of Moses Maimonides. JAMA 194(13):1352–1354. https://doi.org/10.1001/jama.1965.03090260012003
Roybal K, Theobold D, Graham A, DiNieri JA, Russo SJ, Krishnan V, Chakravarty S, Peevey J, Oehrlein N, Birnbaum S, Vitaterna MH, Orsulak P, Takahashi JS, Nestler EJ, Carlezon WA Jr, McClung CA (2007) Mania-like behavior induced by disruption of CLOCK. Proc Natl Acad Sci U S A 104(15):6406–6411. https://doi.org/10.1073/pnas.0609625104
Sakai K, Crochet S (2003) A neural mechanism of sleep and wakefulness. Sleep Biol Rhythms 1(1):29–42. https://doi.org/10.1046/j.1446-9235.2003.00004.x
Sakai K, Sastre JP, Salvert D, Touret M, Tohyama M, Jouvet M (1979) Tegmentoreticular projections with special reference to the muscular atonia during paradoxical sleep in the cat: an HRP study. Brain Res 176(2):233–254. https://doi.org/10.1016/0006-8993(79)90981-8
Sakurai N, Sasaki M (1998) An activity monitor study on the sleep-wake rhythm of healthy aged people residing in their homes. Psychiatry Clin Neurosci 52(2):253–255. https://doi.org/10.1111/j.1440-1819.1998.tb01060.x
Sallanon M, Denoyer M, Kitahama K, Aubert C, Gay N, Jouvet M (1989) Long-lasting insomnia induced by preoptic neuron lesions and its transient reversal by muscimol injection into the posterior hypothalamus in the cat. Neuroscience 32(3):669–683. https://doi.org/10.1016/0306-4522(89)90289-3
Saper CB, Scammell TE (2013) Emerging therapeutics in sleep. Ann Neurol 74(3):435–440. https://doi.org/10.1002/ana.24000
Saper CB, Cano G, Scammell TE (2005) Homeostatic, circadian, and emotional regulation of sleep. J Comp Neurol 493(1):92–98. https://doi.org/10.1002/cne.20770
Sassin JF, Frantz AG, Weitzman ED, Kapen S (1972) Human prolactin: 24-hour pattern with increased release during sleep. Science 177(4055):1205–1207. https://doi.org/10.1126/science.177.4055.1205
Scammell T, Gerashchenko D, Urade Y, Onoe H, Saper C, Hayaishi O (1998) Activation of ventrolateral preoptic neurons by the somnogen prostaglandin D2. Proc Natl Acad Sci U S A 95(13):7754–7759. https://doi.org/10.1073/pnas.95.13.7754
Scammell TE, Arrigoni E, Lipton JO (2017) Neural circuitry of wakefulness and sleep. Neuron 93(4):747–765. https://doi.org/10.1016/j.neuron.2017.01.014
Schenck CH, Mahowald MW (2002) REM sleep behavior disorder: clinical, developmental, and neuroscience perspectives 16 years after its formal identification in SLEEP. Sleep 25(2):120–138. https://doi.org/10.1093/sleep/25.2.120
Schmidt MH, Schmidt HS (2004) Sleep-related erections: neural mechanisms and clinical significance. Curr Neurol Neurosci Rep 4(2):170–178. https://doi.org/10.1007/s11910-004-0033-5
Schredl M (2010) Dreams. Foundations of psychiatric sleep medicine. Cambridge University Press, Cambridge. https://doi.org/10.1017/cbo9780511777493.007
Schredl M, Hofmann F (2003) Continuity between waking activities and dream activities. Conscious Cogn 12(2):298–308. https://doi.org/10.1016/S1053-8100(02)00072-7
Schultz W (2010) Subjective neuronal coding of reward: temporal value discounting and risk. Eur J Neurosci 31(12):2124–2135. https://doi.org/10.1111/j.1460-9568.2010.07282.x
Schwartz S, Maquet P, Frith C (2002) Neural correlates of perceptual learning: a functional MRI study of visual texture discrimination. Proc Natl Acad Sci U S A 99(26):17137–17142. https://doi.org/10.1073/pnas.242414599
Schwartz AR, Patil SP, Laffan AM, Polotsky V, Schneider H, Smith PL (2008) Obesity and obstructive sleep apnea: pathogenic mechanisms and therapeutic approaches. Proc Am Thorac Soc 5(2):185–192. https://doi.org/10.1513/pats.200708-137MG
Siegel JM (2009) The neurobiology of sleep. Semin Neurol 29(4):277–296. https://doi.org/10.1055/s-0029-1237118

Siegel J (2002) The neural control of sleep and waking. Springer-Verlag, New York. https://link.springer.com/content/pdf/bfm%3A978-0-387-21726-0%2F1.pdf

Silber M, Krahn L, Morgenthaler T (2010) Sleep medicine in clinical practice, 2nd edn. CRC Press, Boca Raton, FL. https://doi.org/10.3109/9781616310059

Solms M (2000) Dreaming and REM sleep are controlled by different brain mechanisms. Sleep and dreaming. Cambridge University Press, Cambridge. https://doi.org/10.1017/cbo9780511615511.004

Spiegel R (2000) Mythe et histoire, images et textes: "les vies de dominique-vivant denon". Z Kunstgesch 63(4):562. https://doi.org/10.2307/1594964

Srividya R, Mallick HN, Kumar VM (2006) Differences in the effects of medial and lateral preoptic lesions on thermoregulation and sleep in rats. Neuroscience 139(3):853–864. https://doi.org/10.1016/j.neuroscience.2006.01.003

Stein MD, Friedmann PD (2006) Disturbed sleep and its relationship to alcohol use. Subst Abus 26(1):1–13. https://doi.org/10.1300/J465v26n01_01

Stenberg D (2007) Neuroanatomy and neurochemistry of sleep. Cell Mol Life Sci 64(10):1187–1204. https://doi.org/10.1007/s00018-007-6530-3

Steriade M (2004) Acetylcholine systems and rhythmic activities during the waking–sleep cycle. Prog Brain Res. https://doi.org/10.1016/s0079-6123(03)45013-9

Steriade M, McCarley RW (1990) Brainstem control of wakefulness and sleep. Springer US, New York, NY. https://doi.org/10.1007/978-1-4757-4669-3

Steriade M, McCarley R (2005) Brain control of wakefulness and sleep, 2nd edn. Springer, New York, NY

Steriade M, Domich L, Oakson G, Deschenes M (1987) The deafferented reticular thalamic nucleus generates spindle rhythmicity. J Neurophysiol 57(1):260–273. https://doi.org/10.1152/jn.1987.57.1.260

Steriade M, McCormick DA, Sejnowski TJ (1993) Thalamocortical oscillations in the sleeping and aroused brain. Science 262(5134):679–685. https://doi.org/10.1126/science.8235588

Sterpenich V, Albouy G, Boly M, Vandewalle G, Darsaud A, Balteau E, Dang-Vu TT, Desseilles M, D'Argembeau A, Gais S, Rauchs G, Schabus M, Degueldre C, Luxen A, Collette F, Maquet P (2007) Sleep-related hippocampo-cortical interplay during emotional memory recollection. PLoS Biol 5(11):e282. https://doi.org/10.1371/journal.pbio.0050282

Sterpenich V, Albouy G, Darsaud A, Schmidt C, Vandewalle G, Dang Vu TT, Desseilles M, Phillips C, Degueldre C, Balteau E, Collette F, Luxen A, Maquet P (2009) Sleep promotes the neural reorganization of remote emotional memory. J Neurosci 29(16):5143–5152. https://doi.org/10.1523/JNEUROSCI.0561-09.2009

Strecker RE, Nalwalk J, Dauphin LJ, Thakkar MM, Chen Y, Ramesh V, Hough LB, McCarley RW (2002) Extracellular histamine levels in the feline preoptic/anterior hypothalamic area during natural sleep–wakefulness and prolonged wakefulness: an in vivo microdialysis study. Neuroscience 113(3):663–670. https://doi.org/10.1016/s0306-4522(02)00158-6

Suntsova N, Szymusiak R, Alam MN, Guzman-Marin R, McGinty D (2002) Sleep-waking discharge patterns of median preoptic nucleus neurons in rats. J Physiol 543(Pt 2):665–677. https://doi.org/10.1113/jphysiol.2002.023085

Szymusiak R, Alam N, Steininger TL, McGinty D (1998) Sleep–waking discharge patterns of ventrolateral preoptic/anterior hypothalamic neurons in rats. Brain Res 803(1–2):178–188. https://doi.org/10.1016/s0006-8993(98)00631-3

Takahashi K, Lin JS, Sakai K (2006) Neuronal activity of histaminergic tuberomammillary neurons during wake-sleep states in the mouse. J Neurosci 26(40):10292–10298. https://doi.org/10.1523/JNEUROSCI.2341-06.2006

Takahashi K, Lin JS, Sakai K (2008) Neuronal activity of orexin and non-orexin waking-active neurons during wake-sleep states in the mouse. Neuroscience 153(3):860–870. https://doi.org/10.1016/j.neuroscience.2008.02.058

Takahashi K, Lin JS, Sakai K (2009) Characterization and mapping of sleep-waking specific neurons in the basal forebrain and preoptic hypothalamus in mice. Neuroscience 161(1):269–292. https://doi.org/10.1016/j.neuroscience.2009.02.075

Talamini LM, Bringmann LF, de Boer M, Hofman WF (2013) Sleeping worries away or worrying away sleep? Physiological evidence on sleep-emotion interactions. PLoS One 8(5):e62480. https://doi.org/10.1371/journal.pone.0062480

Tergau F, Wischer S, Paulus W (1999) Motor system excitability in patients with restless legs syndrome. Neurology 52(5):1060–1063. https://doi.org/10.1212/wnl.52.5.1060
Thannickal TC, Moore RY, Nienhuis R, Ramanathan L, Gulyani S, Aldrich M, Cornford M, Siegel JM (2000) Reduced number of hypocretin neurons in human narcolepsy. Neuron 27(3):469–474. https://doi.org/10.1016/s0896-6273(00)00058-1
Thompson JL, Borgland SL (2011) A role for hypocretin/orexin in motivation. Behav Brain Res 217(2):446–453. https://doi.org/10.1016/j.bbr.2010.09.028
Thompson DF, Pierce DR (1999) Drug-induced nightmares. Ann Pharmacother 33(1):93–98. https://doi.org/10.1345/aph.18150
Toh KL, Jones CR, He Y, Eide EJ, Hinz WA, Virshup DM, Ptacek LJ, Fu YH (2001) An hPer2 phosphorylation site mutation in familial advanced sleep phase syndrome. Science 291(5506):1040–1043. https://doi.org/10.1126/science.1057499
Tononi G, Cirelli C (2011) The neurobiology of sleep. Neurobiology of mental illness. Oxford University Press, Oxford. https://doi.org/10.1093/med/9780199798261.003.0085
Trinder J, Kleiman J, Carrington M, Smith S, Breen S, Tan N, Kim Y (2001) Autonomic activity during human sleep as a function of time and sleep stage. J Sleep Res 10(4):253–264. https://doi.org/10.1046/j.1365-2869.2001.00263.x
Trulson ME, Jacobs BL (1979) Raphe unit activity in freely moving cats: correlation with level of behavioral arousal. Brain Res 163(1):135–150. https://doi.org/10.1016/0006-8993(79)90157-4
Tung A, Lynch JP, Mendelson WB (2001) Prolonged sedation with propofol in the rat does not result in sleep deprivation. Anesth Analg 92:1232–1236. https://doi.org/10.1097/00000539-200105000-00028
Ueno R, Ishikawa Y, Nakayama T, Hayaishi O (1982) Prostaglandin D2 induces sleep when microinjected into the preoptic area of conscious rats. Biochem Biophys Res Commun 109(2):576–582. https://doi.org/10.1016/0006-291x(82)91760-0
Van Cauter E, Spiegel K, Tasali E, Leproult R (2008) Metabolic consequences of sleep and sleep loss. Sleep Med 9(Suppl 1):S23–S28. https://doi.org/10.1016/S1389-9457(08)70013-3
Verret L, Leger L, Fort P, Luppi PH (2005) Cholinergic and noncholinergic brainstem neurons expressing Fos after paradoxical (REM) sleep deprivation and recovery. Eur J Neurosci 21(9):2488–2504. https://doi.org/10.1111/j.1460-9568.2005.04060.x
Verrier RL, Muller JE, Hobson JA (1996) Sleep, dreams, and sudden death: the case for sleep as an autonomic stress test for the heart. Cardiovasc Res 31(2):181–211. https://doi.org/10.1016/0008-6363(95)00211-1
Vertes RP (1984) Brainstem control of the events of REM sleep. Prog Neurobiol 22(3):241–288. https://doi.org/10.1016/0301-0082(84)90020-0
Vertes RP (2006) Interactions among the medial prefrontal cortex, hippocampus and midline thalamus in emotional and cognitive processing in the rat. Neuroscience 142(1):1–20. https://doi.org/10.1016/j.neuroscience.2006.06.027
Vetrugno R, Manconi M, Ferini-Strambi L, Provini F, Plazzi G, Montagna P (2006) Nocturnal eating: sleep-related eating disorder or night eating syndrome? A videopolysomnographic study. Sleep 29(7):949–954. https://doi.org/10.1093/sleep/29.7.949
Vyazovskiy VV, Cirelli C, Pfister-Genskow M, Faraguna U, Tononi G (2008) Molecular and electrophysiological evidence for net synaptic potentiation in wake and depression in sleep. Nat Neurosci 11(2):200–208. https://doi.org/10.1038/nn2035
Wagner U, Born J (2008) Memory consolidation during sleep: interactive effects of sleep stages and HPA regulation. Stress 11(1):28–41. https://doi.org/10.1080/10253890701408822
Wagner U, Gais S, Born J (2001) Emotional memory formation is enhanced across sleep intervals with high amounts of rapid eye movement sleep. Learn Mem 8(2):112–119. https://doi.org/10.1101/lm.36801
Wagner U, Gais S, Haider H, Verleger R, Born J (2004) Sleep inspires insight. Nature 427(6972):352–355. https://doi.org/10.1038/nature02223
Walker MP, Stickgold R (2006) Sleep, memory, and plasticity. Annu Rev Psychol 57(1):139–166. https://doi.org/10.1146/annurev.psych.56.091103.070307

Walker MP, Liston C, Hobson JA, Stickgold R (2002) Cognitive flexibility across the sleep–wake cycle: REM-sleep enhancement of anagram problem solving. Cogn Brain Res 14(3):317–324. https://doi.org/10.1016/s0926-6410(02)00134-9

Wang SH, Morris RG (2010) Hippocampal-neocortical interactions in memory formation, consolidation, and reconsolidation. Annu Rev Psychol 61(1):49–79., C41–44. https://doi.org/10.1146/annurev.psych.093008.100523

Watanabe T, Taguchi Y, Shiosaka S, Tanaka J, Kubota H, Terano Y, Tohyama M, Wada H (1984) Distribution of the histaminergic neuron system in the central nervous system of rats; a fluorescent immunohistochemical analysis with histidine decarboxylase as a marker. Brain Res 295(1):13–25. https://doi.org/10.1016/0006-8993(84)90811-4

Weber F (2017) Modeling the mammalian sleep cycle. Curr Opin Neurobiol 46:68–75. https://doi.org/10.1016/j.conb.2017.07.009

Wehr TA (1991) The durations of human melatonin secretion and sleep respond to changes in daylength (photoperiod). J Clin Endocrinol Metab 73(6):1276–1280. https://doi.org/10.1210/jcem-73-6-1276

Welsh DK, Takahashi JS, Kay SA (2010) Suprachiasmatic nucleus: cell autonomy and network properties. Annu Rev Physiol 72(1):551–577. https://doi.org/10.1146/annurev-physiol-021909-135919

Wilhelm I, Diekelmann S, Molzow I, Ayoub A, Molle M, Born J (2011) Sleep selectively enhances memory expected to be of future relevance. J Neurosci 31(5):1563–1569. https://doi.org/10.1523/JNEUROSCI.3575-10.2011

Williams JA, Reiner PB (1993) Noradrenaline hyperpolarizes identified rat mesopontine cholinergic neurons in vitro. J Neurosci 13(9):3878–3883. https://doi.org/10.1523/jneurosci.13-09-03878.1993

Winkelman JW (2006) Sleep-related eating disorder and night eating syndrome: sleep disorders, eating disorders, or both? Comment on Vetrugno R; Manconi M; Ferini-Strambi L et al. Nocturnal eating: sleep-related eating disorder or night eating syndrome? A videopolysomnographic study. Sleep 29(7):949–954. https://doi.org/10.1093/sleep/29.7.876

Wirz-Justice A, Tobler I, Kafka MS, Naber D, Marangos PJ, Borbely AA, Wehr TA (1981) Sleep deprivation: effects on circadian rhythms of rat brain neurotransmitter receptors. Psychiatry Res 5(1):67–76. https://doi.org/10.1016/0165-1781(81)90062-7

Wirz-Justice A, Benedetti F, Terman M (2009) Chronobiology in everyday life. Chronotherapeutics for affective disorders. Karger, Basel. https://doi.org/10.1159/000210076

Wurts SW, Edgar DM (2000) Circadian and homeostatic control of rapid eye movement (REM) sleep: promotion of REM tendency by the suprachiasmatic nucleus. J Neurosci 20(11):4300–4310. https://doi.org/10.1523/jneurosci.20-11-04300.2000

Xi MC, Morales FR, Chase MH (2004) Interactions between GABAergic and cholinergic processes in the nucleus pontis oralis: neuronal mechanisms controlling active (rapid eye movement) sleep and wakefulness. J Neurosci 24(47):10670–10678. https://doi.org/10.1523/JNEUROSCI.1987-04.2004

Xie L, Kang H, Xu Q, Chen MJ, Liao Y, Thiyagarajan M, O'Donnell J, Christensen DJ, Nicholson C, Iliff JJ, Takano T, Deane R, Nedergaard M (2013) Sleep drives metabolite clearance from the adult brain. Science 342(6156):373–377. https://doi.org/10.1126/science.1241224

Zarcone VP Jr, Benson KL, Berger PA (1987) Abnormal rapid eye movement latencies in schizophrenia. Arch Gen Psychiatry 44(1):45–48. https://doi.org/10.1001/archpsyc.1987.01800130047007

Zee PC, Manthena P (2007) The brain's master circadian clock: implications and opportunities for therapy of sleep disorders. Sleep Med Rev 11(1):59–70. https://doi.org/10.1016/j.smrv.2006.06.001

Zeitzer JM, Buckmaster CL, Parker KJ, Hauck CM, Lyons DM, Mignot E (2003) Circadian and homeostatic regulation of hypocretin in a primate model: implications for the consolidation of wakefulness. J Neurosci 23(8):3555–3560. https://doi.org/10.1523/jneurosci.23-08-03555.2003

Genetics and Behaviour

9

Eleni Parlapani, Zoe Nasika, Odysseas Kyriazis, and Ioannis Nimatoudis

9.1 Introduction

Behavioural genetics constitutes a wide research field focusing on genetic, as well as environmental contributors to phenotypic variance of behaviour.

There are different research methods to explore genetic contribution to behaviour:

(a) Animal studies: these are based on animal models simulating human behaviour, e.g. aggressive, impulsive, etc. Selective breeding or genetic manipulation allows studying gene effects on brain development and behaviour.
(b) Heritability studies: these involve quantitative research techniques aiming at determining heritability, e.g. genome-wide linkage studies of multiple alleles segregating within family members. Heritability studies include family, twin and adoption studies.
(c) Molecular genetic studies: these focus on the identification of susceptibility genes, e.g. association between DNA polymorphisms and behavioural traits. The most promising susceptibility alleles are the ones altering gene expression and thus protein levels. Molecular genetic studies include linkage and

E. Parlapani (✉)
1st Department of Psychiatry, General Hospital "Papageorgiou", Faculty of Medicine, School of Health Sciences, Aristotle University of Thessaloniki, Thessaloniki, Greece
e-mail: parlapani@med.auth.gr

Z. Nasika
NHS, Psychiatric Hospital of Thessaloniki, Thessaloniki, Greece

I. Nimatoudis · O. Kyriazis
3rd Department of Psychiatry, University General Hospital AHEPA, Faculty of Medicine, School of Health Sciences, Aristotle University of Thessaloniki, Thessaloniki, Greece

K. N. Fountoulakis, I. Nimatoudis (eds.), *Psychobiology of Behaviour*,
https://doi.org/10.1007/978-3-030-18323-3_9

association studies. The latter constitute a common approach in the field of behavioural genetics. Roughly, they search for significant differences in genetic variant frequencies between individuals characterised by a specific behaviour and healthy controls.

(d) Mechanism studies: these focus on the investigation of the underlying biological mechanisms triggered by gene polymorphisms.

Still, it is almost impossible to base phenotypic expression of behaviour strictly on genes. A broad range of environmental factors is also influential. Thus, it is widely accepted that genetic factors (nature) and environmental factors (nurture) interact with each other. As a result, environmental factors can trigger behaviours for which there is a genetic predisposition (Caspi et al. 2002; Reif et al. 2007; Cicchetti et al. 2012). Gene-environment interaction or genotype-environment correlation (rGE) may be (a) passive, in which there is an association between inherited genotype and childhood environment, e.g. antisocial parents provide both genetic component and an environment promoting the development of antisocial behaviour, and (b) evocative, in which one factor promotes the other, e.g. a child genetically predisposed for aggression will manifest aggressive behaviour provoking others' harsh responses, which evoke in turn further manifestation of child's aggressive behaviour (Jaffee and Price 2008).

Behaviour varies between individuals, a phenomenon called "population variance". Such variance is statistically represented by a bell curve, depicting a "normal distribution". The tails of normal distribution represent individuals with low or high extremes of variance. The term "heritability" is used to describe the proportion of phenotypic variance that may be attributed to genetic factors, e.g. "heritability of 0.50" means that 50% of the variance across the population is explained by non-specific genotype differences. On the other hand, the term "environmental variance" is used to describe the proportion of phenotypic variance that can be explained by environmental factors.

The aim of this chapter is to provide recent research evidence for the association between genes and behaviour. In this first section, a short introduction revises basic knowledge of genetics and provides descriptive information about behaviours and personality traits studied in relation with genetics.

9.1.1 Genetic Variation

Humans are diploid organisms with 23 pairs of chromosomes, 22 somatic and 1 pair of sex chromosomes (female, XX; male, XY). Chromosomes contain all genetic information, coded for by DNA. DNA consists of two strands. Each strand consists of nucleotides. There are four main nitrogenous nucleotide bases, adenine (A), thymine (T), guanine (G) and cytosine (C). Purines consist of A and G while pyrimidines of T and C. These bases couple with each other forming base pairs (bp). Adenine pairs with thymine and guanine with cytosine, in such the two DNA strands are held together forming a double helix, which resembles graphically a ladder.

A DNA segment coding for a specific protein is called "gene", while the sum of genes forms one's "genome". Gene DNA sequences are not fully identical among all people. For instance, at a particular DNA position, one may have adenine whereas another guanine. This DNA variation is called "allele", resulting in wide phenotypic differences among humans. Each individual is a carrier of two alleles, since chromosomes are paired. One allele is of maternal and the other of paternal origin. An individual is homozygous when alleles are identical. When alleles are different, then there is heterozygosity for that particular gene allele. In most cases, individuals are heterozygous.

Alleles detected in at least one out of a hundred (1%) in a population constitute "DNA polymorphisms". Polymorphisms are the result of DNA damage or incorrect DNA replication. In cases of DNA alterations with deleterious effects, i.e. mutations, severe diseases may emerge. Such mutations are usually rapidly eliminated from a population, since mutant carriers are less likely to reproduce. In a broader sense, DNA polymorphisms are not associated with emerging pathology. They are rather the reason why each person is unique. Depending though on polymorphism, protein amount or protein structure may be altered. In this case, the polymorphism is considered "functional". Functional polymorphisms mapping at important sites may be associated with human traits and disorders. The most common polymorphisms are (a) single nucleotide polymorphisms (SNPs): these are formed by variation in a single nucleotide at a specific position. When a purine is changed into another purine or a pyrimidine into another pyrimidine, the polymorphism is called "transition". When a purine is changed into a pyrimidine, the polymorphism is called "transversion". Although there are some triallelic SNPs (i.e. three different base variations), for most SNPs, there are only two different alleles. Single nucleotide polymorphisms may be found at different locations within the genome, and most of them are "silent", caused by synonymous nucleotide changes leaving protein amino acid sequences unaltered (non-synonymous nucleotide changes are the ones altering amino acid sequence and possibly protein's function). Still, a SNP located at a gene's promoter region could be functional, affecting gene expression; (b) short insertion and deletion polymorphisms (INDELs): these constitute insertions (i.e. nucleotide gain, lengthening overall fragment) or deletions (i.e. nucleotide loss, shortening overall fragment) of up to 50 nucleotides at a single locus; and (c) variable-number tandem repeats (VNTRs): these are formed by genetic elements repeated in tandem arrays. They include micro- and minisatellites, primarily distinguished based on size and repeat pattern. Microsatellites or short tandem repeats (STRs) or simple sequence repeats (SSRs) consist of repeated nucleotide sequences ranging from 2 to 6 bp. Minisatellites consist of repeated sequences ranging from 11 to 65 bp.

In case of behavioural genetics, the term "risk allele" is used to describe polymorphic alleles associated with a specific behavioural phenotype. Still, it is extremely difficult to associate a specific behaviour strictly with a certain allele. Behavioural traits normally distributed in a population are attributed to multiple genes interacting with each other. Each gene shows relative limited effects, either enforcing or limiting a trait. Furthermore, gene expression is based on many other expressed genes. Altogether, "polygenic inheritance" (i.e. a complex gene pattern, in some cases across different chromosomes) is involved in the phenotypic expression of human behaviour and personality. Lastly, "linkage disequilibrium" (LD) refers to the non-random association of alleles at two or more loci. Neighbouring alleles tend to co-segregate, i.e. be inherited together. Table 9.1 includes a brief glossary of genetic terms and abbreviations used in this chapter.

Table 9.1 Brief glossary of genetic terms and abbreviations

Terms and abbreviations	Definition
Diploid	A cell/organism with two sets of chromosomes, one of maternal and one of paternal origin
Chromosome	Structure located in the cell nucleus and formed by DNA coiled tightly around histones, the proteins supporting chromosome's structure
XX	Female
XY	Male
DNA	Deoxyribonucleic acid composed of nucleotides, encoding genetic information carried by chromosomes
DNA replication	Production of two identical DNA replicas from one original DNA molecule
Nitrogenous nucleotide bases	Adenine (A), thymine (T), guanine (G), cytosine (C)
Purines	Adenine (A) and guanine (G)
Pyrimidines	Thymine (T) and cytosine (C)
bp	Base pair, formed by nucleotide bases' coupling (adenine pairs with thymine and guanine with cytosine)
kb	Kilobase, unit equal to 1000 bp
Double helix	Two DNA strands coiled around each other
Genome	Sum of genes in an individual
Gene	DNA segment coding for a specific protein
Locus	The exact position of a gene on a chromosome
Gene's promoter region	DNA region initiating gene transcription (i.e. the first step of gene expression), in which a DNA segment is copied into messenger RNA (mRNA)
Gene's coding region	DNA sequence composed of exons, coding for a protein
Exon	DNA sequence encoding RNA, produced after introns have been removed via RNA splicing
Intron	DNA sequence removed by RNA splicing
Codon	Nucleotide triplet coding for an amino acid
Initiation or start codon	Nucleotide triplet defining the beginning of translation, i.e. protein formation. The most common start codon is AUG, coding for amino acid methionine
Amino acid	Structural units forming proteins
Allele	Gene variation due to different nucleotide arrangement, resulting in wide phenotypic differences (each individual carries one allele of maternal and one of paternal origin)
Major allele	The most common allele in a population; in most cases it is the ancestral, also called "wild" allele
Minor allele	The second most common allele in a population

Table 9.1 (continued)

Terms and abbreviations	Definition
Genotype	The combination of two alleles at a specific locus
Homo-/heterozygous	Due to chromosomal diploidy, there are two alleles for any given gene. Carriers of identical alleles at a specific position are homozygous, whereas carriers of different alleles are heterozygous for that particular gene
Alleles in phase	In diploids, "alleles in phase" or "gametic phase" refers to allele combination at different loci on the same chromosome, representing the original combination of maternal and paternal alleles
Risk allele	Allele associated with a specific behaviour or personality trait
Differential susceptibility	Vulnerability or protective effects provided by a gene under specific environmental conditions
DNA polymorphism	Allele detected in at least one out of a hundred in a population
SNP	Single nucleotide polymorphism formed by variation in a single nucleotide at a specific position
Transition	SNP, in which a purine is changed into another purine or a pyrimidine into another pyrimidine
Transversion	SNP, in which a purine is changed into a pyrimidine
Silent SNP	SNP caused by synonymous nucleotide changes leaving amino acid sequence unaltered
Functional SNP	SNP caused by non-synonymous nucleotide changes, altering amino acid sequence and possibly protein's function SNP located at gene's promoter region, affecting gene expression
INDEL	Short insertion and deletion polymorphism, including insertions or deletions of up to 50 nucleotides at a single locus
VNTR	Variable-number tandem repeat formed by genetic elements repeated in tandem arrays
Microsatellite	Short tandem repeat (STR) or simple sequence repeat (SSR) consisting of repeated nucleotide sequences of 2–6 bp
Minisatellite	VNTR consisting of repeated nucleotide sequences of 11–65 bp
Polygenic inheritance	Complex gene pattern, in some cases across different chromosomes, affecting phenotypic expression
LD	Linkage disequilibrium, i.e. non-random association of co-segregating alleles at two or more loci

9.1.2 Behaviour and Personality

9.1.2.1 Behaviour

Behaviour is defined by the way an individual acts and functions in response to internal or external stimuli and under specific circumstances. Behaviours are divided into "innate" and "learned". Innate behaviours are governed by genes. An example of human innate behaviour, or fixed action pattern, is a baby's smile when it's looked at. Such a response makes it attractive and maximises the chance of gaining parental care. Although learned behaviours have a genetic background as well, they are also determined by experience and environmental influences. Examples of learned behaviours include habituation, imprinting, classical conditioning, operant conditioning, observational learning, play and insight learning (reasoning).

Different behaviours have been studied in relation with gene polymorphisms, among which impulsive, suicidal, aggressive, antisocial and criminal. Behaviours laying at the extremes of normal distribution are maladaptive and often associated with psychiatric disorders.

Impulsive Behaviour

Impulsive behaviour is characterised by acting without foresight and is associated with a preference for immediate reward, decision-making without realising risky aspects of a decision and poor volitional control (Evenden 1999). In other words, different aspects of impulsivity include non-planning impulsiveness (i.e. behaving without taking future consequences into consideration), cognitive or attentional impulsiveness (i.e. deciding rapidly without focusing on an assignment) and motor impulsiveness (i.e. acting without thinking) (Patton et al. 1995).

Impulsivity as a personality trait is associated with impulsive behaviour. "Adaptive" impulsivity may be a positive personality trait when there is a demand for immediate confrontation with a crucial situation. In short, impulsive actions that turn out beneficial may be characterised as "spontaneous", "unconventional" or "bold", although their true nature remains impulsive. On the other hand, there is "maladaptive" impulsivity, which is dysfunctional, leads to negative consequences and may be associated with aggressive behaviour.

Heritability of trait impulsivity was estimated at around 45–50% (Pedersen et al. 1988; Hur and Bouchard Jr. 1997). The role of heritability in the manifestation of impulsivity was shown to increase with age. Additionally, heritability of risk-taking was shown to increase with age, though only in males (Anokhin et al. 2009). Behavioural expression of impulsivity was associated with suicidal, aggressive and antisocial behaviour. Additionally, impulsive behaviour constitutes a shared phenotype among different psychiatric disorders, such as attention deficit hyperactivity disorder (ADHD), impulse control and addictive disorders.

Suicidal Behaviour

Suicidal behaviour or suicidality includes thoughts and actions aiming at causing own death. *S*uicide ideation always precedes a suicide attempt and may be originally expressed by a vague death wish. As suicide ideation becomes more intense, a suicide plan, including suicide method, place and time, may be organised. A suicide attempt is an intentional act aiming at causing own death, without succeeding. Lastly, suicide is defined as an intentional fatal act causing own death.

Suicidal behaviour is more common within the context of psychiatric disorders, such as major depressive disorder and schizophrenia (Nock et al. 2008). There seems to be a genetic contribution to suicidal behaviour independent of the genetic contribution to psychiatric disorders. Many genes were implicated in the manifestation of suicidal behaviour (Clayden et al. 2012). Family history of suicidal behaviour is a severe risk factor for suicidal behaviour, while heritability of serious suicidality was estimated at around 55% (Mann et al. 2001). According to some, suicidal behaviour may constitute another form of aggressive behaviour, in which aggression is directed towards one's self. Suicide attempts may be more related to impulsivity, while completed suicide may be more related to aggression.

Aggressive Behaviour

Aggressive behaviour constitutes a broad phenotype, ranging from verbal to physical aggression, including offensive behaviour against others and objects' destruction. Aggression is categorised into two major subtypes: (a) proactive or instrumental. This form is associated with premeditated, controlled assault. Therefore, it is considered as predatory, offensive aggression, aiming at gaining something or achieving a goal, and is related to psychopathy. Proactive aggressive behaviour is more closely related to antisocial and criminal behaviour; (b) reactive or impulsive. In this case, aggressive behaviour is neither planned nor well considered and is almost always accompanied by intense negative emotions, e.g. fear, anxiety, anger, hostility, increased psychomotor activity and excitation of autonomic nervous system. Reactive aggression is usually provoked by external stimuli, such as insults, threats, physical attacks, and is more closely related to affect dysregulation and impulsive behaviour (Vitiello and Stoff 1997). When reactive aggression constitutes a response to threats posed by others, then it is considered "defensive" and thus beneficial. On the contrary, "dysfunctional" reactive aggression is often disproportionate to the stimulus, associated with reduced control over aggressive impulses and closely related to frustration (Crick and Dodge 1996; Raine et al. 2006).

Aggressive behaviour is far more common in males compared with females. Heritability, which was confirmed by meta-analyses of twin and adoption studies, may be more pronounced during adulthood compared with childhood, during which environmental factors are probably equally important. The genetic component of aggressive behaviour was shown to account for 40–60% of the variance (Miles and Carey 1997; Rhee and Waldman 2002; Craig and Halton 2009).

Antisocial Behaviour

Antisocial behaviour constitutes a broad phenotype, characterised among others by aggressive behaviour, criminal behaviour, delinquency and psychopathy (Baker et al. 2007). Antisocial behaviour can be studied under a different perspective, depending on its definition and measurement tools, and is best considered a dimensional phenomenon, a continuum, within which different manifestation and diverse severity may be observed. Altogether, there seems to be a strong association between the clinical manifestation (e.g. aggressive behaviour), the legal aspects (e.g. criminal behaviour, court conviction) and the personality traits (e.g. hostility, dishonesty, psychopathy) related to antisocial behaviour.

Antisocial personality traits include beliefs and attitudes aiming at using or harming others. Still, someone with antisocial personality traits does not necessarily perform aggressive or illegal acts. Since the relation between antisocial personality traits and aggressive or criminal behaviour is not bidirectional, genetic studies should differentiate aggressive versus non-aggressive antisocial behaviour.

Antisocial behaviour is far more common in males compared with females (Craig and Halton 2009). Still, the effect of genetic factors on the manifestation of antisocial behaviour seems to be higher in females during childhood. This sex difference disappears throughout adolescence and in adulthood, since the effect of genetic and environmental factors becomes roughly the same in both sexes (Jacobson et al. 2002). Quantitative genetic studies yielded diverse results, possibly due to heterogeneous definition and assessment of antisocial behaviour. Roughly, the genetic component of antisocial behaviour was shown to account for 40–60% of the variance (Gunter et al. 2010; Fergusson et al. 2011).

Criminal Behaviour

Criminal behaviour, defined as "an act violating public law" or as "failure to act according to public law", was mainly studied within the context of antisocial behaviour/antisocial personality disorder.

9.1.2.2 Personality

Personality is formed by a pattern of relatively permanent traits and unique mental, emotional and behavioural characteristics providing consistency and individuality to one's behaviour. A personality "trait" is a relatively stable characteristic, enduring and consistent across a variety of situations, as well as typical for an individual. Traits are considered to predispose a person to respond in a certain way, regardless of the situation. For instance, an individual with high trait anxiety is prone to interpreting ambiguous stimuli as more threatening, while an individual with high trait anger is prone to reacting with anger towards situations that are least provoking. On the other hand, "state" is a temporary emotional-personality change, constituting a reaction to different stimuli.

Cloninger introduced the Temperament and Character Inventory (TCI), a self-report personality questionnaire, based on his "psychobiological model of personality". According to Cloninger, personality consists of temperament (i.e. heritable-stable traits) and character (i.e. traits influenced by learning and experience, maturing throughout life). Temperament has four dimensions: (a) novelty seeking, (b) harm avoidance, (c) reward dependence and (d) persistence. Character consists of three dimensions: (a) self-directedness, (b) cooperativeness and (c) self-transcendence (Raeymaekers and Van Broeckhoven 1998).

Currently, another widely used taxonomy of personality traits identifies five major personality dimensions: (a) neuroticism, (b) introversion-extraversion, (c) agreeableness, (d) conscientiousness and (e) openness to experience. Each personality trait is represented by a normal distribution. Based on twin studies, heritability of aforementioned dimensions was estimated at around 40% (Borkenau et al. 2001). Neuroticism constitutes a dimensional trait with six different facets: (a) anxiety, (b) depression, (c) hostility, (d) self-consciousness, (e) impulsiveness and (f) vulnerability (Miller et al. 2009). Altogether, twin studies revealed that genetic factors contribute to phenotypic expression of all aforementioned personality facets, accounting for 41–61% of the variance (Jang et al. 1996).

Personality traits laying at the extremes of normal distribution are putatively maladaptive and may predispose to the manifestation of psychiatric disorders. For instance, anxiety-related traits increase vulnerability for the development of anxiety disorders. Different personality traits have been studied in relation with gene polymorphisms, among which trait anxiety, trait impulsivity, trait anger, novelty seeking, sensation seeking, harm avoidance and psychopathy.

Novelty Seeking

High novelty seeking individuals appear quick-tempered, impulsive, curious, exploratory, enthusiastic-excitable, disorderly and extravagant. According to Cloninger's Tridimensional Personality Questionnaire (TPQ, the old version of TCI, measuring novelty seeking, harm avoidance and reward dependence), individuals scoring high in novelty seeking are characterised by impulsive and exploratory behaviour, while those scoring low tend to be rigid, frugal, reflective, stoic and low-tempered. Novelty seeking was correlated with extraversion (Lepine et al. 1994; Tsuchimine et al. 2009). Novelty seeking and extraversion were correlated in turn with suicide ideation and attempt (Brezo et al. 2006). Heritability of novelty seeking was shown to account for 36% of the variance (Heiman et al. 2004).

Sensation Seeking

Sensation seeking is defined as an individual's need for novel, varied, sensory and mental experiences (Zuckerman et al. 1972) and is expressed by four facets: (a) experience seeking, characterised by an attraction towards new experiences (e.g. through travel or various lifestyle choices); (b) disinhibition, characterised by a tendency towards sensation pursuit (e.g. alcohol or sexual intercourse); (c) thrill/adventure seeking, characterised by an engagement in adventurous activities (e.g. extreme sports, reckless driving); and (d) boredom susceptibility, characterised by an aversion to boredom, routine and repetition as well agitation when there is lack of a variety of stimuli.

Sensation seekers pursue new sensations, feelings and experiences in all life aspects, including personal and professional life. Although risk is not the driving force, risk can be underestimated or even considered an additional stimulant for acquiring desired sensations. Behaviours associated with thrill and adventure components of sensation seeking are referred to as "thrill" or "adrenaline seeking behaviours". Individuals exhibiting dysfunctional sensation seeking may engage in sexual risk-taking and gambling.

It was shown that sensation seeking is a highly heritable trait. In males, highly heritable sensation seeking facets included experience seeking (60%) and disinhibition (59%), while thrill/adventure seeking was the least heritable facet (34%). In females, highly heritable facets included thrill/adventure seeking (62%) and disinhibition (52%), while boredom susceptibility was the least heritable facet, accounting for 29% of the variance (Stoel et al. 2006). Sensation seeking was correlated with novelty seeking (McCourt et al. 1993).

Harm Avoidance

Harm avoidance is characterised by anticipatory worry, fear of the unknown, cautiousness, self-doubt, shyness and fatigability. Individuals exhibiting this trait are often characterised as cautious, fearful, discouraged, insecure, negativistic-pessimistic, asthenic and reserved with strangers (Cheung 2007). Harm avoidance was positively correlated with neuroticism and negatively correlated with novelty and sensation seeking (Cloninger 1986; McCourt et al. 1993). Harm avoidance and neuroticism were associated with suicide ideation and attempts (Brezo et al. 2006). Heritability of harm avoidance was shown to account for 36% of the variance (Heiman et al. 2004).

Psychopathy

Psychopathy is a personality trait characterised by lack of empathy (i.e. response congruent to the other's emotional state) and remorse. Psychopathy is related to antisocial behaviour and was shown to have a genetic component ranging from 50% to 80% (Retz et al. 2004; Gunter et al. 2010).

9.2 Genes and Behaviour

This second section focuses on the association between different gene polymorphisms and behaviour/personality. Emphasis was given on human studies, since data from animal studies cannot be easily interpreted in relation to humans. Some animal studies will be presented when there is no available data on humans. Due to the large amount of literature, emphasis was drawn away from genes that have not been extensively studied (Clayden et al. 2012). For instance, although the norepinephrine system has been associated with aggression, genes for which there is lack of consistent evidence, e.g. norepinephrine transporter gene (Kim et al. 2006a), will not be described. Overall, focus was placed on meta-analytic studies, whenever these were available.

9.2.1 Serotonergic System

The serotonergic system is involved, among others, in mood and behaviour regulation. There is evidence for the implication of the serotonergic system in the manifestation of impulsive (Bevilacqua and Goldman 2013), suicidal (Mann et al. 2001), aggressive and antisocial behaviour (Lesch and Merschdorf 2000).

Specifically, it was serotonergic system hypofunction that was associated with impulsive and risk-taking behaviour (Mann 2003). Low 5-hydroxyindoleacetic acid (5-HIAA) levels in cerebrospinal fluid (CSF), indicative of serotonergic system deficiency, were associated with suicidal behaviour within the context of different psychiatric disorders, independent of diagnosis (Asberg et al. 1986; Olivier and van Oorschot 2005). Furthermore, different serotonergic gene polymorphisms were associated with suicidal behaviour (Arango et al. 2003).

The serotonergic system was particularly implicated in the manifestation of aggressive behaviour (Craig and Halton 2009). The theory of a negative association between serotonergic activity and aggression, i.e. low serotonin activity is related to increased aggression levels (Olivier and van Oorschot 2005), was supported by the anti-aggressive effects of selective serotonin reuptake inhibitors (SSRIs) (Fuller 1996; Reist et al. 2003) and other anti-aggressive drugs with serotonergic function, called "serenics", which are under investigation (Miczek et al. 2002; Olivier and van Oorschot 2005). Still, such a direct association may be an oversimplified hypothesis due to serotonin receptors' wide variety, as well as serotonin system's complex regulation (de Almeida et al. 2005).

Lastly, it should be mentioned that research evidence relating different behavioural phenotypes to serotonergic system's dysfunction emphasises the detrimental effects of environmental stressors on the serotonergic system. Tables 9.2, 9.3, 9.4, and 9.5 present meta-analyses of studies investigating serotonergic genes in relation with behaviour and personality.

Table 9.2 MAOA gene polymorphism and behaviour

Gene (chromosome location)	Encoded protein	Polymorphism	Alleles	Behavioural phenotype	Meta-analyses
MAOA (Xp11.3)	Monoamine oxidase A	MAOA-uVNTR, 30 bp repeat element	MAOA-H: high-activity allele, transcribed 2–10 times more efficiently MAOA-L: low-activity allele	Suicidal behaviour	Meta-analysis of five studies (862 suicidal cases versus 1239 healthy controls) Results: no association (Clayden et al. 2012)
				Aggressive/antisocial behaviour	Meta-analysis of five studies (around 2570 males) Results: childhood adversity/abuse was associated with antisocial behaviour in male MAOA-L genotype carriers (Kim-Cohen et al. 2006)
					Meta-analysis of 31 studies (case and healthy control number not mentioned) Results: MAOA-L was proven a risk allele for broadly defined antisocial behaviour (Ficks and Waldman 2014)
					1. Meta-analysis of 20 studies of strictly male or mixed male-female, mainly non-clinical populations (11,064 participants) Results: male MAOA-L carriers with a background of childhood maltreatment (e.g. domestic violence, physical/sexual abuse and parental neglect) exhibited higher levels of aggression/violence, non-violent antisocial behaviour as well as aggressive/violent antisocial behaviour during childhood/adolescence, as well as during adulthood 2. Meta-analysis of 12 studies of female, mainly non-clinical populations (7588 participants) Results: an interaction was observed between MAOA-H genotype and childhood maltreatment (e.g. domestic violence, physical/sexual abuse and parental neglect) in relation with antisocial outcomes. This finding depended though on meta-analysis' study inclusion and requires further investigation (Byrd and Manuck 2014)

Table 9.3 SLC6A4 gene polymorphisms and behaviour/personality

Gene (chromosome location)	Encoded protein	Polymorphisms	Alleles	Behavioural phenotype	Meta-analyses
SLC6A4 (17q11.1–q12)	Serotonin transporter	5-HTTLPR, 20–23 bp repeat element	L: long allele; it was associated with increased gene transcriptional efficiency and exists as LA and LG (low-expressing) S: short allele; it was associated with reduced gene transcriptional activity and is therefore considered low-expressing	Suicidal behaviour	Meta-analysis of 12 studies (three studies of completed/nine studies of attempted suicide; 1168 suicidal cases versus 1371 healthy controls) Results: S allele was proven a risk factor for suicide attempt (Anguelova et al. 2003)
					1. Meta-analysis of 18 studies (1521 suicide attempters and completers versus 2429 healthy controls) Results: no association; subanalysis of 15 studies of only Caucasian populations revealed again no association
					2. Meta-analysis of two studies of patients with a schizophrenia spectrum disorder (146 suicide attempters and 374 non-attempters) and two studies of patients with alcohol dependence (107 suicide attempters and 166 non attempters), leading to a meta-analysis of four studies altogether (258 suicide attempters versus 291 non-attempters) Results: S allele was proven a risk factor for suicide attempt in patients diagnosed with the same psychiatric disorders (the association was significant only in patients with alcohol dependence, when diagnostic categories were considered separately)
					3. Meta-analysis of five studies (190 violent attempters/suicide completers versus 733 healthy controls) Results: association between S allele and violent suicide behaviour (no association when non-violent attempters were compared with healthy controls) (Lin and Tsai 2004)

Gene (chromosome location)	Encoded protein	Polymorphisms	Alleles	Behavioural phenotype	Meta-analyses
					Meta-analysis of 39 studies (3096 suicidal cases versus 5936 healthy controls) Results: overall association between 5-HTTLPR polymorphism and suicidal behaviour, independent of psychiatric diagnoses (Li and He 2007)
					Meta-analysis of 31 studies (6324 suicidal cases versus 10,285 healthy controls) Results: no overall association Subanalysis of 25 studies of suicide attempters revealed an association between S allele and suicide attempt (S allele increased risk for suicide attempt by 13%) Subanalysis of six studies of suicide completers revealed no association (Clayden et al. 2012)
				Aggressive/antisocial behaviour	Meta-analysis of 18 studies (case and healthy control number not mentioned) Results: S was proven a risk allele for broadly defined antisocial behaviour (Ficks and Waldman 2014)
				Trait anxiety	Meta-analysis of 26 studies (7657 participants) Results: no association (slight association, provided the fact that trait anxiety was assessed by a particular scale addressing the five-factor model of personality) (Schinka et al. 2004)
		STin2, 17 bp repeat element	STin2.9: contains nine copies STin2.10: contains ten copies STin2.12: contains 12 copies and constitutes a more potent positive transcriptional regulator	Suicidal behaviour	Meta-analysis of 10 studies (case and healthy control number not mentioned) Results: no association (Li and He 2007)

Table 9.4 Serotonin receptor gene polymorphisms and behaviour

Gene (chromosome location)	Encoded protein	Polymorphisms	Alleles	Behavioural phenotype	Meta-analyses
5-HTR1A (5q11.2–13)	Serotonin 1A receptor	rs6295: C(-1019)G SNP	C G: it was associated with higher receptor expression	Suicidal behaviour	Meta-analysis of four studies (three Caucasian/one Asian population; 957 suicidal cases versus 957 healthy controls) Results: no association, even when only suicide completers were included in analysis (Angles et al. 2012)
					Meta-analysis of six studies (2022 suicidal cases versus 2135 healthy controls) Results: no association (Clayden et al. 2012)
					Meta-analysis of nine studies (seven Caucasian/one Asian/one Mexican population; 2366 suicidal cases versus 2943 healthy controls) Results: no association, even when only Caucasians were included in analysis (Gonzalez-Castro et al. 2013a, b)
5-HTR1B (6q14.1)	Serotonin 1B receptor	rs6296: G861C SNP	G: G/G genotype was associated with higher receptor binding C	Suicidal behaviour	Meta-analysis of seven studies (789 suicidal cases versus 1247 healthy controls) Results: no association (Kia-Keating et al. 2007)
					Meta-analysis of ten studies (2947 suicidal cases versus 4066 healthy controls) Results: no association, even when suicide completers were excluded from analysis (Clayden et al. 2012)

5-HTR2A (13q14–q21)	Serotonin 2A receptor	rs6313: T102C SNP	T C	Suicidal behaviour	Meta-analysis of nine studies (three studies of suicide attempters; 596 suicidal cases versus 1003 healthy controls) Results: no association (Anguelova et al. 2003)
					Meta-analysis of 25 studies (1954 suicidal cases versus 2860 healthy controls) Results: no association, even when analysing different subgroups, e.g. Europeans, Asians, suicidal ideation versus healthy controls, suicide attempt versus healthy controls, violent versus non-violent, etc. (Li et al. 2006)
					Meta-analysis of 18 studies (3759 suicidal cases versus 5692 healthy controls) Results: no association, even when suicide completers were excluded from analysis (Clayden et al. 2012)
					Meta-analysis of 23 studies [2566 suicide attempters and completers versus 3989 healthy controls, as well as 612 suicidal cases and 1129 healthy controls included in a previous meta-analysis (Li et al. 2006); 13 Caucasian/six Asian/four populations of other ethnic origin] Results: no association, even when Caucasian and Asian populations, as well as schizophrenia patients, were analysed separately (Gonzalez-Castro et al. 2013a)
					Meta-analysis of 13 studies [1729 suicide attempters diagnosed with a psychiatric disorder (710 Asians/1019 European-Americans); 1794 non-suicide attempters diagnosed with a psychiatric disorder (759 Asians/920 European-Americans); 2398 healthy controls (906 Asians/1492 European-Americans)] Results: no association when suicide attempters were compared with healthy controls, even when data were analysed separately based on ethnicity No association when suicide attempters were compared with non-attempters diagnosed with the same psychiatric disorders, even when data were analysed separately based on ethnicity Analysis taking psychiatric diagnosis into consideration revealed an association between C/C genotype and suicide attempt in schizophrenia patients. Genotype C/C was not proven a risk factor for suicide attempt in bipolar and in patients with alcohol dependence (Wang et al. 2015)
		rs6311: G-1438A SNP	G A	Suicidal behaviour	Meta-analysis of seven studies (six Asian populations, further data not shown) Results: genotypic analysis with allele A combined [(AA+AG)/GG] revealed an association with suicidal behaviour (Li et al. 2006)
5-HTR2C (Xq23)	Serotonin 2C receptor	rs6318: C68G SNP	C G	Suicidal behaviour	Meta-analysis of seven studies (2297 suicidal cases versus 3431 healthy controls) Results: no association (Clayden et al. 2012)

Table 9.5 TPH1 gene polymorphisms and behaviour

Gene (chromosome location)	Encoded protein	Polymorphisms	Alleles	Behavioural phenotype	Meta-analyses
TPH1 (11p15.1)	Tryptophan hydroxylase, isoform 1	rs1800532: A218C SNP	218A (U allele) 218C (L allele)	Suicidal behaviour	Meta-analysis of seven studies (Caucasian populations only; 898 suicide attempters and completers versus 1179 healthy controls) Results: association between 218A allele and suicide-related behaviours (Rujescu et al. 2003b)
					Meta-analysis of seven studies (Caucasian populations only; 860 suicidal cases versus 1279 healthy controls) Results: association between 218A allele and suicide-related behaviours (Bellivier et al. 2004)
					Meta-analysis of 21 studies (4829 suicidal cases versus 7945 healthy controls) Results: no overall association; association between 218A allele and suicide attempt (Clayden et al. 2012)
		rs1799913: A779C SNP	779A (U allele): it was associated with lower CSF 5-HIAA levels 779C (L allele)	Suicidal behaviour	Meta-analysis of eight studies (1512 suicidal cases versus 3408 healthy controls) Results: no association (Clayden et al. 2012)
		rs1800532 (A218C SNP) together with rs1799913 (A779C SNP)	Described above	Suicidal behaviour	1. Meta-analysis of 15 studies of A218C and/or A779C (3585 suicide attempters and completers versus 2295 healthy controls) Results: no overall association 2. Meta-analysis of nine studies of psychiatric patients (two alcohol dependence/seven major depression, bipolar disorder and schizophrenia; 625 suicide attempters versus 1475 non-attempters) Results: no association (Lalovic and Turecki 2002)
					Meta-analysis of 34 studies of A218C and/or A779C (3922 suicidal cases versus 6700 healthy controls) Results: overall association between A218C/A779C SNPs and suicidal behaviour (different alleles implicated, based on different study characteristics) (Li and He 2006)
					Meta-analysis of 13 studies of psychiatric patients (three schizophrenia/two bipolar disorder/two major depression/three alcohol dependence/one borderline personality disorder/two mixed diagnoses; 1272 suicide attempters versus 1727 non-suicide attempters) Results: no association, independent of mental health status (Saetre et al. 2010)

9.2.1.1 Monoamine Oxidase A Gene (MAOA)

The MAOA Gene Polymorphism

The monoamine oxidase A, MAO-A, is a mitochondrial enzyme in neuronal presynaptic terminals, implicated in the degradation of biogenic amines, i.e. the neurotransmitters dopamine, serotonin (5-hydroxytryptamine, 5-HT) and norepinephrine, after reuptake from the neuronal synaptic cleft. Decreased brain MAO-A levels were associated with impulsive aggression and mental retardation in a study of a Dutch family and specifically in male carriers of a MAOA gene mutation (C936T MAOA stop codon) of maternal origin (Brunner et al. 1993a,b). Additionally, low MAO-A activity in different cortical and subcortical regions detected by positron emission tomography was related to trait aggression, accounting for more than 30% of variance (Alia-Klein et al. 2008).

The MAOA gene, linked to the short arm of X chromosome (Xp11.3), is mainly expressed in catecholaminergic neurons (Hunter 2010). There is a functional polymorphism within the gene's promoter region, located 1.2 kb upstream of the MAOA coding region. This length polymorphic region (MAOA-LPR), a VNTR (MAOA-uVNTR) consisting of a specific 30 bp nucleotide sequence repeated 2, 3, 3.5, 4 or 5 times, was shown to influence gene's transcriptional activity. It was reported that high-activity MAOA gene variants (MAOA-H), containing alleles with 3.5 or 4 copies, were transcribed 2–10 times more efficiently than low-activity MAOA gene variants (MAOA-L), containing alleles with 2, 3 or 5 copies. Thus, the nucleotide repeat number was shown to affect MAO-A enzyme abundance, since MAOA-H genotype was associated with higher, while MAOA-L genotype with reduced MAO-A expression and therefore increased synaptic serotonin levels (Sabol et al. 1998; Buckholtz and Meyer-Lindenberg 2008; Guo et al. 2008).

The MAOA-uVNTR Polymorphism and Behaviour/Personality

The MAOA-L allele was associated with personality traits related to aggressiveness and impulsivity in a community male sample (Manuck et al. 2000). Furthermore, it was suggested that MAOA-L allele association with trait aggression in both healthy males and females may be mediated by increased sensitivity towards negative experiences (Eisenberger et al. 2007). The MAOA-L allele was also associated with antisocial and aggressive/violent behaviour in different psychiatric populations, such as in patients diagnosed with substance abuse and cluster B personality disorders (Reif et al. 2007), as well as in male criminal alcoholics with increased CSF testosterone levels (Sjoberg et al. 2008).

The MAOA genotype was shown to moderate the effects of childhood maltreatment on the manifestation of antisocial behaviour in adulthood. Specifically, male MAOA-L carriers with a background of childhood maltreatment were more prompt to the development of antisocial behaviour compared with male MAOA-H carriers. This was the outcome of a core longitudinal study, emphasising the combined effect of genetic and environmental factors on behavioural outcome (Caspi et al. 2002). Further studies of gene-environment interactions showed that the MAOA-L genotype predicted aggressive behaviour in males with a history of traumatic life events, e.g. separation from parents or family violence, especially during the first 15 years of life (Frazzetto et al. 2007), the presence of at least one short MAOA allele (MAOA-L, 2 or 3 copies) in males and one or two long MAOA alleles (MAOA-H, 3.5, 4 or 5 copies) in females was associated with higher levels of delinquency in adolescents with a history of maltreatment (Aslund et al. 2011), and MAOA-L genotype was associated with aggressive behaviour in males under high provocation (McDermott et al. 2009).

On the other hand, MAOA-L was proven a risk allele for violent behaviour in a population of psychiatric patients, independent of childhood adversity (Reif et al. 2007), while others found no interaction between MAOA genotype and childhood adversity (Haberstick et al. 2005; Huizinga et al. 2006). Lastly, there were studies that did not find an association between MAOA genotype and aggressive/antisocial behaviour at all (Haberstick et al. 2005; Jacob et al. 2005; Widom and Brzustowicz 2006; Prichard et al. 2007a; Weder et al. 2009).

Altogether, different meta-analyses revealed an association between MAOA-L allele and aggressive/violent/antisocial behaviour, while MAOA-H was shown to be the low-risk allele. The first one, a meta-analysis of five independent studies (around 2570 males), reported that childhood adversity (domestic violence, parental neglect, physical or sexual abuse, harsh discipline, etc.) was associated with antisocial behaviour in male MAOA-L genotype carriers (Kim-Cohen et al. 2006). A subsequent meta-analysis of 31 studies confirmed MAOA-L as a risk allele for aggressive and antisocial behaviour (Ficks and Waldman 2014). Lastly, a more recent extended meta-analysis investigated the effect of the interaction between childhood adversity and MAOA genotype on later manifestation of aggressive-antisocial behaviour in both males and females. The meta-analysed studies were mainly conducted in non-clinical populations. Firstly, 20 studies of strictly male or mixed male-female samples (11,064 participants) were meta-analysed. Results revealed that male MAOA-L carriers with a background of childhood maltreatment and other adversities exhibited higher levels of aggression/violence, non-violent antisocial behaviour as well as aggressive/violent antisocial behaviour during childhood/adolescence, as well as during adulthood. When early environmental adversity was more closely investigated, it was shown that MAOA-L genotype increased the risk for aforementioned behavioural outcomes specifically in cases of childhood maltreatment (e.g. domestic violence, physical/sexual abuse and parental neglect), and not in cases of other childhood adversities (e.g. separation, marital difficulties, parental psychopathology). Secondly, 12 studies of female populations (7588 participants) were meta-analysed. Overall, results did not reveal any significant interaction between MAOA genotype and maltreatment/other adversities, predicting aggressive/antisocial outcomes. An interaction was observed between MAOA-H genotype and childhood maltreatment in relation with antisocial outcomes. This association depended though on study inclusion and therefore requires further investigation (Byrd and Manuck 2014).

On the other hand, a meta-analysis of five studies (862 suicidal cases versus 1239 healthy controls) investigating the association between MAOA-L (three copies), MAOA-H (four copies) and suicidal behaviour did not reveal any significant associations (Clayden et al. 2012) (Table 9.2).

9.2.1.2 Serotonin Transporter Gene (SLC6A4)

The SLC6A4 Gene Polymorphisms

The serotonin transporter, SERT or 5HTT, is a monoamine transporter protein at serotonergic neurons' cell body and synaptic terminals, regulating synaptic signalling by removing serotonin from synaptic cleft back to presynaptic neurons.

The transporter is coded for by the gene "solute carrier family 6 member 4" (SLC6A4), located at chromosome 17 (17q11.1–q12) (Nakamura et al. 2000). The polymorphism "5-HTTLPR" constitutes a 20–23 bp repeat element in gene's promoter region (44 bp insertion/deletion), giving rise to up to 14 different alleles (Nakamura et al. 2000). The most common and best studied alleles include the "long" ("L", 16 copies, 528 bp) and the "short" ("S", 14 copies, 484 bp) variant (Heils et al. 1996; Hariri and Weinberger 2003). Homozygosity for the long variant (L/L) was associated with increased transcriptional efficiency compared with the L/S and S/S genotypes (Heils et al. 1996; Cadoret et al. 2003), resulting in higher amounts of serotonin transporter and thus higher serotonin reuptake in blood platelets and lymphoblasts. Contrary, the S allele was associated with reduced SLC6A4 gene transcriptional activity (Lesch et al. 1996; Greenberg et al. 1999), since the S/S genotype corresponded to almost half of SERT protein levels compared with the L/L genotype (Collier et al. 1996). Still, another study reported no association between the 5-HTTLPR genotype and serotonin transporter binding (Jacobsen et al. 2000). Lately, a common polymorphism (rs25531), a single-base substitution (SNP A/G), was reported within the first of two 22 bp imperfect repeats of the L allele. This SNP gives rise to the LA and the LG allele. Thus, together with the S allele, the locus is considered triallelic. The LG allele carries a binding site for a transcription factor (AP2) that possibly suppresses gene expression. Thus, similarly to the S allele, the LG was considered a low-expressing allele (Hu et al. 2006).

Another SLC6A4 gene polymorphism is a 17 bp VNTR in the second intron (STin2), a triallelic polymorphism existing in the form of 9 (STin2.9), 10 (STin2.10) and 12 (STin2.12) copies. Allele STin2.12 was proven a more potent positive transcriptional regulator compared with STin2.10 (MacKenzie and Quinn 1999).

The 5-HTTLPR Polymorphism and Behaviour/Personality

Studies of 5-HTTLPR polymorphism in a Japanese (Sakado et al. 2003) and a Korean male population (Lee et al. 2003b) showed that the S/S genotype was associated with higher levels of trait impulsivity compared with the L/S and the L/L genotype. Similarly, in another study of a non-clinical Caucasian population recruited from the University of Oslo, S/S carriers showed higher levels of impulsivity compared with L/S carriers displaying intermediate and L/L carriers displaying lower levels of impulsiveness. The effect of S allele on the expression of impulsivity was sex-dependent, since males were more likely to exhibit impulsiveness (Walderhaug et al. 2010). On the other hand, a study of a non-clinical Caucasian-Brazilian population did not find any association between the 5-HTTLPR polymorphism and impulsivity (Lage et al. 2011).

A study of German suicide completers with undefined psychiatric diagnoses reported an association between the S allele and violent suicide (Bondy et al. 2000a). The same association was reported in another study of suicide attempters (40 with and 11 without a history of a major psychiatric disorder) (Courtet et al. 2001). Contradictory reports supported an association between the L allele and suicidal behaviour within the context of affective disorders (Du et al. 1999), while other studies did not find any association between the 5-HTTLPR polymorphism and suicidal behaviour (Anguelova et al. 2003). A meta-analysis including nine studies of suicide attempters and three studies of suicide completers analysed a total number of 1168 suicidal cases in comparison with 1371 healthy controls. The results supported an association between the S allele and suicidal behaviour. Still, further analysis showed that this association was credited to studies of suicide attempters only, since meta-analysis of 991 suicide attempters confirmed an association between suicidal behaviour and the S allele, while meta-analysis of 177 suicide completers did not (Anguelova et al. 2003). The following meta-analysis of 18 studies (1521 suicidal cases versus 2429 healthy controls) did not find any association between the 5-HTTLPR polymorphism and suicidal behaviour. In order to rule out confounding effects of ethnicity, a separate analysis included only Caucasians. Again, no significant association was revealed. The same meta-analysis focused only on psychiatric patients, comparing patients with and without suicidal behaviour (4 studies of mood disorder patients, 258 suicide attempters versus 291 non-attempters; 2 studies of schizophrenia/schizoaffective patients, 146 suicide attempters versus 374 non-attempters; 2 studies of alcohol-dependent patients, 107 suicide attempters versus 166 non-attempters). The results revealed a significant association between the S allele and suicide attempts in patients with a positive psychiatric history. Lastly, meta-analysis of five studies (190 cases of violent suicide attempters or completers versus 733 healthy controls) revealed a significant association between the S allele and violent suicidal behaviour. Additionally, S allele frequency was higher in violent compared with non-violent suicide attempters (Lin and Tsai 2004). A subsequent meta-analysis of 39 studies (3096 suicidal cases versus 5936 healthy controls) confirmed overall association of the 5-HTTLPR polymorphism and suicidal behaviour, suggesting though that based on study design and genotypic analysis, both the S and the L allele contribute to risk (Li and He 2007). Lastly, an even more recent meta-analysis of 31 studies (6324 suicidal individuals versus 10,285 healthy controls; 25 studies of suicide attempters, 6 studies of suicide completers) confirmed an association between the S allele and suicidal behaviour in attempted suicide only, emphasising once more the phenotypic heterogeneity between suicide attempt and completed suicide. Based on results, the S allele increased risk for suicide attempt by 13% (Clayden et al. 2012).

In regard to aggression, a study reported a significant association between genotypes with low-expressing alleles (S/S, S/LG and LG/LG) and aggressive behaviour in children aged 5–15 years (Beitchman et al. 2006). Another study emphasised the importance of gene-environment interaction, since a highly adverse childhood environment was associated with later manifestation of violent behaviour only in psychiatric patients being S/S and S/L carriers. Vice versa, the L/L genotype was considered a protective factor against manifestation of violence in adults with a high childhood adverse environment index (Reif et al. 2007). This finding was confirmed in male Caucasian offenders with a history of childhood ADHD. The S allele and the S/S genotype were associated with violent behaviour, explaining 5% of the variance of violent behaviour (Retz et al. 2004). Lastly, a study of a Chinese population of convicted criminals reported an association between the S/S genotype and violent crime, but not antisocial personality disorder (Liao et al. 2004). On the other hand, there were studies that reported no association between the 5-HTTLPR genotype and aggressive behaviour in children (Davidge et al. 2004). Altogether, a recent meta-analysis of 18 studies of the 5-HTTLPR polymorphism confirmed an association between the S allele and the increased risk for antisocial behaviour. Still, the authors noted an effect of publication bias on results, since it was shown that there was a trend towards more publications reporting positive associations (Ficks and Waldman 2014).

Lastly, the S allele (both S/S and S/L genotype) was also associated with neuroticism, reflected by increased trait anxiety (Lesch et al. 1996), specifically within the context of cluster C personality disorders (Jacob et al. 2004). In addition, a study reported an association between the S allele and different aspects of neuroticism, such as trait anxiety, affective temperament (depressive, cyclothymic, irritable and anxious), guilt, hostility and somatisation in a non-clinical Hungarian sample (Gonda et al. 2009). Still, an association between the 5-HTTLPR polymorphism and neuroticism was not always confirmed (Ball et al. 1997). A meta-analysis of 26 studies (7657 subjects) did not reveal a significant association between the 5-HTTLPR genotype and trait anxiety. However, a slight association was indicated, provided the fact that anxiety was assessed by a particular scale addressing the five-factor model of personality (Schinka et al. 2004).

The STin2 Polymorphism and Behaviour/Personality

Based on a post-mortem study of Croatia/Southern Slavic suicide victims, the lower activity STin2.10 allele was associated with suicidal behaviour (Jernej et al. 2004). Altogether though, a meta-analysis of ten studies (case and healthy control number were not reported) failed to support an association between the STin2 polymorphism and suicidal behaviour (Li and He 2007).

According to a study of children displaying aggressive behaviour, in several cases within the context of ADHD, oppositional defiant and conduct disorder, STin2.10 allele frequency was significantly lower compared with STin2.12 allele frequency. Still, this difference was not statistically significant when aggressive children were compared with a control population of healthy adults (Davidge et al. 2004).

The STin2 polymorphism was also studied in relation to different personality traits, measured by the Eysenck Personality Inventory and the TCI. Carriers of the 10-repeat allele scored lower in neuroticism and harm avoidance, while they scored higher in extraversion. Contrary to STin2.10 carriers, STin2.12 carriers scored higher in harm avoidance and lower in extraversion and novelty seeking (Kazantseva et al. 2008) (Table 9.3).

9.2.1.3 Serotonin Receptor Genes

The Serotonin Receptor 1A Gene (5-HTR1A)

The serotonin receptor family includes at least 14 different 5-HT receptors (Hoyer et al. 2002). The serotonin receptor 5-HTR1A is a protein regulating serotonin release by functioning both as a presynaptic autoreceptor in dorsal and medial raphe nuclei serotonergic neurons and a postsynaptic heteroreceptor in non-serotonergic neurons.

The 5-HTR1A receptor is encoded by a gene located at chromosome 5 (5q11.2–13) (Kobilka et al. 1987). There is a common functional SNP (rs6295) in 5-HTR1A gene promoter, C(-1019)G (Wu and Comings 1999). The G allele was associated with higher receptor expression, leading to increased negative feedback inhibition in raphe nuclei serotonergic neurons (mediated by 5-HTR1A autoreceptors) and thus decreased serotonergic activity (Lemonde et al. 2003).

In a study of a Hungarian population, G/G carriers displayed significantly higher impulsivity levels, compared with G/C and C/C carriers (Benko et al. 2010).

The G allele was also recognised as a risk factor for completed suicide in a population of French-Canadian origin (Lemonde et al. 2003) and for suicide attempt in a Polish study (Sawiniec et al. 2007). Still, the latter association was not supported in Ukrainian families of suicide attempters (probands and both parents) (Wasserman et al. 2006) and in a Mexican population of suicide attempters (Gonzalez-Castro et al. 2013b).

The first meta-analysis of four studies of the C(-1019)G polymorphism in relation to suicidal behaviour (three Caucasian and one Asian population; 957 suicidal cases versus 957 healthy controls) did not find any association between the G risk allele and suicidal behaviour (Angles et al. 2012). Lack of an association was confirmed in a subsequent meta-analysis of six studies (2022 suicidal cases versus 2135 healthy controls) (Clayden et al. 2012), as well as in a more recent meta-analysis of nine studies (seven Caucasian, one Asian and one Mexican population; 2366 suicidal cases versus 2943 healthy controls) (Gonzalez-Castro et al. 2013b).

Furthermore, the C(-1019)G polymorphism was investigated in relation with personality traits. The G allele was associated with different anxiety- and depression-related personality traits, such as neuroticism and harm avoidance, in a non-clinical German population (Strobel et al. 2003), though such an association was not always supported (Koller et al. 2006). Another study also failed to report an association between the C(-1019)G polymorphism and different personality traits in a German population of suicide attempters and healthy controls, as well as in an Italian population of patients diagnosed with a mood disorder (Serretti et al. 2009).

Lastly, a few studies of other SNPs, such as *Pro16Leu* (rs1800041, amino acid proline is substituted by amino acid leucine at codon 16), Gly272Asp (rs1800042, amino acid glycine is substituted by amino acid aspartic acid at codon 272) (Anguelova et al. 2003) as well as a C to T transition (rs878567) (Gonzalez-Castro et al. 2013b), did not reveal an association with suicidal behaviour.

The Serotonin Receptor 1B Gene (5-HTR1B)

The serotonin receptor 5-HTR1B is a protein functioning both as a presynaptic autoreceptor in serotonergic neurons, as well as a postsynaptic heteroreceptor in non-serotonergic neurons. The activation of 5-HTR1B autoreceptor modulates neuronal function by inhibiting serotonin release, preventing neuron's overstimulation. There is evidence that 5HTR1B heteroreceptors modulate offensive aggression (Olivier and van Oorschot 2005).

The 5-HTR1B receptor is coded for by a short intronless gene located at chromosome 6 (6q14.1). Several gene polymorphisms have been described, among which a G861C SNP (rs6296). Although G861C is a silent SNP, it is in LD with other functional polymorphisms. There is some evidence that G861C, or another allele in LD with G861C, affects receptor binding (Sanders et al. 2002). Based on a post-mortem study, homozygosity for the G allele was associated with higher receptor binding compared with the G/C heterozygous genotype (Huang et al. 1999).

Although a study found an association between the G allele and history of suicidal behaviour within the context of personality disorders (New et al. 2001), other studies of the G861C SNP did not prove 5-HTR1B a risk gene for manifestation of suicidal behaviour in a Japanese (Nishiguchi et al. 2001), a German (Rujescu et al. 2003c) and a Slavic/Croatian population (Stefulj et al. 2004b). Altogether, lack of an association between the G861C polymorphism and suicidal behaviour was confirmed by a meta-analysis of seven studies (789 suicidal cases versus 1247 healthy controls), in which results were not affected by study heterogeneity, age, gender or ethnicity (Kia-Keating et al. 2007), as well as by a meta-analysis of ten studies (2947 suicidal cases versus 4066 healthy controls) (Clayden et al. 2012).

A study of children displaying aggressive behaviour, in several cases within the context of ADHD, oppositional defiant and conduct disorder, reported a trend towards higher C allele frequency in these children compared with a control population of healthy adults (Davidge et al. 2004), an observation that requires though further verification. Increased C allele frequency was also observed in a Finnish cohort of antisocial alcoholics compared with non-antisocial alcoholics and healthy controls (Lappalainen et al. 1998). Contrary, a previous post-mortem study did not reveal an association between the G861C genotype and pathological aggression (Huang et al. 1999).

The Serotonin Receptor 2A Gene (5-HTR2A)

The serotonin receptor 5-HTR2A is a G *protein*-coupled receptor, regulated by many different interacting proteins and distributed in many different central nervous system areas. This receptor has been implicated in the manifestation of affective and cognitive disorders (Zhang and Stackman Jr. 2015).

The receptor is encoded by the 5-HTR2A gene located at chromosome 13 (13q14–q21). There is a silent SNP in exon 1 (rs6313) as a result of a T/C substitution at position 102. The polymorphism T102C is in almost complete LD with the promoter G-1438A SNP (rs6311), which is also non-functional. Still, a post-mortem study of suicide victims and healthy controls showed that the T102C and the G-1438A SNP affected serotonin binding in both study groups. Specifically, the haplotype 102T/-1438A was associated with increased serotonin binding compared with the haplotype 102C/-1438G (Turecki et al. 1999). Other 5-HTR2A gene SNPs include rs6314 (within gene's coding region; amino acid histidine is substituted by amino acid tyrosine at codon 452, His452Tyr), rs7322347 (an intron 2 SNP, T/A), rs643627 (A/G) and rs594242 (C/G).

Studies of the association between the T102C polymorphism and suicidal behaviour were rather controversial, indicating the C as the risk allele (Zhang et al. 1997), T as the risk allele (Gonzalez-Castro et al. 2013a, b) or no association at all (Bondy et al. 2000b). A meta-analysis of nine studies (596 suicidal attempters and completers versus 1003 healthy controls) did not find any association between the T102C polymorphism and suicidal behaviour (Anguelova et al. 2003). A subsequent meta-analysis of 25 studies (1954 suicidal cases versus 2860 healthy controls) investigated the association between the T102C polymorphism and suicidal behaviour performing several subanalyses, e.g. Europeans only, Asians only, suicidal ideation versus healthy controls, suicide attempt versus healthy controls, violent versus non-violent, etc. Overall, study findings did not support an association between the T102C polymorphism and suicidal behaviour (Li et al. 2006). Lack of an association was confirmed by another meta-analysis of 18 studies (3759 suicidal cases versus 5692 healthy controls) (Clayden et al. 2012).

An even more recent meta-analysis of 23 studies, including 2566 suicide attempters and completers versus 3989 healthy controls (13 Caucasian, 6 Asian and 4 populations of other ethnic origin), as well as 612 suicidal cases and 1129 healthy controls included in a previous meta-analysis (Li et al. 2006), reported no association between the T102C polymorphism and suicidal behaviour, after using allelic models for both C and T allele. No associations were found even when Caucasian and Asian populations were analysed separately, as well as when only schizophrenia patients were considered (Gonzalez-Castro et al. 2013a). A latest meta-analysis of 13 studies [1729 suicide attempters diagnosed with a psychiatric disorder (710 Asians/1019 European-Americans); 1794 non-suicide attempters diagnosed with a psychiatric disorder (759 Asians/920 European-Americans); 2398 healthy controls (906 Asians/1492 European-Americans)] conducted two separate analyses. In the first one, suicide attempters were compared with healthy controls. Results indicated that the C/C genotype was not associated with suicide attempt, even when data was analysed separately based on ethnicity. In the second analysis, suicide attempters were compared with non-attempters diagnosed with the same psychiatric disorders. Again, there was no association between the T102C polymorphism and suicide attempt, even when different ethnic groups were analysed separately. Still, when data was analysed separately based on psychiatric diagnosis, the C/C genotype was proven a risk factor for suicide attempt in schizophrenia patients. On the other hand, this association was confirmed neither in bipolar patients nor in patients with alcohol dependence (Wang et al. 2015).

Although a few data on other polymorphisms, such as His452Tyr and G-1438A, did not reveal an association between 5-HTR2A gene polymorphisms and suicidal behaviour (Anguelova et al. 2003), a meta-analysis of seven studies (six Asian populations) conducting genotypic analysis with the G-1438A SNP allele A combined

[(AA+AG)/GG] reported a significant association with suicidal behaviour (Li et al. 2006). The G-1438A polymorphism was also not associated with different personality traits, such as novelty seeking and harm avoidance. Still, an association was found between the A-1438A genotype and impulsive behaviour assessed by a behavioural task (go/no-go task) in healthy Japanese study participants (Nomura et al. 2006), as well as contradictory evidence for an association between the A-1438A genotype and low levels of impulsive behaviour in a German population of patients with alcohol dependence. The latter association was independent of the presence of comorbidity with a personality disorder (Preuss et al. 2001).

Another study of a non-clinical Caucasian Hungarian population investigated the relation between a set of different 5-HTR2A SNPs and aggressive traits. An association was reported between the intronic SNP T/A (rs7322347) and aggressive traits, in such T/T genotype carriers displayed more aggressive traits compared with allele A carriers (Banlaki et al. 2015).

Lastly, a study of German suicide attempters, diagnosed with an affective, schizophrenia spectrum or borderline personality disorder, searched for an association between different 5-HTR2A gene polymorphisms (rs643627, rs594242 and rs6311) and inwardly/outwardly state and trait anger, as well as aggressive behaviour. Results showed that the A-C-T haplotype (polymorphism/allele, rs643627/A, rs594242/C and rs6311/T), the C-T haplotype (polymorphism/allele, rs594242/C and rs6311/T) and the T allele (rs6311) decreased risk for suicidal behaviour. Additionally, the rs6311 SNP was associated with trait anger, in such the risk genotype C/C was related to higher levels of trait anger, specifically anger turned inwards. Additionally, the C allele was associated with decreased aggressive behaviour inhibition (Giegling et al. 2006).

The Serotonin Receptor 2B Gene (5-HTR2B)

The serotonin receptor 5-HTR2B is a G protein-coupled receptor coded for by a gene located at chromosome 2 (2q36.3–q37.1) and expressed, among others, in the brain (Bonaventure et al. 2002). Its function is currently under investigation, although presynaptic 5-HTR2B receptors were shown to regulate serotonin reuptake (Launay et al. 2006) and were also implicated in mesolimbic dopaminergic activity modulation (Auclair et al. 2010).

There is a functional stop codon (C20T, Q20*), most probably limited to the Finnish population, causing RNA decay and 5-HTR2B expression blockage. In a study of a Finnish population of violent criminal offenders, Q20* carriers showed no cognitive deficits and committed crimes mediated by high impulsivity levels (Bevilacqua et al. 2010).

The Serotonin Receptor 2C Gene (5-HTR2C)

The serotonin receptor 5-HTR2C is again a G protein-coupled receptor, implicated among others in mood, anxiety and reproductive behaviour regulation. The receptor is coded for by a gene located at X chromosome (Xq23). There is a SNP (rs6318) in the gene's coding region, C68G, leading to an amino acid substitution (cysteine is substituted by serine at codon 23, Cys23Ser).

Although one study indicated an association between the serine variant and trait impulsiveness in males displaying repeatedly self-harming behaviour (Evans et al. 2000), another study of suicide completers belonging to two different ethnicities, German and Slavic, did not reveal any association between the C68G SNP and suicidal behaviour (Stefulj et al. 2004a). Absence of an association between the C68G polymorphism and suicidal behaviour was confirmed by a recent meta-analysis of seven studies (2297 suicidal cases versus 3431 healthy controls) (Clayden et al. 2012).

Lastly, a study of a German population of suicide attempters and healthy controls, as well as of Italian patients diagnosed with a mood disorder, did not reveal an association between the C68G polymorphism and different personality traits (Serretti et al. 2009) (Table 9.4).

9.2.1.4 Tryptophan Hydroxylase 1 Gene (TPH1)

The TPH1 Gene Polymorphisms

Tryptophan hydroxylase, TPH, is an enzyme that regulates serotonin availability by catalysing the rate-limiting step in serotonin biosynthesis. The isoform 1, TPH1, is coded for by TPH1 gene, located at chromosome 11 (11p15.1), and is expressed in a variety of tissues. Still, there are contradictory results regarding its expression in the brain (Zill et al. 2007; Gutknecht et al. 2009).

There is a SNP (rs1800532) in intron 7, A218C. The 218A allele is also referred to as "upper"/U allele and constitutes the minor allele, while the 218C is also referred to as "lower"/L allele. Originally, the A218C SNP was not shown to alter TPH1 amino acid sequence (Nielsen et al. 1997). Later, and based on a post-mortem study of suicide victims and healthy controls, 218A allele was considered the high activity allele, since it was associated with significantly higher TPH1 immunoreactivity. Still, the A218C polymorphism was not shown to affect only TPH1 production but 5-HTR2A receptor regulation as well, since the 218A allele was associated with decreased 5-HTR2A receptor density. (Ono et al. 2002).

Another SNP (rs1799913) in intron 7 is the A779C transversion. Accordingly, the allele 779A is referred to as U allele and was related to lower CSF 5-HIAA levels in healthy males (Jonsson et al. 1997), while the allele 779C is referred to as L allele. The A779C SNP, which was also not shown to alter TPH1 amino acid sequence (Nielsen et al. 1997), is in almost complete LD with the A218C SNP in Caucasian populations (alleles 218C and 779C are in phase).

The A218C Polymorphism and Behaviour/Personality

Altogether, studies of the A218C SNP in relation to suicidal behaviour led to great discrepancy both in Caucasian, as well as in Asian populations (Rujescu et al. 2003b). The 218A allele was associated with suicidal behaviour within the context of affective (Mann et al. 1997; Souery et al. 2001) and other psychiatric disorders (Abbar et al. 2001), although the latter finding was not confirmed in Caucasian suicide completers of French-Canadian origin (Turecki et al. 2001), in a family-based study of Israeli adolescent suicide attempters diagnosed with different psychiatric disorders (Zalsman et al. 2001) as well as in a German population of suicide attempters diagnosed with different psychiatric disorders (Rujescu et al. 2003b). On the contrary and based on a post-mortem study of Croatia/Southern Slavic suicide victims, it was the lower activity 218C allele that was associated with suicidal behaviour, especially in combination with the lower ten repeat allele of STin2 (SLC6A4 gene) polymorphism (Jernej et al. 2004).

A meta-analysis of seven studies of Caucasian populations only (898 suicidal cases versus 1179 healthy controls) revealed an association between the 218A allele and suicide-related behaviours (Rujescu et al. 2003b). A subsequent study restricted again to Caucasian populations meta-analysed data from seven studies (860 suicidal cases versus 1279 healthy controls), confirming the association between the 218A allele and suicidal behaviour (Bellivier et al. 2004). A more recent meta-analysis of 21 studies (4829 suicidal cases versus 7945 healthy controls) confirmed partly previous outcomes, since results indicated that the 218A allele was associated only with suicide attempt, and not with completed suicide (Clayden et al. 2012).

Furthermore, homozygosity for the 218A allele was associated with higher aggression and more intense tendency towards unprovoked anger in males (Manuck et al. 1999), as well as with higher levels of proactive aggression (Hennig et al. 2005). The 218A allele was also associated with trait anger, state anger and anger temperament in a German population (Rujescu et al. 2002).

The A779C Polymorphism and Behaviour/Personality

Although evidence was provided for an association between the 779C allele and non-impulsive suicide attempts (Nielsen et al. 1998), results regarding the implication of both A779C and A218C polymorphisms in the manifestation of suicidal behaviour were altogether rather controversial. A meta-analysis of 15 studies (1290 suicide attempters/completers versus 2295 healthy control subjects) investigating the A779C and/or the A218C polymorphism in relation to suicidal behaviour revealed no significant associations. The same meta-analysis searched for an association between the two TPH1 SNPs and suicidal behaviour only in patients with psychiatric diagnoses (9 studies; 625 suicide attempters versus 1475 non-attempters), confirming absence of any association between the SNPs and suicidal behaviour (Lalovic and Turecki 2002). Previous results were confirmed by a meta-analysis of 13 studies (1272 suicide attempters, 1727 non-suicide attempters, all participants were diagnosed with a psychiatric disorder), which did not find any association between the A779C/A218C polymorphisms and suicidal behaviour (Saetre et al. 2010).

Contradictory results were provided by a subsequent meta-analysis of 34 studies (3922 suicidal cases versus 6700 healthy controls), which reported a strong overall association, regardless of alleles, between the TPH1 A779C/A218C polymorphisms and suicidal behaviour. The same meta-analysis revealed though no association between the promoter A-6526G SNP (rs4537731) and suicidal behaviour (Li and He 2006). Lastly, a more recent meta-analysis of eight studies (1512 suicidal cases versus 3408 healthy controls) found again no association between the A779C polymorphism and suicidal behaviour (Clayden et al. 2012).

Homozygosity for the 779C allele was associated with low CSF 5-HIAA levels in a Finnish population of impulsive alcoholic violent offenders (Nielsen et al. 1994), as well as with impulsive aggression in males diagnosed with personality disorders (New et al. 1998). Lastly, the 779A allele was associated with trait anger, state anger and anger temperament in a German population (Rujescu et al. 2002) (Table 9.5).

9.2.1.5 Tryptophan Hydroxylase 2 Gene (TPH2)

The tryptophan hydroxylase isoform 2, TPH2, is expressed in brain serotonergic neurons. The enzyme is encoded by a gene located at chromosome 12 (12q21.1) (Zill et al. 2007). There is a TPH2 gene SNP, C1473G, which has been studied only in animal models. Homozygosity for the G allele in mouse strains was associated with reduced TPH2 activity and lower 5-HT levels (Zhang et al. 2004b). On the contrary, mice homozygous for the C allele showed higher TPH2 activity, which was associated in turn with higher levels of inter-male aggression (Kulikov et al. 2005).

Although TPH2 SNPs have been studied in relation with psychiatric disorders (Walitza et al. 2005; Zhang et al. 2005), further investigation is required regarding their association with aggressive behaviour in humans (Zhang et al. 2006).

9.2.2 Dopaminergic System

The dopaminergic system is involved, among others, in motor control, motivation, emotional stability, reward and cognition. Thus, impulsive, compulsive or addictive behaviours could be related to dopaminergic gene dysregulation.

There is evidence for the implication of dopamine in the manifestation of aggression (de Almeida et al. 2005), since antipsychotic drugs display anti-aggressive effects (Groleger 2007). It has been suggested that dopamine is implicated in the initiation of aggressive behaviour, whereas serotonin in its termination (Olivier and van Oorschot 2005). On the contrary, novelty and sensation seeking were associated with lower dopamine system activity (Cloninger 1986). The Table 9.6 presents meta-analyses of studies investigating dopaminergic genes in relation with behaviour and personality.

9.2.2.1 Dopamine Receptor Genes

The Dopamine 2 Receptor Gene (DRD2)

Dopamine's function is mediated by dopamine receptors, among which the dopamine 2 receptor, D2R, also known as the "antipsychotic dopamine receptor", since it constitutes the main target receptor of all antipsychotic drugs.

The D2R receptor is encoded by the DRD2 gene, located at chromosome 11 (11q22–q23). There is a DRD2 gene TaqI restriction fragment length polymorphism (RFLP), giving rise to alleles A1 and A2 (Grandy et al. 1989). Its functional significance has not been fully elucidated yet. Different studies reported an association

Table 9.6 Dopaminergic gene polymorphisms and personality/behaviour

Gene (chromosome location)	Encoded protein	Polymorphisms	Alleles	Behavioural phenotype	Meta-analyses
DRD4 (11p15.5)	Dopamine 4 receptor	48 bp VNTR	S: short alleles, 2R-5R L: long alleles, 6R-10R	Novelty seeking	Meta-analysis of 20 studies (3907 individuals) Results: no association (Kluger et al. 2002)
					1. Meta-analysis of 14 studies (21 separate samples; 2720 individuals) Results: no association between 7R allele and novelty seeking 2. Meta-analysis of ten studies (12 separate samples; 1719 individuals) grouping all long alleles together Results: association between long alleles and novelty seeking (Schinka et al. 2002)
				Novelty seeking Extraversion Trait impulsivity	Meta-analysis of 36 independent non-clinical adult samples (around 5600 individuals); included studies had grouped 48 bp VNTR into short and long alleles Results: no association between long alleles and personality traits, even when European samples were analysed separately (Munafo et al. 2008)
		rs1800955: C-521T SNP	C T	Novelty seeking	Meta-analysis of four studies (677 individuals) Results: association between C/C genotype and novelty seeking (Schinka et al. 2002)
				Novelty seeking Extraversion Trait impulsivity	Meta-analysis of 11 independent non-clinical adult samples (around 1600 individuals) Results: association between C/T and T/T genotype and novelty seeking, as well as trait impulsivity (T allele carriers displayed lower novelty seeking and trait impulsivity levels). No association with extraversion (Munafo et al. 2008)

(continued)

Table 9.6 (continued)

Gene (chromosome location)	Encoded protein	Polymorphisms	Alleles	Behavioural phenotype	Meta-analyses
COMT (22q11.21)	Catechol-o-methyltransferase	rs4680: G472A SNP (Val158Met)	COMT-L: corresponds to methionine and is associated with low enzyme activity COMT-H: corresponds to valine and is associated with high enzyme activity	Suicidal behaviour	Meta-analysis of 6 studies (519 suicide attempters and completers versus 933 healthy controls) Results: association between COMT-L allele and suicidal behaviour (Kia-Keating et al. 2007)
					Meta-analysis of ten studies (1324 suicidal cases versus 1415 healthy controls) Results: no association (Calati et al. 2011)
					Meta-analysis of 12 studies (2723 suicidal cases versus 1886 healthy controls) Results: no association, even when Caucasians and suicide attempters were analysed separately (Tovilla-Zarate et al. 2011)
					Meta-analysis of nine studies (3226 suicidal cases versus 3055 healthy controls) Results: no association (Clayden et al. 2012)
				Aggressive/violent behaviour	Meta-analysis of 15 studies (2370 schizophrenia patients) Results: presence of at least one COMT-L allele increased risk for manifestation of violent behaviour in male schizophrenia patients by around 50% (Singh et al. 2012)

between the A1 allele and reduced D2 receptor activity in different brain areas (Noble et al. 1997), an association between the A1 allele and decreased receptor density (Pohjalainen et al. 1998) as well as an association between the A1 allele and increased dopamine synthesis, perhaps due to decreased D2R receptor expression (Laakso et al. 2005).

A study of more than 2500 adolescents and young adults showed that contrary to the homozygous A1/A1 and A2/A2 genotype, the heterozygous A1/A2 genotype was associated with serious and violent delinquency only in males (Guo et al. 2007). These findings were discussed within the context of "heterosis". Based on this phenomenon, it is heterozygous rather than homozygous individuals manifesting a trait to a greater or lesser extend (Comings and MacMurray 2000). Elsewhere, the A1 genotype (A1/A1 homozygosity or A1/A2 heterozygosity) was related to aggressive-violent behaviour (Chen et al. 2005).

The Dopamine 4 Receptor Gene (DRD4)

The dopamine receptor 4, D4R, is a D2R-like receptor coded for by the DRD4 gene at chromosome 11 (11p15.5). There is a VNTR in exon III, namely, a 48 bp sequence repeated two (2R) to ten (10R) times. Different alleles vary in regard to number of repeats, nucleotide sequence and variant order, leading to different receptor products containing 32–160 amino acids at the corresponding position. Allele frequencies vary greatly between different populations. It was shown that the 4R allele is the most commonly found, together with the 7R and the 2R allele, while alleles 3R, 5R, 6R and 8R are rare. Alleles up to 5R are considered "short", while the rest, including the 7R allele, "long" (Lichter et al. 1993; Chang et al. 1996; Ding et al. 2002). The 7R allele was associated with decreased in vitro gene expression and receptor binding compared with short alleles (Asghari et al. 1994; Asghari et al. 1995).

There is another DRD4 gene polymorphism, namely, a C-521T SNP (rs1800955) in the promoter region, which is in LD with 48 bp VNTR. This SNP was shown to affect gene's transcription, since the T allele was associated with up to 40% lower transcription levels compared with the C allele (Ronai et al. 2001).

Lastly, there is a tandem 120 bp duplication located 1.2 kb upstream from the initiation codon, giving rise to the long (L) and the short (S) allele. It was shown that the S allele was associated with increased transcriptional activity compared with the L allele (D'Souza et al. 2004).

The DRD4 48 bp VNTR and Personality

A study of the DRD4 48 bp VNTR and its association with novelty seeking in a non-clinical Israeli population reported that 7R allele carriers scored significantly higher in novelty seeking (Ebstein et al. 1996).

Another study supported an association between the 48 bp VNTR and novelty seeking in a sample of white Americans, mainly male (95%, mostly male siblings). Instead of examining the 7R allele as previously described, participants were subgrouped into long or short allele carriers, and it was shown that long allele carriers displayed higher levels of trait novelty seeking (Benjamin et al. 1996). The same association was replicated in another healthy female population, recruited from Japan (Ono et al. 1997), as well as in other Asian (Tomitaka et al. 1999; Lee et al. 2003a) and German (Strobel et al. 1999) populations.

Still, there were studies that were not able to confirm aforementioned results (Baron 1998; Lusher et al. 2001). Contradictory findings were also reported, supporting an association between the 5R allele and high novelty seeking in healthy Japanese individuals (Tsuchimine et al. 2009), an association between the short 2R and 5R alleles and high novelty seeking in a Finnish population (Ekelund et al. 1999) as well as no associations at all in a non-clinical Korean (Kim et al. 2006b) and different Japanese populations (Mitsuyasu et al. 2001; Tochigi et al. 2006).

A former meta-analysis of 20 studies (3907 individuals) applied two different meta-analytic methods. The results did not confirm an association between the 48 bp VNTR and novelty seeking (Kluger et al. 2002). Another meta-analysis published shortly after the previous one reviewed data from 14 studies (21 separate samples, 2720 individuals) and revealed no association between the long 7R allele and novelty seeking. The same study meta-analysed data from ten studies (12 separate samples, 1719 individuals) grouping all long alleles together. In this case, results revealed a significant association between long repeat alleles and high novelty seeking (Schinka et al. 2002). Lastly, a more recent meta-analysis included data from 36 independent non-clinical adult samples (around 5600 individuals). The studies included in this meta-analysis had grouped 48 bp VNTR into short and long alleles and investigated them in relation with novelty seeking, extraversion and trait impulsivity. The results showed absence of an association between long alleles and aforementioned traits, even when samples of European origin were analysed separately. Furthermore, this analysis revealed significant heterogeneity between employed studies (Munafo et al. 2008).

In another study of a mixed population consisting of young men recruited from Harvard University, the 7R allele was proven a significant predictor of sensation seeking, including thrill and adventure seeking (Campbell et al. 2010). Similarly, in a Russian study, 7R allele carriers had higher thrill seeking elements, delinquency and short temper. Still, this was observed only in males, while when social parameters (parental monitoring of youths, exposure to violence) were taken into account, the interaction between thrill seeking and the 7R allele, as well as the gender effect, were no longer significant (Dmitrieva et al. 2011).

Lastly, a study of a Japanese population revealed an association between the 2R-4R short alleles and higher neuroticism levels, including anxiety, depression and vulnerability (Tochigi et al. 2006).

The DRD4 C-521T SNP/120 bp Duplication and Personality

A meta-analysis of four studies (677 individuals) confirmed an association between the C-521T SNP C/C genotype and high novelty seeking (Schinka et al. 2002), a finding that was confirmed by a recent meta-analysis of 11 independent non-clinical adult samples (around 1600 individuals), which reported a significant association between the C-521T SNP and novelty seeking, as well as impulsivity. Namely, T allele carriers (C/T or T/T genotype) displayed lower levels of aforementioned traits, while the C-521T SNP accounted for 2% of the phenotypic variance. Still, the latter polymorphism was not related to extraversion (Munafo et al. 2008).

Lastly, a study of the tandem 120 bp duplication in regard to novelty seeking was performed in four different clinical samples, one diagnosed with bipolar disorder, one with alcohol dependence and two with depression. Combined data revealed an association with novelty seeking. Specifically, individuals genotyped as S/S scored higher in novelty seeking, including impulsivity, extravagance and disorderliness. Still, due to the fact that the S allele is rare, one could assume that it would not contribute greatly to population variance (Rogers et al. 2004) (Table 9.6).

9.2.2.2 Dopamine Transporter Gene (SLC6A3)

The SLC6A3 Gene Polymorphism

The dopamine transporter 1, DAT1, is a protein regulating dopamine synaptic levels by limiting dopamine receptor activation and facilitating neuronal dopamine reuptake.

This protein is coded for by the gene "Solute Carrier Family 6, member 3" (SLC6A3), located at chromosome 5 (5p15.3). Several DAT1 gene polymorphisms have been described. Among these, there is a 40 bp VNTR most commonly repeated nine (DAT1*9R) and ten (DAT1*10R) times. Less abundant alleles include 3, 7 and 11 repeats (Vandenbergh et al. 1992). There is evidence that this polymorphism may be functional, altering gene expression (Fuke et al. 2001). Although it was found that healthy DAT1*10R/10R genotype carriers showed lower striatal transporter binding (Jacobsen et al. 2000) and DAT1*9R allele carriers increased striatal dopamine transporter availability (van Dyck et al. 2005), contradictory findings were also reported (Heinz et al. 2000).

The DAT1 40 bp VNTR and Behaviour

Contrary to the DAT1*9R/9R genotype, the DAT1*10R/10R and the DAT1*10R/9R genotypes were associated with serious and violent delinquency in males (Guo et al. 2007), as well as with aggressive-violent behaviour (Chen et al. 2005). Still, not all studies were able to confirm an association between the DAT1 genotype and violent behaviour (Reif et al. 2007).

Lastly, DAT1 gene, together with other dopamine-related genes, was implicated in the aetiology of ADHD, which is characterised by impulsive behaviour (Khan and Faraone 2006).

9.2.2.3 Dopamine Beta-Hydroxylase Gene (DBH)

Dopamine beta-hydroxylase, DBH, is an enzyme involved in norepinephrine synthesis by catalysing dopamine hydroxylation. Previous studies showed that low plasma norepinephrine levels were associated with antisocial behaviour (Rogeness et al. 1982; Gabel et al. 1995).

The DBH gene is located at chromosome 9 (9q34.2). There is a C1021T SNP, accounting for about 35–52% of the variance of DBH plasma levels. A German study reported an association between the C1021T SNP and personality traits.

Patients diagnosed with more than two personality disorders and genotyped as T/T displayed higher neuroticism levels, as well as higher levels of neuroticism's facet "anger hostility". The same individuals displayed higher novelty seeking levels, as well as higher levels of novelty seeking's facets "impulsiveness" and "disorderliness" (Hess et al. 2009).

9.2.2.4 Catechol-O-Methyltransferase Gene (COMT)

The COMT Gene Polymorphism

Catechol-o-methyltransferase (COMT) is an enzyme catalysing catecholamine methylation. This o-methylation constitutes a major degrading pathway of catecholamine neurotransmitters dopamine, epinephrine and norepinephrine. Brain COMT activity regulates active dopamine and norepinephrine amounts.

The gene coding for COMT is located at chromosome 22 (22q11.21). There is a functional SNP in exon 4 (rs4680, G472A transition), resulting in an amino acid change, i.e. valine is substituted by methionine at codon 158 of membrane-bound COMT (Met158Val) and at codon 108 of soluble COMT (Met108Val). The G472A polymorphism was shown to decrease enzyme thermostability and therefore activity three- to fourfold. Specifically, genetically polymorphic COMT enzyme activity may be low (COMT-LL, corresponding to Met/Met genotype), intermediate (COMT-LH, corresponding to Met/Val genotype) or high (COMT-HH, corresponding to Val/Val genotype). Altogether, carriers of the more active form of the enzyme display less dopamine for neurotransmission (Lachman et al. 1996; Stahl 2003).

The G472A Polymorphism and Behaviour/Personality

The COMT-LL genotype was associated with violent suicide attempts (e.g. hanging, shooting, etc.) but not with non-violent (e.g. drug overdose, gas suffocation, etc.) in a German psychiatric population, independent of diagnosis, compared with healthy controls. Additionally, the latter study found a correlation between the COMT genotype and state anger assessed by the State-Trait Anger Expression Inventory. Specifically, the COMT-HH genotype was related to inward anger expression (higher scores in "Anger-In") while the COMT-LL genotype with outward anger expression (higher scores in "Anger-Out") (Rujescu et al. 2003a). Still, an association between the G472A polymorphism and suicidal behaviour was not confirmed in another mixed (mainly Caucasian) (Russ et al. 2000) and in a Mexican population (Tovilla-Zarate et al. 2011).

Altogether, a meta-analysis of six studies (519 suicide attempters and completers versus 933 healthy controls) revealed a significant association between the COMT-L allele and suicidal behaviour, which was not affected by age or ethnicity. Several studies reported though a gender effect on the association between the G472A polymorphism and suicidal behaviour, reporting COMT-L allele to be more abundant in male compared to female suicidal cases. In accordance, aforementioned meta-analytic results were shown to be affected by the proportion of females to males in both suicidal cases and healthy controls, a gender effect that requires further investigation (Kia-Keating et al. 2007). On the contrary, a subsequent meta-analysis of ten studies (1324 suicidal cases versus 1415 healthy controls) considering both allele frequency and genotype did not report an association between the G472A polymorphism and suicidal behaviour. The meta-analysis of three among ten studies, which compared suicide attempters versus non-suicide attempters diagnosed with the same psychiatric disorders (mood disorders and schizophrenia), showed again lack of any association (Calati et al. 2011). A following meta-analysis of 12 studies (2723 suicidal cases versus 1886 healthy controls) did not find any association between the risk COMT-L allele and suicidal behaviour, even when Caucasians and suicide attempters were analysed separately (Tovilla-Zarate et al. 2011). Lastly, a meta-analysis of nine studies (3226 suicidal cases versus 3055 healthy controls) confirmed lack of an association between the G472A polymorphism and suicidal behaviour (Clayden et al. 2012).

In regard to aggressive behaviour, one study of unselected women showed that COMT-LL carriers displayed lower levels of physical aggression compared with COMT-HH carriers (Kulikova et al. 2008). Altogether, although there was discrepancy in study findings, the G472A polymorphism (in particular COMT-HH genotype) may be associated with aggressive traits [for review see Calati et al. (2011)].

There was one meta-analysis that included 15 studies of the G472A polymorphism in relation with violent behaviour within the context of schizophrenia (2370 schizophrenia patients). Based on results, the presence of at least one COMT-L allele increased risk for violent behaviour in male schizophrenia patients by around 50%. No association was found in females or when data from male and female patients was pooled together (Singh et al. 2012).

Other studies focused on the relation between the COMT genotype and personality traits. There were reports of an association between the COMT-HH genotype and high levels of extraversion, the COMT-HH genotype and the "exploratory excitement" component of novelty seeking in healthy individuals (Reuter and Hennig 2005) as well as the COMT-HH genotype and novelty seeking in a non-clinical female Chinese population (Tsai et al. 2004b). In accordance with previous findings, a study of a non-clinical German population revealed an association between the G472A polymorphism and sensation seeking in females. Specifically, female subjects genotyped as COMT-HH showed higher levels of sensation seeking, including disinhibition, boredom susceptibility and thrill and adventure seeking, compared with subjects genotyped as COMT-LH and COMT-LL (Lang et al. 2007).

Since harm avoidance was correlated inversely with novelty and sensation seeking (McCourt et al. 1993), aforementioned results were not supported by a study of a Korean population, reporting a different association between the COMT genotype and harm avoidance again only in females. Specifically, females carrying the COMT-LL genotype showed the lowest harm avoidance levels, whereas the COMT-LH genotype was associated with intermediate and the COMT-HH with higher harm avoidance levels (Kim et al. 2006c). Lastly, there was some evidence for an association between the COMT-LL genotype and neuroticism in females, which was drawn though from a study employing a selected population (high or low scorers) (Eley et al. 2003).

Altogether, the COMT G472A polymorphism was shown to exhibit a sexually dimorphic effect, relating to personality traits only in females. A plausible explanation could be the correlation between the COMT-HH genotype and low oestrogen levels, which were in turn associated with lower thrill and adventure seeking levels in healthy females (Balada et al. 1993). Still, the latter observation is again not in accordance with previously presented studies [for review of the G472A polymorphism studies in relation to personality traits, see Calati et al. (2011)] (Table 9.6).

9.2.3 Genes Involved in Sexual Behaviour

9.2.3.1 Arginine Vasopressin Receptor 1A Gene (AVPR1A)

Arginine vasopressin (AVP), also called "antidiuretic hormone" (ADH), is a hormone regulating body water retain. Recent evidence suggested though that AVP may be also implicated in social, sexual and reproductive behaviour (Insel 2010).

The arginine vasopressin receptor 1 gene, AVPR1A, is located at chromosome 12 (12q14.2), and its polymorphisms have been studied in relation with reproductive behavioural motifs in other mammals (Young 2002). One study extended research of AVPR1A polymorphisms in humans. A large population consisting of 2085 males and females was genotyped at two different gene loci, c.-5518 AVPR1A (TC)x(TG)y and c.-2481 AVPR1A (AGAT)7_16. Alleles were grouped as short, medium and long for each polymorphism, giving rise to six possible genotypes: short/short, short/medium, short/long, medium/medium, medium/long and long/long. It was observed that males genotyped as long/long AVPR1A (TC)x(TG)y and females genotyped as long/long AVPR1A (AGAT)7_16 showed an increased probability of beginning sexual intercourse at an earlier age, i.e. before the age of 15 (Prichard et al. 2007b).

9.2.3.2 Oxytocin Receptor Gene (OXTR)

Oxytocin is a protein involved in attachment processes, such as social, familiar and maternal bonding, as well as in sexual reproduction. Its function is mediated by oxytocin receptor, OXTR, encoded by a gene located at chromosome 3 (3p25.3).

A large population consisting of 2085 males and females was genotyped at gene locus 1170*712 OXTR (CA)10_15. Alleles were grouped as short and long, giving rise to three genotypes: short/short, short/long and long/long. Based on results, it was more probable for females genotyped as long/long not to use oral contraceptives and to have children at a younger age (Prichard et al. 2007b).

9.2.4 Other Genes

9.2.4.1 Nitric Oxide Synthase Gene (NOS1)

Nitric oxide, NO, is an abundant neurotransmitter in emotion-regulating brain areas. The neuronal nitric oxide synthase, NOS-I, is coded for by NOS1 gene, located at chromosome 12 (12q24.3). The NOS1 is a complex gene, containing multiple protein-coding exons, as well as a variable region with multiple first exons (Zhang et al. 2004a).

There is a highly polymorphic CA dinucleotide repeat (180–210 repeats) within the promoter region (exon 1f), termed NOS1 Ex1f VNTR. Alleles containing 180–196 repeats were defined as short/S and were associated with decreased gene expression, while alleles containing 198–210 repeats were defined as long/L and were associated with maximal gene expression. The NOS1 Ex1f VNTR polymorphism was also shown to dysregulate many other different genes (Reif et al. 2009).

A study of more than 3200 probands (mixed population consisting of healthy controls, patients diagnosed with personality disorder or ADHD, suicide attempters and criminal offenders) reported that the S alleles were associated with psychiatric disorders manifesting impulsive behaviour as a common phenotype, such as ADHD and cluster B personality disorders. Additionally, S alleles were associated with suicidal and aggressive/criminal behaviour, phenotypes relating to impulsivity (Reif et al. 2009).

Still, the implication of NOS1 Ex1f VNTR in the manifestation of impulsiveness warrants further research. The same, above-mentioned, research group assessed impulsiveness and empathy in a male population consisting of criminal offenders with a history of a psychiatric disorder but without acute psychopathology. Previous results were not fully replicated, while there were also contradictory outcomes, since it was heterozygous S/L individuals displaying the highest impulsivity levels, a phenomenon previously described as “heterosis”. On the other hand, homozygous S/S carriers displayed the lowest impulsiveness levels and the highest empathy levels. Conflicting results were attributed to behavioural phenotypes and measures being related but not identical, as well as to employment of a different study population, in the latter case with high psychiatric comorbidity (Retz et al. 2010). Another explanation for outcome diversity could be the fact that environmental factors may moderate the effect of gene polymorphisms on behavioural phenotype. In an effort to clear the issue, the same research group conducted a longitudinal study of children (mean age 15 years) followed up as adults (mean age 18 years). Based on study outcomes, a hypothesis was formed. Short alleles were considered risk alleles for trait adaptive impulsivity, i.e. fast decision-making and excitement seeking, while the S/S risk genotype was associated with behavioural measures of impulsivity. These findings applied for male participants. On the other hand, maladaptive impulsivity, i.e. disinhibition and thoughtlessness, was associated again with the S/S risk genotype, though only in participants with a history of stressful life events, specifically increased perceived maternal rejection and beliefs that parents showed lack of love, appreciation and care towards them. Altogether, environmental stressful events could mediate the phenotypic outcome of the risk S/S genotype, turning adaptive impulsivity into dysfunctional (Reif et al. 2011).

9.2.4.2 Androgen Receptor Gene (AR)

The androgen receptors mediate testosterone’s (i.e. hormone implicated in the development of primary male sexual characteristics) and dihydrotestosterone’s (i.e. hormone regulating secondary male characteristics) function. The androgen receptor gene, AR, is located at X chromosome (Xq12) and presents with a triallelic polymorphism [GCA locus or AR_(CAG)n], giving rise to short, medium and long alleles. Caucasian males homozygous for the medium allele showed more severe antisocial traits. However, it should be noted that this association was rather weak (Prichard et al. 2007a).

9.2.4.3 Nuclear Receptor 4A2 Gene (NR4A2)

Nuclear receptor subfamily four group A member 2, NR4A2, is a protein possibly functioning as a transcription factor involved in dopamine neuron development (Sacchetti et al. 2001). It is encoded by a gene located at chromosome 2 (2q24.1). There was some evidence for an association between the long/long NR4A2 (AC) genotype and antisocial traits in females, a finding that requires though further verification (Prichard et al. 2007a).

9.2.4.4 Transcription Factor AP-2 Beta Gene (TFAP2B)

Transcription factor AP-2 beta, TFAP2B, is a protein acting both as a transcriptional activator and repressor, mediating monoaminergic neuron development and regulating gene expression. The corresponding gene is located at chromosome 6 (6p12.3). There are two polymorphisms in strong LD, TFAP2B (AACA) and TFAP2B (TC) (Prichard and Easteal 2006).

The TFAP2B (CAAA) polymorphism, an intron 2 tetranucleotide repeat (CAAA), repeated four or five times, was associated with different personality traits, such as somatic anxiety and indirect aggression (Damberg et al. 2000). In another study, the short/long genotype of TFAP2B (TC locus) was associated with antisocial traits in females (Prichard et al. 2007a).

9.2.4.5 FK506 Binding Protein 5 Gene (FKBP5)

FK506 binding protein 5, FKBP5, is a heat shock protein 90 co-chaperone, regulating the activity of glucocorticoid receptors and as a result the hypothalamic-pituitary-adrenal axis. The corresponding gene, located at chromosome 6 (6p21.31), has been studied in regard to dysregulated stress response in affective and anxiety disorders (Gillespie et al. 2009).

There are four FKBP5 gene SNPs (rs3800373, rs9296158, rs1360780 and rs9470080) in strong LD, giving rise to six haplotypes. Among these, haplotypes H1 and H2 are considered functional. Three possible diplotypes may be formed, derived from the combination of these two functional haplotypes, H1/H1, H1/H2 and H2/H2.

Study of the aforementioned FKBP5 gene SNPs in a population of substance-dependent African-Americans revealed an association between H1/H1 diplotype and suicidal behaviour, though only in participants with a history of childhood trauma (Roy et al. 2010), while another study reported that the less common H2/H2 diplotype was associated with increased risk for aggressive-violent behaviour in male prisoners with a background of physical abuse (Bevilacqua et al. 2012).

9.2.4.6 Brain-Derived Neurotrophic Factor Gene (BDNF)

Brain-derived neurotrophic factor, BDNF, is a member of the neurotrophin superfamily, a nervous system growth factor implicated in neuronal differentiation, growth, survival and death, affecting multiple neurotransmitter systems, among which the serotonergic and the dopaminergic. It is encoded by a gene located at chromosome 11 (11p14.1). There is a functional SNP (rs6265), G196A, resulting in an amino acid substitution (valine is substituted by methionine at codon 66, Val66Met). The more common G allele codes for valine, while the A allele codes for methionine. An in vitro study indicated that the G196A polymorphism was functional, since the A/A genotype was associated with decreased BDNF neuronal secretion (Egan et al. 2003).

The first meta-analysis of 12 studies (1202 suicidal patients diagnosed with psychiatric disorders versus 1699 non-suicidal patients diagnosed with the same psychiatric disorders and 451 healthy controls) investigating the relation between the G196A polymorphism and suicidal behaviour showed that the low-functioning A allele constituted a risk allele. The association between the A allele and suicidal behaviour was more significant in Asian populations (Chinese, Japanese, Korean) as well as when suicide attempters were compared with non-suicide attempters diagnosed with the same psychiatric disorders (eight studies) (Zai et al. 2012). Contrary to previous findings, a subsequent meta-analysis of seven studies (1700 suicidal cases versus 2584 healthy controls) did not reveal a significant association between the G196A polymorphism and suicidal behaviour (Clayden et al. 2012).

A study of a non-clinical German population focused on the association between the G196A polymorphism and different personality traits. Based on results, the BDNF genotype explained 1.9% of the variance of trait anxiety. Specifically, trait anxiety was significantly higher in G/G carriers (Lang et al. 2005).

In another study, the G allele was associated with increased neuroticism levels, explaining 4% of the genetic variance. Specifically, and in regard to six neuroticism facets, the association was confirmed between the G allele and increased anxiety, depression, self-consciousness and vulnerability (Sen et al. 2003). Similarly, G/G genotype carriers within a German population showed higher neuroticism levels compared with G/A and A/A carriers. It should be noted though that the latter findings failed to reach statistical significance (Lang et al. 2005). Another study of a healthy Japanese population reported an association between the A/A genotype and extraversion, though only in females (Itoh et al. 2004), while a recent genome-wide association study confirmed the association between the G196A polymorphism and extraversion (Terracciano et al. 2010).

On the other hand, a study of a healthy female Chinese population failed to confirm any association between the G196A polymorphism and different personality traits (novelty seeking, harm avoidance, reward dependence and persistence) (Tsai et al. 2004a).

Altogether, a meta-analysis of four studies (non-clinical population consisting of 607 individuals) investigated the relation between the G196A polymorphism and harm avoidance. Results showed that the A/A genotype was associated with a trend towards higher harm avoidance. The same meta-analysis investigated the association between the G196A polymorphism and neuroticism pooling five studies (non-clinical population consisting of 1633 individuals). According to the results, the G/G genotype was associated with higher neuroticism levels (Frustaci et al. 2008).

Lastly, a study of a non-clinical Korean population investigated both COMT G472A polymorphism and BDNF G196A polymorphism in relation to sensation seeking and reported no significant associations when the two polymorphisms were considered separately. On the contrary, focus on the combined effect of COMT and BDNF polymorphisms on sensation seeking revealed a significant association with only one sensation seeking facet. Specifically, among female homozygous or heterozygous COMT-L carriers, female BDNF G/G carriers displayed higher boredom susceptibility levels (Kang et al. 2010) (Table 9.7).

9.2.4.7 Nerve Growth Factor Gene (NGF)

The nerve growth factor, NGF, is a neurotrophic factor involved in neuronal growth and survival of basal forebrain cholinergic neurons. Although the NGF complex consists of α, β and γ subunits, it is the β subunits exhibiting NGF stimulating function.

The NGF gene is located at chromosome 1 (1p13.2). There is a non-synonymous C104T SNP (rs6330) giving rise to the more common C allele coding for alanine and the T allele coding for valine. This amino acid substitution (alanine is substituted by valine at codon 35, Ala35Val) may affect NGF secretion (Syed et al. 2007).

A study of a non-clinical German population revealed a gender-specific effect of the C104T polymorphism on state-trait anxiety levels. Specifically, females genotyped as C/C showed increased anxiety levels compared with heterozygous carriers. Contrary to females, males genotyped as C/C displayed lower anxiety levels compared with heterozygous carriers. Still, results were not significant in both females and males, when homozygous C/C were compared with homozygous T/T individuals (Lang et al. 2008).

9.2.4.8 Cholinergic Receptor Nicotinic Alpha 4 Subunit Gene (CHRNA4)

The cholinergic system is involved in neural plasticity and associated with learning. The nicotinic acetylcholine receptors constitute a receptor family, whose members are formed by diverse combinations of five different subunits (α1–α10 and β2–β4). The neuronal receptor α4β2 is the main receptor found in mammalian brain (Gotti et al. 2009). The α4 subunit is encoded by the CHRNA4 gene, located at chromosome 20 (20q13.2–13.3). There is a synonymous SNP (rs1044396) in exon five, a C to T transition, giving rise to two alleles. Although the SNP does not cause any amino acid change, there is evidence that it affects receptor sensitivity (Eggert et al. 2015).

Table 9.7 Brain-derived neurotrophic factor gene polymorphism and behaviour

Gene (chromosome location)	Encoded protein	Polymorphism	Alleles	Behavioural phenotype	Meta-analyses
BDNF (11p14.1)	Brain-derived neurotrophic factor	rs6265: G196A SNP (Val66Met)	G: the more common coding for valine A: coding for methionine; A/A genotype was associated with decreased BDNF neuronal secretion	Suicidal behaviour	Meta-analysis of 12 studies (1202 suicidal patients diagnosed with a psychiatric disorder versus 1699 non-suicidal patients diagnosed with the same psychiatric disorders and 451 healthy controls) Results: association between A allele and suicidal behaviour, which was even more significant in Asian populations (Chinese, Japanese, Korean), as well as when suicide attempters were compared with non-suicide attempters diagnosed with the same psychiatric disorders (Zai et al. 2012)
					Meta-analysis of seven studies (1700 suicidal cases versus 2584 healthy controls) Results: no association (Clayden et al. 2012)

In a study of a non-clinical sample, contrary to T allele carriers, C/C carriers displayed higher levels of negative emotionality, specifically anxiety and emotional instability, in combination with more intense harm avoidance and behavioural inhibition (Markett et al. 2011). A German study of a large population (1673 subjects) obtained from a German multicentre study of nicotine dependence genetics confirmed aforementioned results, reporting an association between the C allele and higher harm avoidance, as well as increased neuroticism levels (Bey et al. 2016).

Lastly, a study of maltreated children showed that the rs1044396 polymorphism moderated the effects of maltreatment on childhood personality outcome. Children carrying the T/T genotype displayed higher neuroticism levels when they had a background of childhood maltreatment. Contrary to maltreated, non-maltreated children carrying the T/T genotype displayed lower neuroticism levels and higher levels of openness to experience (Grazioplene et al. 2013).

9.3 Conclusions

9.3.1 Discrepancy in Study Outcomes

Association studies within the research field of behavioural genetics have provided in several cases controversial or inconsistent results. There are different issues in regard to study design, leading to outcome discrepancy and difficulties in interpreting findings. Inadequate study strategies may lead to false-positive, probably on chance, or false-negative results. Several factors should be taken into consideration when studying genetic background of behaviour and personality:

1. *Ethnicity*: studies of diverse ethnic groups, showing different allele variant frequency, should pay considerable effort in sample selection so as to avoid bias (Li and He 2007). For instance, research outcomes of MAOA, one of the most studied genes in relation to aggression, were characterised by both discrepancy and inconsistency, partly due to different MAOA allele frequency in diverse ethnic groups (Lea and Chambers 2007). Furthermore, it should be noted that the risk MAOA-L allele is not uncommon, occurring in 40% of the population (Brunner et al. 1993a, b; Hunter 2010).

   Additionally, different allele frequencies between cases and healthy controls in case-control study designs may also confound results. For instance, the frequency of the 5-HTTLPR S allele was shown to be significantly different between Caucasian and Asian healthy controls. In this sense, family-based designs could provide a more appropriate study approach (Lin and Tsai 2004).

   Lastly, the discovery of population-specific mutations associating with behavioural characteristics emphasises the importance of exact description of study populations' genetic background (Brunner et al. 1993a, b; Kelsoe 2010; Zai et al. 2012).

2. *Demographics*: in case of genes with a sexually dimorphic effect, such as the COMT gene, or X-linked genes, such as the MAOA gene, several studies recruited selected samples in regard to sex, possibly limiting results to these particular populations (Manuck et al. 2000; Tsai et al. 2004b; Kulikova et al. 2008). Specifically, MAOA-H/MAOA-L heterozygosity is only present in females, while males are always homozygous, due to the presence of one X chromosome. Since MAOA expression in heterozygous cases has not been fully elucidated yet, many studies excluded all females or included strictly female homozygous allele carriers (Kim-Cohen et al. 2006; Alia-Klein et al. 2008; Derringer et al. 2010; Ficks and Waldman 2014). Furthermore, only a few researchers have provided data on allele frequency in males and females separately, hindering analysis of a plausible gender effect on study outcomes (Lin and Tsai 2004).

   Age is another important variable, specifically in research of gene polymorphisms and personality traits. For instance, novelty seeking diminishes with age, as most personality traits do. In order to avoid bias, recruited individuals should be preferably under the age of 45 (Lusher et al. 2001; Lang et al. 2005). Furthermore, socio-economic or cultural differences may also affect study outcomes in regard to personality traits (Campbell et al. 2010; Kang et al. 2010).
3. *Methodological issues*: studies employing large populations may be at risk of finding false associations, while studies of small populations may fail in revealing statistically significant results. Furthermore, contrary to studies of unselected populations, studies employing selected samples may lead to overestimation of a genetic association, since they focus only on the extremes of normal distribution (Munafo et al. 2008).

   Different definition of a behavioural phenotype may also result in inconsistent findings across studies. For instance, it should be noted that self-report measures of aggression may reflect trait aggression, without reflecting aggressive acts. Additionally, most behaviours and personality traits constitute a continuum that may be defined by a different genetic background. For instance, animal studies suggested that trait aggression may be positively associated with serotonergic activity, whereas impulsive/violent state aggression may be negatively associated with serotonergic activity (Olivier and van Oorschot 2005). Another example is the behavioural spectrum of suicidal behaviour, ranging from death wish to completed suicide. Within this spectrum, suicide attempt constitutes also a broad phenotype, including failed suicide (strong intent of dying, usually careful planning, lethal/violent suicide methods) and suicide gesture (less intent of dying, usually a reaction to acute interpersonal conflicts, less lethal/non-violent suicide methods) (Mann 1998). In this sense, genetic background of severe/violent suicidal behaviour may be different from genetic background in cases of milder/non-violent suicidal manifestations (Lin and Tsai 2004). Accordingly, suicide completers may constitute a distinct group from suicide attempters (Clayden et al. 2012). Lastly, diversity in assessment tools, employed for measuring a specific behaviour or a personality trait, was also shown to moderate results (Schinka et al. 2004). Altogether, each study should describe clearly the behaviour being studied in a particular population, by defining assessment tools and what these measure.

4. *Common pathophysiology*: the possibility of a common pathophysiology underlying extreme manifestation of a specific behaviour and psychiatric disorders should not be overlooked. For instance, alcohol dependence and anorexia/bulimia nervosa are examples of psychiatric disorders characterised by impulsive behaviour. Both disorders were associated with the G-1438A 5-HT2A polymorphism. Additionally, there was evidence that the COMT gene may be a risk gene for schizophrenia. Based on a meta-analysis, presence of at least one COMT-L allele (G472A polymorphism) increased risk for violent behaviour in male schizophrenia patients by around 50% (Williams et al. 2007; Singh et al. 2012).

   Similarly, suicidal behaviour has been mostly studied within the context of different psychiatric disorders, e.g. affective, schizophrenia spectrum, personality and substance use disorders, since 90–95% of suicidal individuals are diagnosed with at least one psychiatric disorder (Gonzalez-Castro et al. 2013a, b). Research has provided evidence for a genetic component for most psychiatric disorders that usually involves the serotonergic and the dopaminergic system. Therefore, it cannot be excluded that risk genes for impulsive, aggressive or suicidal behaviour overlap with susceptibility genes for psychiatric disorders, confounding results. Others suggested that serotonergic system dysfunction may predispose to both suicidal behaviour and psychiatric disorders. In such a case, only psychiatric patients carrying a particular risk allele would manifest suicidal behaviour (Lin and Tsai 2004). Still, although suicidal behaviour has been associated with both TPH1 (Courtet et al. 2005) and SLC6A4 gene (Li and He 2007) independent of psychiatric diagnosis, others failed to confirm this association, claiming that psychiatric history is a major confounding factor. The latter assumption was based on the fact that genetic associations were not confirmed when suicidal versus non-suicidal psychiatric patients were compared (Saetre et al. 2010).

   Altogether, due to comorbidity between specific behaviours, e.g. suicidal and aggressive behaviour, and psychiatric disorders, case-control analyses carry mental health status as a confounding factor. Thus, differentiating the genetic component of a behaviour from the genetic component of a psychiatric disorder may prove a difficult challenge (Schild et al. 2013).

5. *Common comorbidity*: common comorbidity between psychiatric disorders suggests that a certain biological substrate may be shared. Thus, elucidating genetics of a behavioural phenotype present within the context of different diagnostic categories may prove difficult, since the genetic contribution to the behavioural phenotype may be masked by the genetic contributor to comorbid psychiatric disorders.
6. *Environmental factors*: study results have not always supported the association between a gene polymorphism and a particular behaviour or personality trait, such as in case of the MAOA-uVNTR polymorphism and aggressive/antisocial behaviour (Jacob et al. 2005; Huizinga et al. 2006; Widom and Brzustowicz 2006). This may be partly attributed to the fact that the effects of the MAOA

gene polymorphism on the manifestation of aggressive/antisocial behaviour were shown to be moderated by environmental factors, specifically childhood maltreatment (Caspi et al. et al. 2002; Hart and Marmorstein 2009; Derringer et al. 2010). Therefore, studies investigating gene effects without considering history of stressful environmental factors may not be able to reveal an association between a gene polymorphism and a particular behaviour.

7. *Allele grouping:* in case of the DRD4 48 bp VNTR, most studies grouped alleles into short (up to 5) and long (6, 7, 8) repeats. Still, the 7R allele is evolutionary younger than the common 4R allele and has increased in frequency due to positive selection. Thus, it may not be so simply related to other long alleles. Additionally, the 7R allele is extremely rare in Asians; thus it could not possibly contribute to the manifestation of novelty seeking (Kluger et al. 2002).

In regard to the TPH1 gene polymorphisms, diverse A218C and A779C polymorphism alleles have been associated with the manifestation of suicidal behaviour. Therefore it cannot be excluded that the "causative" risk allele is another one, until now unknown, which is not in strong LD with the TPH1 genetic variants implicated so far in suicidal behaviour. As a result, the TPH1 genetic variant associated with suicidal behaviour may depend on the original haplotype carrying the unknown risk allele (Lalovic and Turecki 2002).

Furthermore, several SNPs are not functional and may be in strong LD with other "causative" polymorphisms, located perhaps at different chromosomes. Additionally, a single polymorphism probably accounts for a small proportion of the variance, contributing minorly to the manifestation of a behaviour or expression of a personality trait. Thus, a haplotype approach could be more elucidating (Kluger et al. 2002). For instance, a study of suicide completers investigated the association between three different TPH1 SNPs (A218C; promoter region A-6526G; promoter region G-5806T) and the more extreme manifestation of suicidal behaviour. These three SNPs were in strong LD and were not found to be related to suicidal behaviour when they were analysed separately. On the contrary, when haplotype analysis was conducted, considering all three SNPs simultaneously, the haplotype -6526G-5806T-218C was more abundant in violent suicide completers compared with healthy controls. Based on results, haplotype analysis provided increased statistical power for the identification of a risk locus. Furthermore, although the A218C SNP alone does not seem to be functional, the risk haplotype could affect protein binding (Turecki et al. 2001).

9.3.2 Meta-analyses

Individual studies may lack sufficient statistical power to detect small gene effects. As a result, meta-analyses have been extensively applied in the field of behavioural genetics to investigate a global effect with a greater statistical power, based on studies using smaller sample sizes. Meta-analyses are additionally used to search for sources of heterogeneity between different studies, excluding studies with significant contribution to the magnitude of an association. Although they were proven a useful tool for quantitative summarisation of heterogeneous and/or small sample size studies, meta-analyses are prone to false-positive results due to publication bias (Anguelova et al. 2003). Publication bias is attributed to the fact that studies reporting positive associations are more likely to get published compared with studies reporting negative outcomes (Bellivier et al. 2004). Inclusion of less negative studies, as well as unpublished negative studies, may lead to meta-analyses' undersampling. Thus, studies reporting non-significant associations are equally important as studies with positive findings, in order to avoid overestimation of an association (Munafo et al. 2008). For instance, there was a meta-analysis of the TPH1 A218C polymorphism in relation to suicidal behaviour that pooled data from only one study reporting a positive and six studies reporting a negative association. Overall results revealed a positive association (Bellivier et al. 2004). The one research field that does not seem to be affected by publication bias is genetics of suicidal behaviour (Anguelova et al. 2003).

Still, there were cases of similar meta-analyses, reporting contradictory results, such as in the case of the association between the BDNF G196A polymorphism and suicidal behaviour (Clayden et al. 2012; Zai et al. 2012). Contradictory results may be explained by factors similar to the ones causing heterogeneity between individual studies:

1. *Ethnicity*: pooling samples from different ethnicities is a source of heterogeneity. For instance, genetic distinct populations, such as the Ashkenazi population, were shown to be a significant source of heterogeneity in meta-analyses (Bellivier et al. 2004). Therefore, there were meta-analyses that took ethnicity into consideration, pooling data from studies of Caucasian populations only (Rujescu et al. 2003b; Bellivier et al. 2004).
2. *Methodological issues*: diversity in assessment tools of behaviour and personality was shown to moderate results (Schinka et al. 2004). Therefore, meta-analyses should include studies applying the same measuring tools. For instance, a meta-analysis of the DRD4 48 bp VNTR in relation to novelty seeking based on studies assessing novelty seeking explicitly by Cloninger's TPQ questionnaire (Kluger et al. 2002) revealed different outcomes from a meta-analysis pooling studies independent of assessment scales (Schinka et al. 2002). Lastly, different meta-analytic methods may also contribute to contradictory findings. There were meta-analyses applying fixed effects model, which assumes the same true genetic effects, while other meta-analyses used the random effects model, which assumes normally distributed effects. The latter model may be more advantageous, considering and parameterising in between study variability (Lin and Tsai 2004).
3. *Non-homogeneous phenotypes*: as previously mentioned, human behaviour and personality constitute a spectrum. For instance, suicidal behaviour includes both non-violent and violent suicide. Pooling all suicidal cases together may be a reason why there were negative results among meta-analyses pooling suicide attempters and completers together, instead of analysing them separately (Lalovic and Turecki 2002; Kia-Keating et al. 2007). Contrary to suicidal behav-

iour, another meta-analysis did not prove broad definition of antisocial behaviour to be a cause of heterogeneity among studies (Ficks and Waldman 2014).

4. *Different psychiatric history*: a meta-analysis clearly showed that psychiatric diagnosis (mixed psychiatric populations or psychiatric populations with unclear diagnosis) is a significant source of heterogeneity (Wang et al. 2015). For instance, inclusion of suicidal bipolar patients was shown to moderate meta-analytic results, while inclusion of suicidal patients with different psychiatric diagnoses was not (Bellivier et al. 2004). Altogether, psychiatric history may contribute to outcome diversity, at least in case of genetic studies of suicidal behaviour (Tovilla-Zarate et al. 2011).
5. *Environmental factors*: since behavioural and personality outcome is not affected by genetic factors alone, moderating effects of environmental factors should be taken into consideration. Meta-analyses of studies of gene-environment interactions probably constitute a crucial step towards drawing more certain conclusions (Taylor and Kim-Cohen 2007; Ficks and Waldman 2014).

9.3.3 Closing Remarks

Psychopathology may be viewed as the extreme manifestation of behaviours and personality traits normally distributed in a population. A future aim of behavioural genetics is prediction of the probability that an individual with a specific genetic variation will manifest impulsive, suicidal, aggressive or antisocial/criminal behaviour, behaviours often expressed within the context of different psychiatric disorders. Accordingly, elucidation of the genetics of personality could contribute to early diagnosis and intervention in case of personality disorders (Mulder et al. 1999). Lastly, a wedge issue in regard to behavioural genetics would be adducing evidence of genotypic data explaining aggressive/violent behaviour in legal trials of criminal defendants (Bernet et al. 2007). Genetic studies of criminal behaviour raised severe legal issues. In 2009, a court in Italy lightened the sentence of a man convicted for murder by 1 year, based on scientific evidence linking his violent behaviour to "abnormal genes". Others suggested though tougher sentences in cases of evidence linking criminal behaviour with genetic background, arguing that a genetically determined behaviour will probably be repeated and cannot be treated.

Still, research is a long way before contributing to aforementioned goals. To date, reported associations between gene polymorphisms and behavioural/personality outcomes were proven relatively weak. Furthermore, the influential effect of different study inclusion on meta-analytic outcomes underlines discontinuity in results (Schild et al. 2013; Vassos et al. 2014). Associating a single gene polymorphism with behaviours and personality traits was proven a non-accomplishable task until now. Both human behaviour and personality are of complex origin, the result of genes interacting with each other, as well as with environmental factors. Additionally, behaviour and personality traits manifest themselves in a continuum. Complex phenotypes lying on a continuum are most probably attributed to polygenic inheritance.

Altogether, studies of the association between specific alleles and behaviours/personality traits have mostly contributed to exclude the theory of a single gene's major contribution. "Missing heritability" has been attributed to variants, possibly rare variants with greater effects, which have not been discovered yet, or to gene-gene interactions (Manolio et al. 2009; Zuk et al. 2012). Since a research approach focusing on separate risk alleles was proven unfruitful, genome-wide association studies of large populations (thousands) assaying multiple SNPs could constitute a more effective approach for the identification of candidate genes (Schild et al. 2013; Vassos et al. 2014).

Lastly, behavioural genetics alone are in danger of failing in contributing substantially to effective drug treatment. A certain "risk" allele may not always correspond with high or low protein activity. As a result, the mechanism mediating the effects of a risk genotype remains in several cases unknown. Thus, supplementary studies, enlightening the effects of gene polymorphisms on pathophysiology and focusing on epigenetic regulation, are required.

References

Abbar M, Courtet P, Bellivier F, Leboyer M, Boulenger JP, Castelhau D, Ferreira M, Lambercy C, Mouthon D, Paoloni-Giacobino A, Vessaz M, Malafosse A, Buresi C (2001) Suicide attempts and the tryptophan hydroxylase gene. Mol Psychiatry 6(3):268–273

Alia-Klein N, Goldstein RZ, Kriplani A, Logan J, Tomasi D, Williams B, Telang F, Shumay E, Biegon A, Craig IW, Henn F, Wang GJ, Volkow ND, Fowler JS (2008) Brain monoamine oxidase A activity predicts trait aggression. J Neurosci 28(19):5099–5104

de Almeida RM, Ferrari PF, Parmigiani S, Miczek KA (2005) Escalated aggressive behavior: dopamine, serotonin and GABA. Eur J Pharmacol 526(1-3):51–64

Angles MR, Ocana DB, Medellin BC, Tovilla-Zarate C (2012) No association between the HTR1A gene and suicidal behavior: a meta-analysis. Rev Bras Psiquiatr 34(1):38–42

Anguelova M, Benkelfat C, Turecki G (2003) A systematic review of association studies investigating genes coding for serotonin receptors and the serotonin transporter: II. Suicidal behavior. Mol Psychiatry 8(7):646–653

Anokhin AP, Golosheykin S, Grant J, Heath AC (2009) Heritability of risk-taking in adolescence: a longitudinal twin study. Twin Res Hum Genet 12(4):366–371

Arango V, Huang YY, Underwood MD, Mann JJ (2003) Genetics of the serotonergic system in suicidal behavior. J Psychiatr Res 37(5):375–386

Asberg M, Nordstrom P, Traskman-Bendz L (1986) Cerebrospinal fluid studies in suicide. An overview. Ann N Y Acad Sci 487:243–255

Asghari V, Schoots O, van Kats S, Ohara K, Jovanovic V, Guan HC, Bunzow JR, Petronis A, Van Tol HH (1994) Dopamine D4 receptor repeat: analysis of different native and mutant forms of the human and rat genes. Mol Pharmacol 46(2):364–373

Asghari V, Sanyal S, Buchwaldt S, Paterson A, Jovanovic V, Van Tol HH (1995) Modulation of intracellular cyclic AMP levels by different human dopamine D4 receptor variants. J Neurochem 65(3):1157–1165

Aslund C, Nordquist N, Comasco E, Leppert J, Oreland L, Nilsson KW (2011) Maltreatment, MAOA, and delinquency: sex differences in gene-environment interaction in a large population-based cohort of adolescents. Behav Genet 41(2):262–272

Auclair AL, Cathala A, Sarrazin F, Depoortere R, Piazza PV, Newman-Tancredi A, Spampinato U (2010) The central serotonin 2B receptor: a new pharmacological target to modulate the mesoaccumbens dopaminergic pathway activity. J Neurochem 114(5):1323–1332

Baker LA, Jacobson KC, Raine A, Lozano DI, Bezdjian S (2007) Genetic and environmental bases of childhood antisocial behavior: a multi-informant twin study. J Abnorm Psychol 116(2):219–235

Balada F, Torrubia R, Arque JM (1993) Gonadal hormone correlates of sensation seeking and anxiety in healthy human females. Neuropsychobiology 27(2):91–96

Ball D, Hill L, Freeman B, Eley TC, Strelau J, Riemann R, Spinath FM, Angleitner A, Plomin R (1997) The serotonin transporter gene and peer-rated neuroticism. Neuroreport 8(5):1301–1304

Banlaki Z, Elek Z, Nanasi T, Szekely A, Nemoda Z, Sasvari-Szekely M, Ronai Z (2015) Polymorphism in the serotonin receptor 2a (HTR2A) gene as possible predisposal factor for aggressive traits. PLoS One 10(2):e0117792

Baron M (1998) Mapping genes for personality: is the saga sagging? Mol Psychiatry 3(2):106–108

Beitchman JH, Baldassarra L, Mik H, De Luca V, King N, Bender D, Ehtesham S, Kennedy JL (2006) Serotonin transporter polymorphisms and persistent, pervasive childhood aggression. Am J Psychiatry 163(6):1103–1105

Bellivier F, Chaste P, Malafosse A (2004) Association between the TPH gene A218C polymorphism and suicidal behavior: a meta-analysis. Am J Med Genet B Neuropsychiatr Genet 124b(1):87–91

Benjamin J, Li L, Patterson C, Greenberg BD, Murphy DL, Hamer DH (1996) Population and familial association between the D4 dopamine receptor gene and measures of novelty seeking. Nat Genet 12(1):81–84

Benko A, Lazary J, Molnar E, Gonda X, Tothfalusi L, Pap D, Mirnics Z, Kurimay T, Chase D, Juhasz G, Anderson IM, Deakin JF, Bagdy G (2010) Significant association between the C(-1019)G functional polymorphism of the HTR1A gene and impulsivity. Am J Med Genet B Neuropsychiatr Genet 153b(2):592–599

Bernet W, Vnencak-Jones CL, Farahany N, Montgomery SA (2007) Bad nature, bad nurture, and testimony regarding MAOA and SLC6A4 genotyping at murder trials. J Forensic Sci 52(6):1362–1371

Bevilacqua L, Goldman D (2013) Genetics of impulsive behaviour. Philos Trans R Soc Lond B Biol Sci 368(1615):20120380

Bevilacqua L, Doly S, Kaprio J, Yuan Q, Tikkanen R, Paunio T, Zhou Z, Wedenoja J, Maroteaux L, Diaz S, Belmer A, Hodgkinson CA, Dell'osso L, Suvisaari J, Coccaro E, Rose RJ, Peltonen L, Virkkunen M, Goldman D (2010) A population-specific HTR2B stop codon predisposes to severe impulsivity. Nature 468(7327):1061–1066

Bevilacqua L, Carli V, Sarchiapone M, George DK, Goldman D, Roy A, Enoch MA (2012) Interaction between FKBP5 and childhood trauma and risk of aggressive behavior. Arch Gen Psychiatry 69(1):62–70

Bey K, Lennertz L, Markett S, Petrovsky N, Gallinat J, Grunder G, Spreckelmeyer KN, Wienker TF, Mobascher A, Dahmen N, Thuerauf N, Kornhuber J, Kiefer F, Toliat MR, Nurnberg P, Winterer G, Wagner M (2016) Replication of the association between CHRNA4 rs1044396 and harm avoidance in a large population-based sample. Eur Neuropsychopharmacol 26(1):150–155

Bonaventure P, Guo H, Tian B, Liu X, Bittner A, Roland B, Salunga R, Ma XJ, Kamme F, Meurers B, Bakker M, Jurzak M, Leysen JE, Erlander MG (2002) Nuclei and subnuclei gene expression profiling in mammalian brain. Brain Res 943(1):38–47

Bondy B, Erfurth A, de Jonge S, Kruger M, Meyer H (2000a) Possible association of the short allele of the serotonin transporter promoter gene polymorphism (5-HTTLPR) with violent suicide. Mol Psychiatry 5(2):193–195

Bondy B, Kuznik J, Baghai T, Schule C, Zwanzger P, Minov C, de Jonge S, Rupprecht R, Meyer H, Engel RR, Eisenmenger W, Ackenheil M (2000b) Lack of association of serotonin-2A receptor gene polymorphism (T102C) with suicidal ideation and suicide. Am J Med Genet 96(6):831–835

Borkenau P, Riemann R, Angleitner A, Spinath FM (2001) Genetic and environmental influences on observed personality: evidence from the German Observational Study of Adult Twins. J Pers Soc Psychol 80(4):655–668

Brezo J, Paris J, Turecki G (2006) Personality traits as correlates of suicidal ideation, suicide attempts, and suicide completions: a systematic review. Acta Psychiatr Scand 113(3):180–206

Brunner HG, Nelen M, Breakefield XO, Ropers HH, van Oost BA (1993a) Abnormal behavior associated with a point mutation in the structural gene for monoamine oxidase A. Science 262(5133):578–580

Brunner HG, Nelen MR, van Zandvoort P, Abeling NG, van Gennip AH, Wolters EC, Kuiper MA, Ropers HH, van Oost BA (1993b) X-Linked borderline mental retardation with prominent behavioral disturbance: phenotype, genetic localization, and evidence for disturbed monoamine metabolism. Am J Hum Genet 52(6):1032–1039

Buckholtz JW, Meyer-Lindenberg A (2008) MAOA and the neurogenetic architecture of human aggression. Trends Neurosci 31(3):120–129

Byrd AL, Manuck SB (2014) MAOA, childhood maltreatment, and antisocial behavior: meta-analysis of a gene-environment interaction. Biol Psychiatry 75(1):9–17

Cadoret RJ, Langbehn D, Caspers K, Troughton EP, Yucuis R, Sandhu HK, Philibert R (2003) Associations of the serotonin transporter promoter polymorphism with aggressivity, attention deficit, and conduct disorder in an adoptee population. Compr Psychiatry 44(2):88–101

Calati R, Porcelli S, Giegling I, Hartmann AM, Moller HJ, De Ronchi D, Serretti A, Rujescu D (2011) Catechol-o-methyltransferase gene modulation on suicidal behavior and personality traits: review, meta-analysis and association study. J Psychiatr Res 45(3):309–321

Campbell BC, Dreber A, Apicella CL, Eisenberg DT, Gray PB, Little AC, Garcia JR, Zamore RS, Lum JK (2010) Testosterone exposure, dopaminergic reward, and sensation-seeking in young men. Physiol Behav 99(4):451–456

Caspi A, McClay J, Moffitt TE, Mill J, Martin J, Craig IW, Taylor A, Poulton R (2002) Role of genotype in the cycle of violence in maltreated children. Science 297(5582):851–854

Chang FM, Kidd JR, Livak KJ, Pakstis AJ, Kidd KK (1996) The world-wide distribution of allele frequencies at the human dopamine D4 receptor locus. Hum Genet 98(1):91–101

Chen TJ, Blum K, Mathews D, Fisher L, Schnautz N, Braverman ER, Schoolfield J, Downs BW, Comings DE (2005) Are dopaminergic genes involved in a predisposition to pathological aggression? Hypothesizing the importance of "super normal controls" in psychiatricgenetic research of complex behavioral disorders. Med Hypotheses 65(4):703–707

Cheung G (2007) Stability of the harm avoidance personality trait in late-life depression. Int Psychogeriatr 19(4):778–780

Cicchetti D, Rogosch FA, Thibodeau EL (2012) The effects of child maltreatment on early signs of antisocial behavior: genetic moderation by tryptophan hydroxylase, serotonin transporter, and monoamine oxidase A genes. Dev Psychopathol 24(3):907–928

Clayden RC, Zaruk A, Meyre D, Thabane L, Samaan Z (2012) The association of attempted suicide with genetic variants in the SLC6A4 and TPH genes depends on the definition of suicidal behavior: a systematic review and meta-analysis. Transl Psychiatry 2:e166

Cloninger CR (1986) A unified biosocial theory of personality and its role in the development of anxiety states. Psychiatr Dev 4(3):167–226

Collier DA, Stober G, Li T, Heils A, Catalano M, Di Bella D, Arranz MJ, Murray RM, Vallada HP, Bengel D, Muller CR, Roberts GW, Smeraldi E, Kirov G, Sham P, Lesch KP (1996) A novel functional polymorphism within the promoter of the serotonin transporter gene: possible role in susceptibility to affective disorders. Mol Psychiatry 1(6):453–460

Comings DE, MacMurray JP (2000) Molecular heterosis: a review. Mol Genet Metab 71(1-2):19–31

Courtet P, Baud P, Abbar M, Boulenger JP, Castelnau D, Mouthon D, Malafosse A, Buresi C (2001) Association between violent suicidal behavior and the low activity allele of the serotonin transporter gene. Mol Psychiatry 6(3):338–341

Courtet P, Jollant F, Castelnau D, Buresi C, Malafosse A (2005) Suicidal behavior: relationship between phenotype and serotonergic genotype. Am J Med Genet C Semin Med Genet 133c(1):25–33

Craig IW, Halton KE (2009) Genetics of human aggressive behaviour. Hum Genet 126(1):101–113

Crick NR, Dodge KA (1996) Social information-processing mechanisms in reactive and proactive aggression. Child Dev 67(3):993–1002

D'Souza UM, Russ C, Tahir E, Mill J, McGuffin P, Asherson PJ, Craig IW (2004) Functional effects of a tandem duplication polymorphism in the 5′flanking region of the DRD4 gene. Biol Psychiatry 56(9):691–697

Damberg M, Garpenstrand H, Alfredsson J, Ekblom J, Forslund K, Rylander G, Oreland L (2000) A polymorphic region in the human transcription factor AP-2beta gene is associated with specific personality traits. Mol Psychiatry 5(2):220–224

Davidge KM, Atkinson L, Douglas L, Lee V, Shapiro S, Kennedy JL, Beitchman JH (2004) Association of the serotonin transporter and 5HT1Dbeta receptor genes with extreme, persistent and pervasive aggressive behaviour in children. Psychiatr Genet 14(3):143–146

Derringer J, Krueger RF, Irons DE, Iacono WG (2010) Harsh discipline, childhood sexual assault, and MAOA genotype: an investigation of main and interactive effects on diverse clinical externalizing outcomes. Behav Genet 40(5):639–648

Ding YC, Chi HC, Grady DL, Morishima A, Kidd JR, Kidd KK, Flodman P, Spence MA, Schuck S, Swanson JM, Zhang YP, Moyzis RK (2002) Evidence of positive selection acting at the human dopamine receptor D4 gene locus. Proc Natl Acad Sci U S A 99(1):309–314

Dmitrieva J, Chen C, Greenberger E, Ogunseitan O, Ding YC (2011) Gender-specific expression of the DRD4 gene on adolescent delinquency, anger and thrill seeking. Soc Cogn Affect Neurosci 6(1):82–89

Du L, Faludi G, Palkovits M, Demeter E, Bakish D, Lapierre YD, Sotonyi P, Hrdina PD (1999) Frequency of long allele in serotonin transporter gene is increased in depressed suicide victims. Biol Psychiatry 46(2):196–201

van Dyck CH, Malison RT, Jacobsen LK, Seibyl JP, Staley JK, Laruelle M, Baldwin RM, Innis RB, Gelernter J (2005) Increased dopamine transporter availability associated with the 9-repeat allele of the SLC6A3 gene. J Nucl Med 46(5):745–751

Ebstein RP, Novick O, Umansky R, Priel B, Osher Y, Blaine D, Bennett ER, Nemanov L, Katz M, Belmaker RH (1996) Dopamine D4 receptor (D4DR) exon III polymorphism associated with the human personality trait of Novelty Seeking. Nat Genet 12(1):78–80

Egan MF, Kojima M, Callicott JH, Goldberg TE, Kolachana BS, Bertolino A, Zaitsev E, Gold B, Goldman D, Dean M, Lu B, Weinberger DR (2003) The BDNF val66met polymorphism affects activity-dependent secretion of BDNF and human memory and hippocampal function. Cell 112(2):257–269

Eggert M, Winterer G, Wanischeck M, Hoda JC, Bertrand D, Steinlein O (2015) The nicotinic acetylcholine receptor alpha 4 subunit contains a functionally relevant SNP Haplotype. BMC Genet 16:46

Eisenberger NI, Way BM, Taylor SE, Welch WT, Lieberman MD (2007) Understanding genetic risk for aggression: clues from the brain's response to social exclusion. Biol Psychiatry 61(9):1100–1108

Ekelund J, Lichtermann D, Jarvelin MR, Peltonen L (1999) Association between novelty seeking and the type 4 dopamine receptor gene in a large Finnish cohort sample. Am J Psychiatry 156(9):1453–1455

Eley TC, Tahir E, Angleitner A, Harriss K, McClay J, Plomin R, Riemann R, Spinath F, Craig I (2003) Association analysis of MAOA and COMT with neuroticism assessed by peers. Am J Med Genet B Neuropsychiatr Genet 120b(1):90–96

Evans J, Reeves B, Platt H, Leibenau A, Goldman D, Jefferson K, Nutt D (2000) Impulsiveness, serotonin genes and repetition of deliberate self-harm (DSH). Psychol Med 30(6):1327–1334

Evenden JL (1999) Varieties of impulsivity. Psychopharmacology (Berl) 146(4):348–361

Fergusson DM, Boden JM, Horwood LJ, Miller AL, Kennedy MA (2011) MAOA, abuse exposure and antisocial behaviour: 30-year longitudinal study. Br J Psychiatry 198(6):457–463

Ficks CA, Waldman ID (2014) Candidate genes for aggression and antisocial behavior: a meta-analysis of association studies of the 5HTTLPR and MAOA-uVNTR. Behav Genet 44(5):427–444

Frazzetto G, Di Lorenzo G, Carola V, Proietti L, Sokolowska E, Siracusano A, Gross C, Troisi A (2007) Early trauma and increased risk for physical aggression during adulthood: the moderating role of MAOA genotype. PLoS One 2(5):e486

Frustaci A, Pozzi G, Gianfagna F, Manzoli L, Boccia S (2008) Meta-analysis of the brain-derived neurotrophic factor gene (BDNF) Val66Met polymorphism in anxiety disorders and anxiety-related personality traits. Neuropsychobiology 58(3-4):163–170

Fuke S, Suo S, Takahashi N, Koike H, Sasagawa N, Ishiura S (2001) The VNTR polymorphism of the human dopamine transporter (DAT1) gene affects gene expression. Pharmacogenomics J 1(2):152–156

Fuller RW (1996) The influence of fluoxetine on aggressive behavior. Neuropsychopharmacology 14(2):77–81

Gabel S, Stadler J, Bjorn J, Shindledecker R (1995) Homovanillic acid and dopamine-beta-hydroxylase in male youth: relationships with paternal substance abuse and antisocial behavior. Am J Drug Alcohol Abuse 21(3):363–378

Giegling I, Hartmann AM, Moller HJ, Rujescu D (2006) Anger- and aggression-related traits are associated with polymorphisms in the 5-HT-2A gene. J Affect Disord 96(1-2):75–81

Gillespie CF, Phifer J, Bradley B, Ressler KJ (2009) Risk and resilience: genetic and environmental influences on development of the stress response. Depress Anxiety 26(11):984–992

Gonda X, Fountoulakis KN, Juhasz G, Rihmer Z, Lazary J, Laszik A, Akiskal HS, Bagdy G (2009) Association of the s allele of the 5-HTTLPR with neuroticism-related traits and temperaments in a psychiatrically healthy population. Eur Arch Psychiatry Clin Neurosci 259(2):106–113

Gonzalez-Castro TB, Tovilla-Zarate C, Juarez-Rojop I, Pool Garcia S, Velazquez-Sanchez MP, Genis A, Nicolini H, Lopez Narvaez L (2013a) Association of the 5HTR2A gene with suicidal behavior: case-control study and updated meta-analysis. BMC Psychiatry 13:25

Gonzalez-Castro TB, Tovilla-Zarate CA, Juarez-Rojop I, Pool Garcia S, Genis A, Nicolini H, Lopez Narvaez L (2013b) Association of 5HTR1A gene variants with suicidal behavior: case-control study and updated meta-analysis. J Psychiatr Res 47(11):1665–1672

Gotti C, Clementi F, Fornari A, Gaimarri A, Guiducci S, Manfredi I, Moretti M, Pedrazzi P, Pucci L, Zoli M (2009) Structural and functional diversity of native brain neuronal nicotinic receptors. Biochem Pharmacol 78(7):703–711

Grandy DK, Litt M, Allen L, Bunzow JR, Marchionni M, Makam H, Reed L, Magenis RE, Civelli O (1989) The human dopamine D2 receptor gene is located on chromosome 11 at q22–q23 and identifies a TaqI RFLP. Am J Hum Genet 45(5):778–785

Grazioplene RG, Deyoung CG, Rogosch FA, Cicchetti D (2013) A novel differential susceptibility gene: CHRNA4 and moderation of the effect of maltreatment on child personality. J Child Psychol Psychiatry 54(8):872–880

Greenberg BD, Tolliver TJ, Huang SJ, Li Q, Bengel D, Murphy DL (1999) Genetic variation in the serotonin transporter promoter region affects serotonin uptake in human blood platelets. Am J Med Genet 88(1):83–87

Groleger U (2007) Off-label use of antipsychotics: rethinking "off-label". Psychiatr Danub 19(4):350–353

Gunter TD, Vaughn MG, Philibert RA (2010) Behavioral genetics in antisocial spectrum disorders and psychopathy: a review of the recent literature. Behav Sci Law 28(2):148–173

Guo G, Roettger ME, Shih JC (2007) Contributions of the DAT1 and DRD2 genes to serious and violent delinquency among adolescents and young adults. Hum Genet 121(1):125–136

Guo G, Ou XM, Roettger M, Shih JC (2008) The VNTR 2 repeat in MAOA and delinquent behavior in adolescence and young adulthood: associations and MAOA promoter activity. Eur J Hum Genet 16(5):626–634

Gutknecht L, Kriegebaum C, Waider J, Schmitt A, Lesch KP (2009) Spatio-temporal expression of tryptophan hydroxylase isoforms in murine and human brain: convergent data from Tph2 knockout mice. Eur Neuropsychopharmacol 19(4):266–282

Haberstick BC, Lessem JM, Hopfer CJ, Smolen A, Ehringer MA, Timberlake D, Hewitt JK (2005) Monoamine oxidase A (MAOA) and antisocial behaviors in the presence of childhood and adolescent maltreatment. Am J Med Genet B Neuropsychiatr Genet 135b(1):59–64

Hariri AR, Weinberger DR (2003) Functional neuroimaging of genetic variation in serotonergic neurotransmission. Genes Brain Behav 2(6):341–349

Hart D, Marmorstein NR (2009) Neighborhoods and genes and everything in between: understanding adolescent aggression in social and biological contexts. Dev Psychopathol 21(3):961–973

Heils A, Teufel A, Petri S, Stober G, Riederer P, Bengel D, Lesch KP (1996) Allelic variation of human serotonin transporter gene expression. J Neurochem 66(6):2621–2624

Heiman N, Stallings MC, Young SE, Hewitt JK (2004) Investigating the genetic and environmental structure of Cloninger's personality dimensions in adolescence. Twin Res 7(5):462–470
Heinz A, Goldman D, Jones DW, Palmour R, Hommer D, Gorey JG, Lee KS, Linnoila M, Weinberger DR (2000) Genotype influences in vivo dopamine transporter availability in human striatum. Neuropsychopharmacology 22(2):133–139
Hennig J, Reuter M, Netter P, Burk C, Landt O (2005) Two types of aggression are differentially related to serotonergic activity and the A779C TPH polymorphism. Behav Neurosci 119(1):16–25
Hess C, Reif A, Strobel A, Boreatti-Hummer A, Heine M, Lesch KP, Jacob CP (2009) A functional dopamine-beta-hydroxylase gene promoter polymorphism is associated with impulsive personality styles, but not with affective disorders. J Neural Transm (Vienna) 116(2):121–130
Hoyer D, Hannon JP, Martin GR (2002) Molecular, pharmacological and functional diversity of 5-HT receptors. Pharmacol Biochem Behav 71(4):533–554
Hu XZ, Lipsky RH, Zhu G, Akhtar LA, Taubman J, Greenberg BD, Xu K, Arnold PD, Richter MA, Kennedy JL, Murphy DL, Goldman D (2006) Serotonin transporter promoter gain-of-function genotypes are linked to obsessive-compulsive disorder. Am J Hum Genet 78(5):815–826
Huang YY, Grailhe R, Arango V, Hen R, Mann JJ (1999) Relationship of psychopathology to the human serotonin1B genotype and receptor binding kinetics in postmortem brain tissue. Neuropsychopharmacology 21(2):238–246
Huizinga D, Haberstick BC, Smolen A, Menard S, Young SE, Corley RP, Stallings MC, Grotpeter J, Hewitt JK (2006) Childhood maltreatment, subsequent antisocial behavior, and the role of monoamine oxidase A genotype. Biol Psychiatry 60(7):677–683
Hunter P (2010) The psycho gene. EMBO Rep 11(9):667–669
Hur YM, Bouchard TJ Jr (1997) The genetic correlation between impulsivity and sensation seeking traits. Behav Genet 27(5):455–463
Insel TR (2010) The challenge of translation in social neuroscience: a review of oxytocin, vasopressin, and affiliative behavior. Neuron 65(6):768–779
Itoh K, Hashimoto K, Kumakiri C, Shimizu E, Iyo M (2004) Association between brain-derived neurotrophic factor 196 G/A polymorphism and personality traits in healthy subjects. Am J Med Genet B Neuropsychiatr Genet 124b(1):61–63
Jacob CP, Strobel A, Hohenberger K, Ringel T, Gutknecht L, Reif A, Brocke B, Lesch KP (2004) Association between allelic variation of serotonin transporter function and neuroticism in anxious cluster C personality disorders. Am J Psychiatry 161(3):569–572
Jacob CP, Muller J, Schmidt M, Hohenberger K, Gutknecht L, Reif A, Schmidtke A, Mossner R, Lesch KP (2005) Cluster B personality disorders are associated with allelic variation of monoamine oxidase A activity. Neuropsychopharmacology 30(9):1711–1718
Jacobsen LK, Staley JK, Zoghbi SS, Seibyl JP, Kosten TR, Innis RB, Gelernter J (2000) Prediction of dopamine transporter binding availability by genotype: a preliminary report. Am J Psychiatry 157(10):1700–1703
Jacobson KC, Prescott CA, Kendler KS (2002) Sex differences in the genetic and environmental influences on the development of antisocial behavior. Dev Psychopathol 14(2):395–416
Jaffee SR, Price TS (2008) Genotype-environment correlations: implications for determining the relationship between environmental exposures and psychiatric illness. Psychiatry 7(12):496–499
Jang KL, Livesley WJ, Vernon PA (1996) Heritability of the big five personality dimensions and their facets: a twin study. J Pers 64(3):577–591
Jernej B, Stefulj J, Hranilovic D, Balija M, Skavic J, Kubat M (2004) Intronic polymorphism of tryptophan hydroxylase and serotonin transporter: indication for combined effect in predisposition to suicide. J Neural Transm (Vienna) 111(6):733–738
Jonsson EG, Goldman D, Spurlock G, Gustavsson JP, Nielsen DA, Linnoila M, Owen MJ, Sedvall GC (1997) Tryptophan hydroxylase and catechol-O-methyltransferase gene polymorphisms: relationships to monoamine metabolite concentrations in CSF of healthy volunteers. Eur Arch Psychiatry Clin Neurosci 247(6):297–302
Kang JI, Song DH, Namkoong K, Kim SJ (2010) Interaction effects between COMT and BDNF polymorphisms on boredom susceptibility of sensation seeking traits. Psychiatry Res 178(1):132–136

Kazantseva AV, Gaysina DA, Faskhutdinova GG, Noskova T, Malykh SB, Khusnutdinova EK (2008) Polymorphisms of the serotonin transporter gene (5-HTTLPR, A/G SNP in 5-HTTLPR, and STin2 VNTR) and their relation to personality traits in healthy individuals from Russia. Psychiatr Genet 18(4):167–176

Kelsoe JR (2010) Behavioural neuroscience: a gene for impulsivity. Nature 468(7327):1049–1050

Khan SA, Faraone SV (2006) The genetics of ADHD: a literature review of 2005. Curr Psychiatry Rep 8(5):393–397

Kia-Keating BM, Glatt SJ, Tsuang MT (2007) Meta-analyses suggest association between COMT, but not HTR1B, alleles, and suicidal behavior. Am J Med Genet B Neuropsychiatr Genet 144b(8):1048–1053

Kim CH, Hahn MK, Joung Y, Anderson SL, Steele AH, Mazei-Robinson MS, Gizer I, Teicher MH, Cohen BM, Robertson D, Waldman ID, Blakely RD, Kim KS (2006a) A polymorphism in the norepinephrine transporter gene alters promoter activity and is associated with attention-deficit hyperactivity disorder. Proc Natl Acad Sci U S A 103(50):19164–19169

Kim SJ, Kim YS, Kim CH, Lee HS (2006b) Lack of association between polymorphisms of the dopamine receptor D4 and dopamine transporter genes and personality traits in a Korean population. Yonsei Med J 47(6):787–792

Kim SJ, Kim YS, Kim SY, Lee HS, Kim CH (2006c) An association study of catechol-O-methyltransferase and monoamine oxidase A polymorphisms and personality traits in Koreans. Neurosci Lett 401(1-2):154–158

Kim-Cohen J, Caspi A, Taylor A, Williams B, Newcombe R, Craig IW, Moffitt TE (2006) MAOA, maltreatment, and gene-environment interaction predicting children's mental health: new evidence and a meta-analysis. Mol Psychiatry 11(10):903–913

Kluger AN, Siegfried Z, Ebstein RP (2002) A meta-analysis of the association between DRD4 polymorphism and novelty seeking. Mol Psychiatry 7(7):712–717

Kobilka BK, Frielle T, Collins S, Yang-Feng T, Kobilka TS, Francke U, Lefkowitz RJ, Caron MG (1987) An intronless gene encoding a potential member of the family of receptors coupled to guanine nucleotide regulatory proteins. Nature 329(6134):75–79

Koller G, Bondy B, Preuss UW, Zill P, Soyka M (2006) The C(-1019)G 5-HT1A promoter polymorphism and personality traits: no evidence for significant association in alcoholic patients. Behav Brain Funct 2:7

Kulikov AV, Osipova DV, Naumenko VS, Popova NK (2005) Association between Tph2 gene polymorphism, brain tryptophan hydroxylase activity and aggressiveness in mouse strains. Genes Brain Behav 4(8):482–485

Kulikova MA, Maluchenko NV, Timofeeva MA, Shlepzova VA, Schegolkova JV, Sysoeva OV, Ivanitsky AM, Tonevitsky AG (2008) Effect of functional catechol-O-methyltransferase Val158Met polymorphism on physical aggression. Bull Exp Biol Med 145(1):62–64

Laakso A, Pohjalainen T, Bergman J, Kajander J, Haaparanta M, Solin O, Syvalahti E, Hietala J (2005) The A1 allele of the human D2 dopamine receptor gene is associated with increased activity of striatal L-amino acid decarboxylase in healthy subjects. Pharmacogenet Genomics 15(6):387–391

Lachman HM, Papolos DF, Saito T, Yu YM, Szumlanski CL, Weinshilboum RM (1996) Human catechol-O-methyltransferase pharmacogenetics: description of a functional polymorphism and its potential application to neuropsychiatric disorders. Pharmacogenetics 6(3):243–250

Lage GM, Malloy-Diniz LF, Matos LO, Bastos MA, Abrantes SS, Correa H (2011) Impulsivity and the 5-HTTLPR polymorphism in a non-clinical sample. PLoS One 6(2):e16927

Lalovic A, Turecki G (2002) Meta-analysis of the association between tryptophan hydroxylase and suicidal behavior. Am J Med Genet 114(5):533–540

Lang UE, Hellweg R, Kalus P, Bajbouj M, Lenzen KP, Sander T, Kunz D, Gallinat J (2005) Association of a functional BDNF polymorphism and anxiety-related personality traits. Psychopharmacology (Berl) 180(1):95–99

Lang UE, Bajbouj M, Sander T, Gallinat J (2007) Gender-dependent association of the functional catechol-O-methyltransferase Val158Met genotype with sensation seeking personality trait. Neuropsychopharmacology 32(9):1950–1955

Lang UE, Hellweg R, Bajbouj M, Gaus V, Sander T, Gallinat J (2008) Gender-dependent association of a functional NGF polymorphism with anxiety-related personality traits. Pharmacopsychiatry 41(5):196–199
Lappalainen J, Long JC, Eggert M, Ozaki N, Robin RW, Brown GL, Naukkarinen H, Virkkunen M, Linnoila M, Goldman D (1998) Linkage of antisocial alcoholism to the serotonin 5-HT1B receptor gene in 2 populations. Arch Gen Psychiatry 55(11):989–994
Launay JM, Schneider B, Loric S, Da Prada M, Kellermann O (2006) Serotonin transport and serotonin transporter-mediated antidepressant recognition are controlled by 5-HT2B receptor signaling in serotonergic neuronal cells. FASEB J 20(11):1843–1854
Lea R, Chambers G (2007) Monoamine oxidase, addiction, and the "warrior" gene hypothesis. N Z Med J 120(1250):U2441
Lee HJ, Lee HS, Kim YK, Kim L, Lee MS, Jung IK, Suh KY, Kim S (2003a) D2 and D4 dopamine receptor gene polymorphisms and personality traits in a young Korean population. Am J Med Genet B Neuropsychiatr Genet 121b(1):44–49
Lee JH, Kim HT, Hyun DS (2003b) Possible association between serotonin transporter promoter region polymorphism and impulsivity in Koreans. Psychiatry Res 118(1):19–24
Lemonde S, Turecki G, Bakish D, Du L, Hrdina PD, Bown CD, Sequeira A, Kushwaha N, Morris SJ, Basak A, Ou XM, Albert PR (2003) Impaired repression at a 5-hydroxytryptamine 1A receptor gene polymorphism associated with major depression and suicide. J Neurosci 23(25):8788–8799
Lepine JP, Pelissolo A, Teodorescu R, Teherani M (1994) Evaluation of the psychometric properties of the French version of the Tridimensional Personality Questionnaire (TPQ). Encéphale 20(6):747–753
Lesch KP, Merschdorf U (2000) Impulsivity, aggression, and serotonin: a molecular psychobiological perspective. Behav Sci Law 18(5):581–604
Lesch KP, Bengel D, Heils A, Sabol SZ, Greenberg BD, Petri S, Benjamin J, Muller CR, Hamer DH, Murphy DL (1996) Association of anxiety-related traits with a polymorphism in the serotonin transporter gene regulatory region. Science 274(5292):1527–1531
Li D, He L (2006) Further clarification of the contribution of the tryptophan hydroxylase (TPH) gene to suicidal behavior using systematic allelic and genotypic meta-analyses. Hum Genet 119(3):233–240
Li D, He L (2007) Meta-analysis supports association between serotonin transporter (5-HTT) and suicidal behavior. Mol Psychiatry 12(1):47–54
Li D, Duan Y, He L (2006) Association study of serotonin 2A receptor (5-HT2A) gene with schizophrenia and suicidal behavior using systematic meta-analysis. Biochem Biophys Res Commun 340(3):1006–1015
Liao DL, Hong CJ, Shih HL, Tsai SJ (2004) Possible association between serotonin transporter promoter region polymorphism and extremely violent crime in Chinese males. Neuropsychobiology 50(4):284–287
Lichter JB, Barr CL, Kennedy JL, Van Tol HH, Kidd KK, Livak KJ (1993) A hypervariable segment in the human dopamine receptor D4 (DRD4) gene. Hum Mol Genet 2(6):767–773
Lin PY, Tsai G (2004) Association between serotonin transporter gene promoter polymorphism and suicide: results of a meta-analysis. Biol Psychiatry 55(10):1023–1030
Lusher JM, Chandler C, Ball D (2001) Dopamine D4 receptor gene (DRD4) is associated with novelty seeking (NS) and substance abuse: the saga continues. Mol Psychiatry 6(5):497–499
MacKenzie A, Quinn J (1999) A serotonin transporter gene intron 2 polymorphic region, correlated with affective disorders, has allele-dependent differential enhancer-like properties in the mouse embryo. Proc Natl Acad Sci U S A 96(26):15251–15255
Mann JJ (1998) The neurobiology of suicide. Nat Med 4(1):25–30
Mann JJ (2003) Neurobiology of suicidal behaviour. Nat Rev Neurosci 4(10):819–828
Mann JJ, Malone KM, Nielsen DA, Goldman D, Erdos J, Gelernter J (1997) Possible association of a polymorphism of the tryptophan hydroxylase gene with suicidal behavior in depressed patients. Am J Psychiatry 154(10):1451–1453
Mann JJ, Brent DA, Arango V (2001) The neurobiology and genetics of suicide and attempted suicide: a focus on the serotonergic system. Neuropsychopharmacology 24(5):467–477

Manolio TA, Collins FS, Cox NJ, Goldstein DB, Hindorff LA, Hunter DJ, McCarthy MI, Ramos EM, Cardon LR, Chakravarti A, Cho JH, Guttmacher AE, Kong A, Kruglyak L, Mardis E, Rotimi CN, Slatkin M, Valle D, Whittemore AS, Boehnke M, Clark AG, Eichler EE, Gibson G, Haines JL, Mackay TF, McCarroll SA, Visscher PM (2009) Finding the missing heritability of complex diseases. Nature 461(7265):747–753

Manuck SB, Flory JD, Ferrell RE, Dent KM, Mann JJ, Muldoon MF (1999) Aggression and anger-related traits associated with a polymorphism of the tryptophan hydroxylase gene. Biol Psychiatry 45(5):603–614

Manuck SB, Flory JD, Ferrell RE, Mann JJ, Muldoon MF (2000) A regulatory polymorphism of the monoamine oxidase-A gene may be associated with variability in aggression, impulsivity, and central nervous system serotonergic responsivity. Psychiatry Res 95(1):9–23

Markett S, Montag C, Reuter M (2011) The nicotinic acetylcholine receptor gene CHRNA4 is associated with negative emotionality. Emotion 11(2):450–455

McCourt WF, Gurrera RJ, Cutter HS (1993) Sensation seeking and novelty seeking. Are they the same? J Nerv Ment Dis 181(5):309–312

McDermott R, Tingley D, Cowden J, Frazzetto G, Johnson DD (2009) Monoamine oxidase A gene (MAOA) predicts behavioral aggression following provocation. Proc Natl Acad Sci U S A 106(7):2118–2123

Miczek KA, Fish EW, De Bold JF, De Almeida RM (2002) Social and neural determinants of aggressive behavior: pharmacotherapeutic targets at serotonin, dopamine and gamma-aminobutyric acid systems. Psychopharmacology (Berl) 163(3-4):434–458

Miles DR, Carey G (1997) Genetic and environmental architecture of human aggression. J Pers Soc Psychol 72(1):207–217

Miller DJ, Vachon DD, Lynam DR (2009) Neuroticism, negative affect, and negative affect instability: establishing convergent and discriminant validity using ecological momentary assessment. Pers Individ Differ 47(8):873–877

Mitsuyasu H, Hirata N, Sakai Y, Shibata H, Takeda Y, Ninomiya H, Kawasaki H, Tashiro N, Fukumaki Y (2001) Association analysis of polymorphisms in the upstream region of the human dopamine D4 receptor gene (DRD4) with schizophrenia and personality traits. J Hum Genet 46(1):26–31

Mulder RT, Joyce PR, Sullivan PF, Bulik CM, Carter FA (1999) The relationship among three models of personality psychopathology: DSM-III-R personality disorder, TCI scores and DSQ defences. Psychol Med 29(4):943–951

Munafo MR, Yalcin B, Willis-Owen SA, Flint J (2008) Association of the dopamine D4 receptor (DRD4) gene and approach-related personality traits: meta-analysis and new data. Biol Psychiatry 63(2):197–206

Nakamura M, Ueno S, Sano A, Tanabe H (2000) The human serotonin transporter gene linked polymorphism (5-HTTLPR) shows ten novel allelic variants. Mol Psychiatry 5(1):32–38

New AS, Gelernter J, Yovell Y, Trestman RL, Nielsen DA, Silverman J, Mitropoulou V, Siever LJ (1998) Tryptophan hydroxylase genotype is associated with impulsive-aggression measures: a preliminary study. Am J Med Genet 81(1):13–17

New AS, Gelernter J, Goodman M, Mitropoulou V, Koenigsberg H, Silverman J, Siever LJ (2001) Suicide, impulsive aggression, and HTR1B genotype. Biol Psychiatry 50(1):62–65

Nielsen SD, Storgaard H, Moesgaard F, Gluud C (1994) Prevalence of alcohol problems among adult somatic in-patients of a Copenhagen hospital. Alcohol Alcohol 29(5):583–590

Nielsen DA, Jenkins GL, Stefanisko KM, Jefferson KK, Goldman D (1997) Sequence, splice site and population frequency distribution analyses of the polymorphic human tryptophan hydroxylase intron 7. Brain Res Mol Brain Res 45(1):145–148

Nielsen DA, Virkkunen M, Lappalainen J, Eggert M, Brown GL, Long JC, Goldman D, Linnoila M (1998) A tryptophan hydroxylase gene marker for suicidality and alcoholism. Arch Gen Psychiatry 55(7):593–602

Nishiguchi N, Shirakawa O, Ono H, Nishimura A, Nushida H, Ueno Y, Maeda K (2001) No evidence of an association between 5HT1B receptor gene polymorphism and suicide victims in a Japanese population. Am J Med Genet 105(4):343–345

Noble EP, Gottschalk LA, Fallon JH, Ritchie TL, Wu JC (1997) D2 dopamine receptor polymorphism and brain regional glucose metabolism. Am J Med Genet 74(2):162–166
Nock MK, Borges G, Bromet EJ, Cha CB, Kessler RC, Lee S (2008) Suicide and suicidal behavior. Epidemiol Rev 30:133–154
Nomura M, Kusumi I, Kaneko M, Masui T, Daiguji M, Ueno T, Koyama T, Nomura Y (2006) Involvement of a polymorphism in the 5-HT2A receptor gene in impulsive behavior. Psychopharmacology (Berl) 187(1):30–35
Olivier B, van Oorschot R (2005) 5-HT1B receptors and aggression: a review. Eur J Pharmacol 526(1-3):207–217
Ono Y, Manki H, Yoshimura K, Muramatsu T, Mizushima H, Higuchi S, Yagi G, Kanba S, Asai M (1997) Association between dopamine D4 receptor (D4DR) exon III polymorphism and novelty seeking in Japanese subjects. Am J Med Genet 74(5):501–503
Ono H, Shirakawa O, Kitamura N, Hashimoto T, Nishiguchi N, Nishimura A, Nushida H, Ueno Y, Maeda K (2002) Tryptophan hydroxylase immunoreactivity is altered by the genetic variation in postmortem brain samples of both suicide victims and controls. Mol Psychiatry 7(10):1127–1132
Patton JH, Stanford MS, Barratt ES (1995) Factor structure of the Barratt impulsiveness scale. J Clin Psychol 51(6):768–774
Pedersen NL, Plomin R, McClearn GE, Friberg L (1988) Neuroticism, extraversion, and related traits in adult twins reared apart and reared together. J Pers Soc Psychol 55(6):950–957
Pohjalainen T, Rinne JO, Nagren K, Lehikoinen P, Anttila K, Syvalahti EK, Hietala J (1998) The A1 allele of the human D2 dopamine receptor gene predicts low D2 receptor availability in healthy volunteers. Mol Psychiatry 3(3):256–260
Preuss UW, Koller G, Bondy B, Bahlmann M, Soyka M (2001) Impulsive traits and 5-HT2A receptor promoter polymorphism in alcohol dependents: possible association but no influence of personality disorders. Neuropsychobiology 43(3):186–191
Prichard Z, Easteal S (2006) Characterization of simple sequence repeat variants linked to candidate genes for behavioral phenotypes. Hum Mutat 27(1):120
Prichard ZM, Jorm AF, Mackinnon A, Easteal S (2007a) Association analysis of 15 polymorphisms within 10 candidate genes for antisocial behavioural traits. Psychiatr Genet 17(5):299–303
Prichard ZM, Mackinnon AJ, Jorm AF, Easteal S (2007b) AVPR1A and OXTR polymorphisms are associated with sexual and reproductive behavioral phenotypes in humans. Hum Mutat 28(11):1150. Mutation in brief no. 981. Online
Raeymaekers P, Van Broeckhoven C (1998) Comment - Genes and temperament, a shortcut for unravelling the genetics of psychopathology? Int J Neuropsychopharmacol 1(2):169–171
Raine A, Dodge K, Loeber R, Gatzke-Kopp L, Lynam D, Reynolds C, Stouthamer-Loeber M, Liu J (2006) The Reactive-Proactive Aggression Questionnaire: differential correlates of reactive and proactive aggression in adolescent boys. Aggress Behav 32(2):159–171
Reif A, Rosler M, Freitag CM, Schneider M, Eujen A, Kissling C, Wenzler D, Jacob CP, Retz-Junginger P, Thome J, Lesch KP, Retz W (2007) Nature and nurture predispose to violent behavior: serotonergic genes and adverse childhood environment. Neuropsychopharmacology 32(11):2375–2383
Reif A, Jacob CP, Rujescu D, Herterich S, Lang S, Gutknecht L, Baehne CG, Strobel A, Freitag CM, Giegling I, Romanos M, Hartmann A, Rosler M, Renner TJ, Fallgatter AJ, Retz W, Ehlis AC, Lesch KP (2009) Influence of functional variant of neuronal nitric oxide synthase on impulsive behaviors in humans. Arch Gen Psychiatry 66(1):41–50
Reif A, Kiive E, Kurrikoff T, Paaver M, Herterich S, Konstabel K, Tulviste T, Lesch KP, Harro J (2011) A functional NOS1 promoter polymorphism interacts with adverse environment on functional and dysfunctional impulsivity. Psychopharmacology (Berl) 214(1):239–248
Reist C, Nakamura K, Sagart E, Sokolski KN, Fujimoto KA (2003) Impulsive aggressive behavior: open-label treatment with citalopram. J Clin Psychiatry 64(1):81–85
Retz W, Retz-Junginger P, Supprian T, Thome J, Rosler M (2004) Association of serotonin transporter promoter gene polymorphism with violence: relation with personality disorders, impulsivity, and childhood ADHD psychopathology. Behav Sci Law 22(3):415–425

Retz W, Reif A, Freitag CM, Retz-Junginger P, Rosler M (2010) Association of a functional variant of neuronal nitric oxide synthase gene with self-reported impulsiveness, venturesomeness and empathy in male offenders. J Neural Transm (Vienna) 117(3):321–324

Reuter M, Hennig J (2005) Association of the functional catechol-O-methyltransferase VAL158MET polymorphism with the personality trait of extraversion. Neuroreport 16(10):1135–1138

Rhee SH, Waldman ID (2002) Genetic and environmental influences on antisocial behavior: a meta-analysis of twin and adoption studies. Psychol Bull 128(3):490–529

Rogeness GA, Hernandez JM, Macedo CA, Mitchell EL (1982) Biochemical differences in children with conduct disorder socialized and undersocialized. Am J Psychiatry 139(3):307–311

Rogers G, Joyce P, Mulder R, Sellman D, Miller A, Allington M, Olds R, Wells E, Kennedy M (2004) Association of a duplicated repeat polymorphism in the 5′-untranslated region of the DRD4 gene with novelty seeking. Am J Med Genet B Neuropsychiatr Genet 126b(1):95–98

Ronai Z, Barta C, Guttman A, Lakatos K, Gervai J, Staub M, Sasvari-Szekely M (2001) Genotyping the -521C/T functional polymorphism in the promoter region of dopamine D4 receptor (DRD4) gene. Electrophoresis 22(6):1102–1105

Roy A, Gorodetsky E, Yuan Q, Goldman D, Enoch MA (2010) Interaction of FKBP5, a stress-related gene, with childhood trauma increases the risk for attempting suicide. Neuropsychopharmacology 35(8):1674–1683

Rujescu D, Giegling I, Bondy B, Gietl A, Zill P, Moller HJ (2002) Association of anger-related traits with SNPs in the TPH gene. Mol Psychiatry 7(9):1023–1029

Rujescu D, Giegling I, Gietl A, Hartmann AM, Moller HJ (2003a) A functional single nucleotide polymorphism (V158M) in the COMT gene is associated with aggressive personality traits. Biol Psychiatry 54(1):34–39

Rujescu D, Giegling I, Sato T, Hartmann AM, Moller HJ (2003b) Genetic variations in tryptophan hydroxylase in suicidal behavior: analysis and meta-analysis. Biol Psychiatry 54(4):465–473

Rujescu D, Giegling I, Sato T, Moller HJ (2003c) Lack of association between serotonin 5-HT1B receptor gene polymorphism and suicidal behavior. Am J Med Genet B Neuropsychiatr Genet 116b(1):69–71

Russ MJ, Lachman HM, Kashdan T, Saito T, Bajmakovic-Kacila S (2000) Analysis of catechol-O-methyltransferase and 5-hydroxytryptamine transporter polymorphisms in patients at risk for suicide. Psychiatry Res 93(1):73–78

Sabol SZ, Hu S, Hamer D (1998) A functional polymorphism in the monoamine oxidase A gene promoter. Hum Genet 103(3):273–279

Sacchetti P, Mitchell TR, Granneman JG, Bannon MJ (2001) Nurr1 enhances transcription of the human dopamine transporter gene through a novel mechanism. J Neurochem 76(5):1565–1572

Saetre P, Lundmark P, Wang A, Hansen T, Rasmussen HB, Djurovic S, Melle I, Andreassen OA, Werge T, Agartz I, Hall H, Terenius L, Jonsson EG (2010) The tryptophan hydroxylase 1 (TPH1) gene, schizophrenia susceptibility, and suicidal behavior: a multi-centre case-control study and meta-analysis. Am J Med Genet B Neuropsychiatr Genet 153b(2):387–396

Sakado K, Sakado M, Muratake T, Mundt C, Someya T (2003) A psychometrically derived impulsive trait related to a polymorphism in the serotonin transporter gene-linked polymorphic region (5-HTTLPR) in a Japanese nonclinical population: assessment by the Barratt impulsiveness scale (BIS). Am J Med Genet B Neuropsychiatr Genet 121b(1):71–75

Sanders AR, Duan J, Gejman PV (2002) DNA variation and psychopharmacology of the human serotonin receptor 1B (HTR1B) gene. Pharmacogenomics 3(6):745–762

Sawiniec J, Borkowski K, Ginalska G, Lewandowska-Stanek H (2007) Association between 5-hydroxytryptamine 1A receptor gene polymorphism and suicidal behavior. Przegl Lek 64(4-5):208–211

Schild AH, Pietschnig J, Tran US, Voracek M (2013) Genetic association studies between SNPs and suicidal behavior: a meta-analytical field synopsis. Prog Neuropsychopharmacol Biol Psychiatry 46:36–42

Schinka JA, Letsch EA, Crawford FC (2002) DRD4 and novelty seeking: results of meta-analyses. Am J Med Genet 114(6):643–648

Schinka JA, Busch RM, Robichaux-Keene N (2004) A meta-analysis of the association between the serotonin transporter gene polymorphism (5-HTTLPR) and trait anxiety. Mol Psychiatry 9(2):197–202

Sen S, Nesse RM, Stoltenberg SF, Li S, Gleiberman L, Chakravarti A, Weder AB, Burmeister M (2003) A BDNF coding variant is associated with the NEO personality inventory domain neuroticism, a risk factor for depression. Neuropsychopharmacology 28(2):397–401

Serretti A, Calati R, Giegling I, Hartmann AM, Moller HJ, Rujescu D (2009) Serotonin receptor HTR1A and HTR2C variants and personality traits in suicide attempters and controls. J Psychiatr Res 43(5):519–525

Singh JP, Volavka J, Czobor P, Van Dorn RA (2012) A meta-analysis of the Val158Met COMT polymorphism and violent behavior in schizophrenia. PLoS One 7(8):e43423

Sjoberg RL, Ducci F, Barr CS, Newman TK, Dell'osso L, Virkkunen M, Goldman D (2008) A non-additive interaction of a functional MAO-A VNTR and testosterone predicts antisocial behavior. Neuropsychopharmacology 33(2):425–430

Souery D, Van Gestel S, Massat I, Blairy S, Adolfsson R, Blackwood D, Del-Favero J, Dikeos D, Jakovljevic M, Kaneva R, Lattuada E, Lerer B, Lilli R, Milanova V, Muir W, Nothen M, Oruc L, Papadimitriou G, Propping P, Schulze T, Serretti A, Shapira B, Smeraldi E, Stefanis C, Thomson M, Van Broeckhoven C, Mendlewicz J (2001) Tryptophan hydroxylase polymorphism and suicidality in unipolar and bipolar affective disorders: a multicenter association study. Biol Psychiatry 49(5):405–409

Stahl SM (2003) Deconstructing psychiatric disorders, part 1. Genotypes, symptom phenotypes, and endophenotypes. J Clin Psychiatry 64(9):982–983

Stefulj J, Buttner A, Kubat M, Zill P, Balija M, Eisenmenger W, Bondy B, Jernej B (2004a) 5HT-2C receptor polymorphism in suicide victims. Association studies in German and Slavic populations. Eur Arch Psychiatry Clin Neurosci 254(4):224–227

Stefulj J, Buttner A, Skavic J, Zill P, Balija M, Eisenmenger W, Bondy B, Jernej B (2004b) Serotonin 1B (5HT-1B) receptor polymorphism (G861C) in suicide victims: association studies in German and Slavic population. Am J Med Genet B Neuropsychiatr Genet 127b(1):48–50

Stoel RD, De Geus EJ, Boomsma DI (2006) Genetic analysis of sensation seeking with an extended twin design. Behav Genet 36(2):229–237

Strobel A, Wehr A, Michel A, Brocke B (1999) Association between the dopamine D4 receptor (DRD4) exon III polymorphism and measures of novelty seeking in a German population. Mol Psychiatry 4(4):378–384

Strobel A, Gutknecht L, Rothe C, Reif A, Mossner R, Zeng Y, Brocke B, Lesch KP (2003) Allelic variation in 5-HT1A receptor expression is associated with anxiety- and depression-related personality traits. J Neural Transm (Vienna) 110(12):1445–1453

Syed Z, Dudbridge F, Kent L (2007) An investigation of the neurotrophic factor genes GDNF, NGF, and NT3 in susceptibility to ADHD. Am J Med Genet B Neuropsychiatr Genet 144b(3):375–378

Taylor A, Kim-Cohen J (2007) Meta-analysis of gene-environment interactions in developmental psychopathology. Dev Psychopathol 19(4):1029–1037

Terracciano A, Sanna S, Uda M, Deiana B, Usala G, Busonero F, Maschio A, Scally M, Patriciu N, Chen WM, Distel MA, Slagboom EP, Boomsma DI, Villafuerte S, Sliwerska E, Burmeister M, Amin N, Janssens AC, van Duijn CM, Schlessinger D, Abecasis GR, Costa PT Jr (2010) Genome-wide association scan for five major dimensions of personality. Mol Psychiatry 15(6):647–656

Tochigi M, Hibino H, Otowa T, Kato C, Marui T, Ohtani T, Umekage T, Kato N, Sasaki T (2006) Association between dopamine D4 receptor (DRD4) exon III polymorphism and Neuroticism in the Japanese population. Neurosci Lett 398(3):333–336

Tomitaka M, Tomitaka S, Otuka Y, Kim K, Matuki H, Sakamoto K, Tanaka A (1999) Association between novelty seeking and dopamine receptor D4 (DRD4) exon III polymorphism in Japanese subjects. Am J Med Genet 88(5):469–471

Tovilla-Zarate C, Juarez-Rojop I, Ramon-Frias T, Villar-Soto M, Pool-Garcia S, Medellin BC, Genis Mendoza AD, Narvaez LL, Humberto N (2011) No association between COMT val-

158met polymorphism and suicidal behavior: meta-analysis and new data. BMC Psychiatry 11:151

Tsai SJ, Hong CJ, Yu YW, Chen TJ (2004a) Association study of a brain-derived neurotrophic factor (BDNF) Val66Met polymorphism and personality trait and intelligence in healthy young females. Neuropsychobiology 49(1):13–16

Tsai SJ, Hong CJ, Yu YW, Chen TJ (2004b) Association study of catechol-O-methyltransferase gene and dopamine D4 receptor gene polymorphisms and personality traits in healthy young chinese females. Neuropsychobiology 50(2):153–156

Tsuchimine S, Yasui-Furukori N, Kaneda A, Saito M, Sugawara N, Kaneko S (2009) Minor genetic variants of the dopamine D4 receptor (DRD4) polymorphism are associated with novelty seeking in healthy Japanese subjects. Prog Neuropsychopharmacol Biol Psychiatry 33(7):1232–1235

Turecki G, Briere R, Dewar K, Antonetti T, Lesage AD, Seguin M, Chawky N, Vanier C, Alda M, Joober R, Benkelfat C, Rouleau GA (1999) Prediction of level of serotonin 2A receptor binding by serotonin receptor 2A genetic variation in postmortem brain samples from subjects who did or did not commit suicide. Am J Psychiatry 156(9):1456–1458

Turecki G, Zhu Z, Tzenova J, Lesage A, Seguin M, Tousignant M, Chawky N, Vanier C, Lipp O, Alda M, Joober R, Benkelfat C, Rouleau GA (2001) TPH and suicidal behavior: a study in suicide completers. Mol Psychiatry 6(1):98–102

Vandenbergh DJ, Persico AM, Hawkins AL, Griffin CA, Li X, Jabs EW, Uhl GR (1992) Human dopamine transporter gene (DAT1) maps to chromosome 5p15.3 and displays a VNTR. Genomics 14(4):1104–1106

Vassos E, Collier DA, Fazel S (2014) Systematic meta-analyses and field synopsis of genetic association studies of violence and aggression. Mol Psychiatry 19(4):471–477

Vitiello B, Stoff DM (1997) Subtypes of aggression and their relevance to child psychiatry. J Am Acad Child Adolesc Psychiatry 36(3):307–315

Walderhaug E, Herman AI, Magnusson A, Morgan MJ, Landro NI (2010) The short (S) allele of the serotonin transporter polymorphism and acute tryptophan depletion both increase impulsivity in men. Neurosci Lett 473(3):208–211

Walitza S, Renner TJ, Dempfle A, Konrad K, Wewetzer C, Halbach A, Herpertz-Dahlmann B, Remschmidt H, Smidt J, Linder M, Flierl L, Knolker U, Friedel S, Schafer H, Gross C, Hebebrand J, Warnke A, Lesch KP (2005) Transmission disequilibrium of polymorphic variants in the tryptophan hydroxylase-2 gene in attention-deficit/hyperactivity disorder. Mol Psychiatry 10(12):1126–1132

Wang JY, Jia CX, Lian Y, Sun SH, Lyu M, Wu A (2015) Association of the HTR2A 102T/C polymorphism with attempted suicide: a meta-analysis. Psychiatr Genet 25(4):168–177

Wasserman D, Geijer T, Sokolowski M, Rozanov V, Wasserman J (2006) The serotonin 1A receptor C(-1019)G polymorphism in relation to suicide attempt. Behav Brain Funct 2:14

Weder N, Yang BZ, Douglas-Palumberi H, Massey J, Krystal JH, Gelernter J, Kaufman J (2009) MAOA genotype, maltreatment, and aggressive behavior: the changing impact of genotype at varying levels of trauma. Biol Psychiatry 65(5):417–424

Widom CS, Brzustowicz LM (2006) MAOA and the "cycle of violence": childhood abuse and neglect, MAOA genotype, and risk for violent and antisocial behavior. Biol Psychiatry 60(7):684–689

Williams HJ, Owen MJ, O'Donovan MC (2007) Is COMT a susceptibility gene for schizophrenia? Schizophr Bull 33(3):635–641

Wu S, Comings DE (1999) A common C-1018G polymorphism in the human 5-HT1A receptor gene. Psychiatr Genet 9(2):105–106

Young LJ (2002) The neurobiology of social recognition, approach, and avoidance. Biol Psychiatry 51(1):18–26

Zai CC, Manchia M, De Luca V, Tiwari AK, Chowdhury NI, Zai GC, Tong RP, Yilmaz Z, Shaikh SA, Strauss J, Kennedy JL (2012) The brain-derived neurotrophic factor gene in suicidal behaviour: a meta-analysis. Int J Neuropsychopharmacol 15(8):1037–1042

Zalsman G, Frisch A, King RA, Pauls DL, Grice DE, Gelernter J, Alsobrook J, Michaelovsky E, Apter A, Tyano S, Weizman A, Leckman JF (2001) Case control and family-based studies of tryptophan hydroxylase gene A218C polymorphism and suicidality in adolescents. Am J Med Genet 105(5):451–457

Zhang G, Stackman RW Jr (2015) The role of serotonin 5-HT2A receptors in memory and cognition. Front Pharmacol 6:225

Zhang HY, Ishigaki T, Tani K, Chen K, Shih JC, Miyasato K, Ohara K, Ohara K (1997) Serotonin2A receptor gene polymorphism in mood disorders. Biol Psychiatry 41(7):768–773

Zhang T, Haws P, Wu Q (2004a) Multiple variable first exons: a mechanism for cell- and tissue-specific gene regulation. Genome Res 14(1):79–89

Zhang X, Beaulieu JM, Sotnikova TD, Gainetdinov RR, Caron MG (2004b) Tryptophan hydroxylase-2 controls brain serotonin synthesis. Science 305(5681):217

Zhang X, Gainetdinov RR, Beaulieu JM, Sotnikova TD, Burch LH, Williams RB, Schwartz DA, Krishnan KR, Caron MG (2005) Loss-of-function mutation in tryptophan hydroxylase-2 identified in unipolar major depression. Neuron 45(1):11–16

Zhang X, Beaulieu JM, Gainetdinov RR, Caron MG (2006) Functional polymorphisms of the brain serotonin synthesizing enzyme tryptophan hydroxylase-2. Cell Mol Life Sci 63(1):6–11

Zill P, Buttner A, Eisenmenger W, Moller HJ, Ackenheil M, Bondy B (2007) Analysis of tryptophan hydroxylase I and II mRNA expression in the human brain: a post-mortem study. J Psychiatr Res 41(1-2):168–173

Zuckerman M, Bone RN, Neary R, Mangelsdorff D, Brustman B (1972) What is the sensation seeker? Personality trait and experience correlates of the Sensation-Seeking Scales. J Consult Clin Psychol 39(2):308–321

Zuk O, Hechter E, Sunyaev SR, Lander ES (2012) The mystery of missing heritability: genetic interactions create phantom heritability. Proc Natl Acad Sci U S A 109(4):1193–1198

Psychobiology and Psychoanalysis

10

Elias Andreoulakis and Ioanna Ierodiakonou-Benou

10.1 To Link or Not to Link?

The question whether psychobiology and psychoanalysis can be bridged goes back to the very birth of psychoanalysis. Sigmund Freud, the founder of psychoanalysis, started his career as a neuroscientist; he was trained as a neuroanatomist and originally studied some—conservatively speaking—neurological disorders such as aphasia and cerebral palsies in children before turning to the "neuro-psychoses of defense" and the "studies on hysteria" and the development of psychoanalysis. Freud strongly believed that psychoanalytic ideas could be eventually corroborated by answers provided by biology. He aimed to put psychoanalysis among natural sciences in order to strengthen the validity of his formulations and gain wider acceptance within the scientific community.

This passionate effort of his is illustrated in his writing the *Project for a Scientific Psychology*. In this work, Freud conceptualized mental phenomena in the framework of hydraulic psychic energy and the phenomena of accumulation of excessive amounts of energy and the consequent pressure to discharge. In this context, he described different types of neurons receiving input from inside or outside and exhibiting excitatory, secretory and inhibitory functions (McCrone 2004; Freud 1966).

Freud's initial enthusiasm for the project was soon followed by intense frustration that made him discard "the project" and refrain himself from publishing it. The *Project for a Scientific Psychology* was eventually published after his death, in 1954. However, in this work, Freud had already introduced fundamental concepts of the psychoanalytic theory.

E. Andreoulakis · I. Ierodiakonou-Benou (✉)
3rd Department of Psychiatry, Faculty of Medicine, School of Health Sciences, Aristotle University of Thessaloniki, Thessaloniki, Greece
e-mail: andreli@med.auth.gr

K. N. Fountoulakis, I. Nimatoudis (eds.), *Psychobiology of Behaviour*,
https://doi.org/10.1007/978-3-030-18323-3_10

After abandoning his initial endeavor, Freud turned to a pure psychological theory. The reason for this seemed to be that neuroscience was not technically sufficient at that time to provide the answers he was seeking for. In addition, strict localization approach that was prevailing at his time did not satisfy him (Schore 1997; Northoff 2012). Nevertheless, he did not give up the belief that such a quest would become fruitful in the near future.

Thereon, psychoanalysis and neuroscience followed a rather divergent course for most of the twentieth century. As Kandel points out, "*psychoanalysis represented a revolution in our thinking about the brain in the first half of the 20th century*" but although "*it still represents the most coherent and intellectually satisfying view of the mind it has not evolved scientifically*" mainly in that it suffers from a lack of objective measures and methods to test its theoretical formulations in an experimental setting rather than the analytic situation itself, which is subjective and thus renders susceptible to observer bias. "*As a result, psychoanalysis enters the twenty-first century with its influence in decline.*" Although Kandel acknowledges that "*we are very much at the beginning*," he argues that the biological foundation of many psychoanalytic concepts might be feasible in the near future, judging from the pace of advance in neuroscience (Kandel 1999). Kandel also argues that much of the unease hampering the dialogue between neuroscience and psychoanalysis might be due to a misconception regarding the way genes work and emphasizes on the mutually influenced interplay between "*nature*" and "*nurture*" by highlighting the *dual function of genes*:

1. Genes demonstrate a *template* function (to ensure reliable replication) which is vulnerable to no other influence (environmental or social) than mutations.
2. However, genes also have a *transcriptional* function (protein production) which can be regulated in response to environmental factors (epigenetic regulation), including learning. In that sense, "*Behaviour itself can also modify gene expression and produce sustained alterations in neuronal connectivity and functioning*" (Kandel 1999).

Kandel regards a dialogue between psychoanalysis and neuroscience as feasible, meaningful, and necessary "*if psychoanalysis is to have any future*" within the evolving sciences of the brain. He has outlined some areas in which neuroscience could contribute to psychoanalysis, such as preconscious and unconscious mental processes, psychological causality and its relation to psychopathology, early experiences and the predisposition for a mental disorder, brain structural changes in response to psychotherapy, and the potential for a synergistic effect of psychopharmacology in adjunction to psychoanalysis (Kandel 1999).

Kandel proposed "*a new intellectual framework*" (Kandel 1999) for a dialogue between neuroscience and psychiatry—applicable both for a dialogue involving psychoanalysis—based on certain principles:

1. "*All mental processes, even the most complex psychological processes, derive from operations of the brain*" and "*even those disturbances that are clearly environmental in origin are disturbances in brain function.*"

2. "*Genes, through their protein products, determine the pattern of neuronal interconnections in the brain and, thus, the functioning of neurons. Consequently, genes influence behaviour. In other words, one component of major mental illnesses, either inherited or acquired, is genetic... Genetic illnesses (e.g., schizophrenia) are expressions of altered genes, whereas illnesses acquired as learned behaviour (neuroses) involve the modulation of gene expression by environmental stimuli, leading to the transcription of a previously inactive gene.*"
3. That said, "*altered genes do not, by themselves, account for all of the variance of a given major mental illness and thus, an important contribution of social or developmental factors should also be acknowledged.*"
4. In addition, "*environmental factors as well as behaviour—learning, in particular—can, in reverse modify the expression of genes and consequently the neuronal connectivity and functioning.*"
5. Kandel regards psychotherapy as "*a learning process that can produce long-term changes through a significant modification on genes expression that results in structural and functional changes in the brain*" (Kandel 1999).

As stated above, psychoanalysis and neuroscience have been in a state of mutual disregard and avoidance during most of the twentieth century, but this appears to be changing over the last two decades. A mutual interest on behalf of both neuroscientists and psychoanalysts regarding the bridging of the two disciplines has emerged (Yovell et al. 2015). Thus, a century after, the interest in continuing Freud's initial endeavor has reemerged, on the ground of recent technological advances that allow in vivo study of the brain in function.

Recent advances in neuroscience have also led to a reframing of brain functioning conceptualization. Strict localization approach has given its place to a function-based and a diffuse dynamic localization approach. The modular approach of highly specialized segregations of neurons has been reconsidered in the light of findings concerning complex interconnections and interactions in the context of neuronal integration, i.e., the hierarchical organization and coordination of neuronal activity across various interconnected brain areas (Northoff 2012). Research focused on brain connectivity—anatomical, functional, and effective (Friston 2011)—is thought to be able to enhance the convergence between psychoanalysis and neuroscience by providing insightful information on the circuitries and mechanisms underlying mental functions (Salone et al. 2016). Northoff proposes that neuroscientific research may now go beyond the function and localization-based approach, moving from the "Neural *Correlates* of Psychodynamics" to the "Neural *Predispositions* of Psychodynamics," which he relates with the brain's *resting-state activity* and its "*spatiotemporal structure*." The latter reflects the two essential characteristics of the resting-state activity, namely, the low-frequency fluctuations and the patterns of functional connectivity. According to Northoff, the "spatiotemporal structure" of the brain's resting-state activity may be parallelized to the Freudian conceptualization of the "psychological structure" of the mental apparatus, in that both concepts reflect a process and a dynamic organization rather than a strictly anatomical entity. In addition, spatiotemporal structure of the resting-state activity may provide a model related with the ego functions and dreams (Northoff 2012). Northoff also suggests a direct link between abnormalities of the "spatiotemporal structure" of the resting state and psychopathological symptoms such as ego disturbances and auditory hallucinations observed in schizophrenia (Northoff 2015).

In the context of this revitalized interest, several researchers have revisited Freud's ideas linking them with current neuroscientific findings (Carhart-Harris and Friston 2010; Carhart-Harris et al. 2008; Fotopoulou et al. 2012; Kandel 1999; Kaplan-Solms 2000), while others challenge this revisiting on Freudian ideas arguing that new insights provided so far are not unique to psychoanalytical concepts but can be embedded in the cognitive, affective, and social neuroscience (Ramus 2013). There are also some psychoanalysts who consider that the time for a rapprochement between neuroscience and psychoanalysis has come (Schore 1997; Fonagy 2003; Kernberg 2004). However, the issue of rapprochement has not been free of confrontation and has given rise to several concerns as will be discussed later on.

This revitalized interest in the quest for a rapprochement between psychoanalysis and neuroscience has led to the emergence of a bridging interdiscipline named *neuropsychoanalysis* (Nersessian and Solms 1999). Neuropsychoanalysis aims to link psychoanalysis with neuroscience and understand the human mind by investigating its functions in a cross-species context (Panksepp and Solms 2012). Although the proponents of neuropsychoanalysis argue that the investigation of neural networks associated with complex behaviors is essential for the thorough understanding of the latter, they refrain themselves from the prevailing reductionism and the temptation to make hasty inferences concerning causality regarding the relationship between human behavior and the functioning in certain brain areas (Panksepp and Solms 2012).

As mentioned above, the meaningfulness of a dialogue between psychoanalysis and neuroscience represents a matter of debate, probably representing a new era of what might lead to the—hopefully fruitful—"controversial discussions of the twenty-first century." Many questions and concerns have risen, such as: Is this bridging legitimate? Is it feasible? Would it be meaningful and beneficial for both disciplines, i.e., neuroscience and psychoanalysis?

A fruitful controversy has begun regarding the aim to engage psychoanalysis and neuroscience in a dialogue within the framework of neuropsychoanalysis, with arguments raised for (Yovell et al. 2015; Canestri 2015; Panksepp and Solms 2012; Pugh 2007; Mancia 2007) and against the newborn interdiscipline (Blass and Carmeli 2007, 2015; Carmeli and Blass 2013; Pulver 2003; Ramus 2013).

While proponents argue that neuropsychoanalysis is consistent with Freud's conceptualization and represents a legitimate continuation of his legacy, opponents challenge such an inference as contradicting the broader spirit of his work (Blass and Carmeli 2015).

Solms and colleagues, leading proponents of bridging psychoanalysis with neuroscience by means of neuropsychoanalysis, argue that such a dialogue should be fostered on the grounds that psychoanalysis and neuroscience essentially study the same entity, which they call as "*mindbrain*," speaking of a "*dual-aspect monism*." At the same time, however, they acknowledge that neither discipline can be sufficient in exploring and describing this entity on its own. They argue in favor of a mutually respectful dialogue and collaboration between the two disciplines that "*does not eliminate each other's unique perspective.*" They admit that psychoanalysis and neurosciences are and will continue to be two inherently different disciplines that employ different methodologies to investigate two complementary aspects of the human mind—the subjective and the objective—"*neither of which is reducible to the other or more real than the other*" (Yovell et al. 2015; Solms and Turnbull 2002). The very essence of "mindbrain" and "dual-aspect monism" has been challenged (Blass and Carmeli 2015; Karlsson 2010), particularly by Blass, Carmeli, and colleagues who have extensively stressed important and plausible concerns (Blass and Carmeli 2007, 2015; Carmeli and Blass 2013). Blass and Carmeli challenge the very essence of "dual-aspect monism" postulated by Solms and colleagues although they acknowledge that there is indeed an inherent tie and a mutually influenced relationship between the mind and the brain and that mental phenomena can be "correlated" with brain functions. They would prefer the term "correlated" as less reductionistic compared to "mapped to a biological substrate." Besides challenging the opinion that neuroscience and psychoanalysis study the same entity, Blass and Carmeli argue that even if this argument is accepted, it could not be a sufficient condition for an in depth and mutually influencing dialogue, as in many other approaches that might share the same study object with psychoanalysis. Exerting further criticism on dual-aspect monism, Blass and Carmeli argue that psychoanalysis and neuroscience do not share the same research question, since psychoanalysis studies the subjective experience, whereas neuroscience studies the objective part (brain function). They argue that "*the analyst not only seeks to discover patient's meaning but also goes beyond this by trying to make them manifest themselves in the analytic setting in order for them to be amenable to working through. To this end, they argue, biology has nothing to say.*"

Blass and Carmeli also consider neuroscientific findings as irrelevant to psychoanalysis using some metaphors to further illustrate and corroborate their view. In specific, they argue that the study of the piano structure and function though related with the music produced would be irrelevant to study of the music itself as the study of chemicals would be irrelevant to the painter. In line with this, they raise plausible concerns against a materialistic and reductionistic approach which would contradict the very essence of Freud's spirit as illustrated in the following passage: "*It will soon be clear what the mental apparatus is; but I must beg you not to ask what material it is constructed of. That is not a subject of psychological interest... We shall leave entirely on one side the material line of approach*" (Freud 1969).

Blass and Carmeli also criticize some of the neuroscientific findings as general psychological but also do not consider the knowledge gained from neuroscience as necessary for psychoanalysis, arguing that the analyst should not ask for external information sources so that he can understand what is happening in the patient's mind and what his/her subjective meanings are. They conclude by arguing that neuroscientific findings are not only irrelevant (as not studying the same entity or addressing the same research question) but even harmful to psychoanalysis, particularly if neuroscientific knowledge is to interfere with the practice of psychoanalysis (Blass and Carmeli 2015). This is indeed a matter of concern to which we will come back later on.

Proponents of the dialogue argue that this could be mutually beneficial; on one hand, the psychoanalytic theory can provide a conceptual framework guiding neuroscientific research, whereas, on the other hand, the neuroscience may contribute to the neurobiological foundation of psychoanalytic theory and concepts. Consequently, this might lead to a fruitful revisiting of both psychoanalytic theory and practice (Yovell et al. 2015).

At this point, some authors—although not entirely against a dialogue—speak of a somehow one-sided benefit as they argue that psychoanalysis might help neuroscience in a better targeted research, whereas the knowledge derived from neuroscience would be meaningful to a to a cognitive neuroscientist but not a psychoanalyst (Mechelli 2010). Others agree that neuroscience could be beneficial for the psychoanalytic theory, in that neuroscience might help psychoanalysis sort and systematize the knowledge within the abundance of psychoanalytic ideas and answer essential questions, such as "how therapy cures" (Fonagy 2003) or by providing evidence that would enable choosing between competing analytic theories whereas it is also stressed that it would be irrelevant and would have nothing to offer to psychoanalytic practice (Pulver 2003).

There is some skepticism concerning the validation of psychoanalytic theory by means of neuroscientific evidence. Pulver as well as Blass and Carmeli acknowledge that psychoanalytic models of explanations cannot be incongruent to the new findings of neuroscience (Blass and Carmeli 2007; Pulver 2003). That said, it is also stressed that neuroscience should not represent the "final court of appeal for psychoanalytic models" (Blass and Carmeli 2007). Similarly, Mitchell (1995) argues that it is communication within the psychoanalytic community that can provide the necessary "testing-ground" for psychoanalytic theory and practice. In addition, Blass and colleagues also plausibly argue that psychoanalysis should not restrict and redirect its interest only in concepts (such as certain drives and motives), the existence of which has been confirmed and mapped biologically (Blass and Carmeli 2007, 2015). Despite the disagreement concerning the role of neuroscience in validating or revisiting (or probably discarding) certain formulations of the psychoanalytic theory, the proponents of neuropsychoanalysis have also refrained themselves from assigning a decisive role to neuroscience. As they state, this is not only because any new findings might be proved to be wrong in the long run but also because any findings leading to a proposal for a revisiting of psychoanalytic theory and practice should always be tested within the analytic situation (Yovell et al. 2015).

There are also some concerns raised in terms of methodology and the feasibility of a meaningful dialogue. Regarding the cross-validation of psychoanalytic concepts, Mechelli plausibly stresses that a necessary assumption is that the concepts under investigation many of which have been so far unique to psychoanalytic thinking have to be defined in such a manner that would make them widely acceptable. In addition, a fully objective interpretation of neuroscientific findings might be flawed by the researcher's theoretical predisposition. Moreover, neuroscientific findings have to be consolidated and freed from controversies in order to serve as a testing-ground knowledge base (Mechelli 2010).

One important barrier in bridging psychoanalytic concepts and neuroscience lies in the arduous task to link the subjective experience of a person (first-person experience) with the observation of his/her brain function from a third-person (investigator) perspective. In an attempt to overcome this barrier, Norhoff and colleagues recommended the employment of a strategy termed as "first-person neuroscience" which aims to "*preserve the individual contents*" of psychological states as obtained from first-person perspective but at the same time, "*make them amenable to objectification and quantification*" in order to be accessible to systematic data gathering and investigation "*without compromising their complexity and richness*." To this end, self-rated instruments such visual analogue scales or semi-structured interviews might be used in order to transcribe the first-person experience into measurable and reliable data. Northoff goes on to point to the need for a further development in assessment tools, namely, "first-person questionnaires" (Northoff and Heinzel 2006; Northoff et al. 2007).

A point of convergence is that opponents of neuropsychoanalysis admit that "*neuroscientific findings can help demarcate the limits of psychoanalysis*," in that they can identify the cases where the benefit of psychoanalysis is limited by brain abnormalities that render a patient less capable of "*meaningful expression*" (Blass and Carmeli 2007). That said, they clarify that this should be not be taken as a categorical argument that "psychoanalysis has nothing to say" in cases of severe brain damage and hold that there always remains a psychic meaning in the patient's behavior no matter how this behavior is related with certain brain lesions (Blass and Carmeli 2015).

Concerning the interference of neuroscientific findings with psychoanalytic practice, Blass and colleagues criticize Solms' and colleagues' argument presented by means of a clinical illustration in which neuroscientific knowledge on the unfeasibility of retrieval of early traumatic memories due to certain neuronal dysfunction is conveyed to the analysand. They consider this as a consolatory and nonanalytical practice despite the relief of anxiety or guilt it might provide to the patient. They argue that "*when information and consolation are introduced into the analytic setting to calm latent anxieties to the neglect of the dynamic sources of the anxieties, the dynamic sources remain untouched*" (Blass and Carmeli 2015). This is the main argument concerning the potential harmfulness of a dialogue between neuroscience and psychoanalysis as stressed by Blass and Carmeli. We agree that acquisition of neuroscientific knowledge on behalf of the analyst is one thing, while conveying such a knowledge to the patient is another.

In response to the necessity of a dialogue between neuropsychoanalysis and psychoanalysis in order to avoid scientific isolation of the latter, Blass and Carmeli (2015) argue that "*the spirit of psychoanalysis should not be distorted in order to belong to the mainstream*".

Another potential benefit proposed by those who argue in favor of neuropsychoanalysis is that neuroscience could help psychoanalysis further demonstrate its therapeutic efficacy in an evidence-based context, by identifying structural or functional changes in the brain in response to the therapy. However, as Mechelli plausibly stresses, neuroanatomical changes are of a quantitative nature and cannot be fully informative about the nature or the quality of the analytic experience or the effectiveness of treatment (Mechelli 2010). We would corroborate and go on to say that effectiveness of treatment is also to be assessed in terms of insight acquired, growth gained, and the improvement achieved in patient's living and relating with others besides the assessment of his/her functioning in various domains of life.

10.2 Neuroscientific Findings in Relation to Psychoanalytic Theory

We will now turn to the presentation of some current neuroscientific findings that are linked to essential aspects of the psychoanalytic theory.

10.2.1 The Unconscious

The study of the unconscious and its dominant role in mental life represents a cornerstone in psychoanalytic theory, which is also unique to psychoanalysis, in that the conceptualization of unconscious (encompassing phantasies, emotions, and motivation) has been fundamentally different from its view in cognitive neuroscience. In that context, some authors prefer to refer to "*non-conscious*" instead of unconscious processes to discriminate cognitive from emotional and motivational aspects of the unconscious (Berlin 2011).

Moving closer to the psychoanalytic conceptualization of unconscious, the interest in unconscious processes has been expanded recently from the study of cognitive functions, such as memory and learning, to the exploration of emotions and motivation within the evolving fields of affective and social neuroscience.

As for the "cognitive" part of the unconscious, it has been found that unconscious perception of subliminal stimuli can induce motor responses or, at a more complex level, activate high-level cortical areas to induce an appropriate behavioral response. In addition, semantic priming and associative learning have been found to take place outside awareness. Notably, besides unconscious cognitive processes, there is sufficient evidence that emotion processing can also occur outside conscious awareness; subliminal (thus not consciously perceived) emotionally valenced stimuli can activate emotional circuits including, but not restricted to, the amygdala (Berlin 2011). It has also been found that conscious interfering in the processing of a stimulus can eliminate the conscious perception of the emotional valence of the stimulus. In other words one could say that conscious interfering, particularly exerted by the prefrontal cortex, can *repress* information toward the unconscious. In addition to the interaction between cognitive and unconscious emotional processing, it has been found that unconscious emotionally valenced stimuli can influence motivation, social judgment, and goal-directed behavior. Decisions seem to be influenced by unconscious processes more than we believed (Berlin 2011). Recent reviews have challenged the so far prevailing role of consciousness on the setting and pursuing personal goals and documented the presence of what has been described as "unconscious will," which initiates the perseverance of personal goals before or even outside conscious attending to them (Custers and Aarts 2010; Dijksterhuis and Aarts 2010). Conscious decision-making is thought to be more precise but with less capacity to integrate larger proportion of information compared to unconscious thinking. Thus, it is proposed that conscious thinking might be more suitable for simple matters, whereas unconscious processing might offer better choices under complex conditions (Berlin 2011).

The increased interest in the study of all the aspects of the unconscious has led to the exploration of its neurological substrate. Schore assigns the neurological substrate of the dynamic unconscious to the right hemisphere, which he considers as the "nest" for sexual and aggressive impulses. He also argues that repression of such impulses an interhemispheric inhibition exerted by the left hemisphere upon the right. However, such an inhibition can be feasible only after the left hemisphere has established dominance over the right, i.e., after the age of 4. This led Schore to conclude that the formation of the unconscious in the early years of life must rely on mechanisms other than repression (Schore 2003). In contrast, Kaplan-Solms and Solms challenge the linking of the right hemisphere with the capacity for repression based on their clinico-anatomical studies in patients with damage in the perisylvian area of the right hemisphere. These patients demonstrated greater ego-deficits, and thus less capacity for repression, compared to their left-hemisphere counterparts (Kaplan-Solms 2000).

10.2.2 Defense Mechanisms

It is stressed that the complex emotional-cognitive interaction underlying the constitution of defense mechanisms can be better explored in the context of neuronal integration (Northoff et al. 2007).

10.2.2.1 Regression

It has been postulated that the *sensorimotor regression* (as typically observed in catatonia) might be related with disruption in the function of a neural circuit involving the orbitofrontal, the premotor, and the medial prefrontal cortices. The orbitofrontal cortex has also been highlighted in constituting defenses that are relatively more mature (Northoff et al. 2007).

10.2.2.2 Repression

Although the neural mechanism underlying *repression* is still unknown, it has been observed that people with longer "utilization brain duration," i.e., the critical time period of neural activation in order for a stimulus to become conscious, might be more prone to repression (Berlin 2011).

10.2.2.3 Suppression

Compared to repression, *suppression*, i.e., the conscious casting unbearable thoughts or emotions away from awareness, has been associated with increased activation of dorsolateral prefrontal cortex and reduced activation of the hippocampus. It has also been found that as long as suppression is constantly repeated, it can evolve to repression (Berlin 2011).

10.2.3 Memory and the Unconscious

It is stressed that the unconscious as a storage of content related—though not limited—to memories can plausibly be linked to the structures involved in memory (Mancia 2006).

Memory encompasses a dual system consisting of overlapping and usually working together subsystems, namely:

(a) *Explicit* or *declarative* memory, which is conscious and encodes autobiographical and other factual information on people, objects, and places. The medial temporal lobe and particularly the hippocampus complex are essential for its function (Milner et al. 1998). However, it has also been stressed that after sufficient time, the storage or retrieval of explicit memory no longer depends on hippocampus but on the neocortex instead (Squire and Alvarez 1995).
(b) *Implicit* memory, which involves unconscious information of many kinds involving different brain areas (following in parenthesis below).

In specific, implicit memory involves (a) *procedural* memory for skills and habits (striatum); (b) *priming* (neocortex); (c) *classical conditioning*, further related to *emotional responses* (amygdala) and the *skeletal musculature* (cerebellum); and, finally, (d) *associated learning* (reflex pathways) (Milner et al. 1998). There occurs a significant interplay between the two subsystems as many learning processes involve both, whereas the repetition of a task can transform explicit memory to implicit, of procedural type (Kandel 1999). In addition, amygdala, involved in implicit memory, also exerts an influence on explicit memory, due to its involvement in emotion processing (Milner et al. 1998).

Linking memory with the unconscious, Kandel postulates that the part of the unconscious that does not contain conflict-evoking or repressed material but is concerned with habits and skills, either perceptual or motor (i.e., what he calls the "procedural unconscious"), can be mapped to the procedural implicit memory (Kandel 1999). On the other hand, the dynamic (repressed) unconscious might be linked with the circuit involved in explicit memory (Kandel 1999; Mancia 2006). Furthermore, given that amygdala matures earlier than the hippocampus, Mancia argues that early infancy experiences cannot be stored in the explicit memory (which is hippocampus-dependent) but are filed in the implicit memory since this is the only one available at that time (Mancia 2006). Besides early experiences of any kind, traumatic experiences (even of later time) cannot be filed in explicit memory either due to a trauma induced loss of hippocampal neurons (McEwen and Sapolsky 1995). Such experiences are also thought to be filed in implicit memory, constituting what Mancia calls a *late unrepressed unconscious,* which is irretrievable by conscious effort though not because of repression (Mancia 2006). Mancia argues that these irretrievable experiences might be accessible through dreaming, as in the case of patients with bilateral hippocampal lesions who were failed to consciously recall the previous training for a memorizing game but mentioned dreaming about it (Stickgold et al. 2000).

10.2.4 Dreaming

In this part, we will refrain from dealing with the neurobiology of dreaming as a psychophysiological phenomenon and will be concerned with the neuroscientific findings that might provide insight regarding the function of dreaming, as the latter reflects the aspect of dreaming that psychoanalysis is concerned with. There are some findings that corroborate the Freudian conceptualization of the dreamwork. For instance, Mancia argues that there are neuroscientific findings that can confirm the "*derepression*" function of dreaming, i.e., bringing repressed material back to light. According to Mancia, such a function can be inferred from the reciprocal pattern of activation/deactivation observed during REM-related dreaming compared to deactivation/activation of the same areas, namely, the hippocampus and the dorsolateral frontal cortex, respectively, during voluntary repression. (It is a matter of terminology whether this should better be called suppression.) In addition, Johnson argues that there is strong evidence corroborating the "*wish-fulfilment*" function of dreams, as stressed by Freud (Johnson 2001).

10.2.5 Empathy

The discovery of the mirror neuron systems in the macaque monkey (di Pellegrino et al. 1992) and the verification of the existence of a similar system in humans (Rizzolatti 2005) have given the opportunity to view human beings no longer as a social and self-centered minds but as capable of interpersonal communion and other-centered participation at the very outset of life (as of infancy) in the context of "intersubjectivity" (Bråten 2007). In addition, the discovery of mirror neurons has enhanced the study of an issue closely related to psychoanalysis such as empathy.

10.2.5.1 Actions

A circuit encompassing the middle temporal gyrus, the rostral inferior parietal lobule, and the ventral premotor cortex has been found to be involved in both the performance of one's own actions and the observation of the same actions by others (Keysers and Gazzola 2006). In addition, important properties of the mirror system, namely, (a) *selectivity* to a certain kind of a precise action as well as (b) a *somatotopical organization*, not only for actions but also *for the related sounds* have also been found in humans (Gazzola et al. 2006). Another interesting finding is that mirroring might also be gender-specific as revealed in a study among male and female dancers (Calvo-Merino et al. 2005). However, in our opinion, it might warrant further clarification whether such a mirroring represents a gender-specific or a role-specific one.

10.2.5.2 Sensations

Concerning sensations, it has been found that the somatosensory cortices are involved in both experiencing touch on oneself and viewing others being touched (Keysers et al. 2004). Similarly, anterior cingulate and insular cortices have been found to be involved in the both the first-person experience of one's own pain and the perception of others in pain (Jackson et al. 2005; Avenanti et al. 2005).

10.2.5.3 Emotions

As for emotions, the anterior insula has been found to involve both in the experience of disgust and in the observation of disgust in others, whereas amygdala has been implicated in both the subjective experience of fear and in the perception of fear in others. However, Keysers and colleagues challenge the opinion that amygdala shows fear selectivity. Instead, they argue that amygdala is involved in processing of all facial expressions and that recognition of fear in fearful facial expressions might be confounded by factors such as focusing on the eye part of the picture or not or on the investigation method, i.e., photo with fearful expressions vs. video (Keysers and Gazzola 2006). They conclude that the role of amygdala in the experience of fear in both a first- and third-person perspective might be indirect, with amygdala representing an information providing part within a circuit involving other brain areas which might be more directly associated with fear than amygdala.

Concerning the understanding of the emotions of others from a third-person perspective, Damasio and colleagues consider a somatosensory representation of the body state of the person observed as a necessary condition (Damasio 2003b). In contrast, Keysers and Gazzola (2006) hold that somatosensory representations are important in the case of others' sensations but may not be central to our understanding of others' emotions and actions.

10.2.6 Empathy and Differentiation Between the Self and Others

Given the existence of a shared circuit involved in both first-person and third-person experience, there emerges the issue of differentiation between the representation of the self vs. the others.

Keysers and colleagues argue that there can be a discrimination between one's own experience and the perception of other's experience on the basis of a different pattern of activations among the (same) areas involved in both (Keysers and Gazzola 2006). Other authors argue that brain areas such as the right inferior parietal lobule and the posterior cingulate gyrus are involved in the explicit differentiation between self and others (Decety and Sommerville 2003; Vogt 2005). Van Veluw and colleagues reviewed relevant studies and concluded that although there are regions of higher order prefrontal cortex involved in both self-representation and the presentation of others, there are also distinct neural circuits with different connectivity patterns separately involved in each kind of representation. In specific, regions involved in both representations were the superior temporal gyrus and the ventromedial prefrontal cortex. The medial prefrontal cortex, the temporoparietal junction bilaterally, the precuneus, and the middle temporal gyrus bilaterally were involved in the representation of others, whereas the right superior temporal gyrus, the right parahippocampal gyrus, the right inferior frontal gyrus/anterior cingulate cortex, and the left inferior parietal lobe were involved in self-representation. According to the authors, these findings may provide new insight into disorders such as autism, schizophrenia, and borderline personality disorder, in which a disruption of the self and other representations is profound (van Veluw and Chance 2014).

10.2.7 The Self

In addition to the finding that the medial prefrontal cortex is selectively activated during self-referential processing (Kelley et al. 2002), there has been a link between different facets of self-referential stimuli processing and certain areas within the cortical midline structures, more specifically, *self-representation* and orbitomedial prefrontal cortex, *self-monitoring* and supragenual anterior cingulate cortex, *self-evaluation* and dorsomedial prefrontal cortex, and *self-integration* and posterior cingulate cortex (Northoff and Bermpohl 2004). Northoff and colleagues conclude that the self is related with the neuronal connectivity and integration among different brain areas. They also postulate that certain areas along the cortical midline structures linked with certain aspects of the self, extending what has been mentioned above (for instance, the right posterior insula and right inferior parietal cortex—linked with self-agency—and the right parietal and ventromedial prefrontal cortex, linked with self-ownership-), might represent the neural correlate of what Damasio has termed as the "*core self*," i.e., the continuous interaction between intero- and exteroceptive stimuli that allows the self to be perceived as a unit (Northoff 2012; Northoff and Bermpohl 2004; Damasio 2003a).

10.2.8 The Ego and the Default Mode Network

Raichle et al. have introduced the notion of default mode network (DMN) to emphasize on the brain's intrinsic activity, i.e., activity not directly related to identifiable sensory or motor events (Raichle and Snyder 2007; Raichle et al. 2001). The DMN includes the medial prefrontal cortex, the posterior cingulated cortex, the inferior parietal lobule, and the medial temporal lobes (Raichle and Snyder 2007; Raichle et al. 2001; Buckner et al. 2008). The major nodes of DMN demonstrate a strong structural and functional connectivity (Greicius et al. 2009) that develops ontogenetically, as, for instance, the medial prefrontal cortex and the posterior cingulated cortex are absent or underdeveloped in infants (Fair et al. 2008; Kelly et al. 2009).

DMN shows a competitive activity—antiphase relationship with reciprocal pattern of activation/deactivation—compared to another large-scale intrinsic network which is activated during cognitive tasks, occasionally referred to as the "*attention system*" (Fox et al. 2005).

Carhart-Harris and colleagues consider DMN as sitting at the top of the hierarchical organization of the brain and parallelize it with what they call "*the core of the ego.*" They postulate that just like ego—by employing the secondary process—tries to bound the free energy of primary process (the latter entailed by the Id), so are higher cortical regions, the DMN and medial prefrontal cortex, in particular, trying to suppress the activity of lower limbic areas, i.e., the medial temporal lobe structures such as hippocampus, parahippocampus, amygdala, and endorhinal cortex, thus exerting a *top-down control*. In sum, Harris parallelizes the Freudian concept of ego and the default mode network based on:

(a) The integrated and compound nature of both.
(b) The parallelism in the course of their development and maturation (progressive development of functional connectivity among DMN parts over time).
(c) The conceptualization of the ego as an agency that acts as a recipient and regulatory agency (top-down control) of endogenous activity coming from the Id (bottom-up) related with drives, memories, and affects similar to the structural and functional connectivity between the cortical nodes of DMN and the lower limbic structures.
(d) The conceptualization of ego as the tonic reservoir of activity similar to the high resting activity and metabolism characterizing the DMN.
(e) The Freudian conceptualization of the ego as the "seat of the sense-of-self" and the involvement of the DMN in self-referential processing.

Carhart-Harris and colleagues conclude that such a synthetic approach could provide insight for disorders like anxiety, depression, and addiction (Carhart-Harris and Friston 2010, 2012).

10.3 Conclusive Thoughts

We think that inevitably, when translating from one language to another, some shades of the original meaning cannot be grasped, no matter the expertise competence of the translators. Similar to this is Bion's postulation regarding "*O*," or "*ultimate reality*," which can never be known; it can tentatively be inferred through its transformations, but there always be something that cannot be grasped. Hence, we think that there will also be something embedded in psychoanalytic understanding that cannot be fully conveyed to neuroscientific findings and vice versa. That said, we believe that a dialogue between the two disciplines remains meaningful. Arguing in favor of a mutually respective and collaborative dialogue between the two perspectives, there come to our minds Bion's formulations regarding the "*language of substitution*" (which we would like to parallelize with the descriptive nature of the neuroscientific language), the "*language of achievement*" (which we would parallelize with the empathic understanding within psychoanalytic thinking and practicing), and his recommendation that the analyst should be sensitive to and employing both (Bion 1984a). In other words, we might say that neuroscience and psychoanalysis could provide insight deriving from different perspectives or "*vertices*" that could synergistically constitute a meaningful "*binocular vision*," to mention another Bion's formulation (Bion 1984a, b) we regard as applicable.

Thus, we think that although psychoanalysis and neuroscience do differ in terms of research question and aim, a dialogue between the two would still be meaningful, despite the limitations. Such a dialogue might not be that threatening for psychoanalysis, since there are always be an ultimate question, concerning the subjective meaning, which could be addressed only by psychoanalysis, no matter the progress in neuroscientific knowledge base. (Not to mention the fundamental difference lying in that psychoanalysis goes even further than addressing a question to dealing with dynamics that any answer evokes.)

In order to illustrate our postulation, we invite you to think of an imaginary dialogue between a patient and a neuroscientist. (At the moment, we put aside the concern regarding the provision of neuroscientific information to the patient, for the sake of the conversation.)

- Why do I feel so guilty?
- It's probably because your "guilt" circuit was activated.
- Ok, but what made it activate in the first place?

Even if the neuroscientist could provide an answer such as "an unconscious phantasy—probably an aggressive one—arose," he could not identify its qualitative characteristics, not restricted to its content, unless neuroscience could visualize one's unconscious thoughts, which is highly unlikely.

Even after overcoming this barrier, the neuroscientist would probably be faced up with another question "…And what made this fantasy come up in here and now?...".

This is a question that cannot be answered unless the dynamics concerning patients relating to others in all three dimensions, i.e., past, present, and transference, are put forward and the meaning the patients attributes to this knowledge gained is explored. Thereon, following the insight gained another question such as "now what?" might emerge, referring to the process of change, i.e., transforming knowledge gained to behavioral change. Another circle of resistances and dynamics to be elucidated and worked through by analysis. Another storm of anxieties to be "*held*" and "*contained*."

What we want to emphasize is that absolute convergence or even assimilation between the two disciplines should not be considered as meaningless or feared as harmful, since it is unfeasible in the first place. This is due to a fundamental difference between psychoanalysis and neuroscience in terms of both the nature of the main question and aim to be addressed as well as the suitability and efficiency of each domain to provide a meaningful answer and working through. Psychoanalysis seeks to come up with answers concerning *why*, i.e., identify the subjective *meaning* of certain thoughts, fantasies, or behaviors; from another vertex, neuroscience can effectively provide some answers concerning *how*, by identifying the neuronal substrate and mechanisms correlated with certain mental phenomena. At this point, we would also like to stress parenthetically that tracking certain behaviors down to their neural substrate does not necessarily reflect a causal relationship. Cross-sectional studies of first-person experience and third-person observation are inherently incapable of providing inferences regarding causality in the first place; that is, in the context of a bidirectional interplay between neurobiology and behaviour, a certain behavior observed in a certain moment might be accounted for by a functional disruption in a certain neuronal circuit, but, conversely, it might also be the case that this behavior is the preceding cause, in that it might had previously modified the relevant neural circuits. Prospective studies would be helpful, but they have to overcome the barrier of setting the "zero" (starting) point of observation, among other complexities.

In line to the relevant arguments postulated in favor of neuropsychoanalysis stressed so far, we think that neuroscience could help elucidate and provide further scientific foundation and wide acceptance for some psychoanalytic concepts and formulations. We could mention "*repetition compulsion*," "*death drive*," "*innate aggression*," and "*primary envy*"; the Kleinian concepts of "*schizoid-paranoid*" and "*depressive position*," "*projective identification*," and "*empathy*"; and Bion's "*alpha function*," "*dreamwork*," the capacity for "*symbolization*," "*psychological thinking*," "*mentalization*," or even the "*transference/countertransference*" itself, to name some among a large list. In addition, neuroscience might provide insight and further validation regarding some aspects of psychoanalytic technique (or even the setting) such as whether the "*free association*" indeed enhances the emergence of unconscious content by eliminating conscious censorship or whether the "*evenly suspended attention*" or the "*reverie*" on behalf of the analyst reflects a brain state where the grasp of patient's unconscious meaning of communication is really enhanced, or even if "lying on the couch" and talking with no visual contact with and not frequent feedback from the listener indeed enhances the unfolding of early memories and anxieties, among others. That said, we do not argue that the abundance of empirical knowledge and clinical experience added so far is not enough to establish the psychoanalytic theory, setting and technique as valid and meaningful; nevertheless, neuroscientific cross-validation might not only provide psychoanalysis wider acceptance beyond the psychoanalytic community but it might also in some cases provide useful insight that might lead to a fruitful revisiting of certain aspects of psychoanalysis. On this ground, neuroscientific knowledge could also contribute to a richer dialogue regarding several issues—hopefully even the controversial ones—within the psychoanalytic community regarding both the theory and the technique of psychoanalysis and contribute to the scientific invigoration and evolution of the latter.

On the other hand, one thing that warrants attention is the use of any insight gained from neuroscience within the analytic practice in the context of the sessions. For instance, we think that including solid neuroscientific findings into psychoanalytic knowledge base is one thing, but conveying such knowledge to the patients is another; therefore we believe that such a practice should be employed with caution, not just in order to keep in line with the Freudian recommendation for "*abstinence*" (Freud 1958). Providing the patient with knowledge about certain limitations identified by the neuroscientific research (as Solms and colleagues showed in their case report, namely, inability to access early traumatic memories) might significantly diverge from the very spirit of psychoanalytic practice if employed thoughtlessly. Although this could ameliorate patient's anxiety, it might also hinder the working-through of the underlying dynamics. We think that any information provision should better not precede the extensive unfolding and working-through of the dynamics underlying any question posed. To take another example, however, tempting it might be for the analysts to provide neuroscientific evidence demonstrating the efficacy of psychoanalytic treatment, such a practice might compromise the working-through of the patient's ambivalence on entering or continuing analysis and engaging in an intimate human relationship—with all kinds of anxiety associated with it—or even addressing the possibility of the patient employing a "*reversible perspective*," as pointed out by Bion (1984a, b).

We generally agree that neuroscience might "*help demarcate the limits of psychoanalysis*" and identify patients most likely to benefit from psychoanalysis or brief psychodynamic psychotherapy; indeed, we think that neuroscience could contribute to a synthesis concerning the criteria employed to assess the applicability and potential benefit of a certain type of treatment among the psychodynamically originated therapies. Nevertheless, we think that the decisive role should be preserved for the clinical criterion and the risk of misuse of neuroscientific evidence that would lead to hasty categorical arguments for analyzable vs. non-analyzable patients should be acknowledged and managed.

As Fonagy points out, science progresses in an interdisciplinary context anyway. Thus, psychoanalysts cannot avoid being aware of recent developments in science, even if they are not consciously engaged in such a dialogue (Fonagy 2003).

Besides, Freud himself and many other prominent psychoanalytic thinkers have always been in a dialogue with philosophy or even the natural sciences. Why not with neuroscience then?

We think that there has emerged an ongoing fruitful dialogue between proponents and opponents of neuropsychoanalysis that can provide the psychoanalytic community with further insight. Both sides have admitted points of convergence and satisfactorily addressed their concerns alongside with a better clarification or even slight modification of their original arguments.

Conclusively, we think that despite the limitations and concerns, a dialogue between psychoanalysis and neuroscience would be meaningful and beneficial for both disciplines provided that a mutually respected attitude is safeguarded. We agree with Yovell et al. (2015) stressing that a deliberate negligence and isolation should be avoided. However, we also believe that analysts should be aware of neuroscientific findings but at the same time they should be able to enter an "*as if not knowing*" state, in line with the recommendations of Freud for a self-imposed "*artificial blindness*" (Freud et al. 1972) and Bion for the "*abandonment of memory and desire*" within the analytic session (Bion 1984a). We think that the analysts should refrain themselves from using neuroscientific knowledge as a "*procrustean bed*" trying to fit patients' meanings into the prevailing theoretical model.

Appendix

Self-Assessment Questions

1. Sigmund Freud, founder of psychoanalysis:
   (a) Was an eminent psychologist who had nothing to do with neurobiology.
   (b) Essentially never gave up the belief that psychoanalytic ideas could eventually find biological evidence.
   (c) Had no interest in establishing psychoanalysis as a widely accepted discipline within medical science.
   (d) None of the above.
2. According to Kandel, neuroscience could contribute to psychoanalysis in:
   (a) Understanding preconscious and unconscious mental processes.
   (b) Clarifying psychological causality and its relation to early experience.
   (c) Identifying brain structural changes in response to psychotherapy.
   (d) All of the above.
3. According to Kandel:
   (a) Psychobiology and psychoanalysis have nothing in common to be studied.
   (b) There is a necessity for a scientific dialogue between neuroscience and psychiatry-psychoanalysis.
   (c) The time for a dialogue between neuroscience and psychoanalysis has not come yet.
   (d) None of the above.
4. Human behavior is determined by genes which are in no way influenced by environmental factors.
   True False.
5. Which of the following is true?
   (a) Psychoanalysis and neuroscience have always been in a state of mutual disregard and conflict.
   (b) Over the last two decades, there has risen a mutual interest of both neuroscientists and psychoanalysts regarding the bridging of the two disciplines.
   (c) Bridging between neuroscience and psychoanalysis should lead to a merging of the two disciplines in one.
   (d) None of the above.

6. "First-person neuroscience" aims to study:
   (a) The subjective experience of mental life by means of carefully designed first-person assessment tools.
   (b) A person's behavior observed by the most suitable (first-person) investigator.
   (c) A person's early experiences with parents and siblings (first-degree relatives).
   (d) None of the above.
7. It is unanimously accepted that neuroscientific findings represents the final "court of appeal" for psychoanalytic theory and practice.
   True False.
8. Neuropsychoanalysis aims:
   (a) To teach psychoanalysts how the brain operates.
   (b) To teach neuroscientists the dynamics that underlie human behavior.
   (c) To provide evidence that could help neuroscience and psychoanalysis engage in a mutually beneficial dialogue.
   (d) None of the above.
9. Dual-aspect monism as postulated by Solms and colleagues refers to.
   (a) The mind and brain seen as one and the same entity investigated from two complementary perspectives.
   (b) The existence of two contradictory opinions on how the brain operates.
   (c) The dual function of genes.
   (d) None of the above.
10. The "spatiotemporal structure" of the brain resting-state activity has been parallelized by Northoff with the Freudian non-anatomical conceptualization of the "psychological structure" of the mental apparatus.
    True False.
11. The discovery of mirror neurons has provided insight regarding:
    (a) Hallucinations.
    (b) Projections.
    (c) Empathy.
    (d) None of the above.
12. The existence in humans of a mirror neurons system similar to that discovered in monkeys is true for:
    (a) Actions.
    (b) Emotions.
    (c) Sensations.
    (d) All of the above.
13. According to current neuroscientific findings:
    (a) The "wish-fulfilling" function of dreaming can no longer be supported.
    (b) The "derepression" function of dreaming can be confirmed.
    (c) The interpretation of dreams renders clinically meaningless.
    (d) None of the above.

14. The default mode network (DMN):
    (a) Is responsible for the default answers brain utilizes when faced up with vague stimuli.
    (b) Is linked to the extrinsic view of brain function.
    (c) Is located in the limbic system.
    (d) None of the above.
15. Certain memories might be unable to enter consciousness due to:
    (a) Repression.
    (b) Acute loss of hippocampal neurons in case of traumas.
    (c) Suppression.
    (d) Any of the above.
16. DMN and the concept of Ego have been parallelized, based on:
    (a) The integrated and compound nature of both.
    (b) Their acting as a regulatory system exerting a top-down control.
    (c) Their representing a tonic reservoir of activity.
    (d) All of the above.
17. The existence of unconscious mental processes has been verified for cognitive but not emotional aspects.
    True False.
18. Among others, one thing that makes the absolute convergence between psychoanalysis and neuroscience unfeasible is that the latter cannot satisfactorily address and handle.
    (a) The how.
    (b) The why.
    (c) The where and what.
    (d) None of the above.
19. A dialogue between neuroscience and psychoanalysis might.
    (a) Help neuroscientists for a better targeted research.
    (b) Provide scientific foundation for some psychoanalytic formulations.
    (c) Provide insight about the effectiveness of certain aspects of psychoanalytic technical tools.
    (d) All of the above.
20. Neuroscientific findings could help clinicians decide between providing patients either psychoanalytic or psychopharmacological treatment.
    True False.

Answers

1	B	6	A	11	C	16	D
2	D	7	False	12	D	17	False
3	B	8	C	13	B	18	B
4	False	9	A	14	D	19	D
5	B	10	True	15	D	20	False

References

Avenanti A, Bueti D, Galati G, Aglioti SM (2005) Transcranial magnetic stimulation highlights the sensorimotor side of empathy for pain. Nat Neurosci 8(7):955–960. https://doi.org/10.1038/nn1481

Berlin HA (2011) The neural basis of the dynamic unconscious. Neuropsychoanalysis 13(1):5–31

Bion WR (1984a) Attention and interpretation. Karnac Books, London

Bion WR (1984b) Learning from Experience. Karnac Books, London

Blass RB, Carmeli Z (2007) On: the case against neuropsychoanalysis – reply to Drs Mancia and Pugh. Int J Psychoanal 88:1068–1070

Blass RB, Carmeli Z (2015) Further evidence for the case against neuropsychoanalysis: How Yovell, Solms, and Fotopoulou's response to our critique confirms the irrelevance and harmfulness to psychoanalysis of the contemporary neuroscientific trend. Int J Psychoanal 96(6):1555–1573. https://doi.org/10.1111/1745-8315.12449

Bråten S (2007) On being moved: from mirror neurons to empathy, vol 68. John Benjamins Publishing, Amsterdam

Buckner RL, Andrews-Hanna JR, Schacter DL (2008) The brain's default network: anatomy, function, and relevance to disease. Ann N Y Acad Sci 1124:1–38. https://doi.org/10.1196/annals.1440.011

Calvo-Merino B, Glaser DE, Grezes J, Passingham RE, Haggard P (2005) Action observation and acquired motor skills: an FMRI study with expert dancers. Cereb Cortex 15(8):1243–1249. https://doi.org/10.1093/cercor/bhi007

Canestri J (2015) The case for neuropsychoanalysis. Int J Psychoanal 96(6):1575–1584. https://doi.org/10.1111/1745-8315.12474

Carhart-Harris RL, Friston KJ (2010) The default-mode, ego-functions and free-energy: a neurobiological account of Freudian ideas. Brain 133(Pt 4):1265–1283. https://doi.org/10.1093/brain/awq010

Carhart-Harris RL, Friston KJ (2012) Free-energy and Freud: an update. In: Photopoulou A, Pfaff D, Conway MA (eds) From the couch to the lab: trends in neuropsychoanalysis: psychology, psychoanalysis and cognitive neuroscience in dialogue. Oxford University Press, Oxford, pp 219–229

Carhart-Harris RL, Mayberg HS, Malizia AL, Nutt D (2008) Mourning and melancholia revisited: correspondences between principles of Freudian metapsychology and empirical findings in neuropsychiatry. Ann General Psychiatry 7:9. https://doi.org/10.1186/1744-859X-7-9

Carmeli Z, Blass R (2013) The case against neuroplastic analysis: a further illustration of the irrelevance of neuroscience to psychoanalysis through a critique of Doidge's The Brain that Changes Itself. Int J Psychoanal 94(2):391–410. https://doi.org/10.1111/1745-8315.12022

Custers R, Aarts H (2010) The unconscious will: how the pursuit of goals operates outside of conscious awareness. Science 329(5987):47–50. https://doi.org/10.1126/science.1188595

Damasio A (2003a) Feelings of emotion and the self. Ann N Y Acad Sci 1001:253–261

Damasio AR (2003b) Looking for spinoza: joy, sorrow, and the feeling brain. Harcourt, Houghton Mifflin

Decety J, Sommerville JA (2003) Shared representations between self and other: a social cognitive neuroscience view. Trends Cogn Sci 7(12):527–533

Dijksterhuis A, Aarts H (2010) Goals, attention, and (un)consciousness. Annu Rev Psychol 61:467–490. https://doi.org/10.1146/annurev.psych.093008.100445

Fair DA, Cohen AL, Dosenbach NU, Church JA, Miezin FM, Barch DM et al (2008) The maturing architecture of the brain's default network. Proc Natl Acad Sci USA 105:4028–4032

Fonagy P (2003) Psychoanalysis today. World Psychiatry 2(2):73–80

Fotopoulou A, Pfaff D, Conway MA (2012) From the couch to the lab: trends in psychodynamic neuroscience. Oxford University Press, Oxford

Fox MD, Snyder AZ, Vincent JL, Corbetta M, Van Essen DC, Raichle ME (2005) The human brain is intrinsically organized into dynamic, anticorrelated functional networks. Proc Natl Acad Sci USA 102:9673–9678

Freud S (1958) Observations on transference-love (further recommendations on the technique of psycho-analysis III). In: The standard edition of the complete psychological works of Sigmund Freud, volume XII (1911–1913): the case of schreber, papers on technique and other works, pp 157–171
Freud S (1966). Project for a scientific psychology) The standard edition of the complete psychological works of Sigmund Freud, vol 1. Hogarth Press, London, pp 283–397
Freud S (1969) The question of lay analysis: conversations with an impartial person. WW Norton & Company, New York, NY
Freud S, Andreas-Salomé L, Pfeiffer E (1972) Sigmund Freud and Lou Andreas-Salomé—letters, vol 89. Chatto & Windus, London
Friston KJ (2011) Functional and effective connectivity: a review. Brain Connect 1(1):13–36. https://doi.org/10.1089/brain.2011.0008
Gazzola V, Aziz-Zadeh L, Keysers C (2006) Empathy and the somatotopic auditory mirror system in humans. Curr Biol 16(18):1824–1829. https://doi.org/10.1016/j.cub.2006.07.072
Greicius MD, Supekar K, Menon V, Dougherty RF (2009) Resting-state functional connectivity reflects structural connectivity in the default mode network. Cereb Cortex 19:72–78
Jackson PL, Meltzoff AN, Decety J (2005) How do we perceive the pain of others? A window into the neural processes involved in empathy. NeuroImage 24(3):771–779. https://doi.org/10.1016/j.neuroimage.2004.09.006
Johnson B (2001) Drug dreams: a neuropsychoanalytic hypothesis. J Am Psychoanal Assoc 49(1):75–96
Kandel ER (1999) Biology and the future of psychoanalysis: a new intellectual framework for psychiatry revisited. Am J Psychiatry 156(4):505–524. https://doi.org/10.1176/ajp.156.4.505
Kaplan-Solms K (2000) Clinical studies in neuro-psychoanalysis: introduction to a depth neuropsychology. Karnac Books, London
Karlsson G (2010) Psychoanalysis in a new light. Cambridge University Press, Cambridge
Kelley WM, Macrae CN, Wyland CL, Caglar S, Inati S, Heatherton TF (2002) Finding the self? An event-related fMRI study. J Cogn Neurosci 14(5):785–794. https://doi.org/10.1162/08989290260138672
Kelly AM, Di Martino A, Uddin LQ, Shehzad Z, Gee DG, Reiss PT et al (2009) Development of anterior cingulate functional connectivity from late childhood to early adulthood. Cereb Cortex 19:640–657
Kernberg O (2004) Psychoanalytic affect theory in the light of contemporary neurobiology. 6th International psychoanalytic symposium, Delphi, Oct 2004
Keysers C, Gazzola V (2006) Towards a unifying neural theory of social cognition. Prog Brain Res 156:379–401. https://doi.org/10.1016/S0079-6123(06)56021-2
Keysers C, Wicker B, Gazzola V, Anton JL, Fogassi L, Gallese V (2004) A touching sight: SII/PV activation during the observation and experience of touch. Neuron 42(2):335–346
Mancia M (2006) Neuropsychoanalysis and neuroscience. Springer Verlag, Milan
Mancia M (2007) On: the case against neuropsychoanalysis. Int J Psychoanal 88(Pt 4):1065–1067. author reply 1068–1070
McCrone J (2004) Freud's neurology. Lancet Neurol 3(5):320. https://doi.org/10.1016/S1474-4422(04)00747-1
McEwen BS, Sapolsky RM (1995) Stress and cognitive function. Curr Opin Neurobiol 5(2):205–216
Mechelli A (2010) Psychoanalysis on the couch: can neuroscience provide the answers? Med Hypotheses 75(6):594–599. https://doi.org/10.1016/j.mehy.2010.07.042
Milner B, Squire LR, Kandel ER (1998) Cognitive neuroscience and the study of memory. Neuron 20(3):445–468
Mitchell SA (1995) Hope and dread in psychoanalysis. Basic Books, New York, NY
Nersessian E, Solms M (1999) Editors' introduction. Neuro-psychoanalysis 1:3–4
Northoff G (2012) Psychoanalysis and the brain – why did Freud abandon neuroscience? Front Psychol 3:71. https://doi.org/10.3389/fpsyg.2012.00071

Northoff G (2015) Is schizophrenia a spatiotemporal disorder of the brain's resting state? World Psychiatry 14(1):34–35. https://doi.org/10.1002/wps.20177
Northoff G, Bermpohl F (2004) Cortical midline structures and the self. Trends Cogn Sci 8(3):102–107. https://doi.org/10.1016/j.tics.2004.01.004
Northoff G, Heinzel A (2006) First-person neuroscience: a new methodological approach for linking mental and neuronal states. Philos Ethics Humanit Med 1(1):1
Northoff G, Bermpohl F, Schoeneich F, Boeker H (2007) How does our brain constitute defense mechanisms? First-person neuroscience and psychoanalysis. Psychother Psychosom 76(3):141–153. https://doi.org/10.1159/000099841
Panksepp J, Solms M (2012) What is neuropsychoanalysis? Clinically relevant studies of the minded brain. Trends Cogn Sci 16(1):6–8. https://doi.org/10.1016/j.tics.2011.11.005
di Pellegrino G, Fadiga L, Fogassi L, Gallese V, Rizzolatti G (1992) Understanding motor events: a neurophysiological study. Exp Brain Res 91(1):176–180
Pugh G (2007) On: the case against neuropsychoanalysis. Int J Psychoanal 88(Pt 4):1067–1068. author reply 1068–1070
Pulver SE (2003) On the astonishing clinical irrelevance of neuroscience. J Am Psychoanal Assoc 51(3):755–772
Raichle ME, Snyder AZ (2007) A default mode of brain function: a brief history of an evolving idea. NeuroImage 37(4):1083–1090.; discussion 1097-1089. https://doi.org/10.1016/j.neuroimage.2007.02.041
Raichle ME, MacLeod AM, Snyder AZ, Powers WJ, Gusnard DA, Shulman GL (2001) A default mode of brain function. Proc Natl Acad Sci U S A 98(2):676–682. https://doi.org/10.1073/pnas.98.2.676
Ramus F (2013) What's the point of neuropsychoanalysis? Br J Psychiatry 203(3):170–171. https://doi.org/10.1192/bjp.bp.113.127217
Rizzolatti G (2005) The mirror neuron system and its function in humans. Anat Embryol 210(5-6):419–421. https://doi.org/10.1007/s00429-005-0039-z
Salone A, Di Giacinto A, Lai C, De Berardis D, Iasevoli F, Fornaro M, De Risio L, Santacroce R, Martinotti G, Giannantonio MD (2016) The interface between neuroscience and neuro-psychoanalysis: focus on brain connectivity. Front Hum Neurosci 10:20. https://doi.org/10.3389/fnhum.2016.00020
Schore AN (1997) A century after Freud's project: is a rapprochement between psychoanalysis and neurobiology at hand? J Am Psychoanal Assoc 45(3):807–840
Schore AN (2003) Affect regulation and the repair of the self (Norton series on interpersonal neurobiology), vol 2. WW Norton & Company, New York, NY
Solms M, Turnbull O (2002) The brain and the inner world: an introduction to the neuroscience of subjective experience. Karnac Books, London
Squire LR, Alvarez P (1995) Retrograde amnesia and memory consolidation: a neurobiological perspective. Curr Opin Neurobiol 5(2):169–177
Stickgold R, Malia A, Maguire D, Roddenberry D, O'Connor M (2000) Replaying the game: hypnagogic images in normals and amnesics. Science 290(5490):350–353
van Veluw SJ, Chance SA (2014) Differentiating between self and others: an ALE meta-analysis of fMRI studies of self-recognition and theory of mind. Brain Imaging Behav 8(1):24–38. https://doi.org/10.1007/s11682-013-9266-8
Vogt BA (2005) Pain and emotion interactions in subregions of the cingulate gyrus. Nat Rev Neurosci 6(7):533–544. https://doi.org/10.1038/nrn1704
Yovell Y, Solms M, Fotopoulou A (2015) The case for neuropsychoanalysis: why a dialogue with neuroscience is necessary but not sufficient for psychoanalysis. Int J Psychoanal 96(6):1515–1553. https://doi.org/10.1111/1745-8315.12332

11 Psychophysiology and Psychosomatics

Kostas N. Fountoulakis

11.1 Psychophysiology

The discipline of psychophysiology concerns the study of the link between psychological processes and somatic physiology, which is the interface between mind and body (Cacioppo et al. 2007). Historically, most of research concerns the physiological responses and the function of the organ systems innervated by the autonomic nervous system, especially under stressful conditions (Lang et al. 2000), but more recent research has focused on the brain with the use of electrophysiological techniques and functional neuroimaging (Hugdahl 1984; Schwartz 1999; Striefel 1999; Bechtereva 2000; Gordon 2001).

Psychophysiology utilizes instruments and devices which record the activity of various physiological systems in the human body and compares these results with measures of psychological states. The main fields of psychophysiology research concern the study of emotions (Ekman et al. 1983; Panconesi and Hautmann 1996; Glynn et al. 2002; Davidson 2003; Corr and Perkins 2006; Critchley 2009; Stemmler and Wacker 2010; Quigley and Barrett 2014; Katz and Greenberg 2015), pain (Handwerker and Kobal 1993; Keefe and Smith 2002; Tiller 2006; Wachholtz et al. 2015), psychopathology (Fowles 1988; Steinhauer and Hakerem 1992; Carroll and Sheffield 1998; Wilhelm and Roth 2001; Corr and Perkins 2006; Tiller 2006; Stemmler and Wacker 2010), but also forensics (Yankee 1995; Furedy 1996).

The term "psychophysiology" comes from the Greek "*ψυχή*" (psichi meaning the soul) plus "*φυσιο-*" (physio—meaning nature) and "*λογια*" (logia meaning the study of things). The term "psychosomatics" comes again from "*ψυχή*" plus "*σωματικός*" (somatikos meaning bodily function).

K. N. Fountoulakis (✉)
3rd Department of Psychiatry, Faculty of Medicine, School of Health Sciences, Aristotle University of Thessaloniki, Thessaloniki, Greece

K. N. Fountoulakis, I. Nimatoudis (eds.), *Psychobiology of Behaviour*,
https://doi.org/10.1007/978-3-030-18323-3_11

11.1.1 Brain Psychophysiology

Brain psychophysiology includes the study of brain function with electroencephalography (EEG), functional magnetic resonance imaging (fMRI), magnetoencephalography (MEG), event-related potentials (ERPs) (Coles 1989), and other similar methods.

- EEG records electrical activity through the electrodes placed along the scalp, corresponding to the brain's spontaneous electrical activity. This activity corresponds to the summation of the synchronous activity of billions of neurons that have similar spatial orientation. Its diagnostic applications include the diagnosis of epilepsy, sleep disorders, coma, encephalopathies, and brain death. A routine clinical EEG lasts 20–30 min plus preparation time and usually uses the 10–20 system for the placing of the electrodes. For the first time, EEG was recorded by Hans Berger (1873–1941) in 1924. ERPs are derivatives of the EEG technique and involve the averaging of the EEG responses to the presentation of a visual, somatosensory, or auditory stimulus (Niedermeyer and da Silva 2004; Schreckenberger et al. 2004).
- MEG records the magnetic fields produced by the brain's electrical activity. For the first time, they were recorded by David Cohen in 1968 (Cohen 1968; Cohen and Cuffin 1983). The most important problem of biomagnetism is that the signal is relatively weak in comparison to the environmental noise and to the sensitivity of sensors. Event potentials can also be recorded with MEG.
- ERPs were recorded for the first time in 1935 by Pauline and Hallowell Davis (1896–1992). Grey Walter discovered the first cognitive ERP component, called the contingent negative variation (CNV) in 1964 (Walter et al. 1964), and the next year, Samuel Sutton, Margery Braren, and Joseph Zubin discovered the P3 (P300) component (Sutton et al. 1965). ERPs are one of the most widely used methods to study the physiological correlates of sensory, perceptual, and cognitive activity as well as the process of information. The pattern and the timing of responses and their components provide an image of brain function and the timing of information processing. After the P3, a number of other components were discovered including the ELAN, the N400, and the P600/SPS (Handy 2005).
- Magnetic resonance imaging (MRI) is used in radiology to image the anatomy and the physiology of the human body (McRobbie 2007). Essentially MRI scanners use strong magnetic fields, radio waves, and field gradients to form images of the body. The technique of the MRI is based upon the property of certain atomic nuclei to absorb and emit radio frequency energy when placed in an external magnetic field. In contrast to simple MRI, the functional MRI (fMRI) is used to record changing neural activity in different parts of the brain as a response to external stimuli (Heeger and Ress 2002). Blood oxygenation level-dependent (BOLD) fMRI measures the hemodynamic response to transient neural activity resulting from a change in the ratio of oxyhemoglobin and deoxyhemoglobin. Increased neural activity increases the demand for oxygen, and subsequently there is an increase in the amount of oxygenated hemoglobin

relative to deoxygenated hemoglobin. Other techniques include the arterial spin labeling (ASL) or weighting the MRI signal by cerebral blood flow (CBF) and cerebral blood volume (CBV) (Singer 1959).

Detailed description of the above methods and techniques is beyond the scope of the current chapter, and it can be found in specialized books and articles. However, it is important to note that in comparison to other imaging techniques (e.g., fMRI), electrophysiological methods have very high temporal resolution, on the order of milliseconds, but rather low spatial resolution on the scalp. However, on the contrary, electrophysiological methods measure the neural activity that occurs only at the surface of the brain and not at deep structures, and unlike PET and MRS, they cannot identify neurotransmitters, drugs, and metabolic indices. In research settings, structural MRI or functional MRI (fMRI) can be combined with EEG (electroencephalography) under the condition that the EEG equipment is MR compatible.

11.1.2 Pupillometry and Eye Movements

The testing of the eye pupil function gives information on the two components of the autonomic nervous system, that is, the sympathetic nervous system and the craniosacral or parasympathetic nervous system. These two parts of the autonomic nervous system exert opposite effects on the pupil (Ropper and Brown 2005).

One of the most interesting features of the eye is the pupil reaction to light, during which, acetylcholine acts on the muscarinic receptors of the sphincter muscle and leads to miosis (decrease of the size of the pupil). The dilator muscle of the pupil has an opposite function. The neurotransmitter released in the neuromuscular junction of the dilator muscle is norepinephrine, which through its effect on a-adrenergic receptors, leads to constriction of the dilator muscle and causes mydriasis. During miosis, acetylcholine is secreted in the dilator as well, inhibiting norepinephrine activity. Therefore the change of the pupil size as a response to a light stimuli is based on the functional balance between parasympathetic and sympathetic nervous systems (balance between acetylcholine and norepinephrine).

In the absence of a documented peripheral disorder, the contribution of peripheral structures is negligible in comparison to that of central ones (Fotiou et al. 2000b). So, the pupil reaction to light may be used either to test peripheral structures of the autonomic nervous system or, in cases in which peripheral structures are intact, to assess acetylcholine and norepinephrine function in the brain.

The assessment of the pupil reflex has been used to study the deficit of the parasympathetic supply to the pupil, among others, in alcoholics (Tan et al. 1984), pain (Chapman et al. 1999), HIV (Maclean and Dhillon 1993), depression (Sokolski and Demet 1996; Fountoulakis et al. 1999), generalized anxiety disorder (Bakes et al. 1990), and Alzheimer's disease (Scinto et al. 1994; Prettyman et al. 1997; Grunberger et al. 1999; Fotiou et al. 2000a; Granholm et al. 2003; Frost et al. 2013).

The critical point in this kind of research is the method used for the assessment of the pupil. Today, parasympatheticomimetic agents like methacholine 2.5% and pilocarpine 0.125%, 0.0625%, or 0.05% or mydriatic agents like tropicamide, cocaine 4%, epinephrine 0.001%, phenylephrine 10%, or hydroxyamphetamine are used. Important drawbacks of the method of direct application of these agents in the eye are that the eye epithelium may manifest different permeability across individuals and disorders and technical problems concerning the stability of the solutions applied and the characteristics measured (should one measure the difference in reaction to a standard dose of an agent-stimulus or the smallest dose that produces the minimum reaction?). It is obvious that the pharmacological methods of testing the pupil reactivity are poor and the results of studies that use these methods should be considered with caution.

On the contrary, optical methods which utilize recording of the pupil reaction to dark and light or the reaction to various stimuli are more reliable and provide a significant number of dynamic information, including speed, acceleration, and size at different time points (Fig. 11.1). These methods are noninvasive, do not depend on the condition of the epithelium or the stability of solutions, and measure a natural phenomenon, not a chemically induced one (Fotiou et al. 2000b). The variance due to the method itself is minimal.

Eye movement refers to both the voluntary and the involuntary movement of the eyes (Wehner 2005). Their function is to help in acquiring, fixating, and tracking visual stimuli (Henderson 2003). Rapid eye movement is a special type which occurs during REM sleep. In humans, there are three types of voluntary eye movement: smooth pursuit (following an object that moves), vergence shifts (movements of the eyes in opposite directions), and saccades (rapid movement of eyes while scanning a visual scene (Findlay 2009)), but also there are involuntary random jittering movements. All voluntary movements are under the control of the frontal lobe (Pierrot-Deseilligny et al. 2004). Eye movements can be recorded via electrooculogram (EOG) and direction-of-gaze methods.

11.1.3 Electrodermal Activity (EDA)

EDA was discovered for the first time in 1849 by Emil du Bois-Reymond (1818–1896). It refers to the continuous variation in the electrical characteristics of the skin. The combined changes between electrodermal resistance and electrodermal potential make up electrodermal activity. There are many terms which were used to denote this characteristic, including skin conductance, galvanic skin response, electrodermal response, psychogalvanic reflex, skin conductance response, sympathetic skin response, and skin conductance level (Bach and Friston 2013; Boucsein 2013).

As shown by Hermann and Luchsinger in 1878, the mechanism behind this variation in skin resistance is mainly sweating which is under the control of the sympathetic nervous system. As such, it is considered to be a measure of arousal. It was Marie Gabriel Romain Vigouroux (1831–1911) who in 1879 was the first to relate

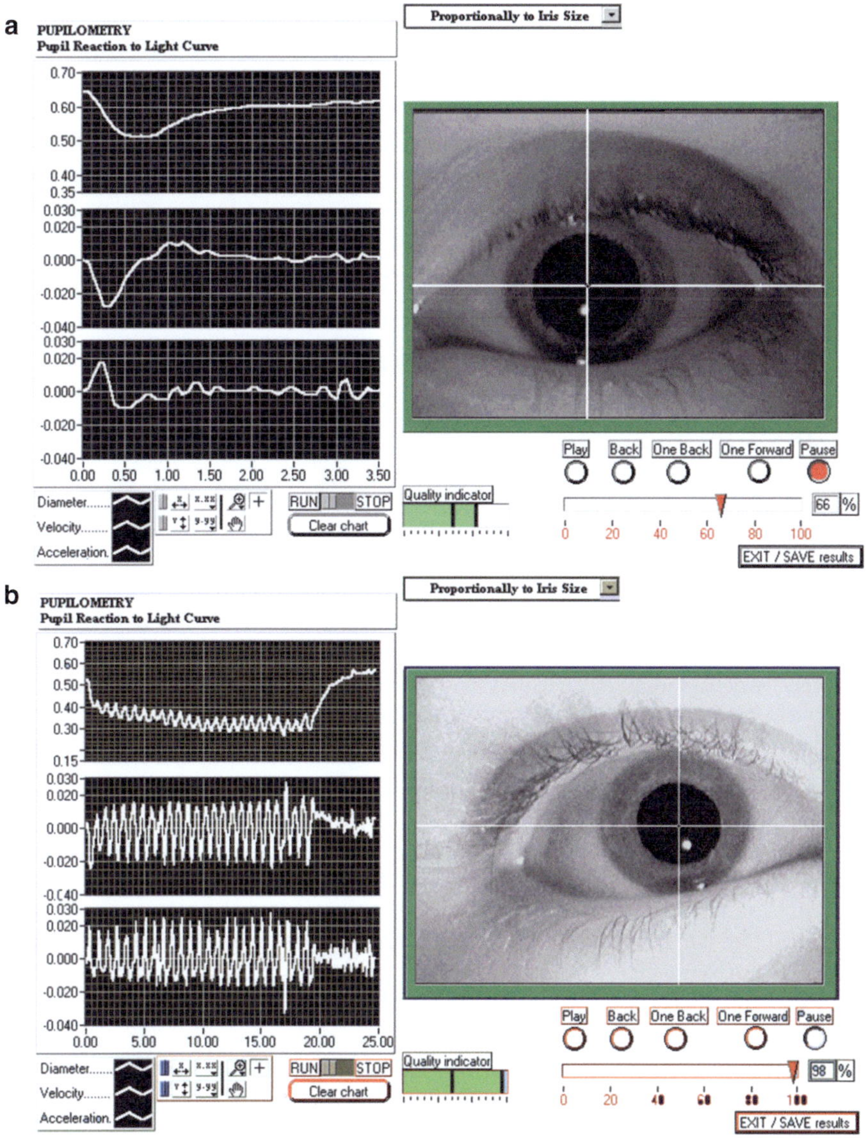

Fig. 11.1 Pupillometry. (**a**) Pupil reaction to a single flash. (**b**) Pupil reaction to multiple flashes with 1 Hz. In both images the upper curve concerns the size change vs. time, the middle concerns the change in velocity, and the lower concerns the change in acceleration

EDA with psychological factors, while in 1888, Charles Samson Féré (1852–1907) showed that EDA could be changed by emotional stimulation and inhibited by drugs. In 1889 Ivan Romanovich Tarkhanov (or Ivane Tarkhnishvili; 1846–1908) observed spontaneous variations in skin electrical potentials and designed a device to measure these oscillations in real time.

EDA was used by the famous psychoanalyst Carl Gustav Jung (1875–1961) to evaluate the emotional sensitivities of patients to lists of words during word association testing and considered it to be a method to access the unconscious.

While EDA is highly responsive to emotions in some people, including fear, anger, startled response, orienting response, and sexual feelings, the relationship is inconsistent across different people and specifically different or flattened in psychopaths. This later fact makes problematic the inclusion of EDA in polygraphic lie detectors. Also external factors including temperature and humidity as well as medications and other internal (somatic) factors can lead to inconsistent results.

11.1.4 Other Psychophysiological Methods

The cardiovascular measures in psychophysiology include the recording of heart rate and its variability and cardiodynamics via impedance cardiography. The recording of somatic muscle activity includes electromyography and thermogram (Ioannou et al. 2014), while the assessment of the gastrointestinal tract includes electrogastrogram (Musial et al. 2008).

The main problem with most of the abovementioned methods is that they are sensitive to artifacts because of any physical activity or motion, they depend on the basal levels of arousal, and responsiveness differs significantly among individuals. Another very important issue is that they are very sensitive to medication status.

11.2 Stress

Stress is an organism's response to a stressor which is often an external factor such as an environmental condition. The term has its origins in physiology and engineering. Stress could be purely psychological in origin, or it could be mainly physiological or biological. Pure forms of stress are rare, and usually although one aspect could be predominant, stress includes both psychological and physiological/biological components (Goldstein and Kopin 2007).

Among scientists but also among lay people, stress is considered to be a major element of life and central to both mental and somatic health and disease. It is the universality of its unpleasant experience that makes stress so important both in science and in culture. These beliefs are dated back to antiquity and to a certain extend have been also incorporated in the theological teachings of major religions.

Homeostasis is a concept which refers to the equilibrium toward which most biochemical processes tend to or strive to maintain. The term was coined for the first time by Walter Cannon (1871–1945) in 1926. He made major contributions concerning the role of the sympathetic nervous system in the stress response. He also focused on more immediate (short-term) responses to stressors (Cannon 1926, 1932). Stressors tend to disrupt homeostasis in an extreme way, and as a response, the organism utilizes compensatory mechanisms and reserves to counterbalance the insult. This compensatory efforts usually consume resources to an unusual extend and therefore could be considered as an additional source of stress. Hans Selye (1907–1982) almost immediately pointed out the ambiguity in defining stress and especially in discovering what is the cause and what is the effect (Selye 1978). He also coined the term "stress" for the first time, and soon after, in 1936, he developed the concept of the general adaptation syndrome, which corresponds to a simplified model of how organisms respond to stress (Fig. 11.2). It includes three phases: a non-specific mobilization phase (mainly sympathetic nervous system activity), a resistance phase (efforts to cope), and a recovery or an exhaustion phase (depletion of physiological resources) (Selye 1950, 1954). It was because of Hans Selye's prolific scholarship that the term "stress" was solidly embedded both in science and in lay culture.

The body reacts to stressful conditions with the activation of the sympathetic nervous system and the initiation of the fight-or-flight response. On the contrary, the role of the parasympathetic nervous system is to return the body to the normal physiological status and to facilitate the restoration of the reserves. The reaction eventually includes the activation of a variety of body systems and mechanisms, including the endocrine and the immune systems. But as Hans Selye argued, stress decreases the ability of the organism to adapt by consuming "adaptation energy" (Selye 1938). This term refers to complex abstract abilities and orientation rather than some specific mechanism or energy in the frame of physics (Gorban et al. 2016).

There is a widespread lay belief but also a number of theories suggesting there is a strong connection between stress and illness, both mental and somatic. According to theoretical approaches, both acute and chronic stress can lead to changes in physiology, which could create vicious cycles especially in combination with personality features or changes in behavior (Cohen et al. 2007; Jeronimus et al. 2014). As a result, the individual could become more susceptible to illnesses like the common cold (Cohen et al. 1997), morbidity, and mortality because of coronary heart disease as well as increased serum lipids (Calderon Jr. et al. 1999) and depression and

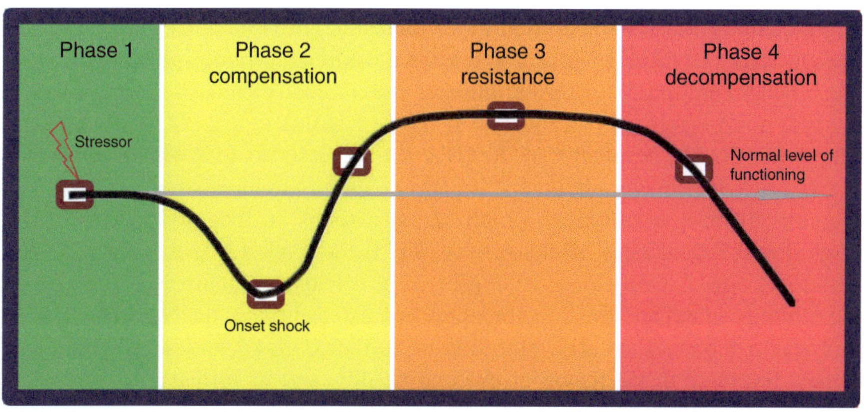

Fig. 11.2 The general adaptation syndrome introduced by Hans Selye (1907–1982)

anxiety (Schlotz et al. 2011). Chronic stress increases the accumulation of visceral fat, which leads to hormonal and metabolic changes, thus causing a variety of health problems, especially cardiovascular (Shively et al. 2009).

It seems that the type of stressful condition in combination with early trauma, age, personality factors, and genetic vulnerabilities determine the overall outcome (Heim et al. 2000; Pace et al. 2006). Especially chronic stress seems to exert a worse effect on global health (Pinquart and Sorensen 2003; Schneiderman et al. 2005; Jeronimus et al. 2013, 2014). Age seems to be an important factor since experiencing of extremely stressful conditions at a young age increases future vulnerabilities (Miller et al. 2009). On the contrary, some people do not show any deterioration of their health even after prolonged periods of stress. This resilience is called "hardiness" (Kobasa 1982). Essentially most healthy individuals will not show any health deterioration even after chronic stress, and a factor which seems to be protective is the belief that stress will not (Keller et al. 2012; Richardson et al. 2012; Nabi et al. 2013). This is a strong evidence that psychological factors and reactions mediate the harmful effects of stress on somatic health. It is interesting to note that the above research findings suggest the presence of an intergenerational effect concerning early chronic stress (Miller et al. 2009).

Animal studies suggest that chronic exposure to a specific stressor leads to the development of tolerance concerning that specific stressor but also to a hypersensitivity toward novel stressors. The exact phenomenon of such a "sensitization" (meaning the enhanced response to the same stress over time) or "desensitization" (meaning the reduced response to the same stress over time) has not been elucidated yet, and it is unknown when, why, and how it might occur. However it is solidly known that the response to stress varies dramatically as a function of frequency and duration of the exposure to the stressful insult.

The mechanism of the response to stress from the periphery of the body is mainly mediated by the hypothalamic-pituitary-adrenal (HPA) axis (Fig. 11.3). It is triggered when the hypothalamus receives input from higher structures (e.g., cerebral cortex, limbic system, visceral organs) concerning the presence of stress and the deterioration of the supposed homeostatic state. In response to stress, the hypothalamus releases the corticotropin-releasing hormone (CRH) which stimulates the anterior pituitary gland, to release the adrenocorticotropic hormone into the bloodstream which in turn stimulates the adrenal glands to secrete cortisol but also norepinephrine (McEwen et al. 1997; Kaufman et al. 2000; Sanders and Straub 2002). The activation of the HPA axis lasts for as long as the stressor is present and insufficient amounts of cortisol exist in the bloodstream. As soon as the cortisol level is sufficiently high, the stressor is no longer present or some kind of new balance has been achieved, and the so-called homeostasis has been restored, cortisol itself causes inhibition of its release by binding on receptors in the pituitary gland and the hypothalamus but also as high as the hippocampus. In this way a feedback creates inhibition of further cortisol release. Cortisol is a steroid hormone of the group of glucocorticoids, and its primary function is to control and redistribute glucose consumption and subsequently energy and metabolism rates in favor of those body regions in need, depending on the

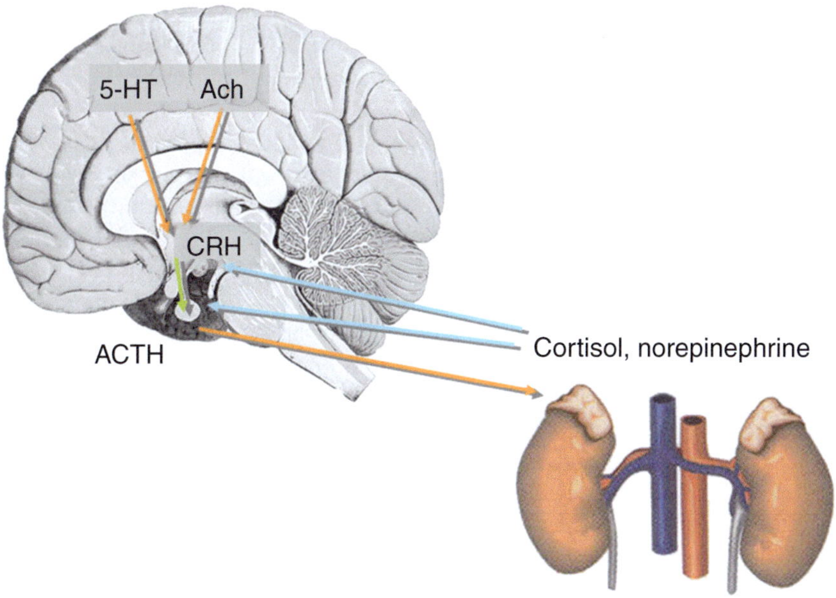

Fig. 11.3 The hypothalamus-pituitary-adrenal (HPA) axis

situation. A second major effect of cortisol is the suppression of the immune system, so that energy resources are reserved for those body organs which take part in the fight or flight process. Energy is also conserved by the inhibition of growth and reproduction mechanisms. Under chronic stress this suppression of systems and functions could be chronic also, leaving the organism vulnerable to infectious insults (Miller 1985; Khansari et al. 1990; Wolf et al. 1994; Jessop et al. 1997; Poliak et al. 1997; Antoni et al. 2002; Karacabey et al. 2005; Adamo 2014; Gur and Bailey 2016).

This is the reason why the first biological marker which had been proposed in psychiatry was the dexamethasone suppression test (DST). It had been described as a biological marker for depression (Carroll et al. 1968; Fountoulakis et al. 2004; Yerevanian et al. 2004), it had also been associated with suicidal behavior, melancholic, and latter atypical features, but essentially it is a biomarker of stress, not of clinical depression. This test mainly reflects the HPA axis and norepinephrine activity (Greene and Dalton 1953; Carroll et al. 1968; Nuller and Ostroumova 1980; Mendlewicz et al. 1984; Stokes et al. 1984; Evans and Golden 1987; The APA Task Force on Laboratory Tests in Psychiatry 1987). The 1 mg DST protocol demands the administration of 1 mg dexamethasone per os at 23.00 of the first day and determination of cortisol serum levels simultaneously and the next day at 16.00 and 23.00. A subject is considered to be non-suppression if cortisol levels are above 5 μg/dl in either second day measurement. Nearly 4–10% of normal persons are reported to be DST non-suppressors (Watson et al. 1987; Sharma et al. 1988; Yeragani 1990). The reason for this is unknown; however, it has been suggested that it is due to an underlying stressful situation since it probably reflects the degree of psychological pressure or discomfort of the subject and not a specific vulnerability or characteristic of depression (Ceulemans et al. 1985; Shuchter et al. 1986).

The effect of stress on the immune system is reported to be profound, although individual differences are significant and predictions at the patient level impossible (Cohen and Hamrick 2003; Irwin 2008). It is of course known that the immune system constitutes the body's defense against invading external pathogens but also against abnormal body cells like cancer cells. It includes a non-specific mechanism-based macrophages and granulocytes which release pro-inflammatory cytokines (tumor necrosis factor-α (TNF-α) and interleukin-6 (IL-6)). They do not demand the identification of the type of insult, and their function is to consume debris and invading organisms. There is also a second mechanism, which is specific and requires the identification of invading organisms and molecules and discriminating them from those of the body. The invading molecules and cells are called "antigens." B cells secrete specific antibodies which attack antigens, while T cells kill those body cells that host pathogens. Both types of immunity are controlled via the secretion of interleukins and cytokines from immune cells. In animals, acute exposure to stressors produces alterations in almost every aspect of immunity in a dose-dependent manner to the severity of the stressor. In humans, acute stress inhibits cellular immune responses and increases markers of inflammation (e.g., IL-6), and the overall impact when the stress is chronic puts the individual at increased risk for infectious diseases (Irwin et al. 2006)

The HPA axis and the autonomic nervous system connect the brain with the immune system. It seems that sympathetic activation suppresses the non-specific (cellular) immune mechanism while it might enhance the specific (humoral) mechanisms. It has also been reported that chronic stress might modify the sympathetic innervation (Sloan et al. 2007), while also disordered sleep and loss of sleep alter inflammatory function and have an adverse effect on disease progression (Lange et al. 2003).

However, it is important to stress that there is significant heterogeneity in the population which could at least be partially attributed to coping and personality as well as to idiosyncratic neuroendocrine and sympathetic activity. In spite of this problematic modeling, dealing with stress is important both for healthy and for patient populations. It is interesting that because the average individual feels comfortable talking about the stress he experiences, this is frequently the first thing discussed during a psychiatric consultation. Unfortunately, this is often more confusing than clarifying. Because of a number of methodological issues, it is almost impossible to scientifically prove that stress can cause any kind of somatic disorder. On the other hand, it is possible, and a number of studies support the suggestion that acute stress acts as a trigger on an underlying subclinical somatic condition. This, in turn, has shifted the focus from the stressor as a malignant insult to the vulnerability of the organism, both in psychological and in organic terms.

11.3 Psychosomatic Medicine

Although the term refers to the mind-body interaction in the development of somatic disorders, the lay interpretation during the last decades referred to the somatization type of disorders, and therefore in 1980, in the DSM-III, the term "psychological factors affecting physical conditions" replaced the previously used "psychophysiological (or psychosomatic) disorders." However, the term continues to be used in research, in journal titles, and in scientific associations, and also in 2003, the American Board of Medical Specialties and the American Board of Psychiatry and Neurology approved the specialty of psychosomatic medicine (Gitlin et al. 2004).

An early psychosomatic concept was proposed by Hippocrates (460–370 BC) who, in the frame of his fluid theory, suggested that fluid imbalance and the resulting disease could be caused by emotional upset. He was also the first to suggest a holistic approach in the treatment. Around the same time, Plato (427–348 BC), in his book *Timaeus*, suggested that trouble in the soul could bring trouble to the body, and, in *Charmides*, he quoted Socrates (470–399 BC) "…neither is it proper to cure the body without the soul." Aristotle (384–322 BC) wrote that emotions affect the body. Aretaeus of Cappadocia (first century BC) proposed that emotional problems constitute one of the six major causes of paralysis. The term "psychosomatic" was first coined in 1818 by Johann Christian August Heinroth (1773–1843) and "psychosomatic medicine" as a term was used for the first time by Felix Deutsch (1884–1964) in the early 1920s (Deutsch 1939).

The modern evolution of the psychosomatic concept followed a much different trajectory. While in the nineteenth century the diagnosis of spinal irritation was frequent in order to label somatization, this was later replaced by "reflex neurosis" which suggested that an affected organ could cause irritation in any other organ in the body through the nervous connections between organs. In the thought of physicians of the time, the uterus was the most important and frequently affected organ and the source of the problem. At the end of the nineteenth century, the reflex theory was abandoned in favor of the concept of a central disorder being responsible for symptoms, paving the way for the concept of the psychological origin of somatic symptoms in the early twentieth century (Shorter 1992). In this frame, George Miller Beard (1839–1883) coined the term "neurasthenia" in 1969, while the term "somatization" came from the false translation of the term "organsprache" coined by Wilhelm Stekel (1868–1940) in the 1920s (Marin and Carron 2002; Mai 2004).

The early twentieth century was marked by the theories of Sigmund Freud (1856–1939). In his early theories, libido and sexuality were utilized as a bridge between the body and the mind. Later, in the 1920s, Karl Abraham (1877–1925) and Sandor Ferenczi (1973–1933) further elaborated on these concepts with Georg Groddeck (1866–1934) going much further to the extremes and attaching a symbolic meaning to fever and hemorrhage in 1929. During the second half of the twentieth century, two main branches emerged, one concerned the contribution of psychological factors on the development of physical disorders (psychosomatics) and the second concerned the development of psychological disorders in somatic patients (consultation-liaison psychiatry and psychological medicine). Psychosomatics were initially dominated by two conflicting theories. The first one, introduced by Franz Alexander (1891–1964) in the 1930s, was a theory of specificity suggesting that specific emotions led to specific cell and tissue damage, while the second was a non-specific one, suggesting that generalized anxiety and stress contributed to a number of un-predetermined somatic conditions. Franz Alexander considered intrapsychic conflict to constitute a stress factor, which even after suppression could lead organic pathology because the fight or flight response becomes chronic in spite of the suppression of the conflict (Alexander 1950). Depending on the specific predisposition the person carries, this prolonged alertness could lead to gastric ulcer, colitis, asthma, migraine, hypertension, or arthritis, but empirical data failed to recognize the same type of conflict behind the same somatic disease, thus making problematic the psychoanalytic approach to psychosomatics.

In 1939, Karen Horney (1885–1952) and, in 1948, James Halliday (1897–1983) shifted the emphasis on culture and its effect on the development of psychosomatic illness, mainly through the mother (Horney 1937).

An important contribution which later was solidly embedded in lay culture was made in 1954 by Helen Flanders Dunbar (1902–1959) who proposed that the ambitious, highly motivated male would be prone to coronary occlusion (Dunbar 1954). In 1959, two cardiologists, Meyer Friedman (1910–2001) and Ray H. Rosenman (1923–2013), proposed the existence of type A personality which is almost identical to the description of Dunbar and includes hostility as an important personality trait (Friedman et al. 1970). According to them, hostility was the main risk factor for the development of coronary artery disease. Around the same time, Harold Wolff (1898–1962) and Stewart George Wolf (1914–2005) reported that chronic hyperfunction or hypofunction activities of the mucosa of the gastrointestinal and respiratory systems could produce pathology and they correlated overfunctioning with hostility and underfunctioning with fear or sadness.

The patient's entire reactive patterning and his or her life history account for whether he or she reacts to stress by hyperfunctioning or hypofunctioning. Further on this theory of non-specific causation, Hans Selye suggested that the hormonal response to stress by the HPA could ultimately lead to the development of a number of organic diseases, which he considered to be a by-product of the attempt to adapt to stress. This was in sharp contrast with the prevailing theoretical model in medicine at that time, which demanded a specific cause to lead to a specific effect.

An important contribution was made in the late 1960s by two psychiatrists, Thomas Holmes and Richard H. Rahe. They developed the Social Readjustment Rating Scale (SRRS) (Holmes and Rahe 1967; Harmon et al. 1970; Rahe et al. 1971) on the basis of the medical records of over 5000 patients. Later they tested it in 2500 US sailors, and they reported a weak but significant correlation between scale scores and somatic illness (Rahe et al. 1970, 1972). An important contribution of the scale and this line of research was the identification of loss of spouse as the most intense stressor.

Already since the dawn of the twentieth century, Adolph Meyer (1866–1950) developed the psychobiological approach by emphasizing the interaction of developmental, psychological, social, and biological factors in mental health and disease.

This concept was further elaborated by Zbigniew Lipowski (1924–1997) (Lipowski 1967a, b, 1968), and eventually a vague but popular "biopsychosocial model" was proposed by George Engel (1913–1999) in 1977 (Engel 1977). Leon Eisenberg (1922–2009) suggested a developmental approach according to which the major brain structures and neural pathways are genetically specified, but the detailed connections are shaped and thus reflect social and interpersonal factors and the external environment in general (Eisenberg 1995). The last major contribution in the field came by Peter Sifneos (1920–2008) and John C. Nemiah (1918–2009) who coined the term alexithymia to describe a developmental arrest in the ability to express affect related to conflict, and as a result, psychosomatic symptoms emerge.

The most compelling data on psychosomatic medicine concern the impact of stress on atherosclerosis and cardiovascular disease in general which is now thought to correlate to an inflammatory process in vascular endothelium. The effect on immunity has profound consequences on a number of diseases including infections, autoimmune disorders, and cancer. The infectious diseases best studied include viral diseases such as herpes simplex, HIV, Epstein-Barr virus infections, and the common cold (Cohen et al. 1991). It seems that under stress, the immunologic suppression by glucocorticoids is not fully achieved, and specifically in rheumatoid arthritis patients, when chronic stress is present, there is an increase in IL-6, which correlates with symptoms of disease including fatigue, pain, and functional limitations (Davis et al. 2008). In cancer, human research linking stress to cancer onset and progression has produced inconsistent results. One recent study suggests there is a link between chronic stress and the expression of tumor-promoting peptides in women with ovarian cancer (Thaker et al. 2006). In cancer survivors, it has been reported that persistent fatigue is associated with elevated levels of markers of inflammation (Bower et al. 2002; Collado-Hidalgo et al. 2006), while treatment for cancer or hepatitis C with high doses of cytokines frequently induces sickness behaviors and depression. Interestingly, pretreatment with antidepressants prevents the manifestation of these symptoms (Dantzer 2001; Reichenberg et al. 2001; Konsman et al. 2002). High-dose interferon-α induces significant activation in the dorsal anterior cingulate cortex, which is responsible for affect generation and with cognitive control during highly demanding cognitive tasks (Reichenberg et al. 2001; Capuron et al. 2003). Such activations could lead to manic or psychotic symptoms and neurocognitive disorder (Fountoulakis et al. 2008).

There has also been a surge of interest in recent years in the phenomenon of emotional (or psychological) resilience (Lazarus 2006). Whereas serious traumatic stressors almost inevitably result in short-term symptoms and a decline in functioning, some individuals rebound especially quickly from such insults and go on to function extremely well, even in the face of ongoing chronic stress. Determination of individual differences in personality that characterize psychological resilience and the neurobiological mechanisms that underlie these observed differences is a subject of intense research scrutiny at the present time (Charney 2004). One intriguing report correlated the serotonin reuptake transporter gene with resilience (Caspi et al. 2003), but subsequent research failed to confirm it (Risch et al. 2009).

11.4 Consultation-Liaison Psychiatry (Psychological Medicine)

After the middle of the twentieth century, the prevalence of various psychiatric problems in medical patients became the focus of interest and research. The importance of assessing and understanding the personality of the medical patient in the long-term success of treatment and the overall outcome was stressed (Lipsitt 2001, 2003). Although the approach was initially psychoanalytic, more practical and operationalized approaches were dominant later. In spite of this, the overall prevalence of reasons for requests for assistance in patient management remained unchanged since the 1960s and included suicide, delirium and psychosis, hostility, depression, uncooperativeness, adverse events of medication, and preparation of patients for major events, including surgery. The first consultation-liaison psychiatry textbook was published in 1968 by John Schwab (1923–2010). However, recently, the focus of research has been on the relationship between chronic medical conditions and psychiatric disorders with a special focus also on economic outcomes. It is widely accepted that the co-occurrence of mental and somatic disorders constitutes a significant public health problem, since mental disorders are both risk factors for the development but also poor prognostic indicators of chronic medical illnesses (Evans and Charney 2003). This coexistence also worsens disability and has a huge impact on quality of life while at the same time increasing healthcare utilization and the overall cost. Another important consequence of mental-somatic comorbidity is poor adherence to all aspects of treatment including medication and lifestyle changes as well as safe behaviors in sex and drug abuse, thus increasing the overall mortality.

It is estimated that between 20% and 67% of medical patients also suffer from a mental disorder, a rate impressively higher from what is seen in the general population, with patients in general hospitals being at the highest risk. Depression and delirium are the most common mental disorders in this population with substance abuse being an additional significant problem. A general rule could be that mental disorders are 2–3 times more common in somatic patients in comparison to the general population.

Historically, the majority of consultation-liaison work is done within a general hospital setting, but recently with the increasing size of outpatient facilities, it embraces a variety of settings and deals with varying needs and problems. New challenges emerge as diagnostic, assessment, and therapeutic tools continue to evolve and change concerning the total field of medicine. On one hand this is extremely helpful, but on the other hand, it makes more difficult the synthesis of findings.

References

Adamo SA (2014) The effects of stress hormones on immune function may be vital for the adaptive reconfiguration of the immune system during fight-or-flight behavior. Integr Comp Biol 54:419–426

Alexander F (1950) Psychosomatic medicine: its principles and application. W. W. Norton, New York, NY

Antoni MH, Cruess DG, Klimas N, Maher K, Cruess S, Kumar M, Lutgendorf S, Ironson G, Schneiderman N, Fletcher MA (2002) Stress management and immune system reconstitution in symptomatic HIV-infected gay men over time: effects on transitional naive T cells (CD4(+) CD45RA(+)CD29(+)). Am J Psychiatry 159:143–145
Bach DR, Friston KJ (2013) Model-based analysis of skin conductance responses: towards causal models in psychophysiology. Psychophysiology 50:15–22
Bakes A, Bradshaw CM, Szabadi E (1990) Attenuation of the pupillary light reflex in anxious patients. Br J Clin Pharmacol 30:377–381
Bechtereva NP (2000) Psychophysiology by the end of the 20th century. Int J Psychophysiol 35:219–236
Boucsein W (2013) Electrodermal activity. Springer Science & Business Media, New York, NY
Bower JE, Ganz PA, Aziz N, Fahey JL (2002) Fatigue and proinflammatory cytokine activity in breast cancer survivors. Psychosom Med 64:604–611
Cacioppo J, Tassinary L, Berntson G (2007) Handbook of psychophysiology, 3rd edn. Cambridge University Press, Cambridge
Calderon R Jr, Schneider RH, Alexander CN, Myers HF, Nidich SI, Haney C (1999) Stress, stress reduction and hypercholesterolemia in African Americans: a review. Ethn Dis 9:451–462
Cannon W (1926) Physiological regulation of normal states: some tentative postulates concerning biological homeostatics. In: Pettit A (ed) A Charles Richet: ses amis, ses collègues, ses élèves. Éditions Médicales, Paris, p 91
Cannon W (1932) The wisdom of the body. W. W. Norton, New York, NY
Capuron L, Raison CL, Musselman DL, Lawson DH, Nemeroff CB, Miller AH (2003) Association of exaggerated HPA axis response to the initial injection of interferon-alpha with development of depression during interferon-alpha therapy. Am J Psychiatry 160:1342–1345
Carroll D, Sheffield D (1998) Social psychophysiology, social circumstances, and health. Ann Behav Med 20:333–337
Carroll BJ, Martin FI, Davies B (1968) Resistance to suppression by dexamethasone of plasma 11-O.H.C.S. levels in severe depressive illness. Br Med J 3:285–287
Caspi A, Sugden K, Moffitt TE, Taylor A, Craig IW, Harrington H, McClay J, Mill J, Martin J, Braithwaite A, Poulton R (2003) Influence of life stress on depression: moderation by a polymorphism in the 5-HTT gene. Science 301:386–389
Ceulemans DL, Westenberg HG, van Praag HM (1985) The effect of stress on the dexamethasone suppression test. Psychiatry Res 14:189–195
Chapman CR, Oka S, Bradshaw DH, Jacobson RC, Donaldson GW (1999) Phasic pupil dilation response to noxious stimulation in normal volunteers: relationship to brain evoked potentials and pain report. Psychophysiology 36:44–52
Charney DS (2004) Psychobiological mechanisms of resilience and vulnerability: implications for successful adaptation to extreme stress. Am J Psychiatry 161:195–216
Cohen D (1968) Magnetoencephalography: evidence of magnetic fields produced by alpha-rhythm currents. Science 161:784–786
Cohen D, Cuffin BN (1983) Demonstration of useful differences between magnetoencephalogram and electroencephalogram. Electroencephalogr Clin Neurophysiol 56:38–51
Cohen S, Hamrick N (2003) Stable individual differences in physiological response to stressors: implications for stress-elicited changes in immune related health. Brain Behav Immun 17:407–414
Cohen S, Tyrrell DA, Smith AP (1991) Psychological stress and susceptibility to the common cold. N Engl J Med 325:606–612
Cohen S, Doyle WJ, Skoner DP, Rabin BS, Gwaltney JM Jr (1997) Social ties and susceptibility to the common cold. JAMA 277:1940–1944
Cohen S, Janicki-Deverts D, Miller GE (2007) Psychological stress and disease. JAMA 298:1685–1687
Coles MG (1989) Modern mind-brain reading: psychophysiology, physiology, and cognition. Psychophysiology 26:251–269

Collado-Hidalgo A, Bower JE, Ganz PA, Cole SW, Irwin MR (2006) Inflammatory biomarkers for persistent fatigue in breast cancer survivors. Clin Cancer Res 12:2759–2766
Corr PJ, Perkins AM (2006) The role of theory in the psychophysiology of personality: from Ivan Pavlov to Jeffrey Gray. Int J Psychophysiol 62:367–376
Critchley HD (2009) Psychophysiology of neural, cognitive and affective integration: fMRI and autonomic indicants. Int J Psychophysiol 73:88–94
Dantzer R (2001) Cytokine-induced sickness behavior: mechanisms and implications. Ann N Y Acad Sci 933:222–234
Davidson RJ (2003) Affective neuroscience and psychophysiology: toward a synthesis. Psychophysiology 40:655–665
Davis MC, Zautra AJ, Younger J, Motivala SJ, Attrep J, Irwin MR (2008) Chronic stress and regulation of cellular markers of inflammation in rheumatoid arthritis: implications for fatigue. Brain Behav Immun 22:24–32
Deutsch F (1939) The choice of organ in organ neurosis. Int J Psychoanal 20:252–262
Dunbar F (1954) Emotions and bodily changes. Columbia University Press, New York, NY
Eisenberg L (1995) The social construction of the human brain. Am J Psychiatry 152:1563–1575
Ekman P, Levenson RW, Friesen WV (1983) Autonomic nervous system activity distinguishes among emotions. Science 221:1208–1210
Engel GL (1977) The need for a new medical model: a challenge for biomedicine. Science 196:129–136
Evans DL, Charney DS (2003) Mood disorders and medical illness: a major public health problem. Biol Psychiatry 54:177–180
Evans D, Golden R (1987) The dexamethasone suppression test: a review. In: Nemeroff C, Loosen P (eds) Handbook of clinical psychoneuroendocrinology. John Wiley and Sons, New York, NY, pp 313–335
Findlay JM (2009) Saccadic eye movement programming: sensory and attentional factors. Psychol Res 73:127–135
Fotiou F, Fountoulakis KN, Tsolaki M, Goulas A, Palikaras A (2000a) Changes in pupil reaction to light in Alzheimer's disease patients: a preliminary report. Int J Psychophysiol 37:111–120
Fotiou F, Fountoulakis KN, Goulas A, Alexopoulos L, Palikaras A (2000b) Automated standardized pupillometry with optical method for purposes of clinical practice and research. Clin Physiol 20:336–347
Fountoulakis K, Fotiou F, Iacovides A, Tsiptsios J, Goulas A, Tsolaki M, Ierodiakonou C (1999) Changes in pupil reaction to light in melancholic patients. Int J Psychophysiol 31:121–128
Fountoulakis K, Iacovides A, Fotiou F, Nimatoudis J, Bascialla F, Ioannidou C, Kaprinis G, Bech P (2004) Neurobiological and psychological correlates of suicidal attempts and thoughts of death in patients with Major Depression. Neuropsychobiology 49:42–52
Fountoulakis KN, Giannakopoulos P, Kovari E, Bouras C (2008) Assessing the role of cingulate cortex in bipolar disorder: neuropathological, structural and functional imaging data. Brain Res Rev 59:9–21
Fowles DC (1988) Psychophysiology and psychopathology: a motivational approach. Psychophysiology 25:373–391
Friedman M, Byers SO, Roseman RH, Elevitch FR (1970) Coronary-prone individuals (type A behavior pattern). Some biochemical characteristics. JAMA 212:1030–1037
Frost S, Kanagasingam Y, Sohrabi H, Bourgeat P, Villemagne V, Rowe CC, Macaulay SL, Szoeke C, Ellis KA, Ames D, Masters CL, Rainey-Smith S, Martins RN, Group AR (2013) Pupil response biomarkers for early detection and monitoring of Alzheimer's disease. Curr Alzheimer Res 10:931–939
Furedy JJ (1996) The North American polygraph and psychophysiology: disinterested, uninterested, and interested perspectives. Int J Psychophysiol 21:97–105
Gitlin DF, Levenson JL, Lyketsos CG (2004) Psychosomatic medicine: a new psychiatric subspecialty. Acad Psychiatry 28:4–11
Glynn LM, Christenfeld N, Gerin W (2002) The role of rumination in recovery from reactivity: cardiovascular consequences of emotional states. Psychosom Med 64:714–726

Goldstein DS, Kopin IJ (2007) Evolution of concepts of stress. Stress 10:109–120
Gorban AN, Tyukina TA, Smirnova EV, Pokidysheva LI (2016) Evolution of adaptation mechanisms: adaptation energy, stress, and oscillating death. J Theor Biol 405:127–139
Gordon E (2001) Integrative psychophysiology. Int J Psychophysiol 42:95–108
Granholm E, Morris S, Galasko D, Shults C, Rogers E, Vukov B (2003) Tropicamide effects on pupil size and pupillary light reflexes in Alzheimer's and Parkinson's disease. Int J Psychophysiol 47:95–115
Greene R, Dalton K (1953) The premenstrual syndrome. Br Med J 1:1007–1014
Grunberger J, Linzmayer L, Walter H, Rainer M, Masching A, Pezawas L, Saletu-Zyhlarz G, Stohr H, Grunberger M (1999) Receptor test (pupillary dilatation after application of 0.01% tropicamide solution) and determination of central nervous activation (Fourier analysis of pupillary oscillations) in patients with Alzheimer's disease. Neuropsychobiology 40:40–46
Gur TL, Bailey MT (2016) Effects of stress on commensal microbes and immune system activity. Adv Exp Med Biol 874:289–300
Handwerker HO, Kobal G (1993) Psychophysiology of experimentally induced pain. Physiol Rev 73:639–671
Handy T (2005) Event related potentials: a methods handbook. Bradford/MIT Press, Cambridge, MA
Harmon DK, Masuda M, Holmes TH (1970) The Social Readjustment Rating Scale: a cross-cultural study of Western Europeans and Americans. J Psychosom Res 14:391–400
Heeger DJ, Ress D (2002) What does fMRI tell us about neuronal activity? Nat Rev Neurosci 3:142–151
Heim C, Newport DJ, Heit S, Graham YP, Wilcox M, Bonsall R, Miller AH, Nemeroff CB (2000) Pituitary-adrenal and autonomic responses to stress in women after sexual and physical abuse in childhood. JAMA 284:592–597
Henderson JM (2003) Human gaze control during real-world scene perception. Trends Cogn Sci 7:498–504
Holmes TH, Rahe RH (1967) The social readjustment rating scale. J Psychosom Res 11:213–218
Horney K (1937) The neurotic personality of our time. W. W. Norton, New York, NY
Hugdahl K (1984) Human psychobiology in Scandinavia: I. Psychophysiology—theory, method and empirical research. Scand J Psychol 25:194–213
Ioannou S, Gallese V, Merla A (2014) Thermal infrared imaging in psychophysiology: potentialities and limits. Psychophysiology 51:951–963
Irwin MR (2008) Human psychoneuroimmunology: 20 years of discovery. Brain Behav Immun 22:129–139
Irwin MR, Wang M, Campomayor CO, Collado-Hidalgo A, Cole S (2006) Sleep deprivation and activation of morning levels of cellular and genomic markers of inflammation. Arch Intern Med 166:1756–1762
Jeronimus BF, Ormel J, Aleman A, Penninx BW, Riese H (2013) Negative and positive life events are associated with small but lasting change in neuroticism. Psychol Med 43:2403–2415
Jeronimus BF, Riese H, Sanderman R, Ormel J (2014) Mutual reinforcement between neuroticism and life experiences: a five-wave, 16-year study to test reciprocal causation. J Pers Soc Psychol 107:751–764
Jessop DS, Douthwaite JA, Conde GL, Lightman SL, Dayan CM, Harbuz MS (1997) Effects of acute stress or centrally injected interleukin-1beta on neuropeptide expression in the immune system. Stress 2:133–144
Karacabey K, Saygin O, Ozmerdivenli R, Zorba E, Godekmerdan A, Bulut V (2005) The effects of exercise on the immune system and stress hormones in sportswomen. Neuro Endocrinol Lett 26:361–366
Katz D, Greenberg M (2015) Adolescent stress reactivity and recovery: examining the relationship between state rumination and the stress response with a school-based group public speaking task. Psychoneuroendocrinology 61:23
Kaufman J, Plotsky PM, Nemeroff CB, Charney DS (2000) Effects of early adverse experiences on brain structure and function: clinical implications. Biol Psychiatry 48:778–790

Keefe FJ, Smith S (2002) The assessment of pain behavior: implications for applied psychophysiology and future research directions. Appl Psychophysiol Biofeedback 27:117–127
Keller A, Litzelman K, Wisk LE, Maddox T, Cheng ER, Creswell PD, Witt WP (2012) Does the perception that stress affects health matter? The association with health and mortality. Health Psychol 31:677–684
Khansari DN, Murgo AJ, Faith RE (1990) Effects of stress on the immune system. Immunol Today 11:170–175
Kobasa SC (1982) The hardy personality: toward a social psychology of stress and health. In: Sanders G, Suls J (eds) Social psychology of health and illness. Lawrence Erlbaum Assoc, Hillsdale, NJ, pp 1–25
Konsman JP, Parnet P, Dantzer R (2002) Cytokine-induced sickness behaviour: mechanisms and implications. Trends Neurosci 25:154–159
Lang PJ, Davis M, Ohman A (2000) Fear and anxiety: animal models and human cognitive psychophysiology. J Affect Disord 61:137–159
Lange T, Perras B, Fehm HL, Born J (2003) Sleep enhances the human antibody response to hepatitis A vaccination. Psychosom Med 65:831–835
Lazarus R (2006) Stress and emotion: a new synthesis. Springer Publishing Company, New York, NY, p 2006
Lipowski ZJ (1967a) Review of consultation psychiatry and psychosomatic medicine. I. General principles. Psychosom Med 29:153–171
Lipowski ZJ (1967b) Review of consultation psychiatry and psychosomatic medicine. II. Clinical aspects. Psychosom Med 29:201–224
Lipowski ZJ (1968) Review of consultation psychiatry and psychosomatic medicine. 3. Theoretical issues. Psychosom Med 30:395–422
Lipsitt DR (2001) Consultation-liaison psychiatry and psychosomatic medicine: the company they keep. Psychosom Med 63:896–909
Lipsitt DR (2003) What do consultation-liaison (C-L) psychiatry and psychosomatic medicine (PM) have in common? Psychiatr Neurol Jpn 105:332–338
Maclean H, Dhillon B (1993) Pupil cycle time and human immunodeficiency virus (HIV) infection. Eye 7(Pt 6):785–786
Mai F (2004) Somatization disorder: a practical review. Can J Psychiatr 49:652–662
Marin C, Carron R (2002) The origin of the concept of somatization. Psychosomatics 43:249–250
McEwen BS, Biron CA, Brunson KW, Bulloch K, Chambers WH, Dhabhar FS, Goldfarb RH, Kitson RP, Miller AH, Spencer RL, Weiss JM (1997) The role of adrenocorticoids as modulators of immune function in health and disease: neural, endocrine and immune interactions. Brain Res Brain Res Rev 23:79–133
McRobbie D (2007) MRI from picture to proton. Cambridge University Press, Cambridge
Mendlewicz J, Hubain P, Koumakis C (1984) Further investigation of the dexamethasone suppression test in affective illness: relationship to clinical diagnosis and therapeutic response. Neuropsychobiology 12:23–26
Miller NE (1985) Effects of emotional stress on the immune system. Pavlov J Biol Sci 20:47–52
Miller G, Chen E, Cole SW (2009) Health psychology: developing biologically plausible models linking the social world and physical health. Annu Rev Psychol 60:501–524
Musial F, Hauser W, Langhorst J, Dobos G, Enck P (2008) Psychophysiology of visceral pain in IBS and health. J Psychosom Res 64:589–597
Nabi H, Kivimaki M, Batty GD, Shipley MJ, Britton A, Brunner EJ, Vahtera J, Lemogne C, Elbaz A, Singh-Manoux A (2013) Increased risk of coronary heart disease among individuals reporting adverse impact of stress on their health: the Whitehall II prospective cohort study. Eur Heart J 34:2697–2705
Niedermeyer E, da Silva FL (2004) Electroencephalography: basic principles, clinical applications, and related fields. Lippincot Williams & Wilkins, Philadelphia, PA
Nuller J, Ostroumova M (1980) Resistance to inhibiting effect of dexamethasone in patients with endogenous depression. Acta Psychiatr Scand 61:169–177

Pace TW, Mletzko TC, Alagbe O, Musselman DL, Nemeroff CB, Miller AH, Heim CM (2006) Increased stress-induced inflammatory responses in male patients with major depression and increased early life stress. Am J Psychiatry 163:1630–1633
Panconesi E, Hautmann G (1996) Psychophysiology of stress in dermatology. The psychobiologic pattern of psychosomatics. Dermatol Clin 14:399–421
Pierrot-Deseilligny C, Milea D, Muri RM (2004) Eye movement control by the cerebral cortex. Curr Opin Neurol 17:17–25
Pinquart M, Sorensen S (2003) Differences between caregivers and noncaregivers in psychological health and physical health: a meta-analysis. Psychol Aging 18:250–267
Poliak S, Mor F, Conlon P, Wong T, Ling N, Rivier J, Vale W, Steinman L (1997) Stress and autoimmunity: the neuropeptides corticotropin-releasing factor and urocortin suppress encephalomyelitis via effects on both the hypothalamic-pituitary-adrenal axis and the immune system. J Immunol 158:5751–5756
Prettyman R, Bitsios P, Szabadi E (1997) Altered pupillary size and darkness and light reflexes in Alzheimer's disease. J Neurol Neurosurg Psychiatry 62:665–668
Quigley KS, Barrett LF (2014) Is there consistency and specificity of autonomic changes during emotional episodes? Guidance from the conceptual act theory and psychophysiology. Biol Psychol 98:82–94
Rahe RH, Mahan JL Jr, Arthur RJ (1970) Prediction of near-future health change from subjects' preceding life changes. J Psychosom Res 14:401–406
Rahe RH, Lundberg U, Bennett L, Theorell T (1971) The social readjustment rating scale: a comparative study of Swedes and Americans. J Psychosom Res 15:241–249
Rahe RH, Biersner RJ, Ryman DH, Arthur RJ (1972) Psychosocial predictors of illness behavior and failure in stressful training. J Health Soc Behav 13:393–397
Reichenberg A, Yirmiya R, Schuld A, Kraus T, Haack M, Morag A, Pollmacher T (2001) Cytokine-associated emotional and cognitive disturbances in humans. Arch Gen Psychiatry 58:445–452
Richardson S, Shaffer JA, Falzon L, Krupka D, Davidson KW, Edmondson D (2012) Meta-analysis of perceived stress and its association with incident coronary heart disease. Am J Cardiol 110:1711–1716
Ropper A, Brown R (2005) Principles of neurology, 8th edn. McGraw Hill, New York, NY
Sanders VM, Straub RH (2002) Norepinephrine, the beta-adrenergic receptor, and immunity. Brain Behav Immun 16:290–332
Schlotz W, Yim IS, Zoccola PM, Jansen L, Schulz P (2011) The Perceived Stress Reactivity Scale: measurement invariance, stability, and validity in three countries. Psychol Assess 23:80–94
Schneiderman N, Ironson G, Siegel SD (2005) Stress and health: psychological, behavioral, and biological determinants. Annu Rev Clin Psychol 1:607–628
Schreckenberger M, Lange-Asschenfeldt C, Lochmann M, Mann K, Siessmeier T, Buchholz HG, Bartenstein P, Grunder G (2004) The thalamus as the generator and modulator of EEG alpha rhythm: a combined PET/EEG study with lorazepam challenge in humans. NeuroImage 22:637–644
Schwartz MS (1999) What is applied psychophysiology? Toward a definition. Appl Psychophysiol Biofeedback 24:3–10
Scinto LF, Daffner KR, Dressler D, Ransil BI, Rentz D, Weintraub S, Mesulam M, Potter H (1994) A potential noninvasive neurobiological test for Alzheimer's disease. Science 266:1051–1054
Selye H (1938) Adaptation energy. Nature 141:926
Selye H (1950) Stress and the general adaptation syndrome. Br Med J 1:1383–1392
Selye H (1954) The alarm reaction, the general adaptation syndrome, and the role of stress and of the adaptive hormones in dental medicine. Oral Surg Oral Med Oral Pathol 7:355–367
Selye H (1978) The stress of life. McGraw-Hill, New York, NY
Sharma RP, Pandey GN, Janicak PG, Peterson J, Comaty JE, Davis JM (1988) The effect of diagnosis and age on the DST: a metaanalytic approach. Biol Psychiatry 24:555–568
Shively CA, Register TC, Clarkson TB (2009) Social stress, visceral obesity, and coronary artery atherosclerosis: product of a primate adaptation. Am J Primatol 71:742–751

Shorter E (1992) From paralysis to fatigue: A history of psychosomatic illness in the modern era. Free Press, New York, NY

Shuchter SR, Zisook S, Kirkorowicz C, Risch C (1986) The dexamethasone suppression test in acute grief. Am J Psychiatry 143:879–881

Singer JR (1959) Blood flow rates by nuclear magnetic resonance measurements. Science 130:1652–1653

Sloan EK, Capitanio JP, Tarara RP, Mendoza SP, Mason WA, Cole SW (2007) Social stress enhances sympathetic innervation of primate lymph nodes: mechanisms and implications for viral pathogenesis. J Neurosci 27:8857–8865

Sokolski KN, Demet EM (1996) Increased pupillary sensitivity to pilocarpine in depression. Prog Neuro-Psychopharmacol Biol Psychiatry 20:253–262

Steinhauer SR, Hakerem G (1992) The pupillary response in cognitive psychophysiology and schizophrenia. Ann N Y Acad Sci 658:182–204

Stemmler G, Wacker J (2010) Personality, emotion, and individual differences in physiological responses. Biol Psychol 84:541–551

Stokes P, Stoll P, Koslow S, Maas J, Davis J, Swann A, Robins E (1984) Pretreatment DST and hypothalamic-pituitary-adrenocortical function in depressed patients and comparison groups. Arch Gen Psychiatry 41:257–267

Striefel S (1999) Is the working definition of applied psychophysiology proposed by Schwartz too narrow/restrictive? Appl Psychophysiol Biofeedback 24:11–19. discussion 43–54

Sutton S, Braren M, Zubin J, John ER (1965) Evoked-potential correlates of stimulus uncertainty. Science 150:1187–1188

Tan ET, Lambie DG, Johnson RH, Whiteside EA (1984) Parasympathetic denervation of the iris in alcoholics with vagal neuropathy. J Neurol Neurosurg Psychiatry 47:61–64

Thaker PH et al (2006) Chronic stress promotes tumor growth and angiogenesis in a mouse model of ovarian carcinoma. Nat Med 12:939–944

The APA Task Force on Laboratory Tests in Psychiatry (1987) The Dexamethasone Suppression Test: an overview of its current status in psychiatry. Am J Psychiatr 144:1253–1262

Tiller WA (2006) Human psychophysiology, macroscopic information entanglement, and the placebo effect. J Altern Complement Med 12:1015–1027

Wachholtz A, Foster S, Cheatle M (2015) Psychophysiology of pain and opioid use: implications for managing pain in patients with an opioid use disorder. Drug Alcohol Depend 146:1–6

Walter WG, Cooper R, Aldridge VJ, McCallum WC, Winter AL (1964) Contingent negative variation: an electric sign of sensorimotor association and expectancy in the human brain. Nature 203:380–384

Watson SJ, Akil H, Young E (1987) Hypothalamic-pituitary-adrenal axis peptides in affective disease: focus on the ACTH/β-endorphin system. In: Nemeroff CB, Loosen PT (eds) Handbook of Clinical Psychoneuroendocrinology. John Wiley and Sons, New York, NY, pp 384–396

Wehner R (2005) Sensory physiology: brainless eyes. Nature 435:157–159

Wilhelm FH, Roth WT (2001) The somatic symptom paradox in DSM-IV anxiety disorders: suggestions for a clinical focus in psychophysiology. Biol Psychol 57:105–140

Wolf TM, Cole B, Fahrion S, Norris P, Coyne L (1994) Age and sex modulate effects of stress on the immune system: a multivariate analysis. Int J Neurosci 79:121–132

Yankee WJ (1995) The current status of research in forensic psychophysiology and its application in the psychophysiological detection of deception. J Forensic Sci 40:63–68

Yeragani VK (1990) The incidence of abnormal dexamethasone suppression in schizophrenia: a review and a meta-analytic comparison with the incidence in normal controls. Can J Psychiatr 35:128–132

Yerevanian B, Feusner J, Koek R, Mintz J (2004) The dexamethasone suppression test as a predictor of suicidal behavior in unipolar depression. J Affect Disord 83:103–108

Ethology, Evolutionary Psychology, Sociobiology, and Evolutionary Psychiatry

12

Kostas N. Fountoulakis

12.1 Introduction

The theory of evolution of species is one of the most celebrated pieces of science through the centuries. Although similar thoughts existed since antiquity, it was Charles Robert Darwin (1809–1882; Fig. 12.1) who established that all species of life have descended over time from common ancestors with his 1859 book *On the Origin of Species* (Darwin 1859). The theory of evolution is important not only because of its mere scientific value; it serves as a paradigm of what a scientific theory is and how it is developed in sharp contrast with religious or ideologically driven beliefs. Its effect on culture and society is so profound that it has been the focus of debate for more than a century and even the subject of a trial (the Scopes Monkey Trial in 1925). It is interesting that currently it is not taught in most medical schools not even in the frame of biology or genetics, while there is a significant number of academics who support the divine or "intelligent" design theory which does not meet the criteria to be considered a scientific theory.

Its scientific importance lies to the fact that it connects cellular biology, physiology, molecular biology, genetics, immunology, anatomy, microbiology, and literally every life science. It provides a conceptual framework which obeys to the laws of science and includes all Earth environment and all disciplines of human knowledge and activity. It suggests that while Earth itself is about 4.6 billion years old, there were molecules with the ability to self-duplicate since 4 billion years ago and single-celled organisms (prokaryotic) since 3.5 billion years ago. Multicellular organisms (eukaryotic) appeared some 1.7 billion years ago, while the first ancestors of humans evolved probably 2 million years ago (Fig. 12.2). It is very interesting to know that the genetic material of modern humans but also of all modern living beings is the end product of ancient natural experiments and loans from

K. N. Fountoulakis (✉)
3rd Department of Psychiatry, Faculty of Medicine, School of Health Sciences, Aristotle University of Thessaloniki, Thessaloniki, Greece

K. N. Fountoulakis, I. Nimatoudis (eds.), *Psychobiology of Behaviour*,
https://doi.org/10.1007/978-3-030-18323-3_12

Fig. 12.1 Charles Robert Darwin (1809–1882)

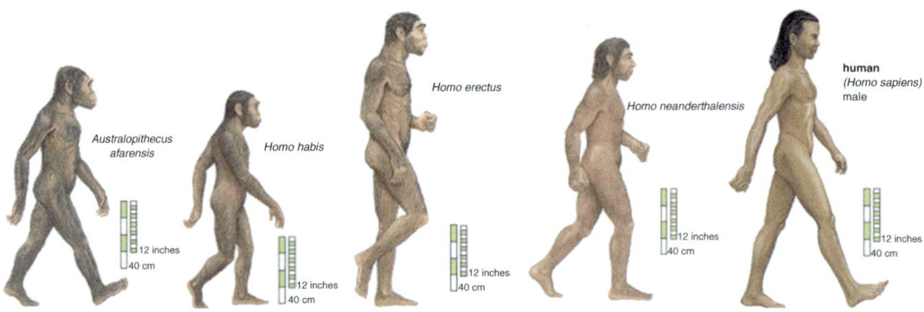

Fig. 12.2 Evolution of modern humans (*Homo sapiens sapiens*). Modified from http://imgarcade.com/1/evolution-of-man-names-of-stages

other species and other kingdoms of life and not simply random mutations as a product of influence of external factors. Approximately 10% of the genes of modern humans come from retroviruses whose genetic material was insinuated into germ cells millions of years ago.

The Darwinian or evolutionary approach is of proven utility when we try to understand how behaviors evolved in parallel with genetic structure. In fact, genetics, environment, and behavior seem to be closely related since genetics determine behavior which serves adaptation to the environment, but the feedback loops between these three domains are extremely complex and dynamic.

One important characteristic of the evolutionary approach to behavior and social structure is that it utilizes a somewhat "cynical" approach to interpret and understand things. This is often in contrast with the ethical-philosophical or even religious beliefs and concepts the average lay person but also probably the majority of the scientific community have. Phenomena like homicide or rape are considered under the concept that in order to exist, there must had been some evolutionary advantage in the far past for those who committed them, or at least they constitute residual out of frame behaviors. These approaches might bother since they seemingly deprive the ethical load of these acts and they seemingly justify and provide support or excuse for those engaging in similar behaviors. The source of individual differences is another such example of sociopolitically sensitive issues. Often individual differences are seen as a reason or a cause of social and hierarchical inequality, which is indeed not unusual in human society. The big question, however, is whether sociopolitical and ideological forces should dictate the outcome of scientific research. The reader should have in mind that the scientific method tries to elucidate the roots and the sources of behaviors without the bias that might come with millennia of social evolution and organization. This is the way science works in all fields, and this is why science was able to make the significant progress that changed human life in the last few centuries, prolonged life expectancy, improved quality of life, and reduced poverty. Also this is the way through which science might be able to reduce the manifestation of these unwanted behaviors and help to improve the humanity. After all, humanity is only partially a matter of genes; modern humans and chimpanzees differ in less than 5% of total genome (Varki and Altheide 2005), yet human and chimpanzee societies are radically different.

It is the cognitive and metacognitive processes that define what we are, how we see ourselves and the others, how we consider and respect them and their rights, and what we believe it is our place in the universe and among other species. These qualities made human beings aware of their existential dilemmas and tragedies and raised them above and beyond the basic instincts and biological pressure.

12.1.1 Ethology

The discipline of ethology concerns the study of animal behavior. It usually considers it to be an evolutionarily adaptive trait. The term "ethology" comes from the Greek word *Ðθος* (ethos meaning "character") and *–λογία* (–logia meaning "the study or knowledge of"). The first to use the term was John Stuart Mill (1806–1873). In his 1843 book *System of Logic*, he advised the development of ethology as a field of science for the study of individual and national differences in character (Robson 1974). As the discipline we know today, it was introduced in 1902 by William Morton Wheeler (1865–1937), an American myrmecologist (he was studying ants) (Matthews and Matthews 2010). Typically, ethology studies a type of behavior across species rather than behaviors specific to a single specific species

Fig. 12.3 1973 Nobel Prize in physiology or medicine

and in relationship to the known animal anatomy, physiology, neurobiology, and phylogenetic history.

Ethology was part of Darwin's work but was advanced mainly through the works of Charles Otis Whitman (1842–1910), Oskar Heinroth (1871–1945), and Wallace Craig (1876–1954) which were ornithologists. The discipline reached its highest fame when three scientists, Nikolaas Tinbergen (1907–1988), Konrad Lorenz (1903–1989), and Karl von Frisch (1886–1982), were awarded the 1973 Nobel Prize in Physiology or Medicine (Fig. 12.3).

12.1.2 Evolutionary Psychology

In this frame, evolutionary psychology studies behavior and psychological function from an evolutionary perspective. It assumes that human psychology and behavior is a direct product of evolutionary processes and constitute evolved adaptations, which follow the laws of Darwinian theory through natural or sexual selection (Williams 1966). This implies that not only the brain but also the mind and behavior follow these laws and rules and therefore evolutionary psychology can serve as a unifying theoretical approach for the behavioral sciences by putting research of psychological function in the frame of the human evolutionary past.

This concept has a number of other implications; it considers psychology to be part of biology rather than theoretical sciences and philosophy, which is true if one considers the fact that modern psychology emerged from anthropology and the study of individual differences in the early nineteenth century. It also views the mind in the frame of computational theory and considers mental processes to be computational operations as responses to the environment. It utilizes elements and concepts from cognitive psychology, artificial intelligence, evolutionary biology, genetics, behavioral ecology and ethology, anthropology, archaeology, and zoology.

This computational approach to mind also suggests that every psychological function has evolved to solve a problem and that the complexity of needs and subsequently of evolved adaptations and solutions to these needs gave birth to broader psychological processes like reason, intelligence, emotion, or impulses and not vice versa as many philosophers imply. It also implies that modern humans have minds, which were evolved and are specifically equipped for behavioral adjustment and survival in the environment humans were facing during the Pleistocene and the Stone Age and not necessarily for the modern environment.

There is a number of behaviors, abilities, and psychological traits which could be good candidates to be considered as results of evolutionary procedures. They occur universally, that is, in all cultures and across the globe, and they concern basic cognitive abilities like expression, recognition, and interpretation of emotions and their behavioral dimensions including facial and hand gestures, discern kin from nonkin, select healthier and more fit mates, and cooperate with other members of the group or tribe. This approach has important although controversial applications in economic and political theory, health, law, and mental health among others (Dunbar and Barret 2007; Buss 2005).

12.1.3 Sociobiology

Sociobiology was introduced as a term in 1975 by Edward Osborne Wilson (1929–) (Wilson 1975). This constituted the highlight of a long process which took place in the late 1960s and early 1970s and attempted to bridge psychology, sociology, ethology, evolutionary and population biology, ecology, anthropology, game theory, and genetics (Barash 2003b). Among other issues, the word "psychobiology" has some political implications, and therefore the alternative term "behavioral ecology" is also used. Essentially, sociobiology differs very little from evolutionary psychology. At times and especially in the beginning, there was severe and vicious ideological criticism which included the broader spectrum of evolutionary psychology and sociobiology.

One major feature in the sociobiological approach is that natural selection occurs among genes, not at the level of groups or species. This has been shown among others by George C. Williams (1926–), William Donald Hamilton (1936–2000), and John Maynard Smith (1920–2004). Their work also introduced the term "inclusive fitness" to denote the sum of accumulated reproductive success of individual genes within family lines by passing the generations.

12.1.4 Evolutionary (Darwinian) Psychiatry

The Darwinian theory had a major influence on how health and disease are conceptualized and what might constitute the best treatment option. A number of medical conditions including obesity, anemias, autoimmune diseases, and hypertension were put in an evolutionary frame (Gluckman et al. 2009). According to this approach, mental health and disease should be understood as the end result of the interaction between the organism and the environment, with the addition that abnormal behaviors and symptoms could reflect either an extreme form of otherwise adaptive behaviors or the triggering of them in an out-of-frame or proportion way and under inappropriate conditions.

This is somehow different from the standard approach in clinical psychiatry and psychopathology where connections with recent events and reaction to recent problems in the frame of the individual patient's and core family are considered (proximate mechanisms). In contrast, evolutionary psychiatry stresses that variation (including variation in behavior) is not only normal but evolutionarily necessary and is concerned with the misfit of preexisting normal coping mechanisms which could cause maladjustment or disease if triggered in an inappropriate way or persist for longer than expected and especially in an environment and with social demands extremely different from those encountered by humans during the Pleistocene and the Stone Age (Fabrega 2002).

12.2 Historical Overview

The theory of evolution has a long history and roots in the antiquity. It exists in the teachings of the pre-Socratic Greek philosophers, especially Anaximander (610–546 BC) and Empedocles (495–430 BC), with his work *Περί φύσεως* (De rerum natura; *On the Nature of Things*). However, in contrast to them, Aristotle (384–322 BC; Fig. 12.4) utilized the idea of fixed natural unchanged and preexisting patterns, known as *μορφή* (morfi meaning form) or *είδος* (idos meaning species). He rejected any idea of changing or evolving forms and species and suggested that all naturally existing living or nonliving things were actually incomplete reflections of these preexisting ideal forms. He also introduced the concept of *ιεραρχική κλίμακα* (scala naturae; hierarchical scale), according to which both living beings and nonliving things are classified on an ideal pyramid with simple nonliving things at the base, plants and simpler animals at the lower levels, and humans at the top (Fig. 12.5). These ideas can be considered to be the early conceptualization of the "intelligent design theory"; they demand some kind of divine top-down cosmic order and in combination with Christianity came to dominate the western world until relatively recently.

With the Renaissance and since the seventeenth century, the method of modern science became gradually dominant and demanded the application of the same physical laws for all visible things without the contribution of any divine cosmic order. An important advance was the classification of plants and animals

Fig. 12.4 Aristotle (384–322 BC)

in 1735 by Carl Linnaeus (1707–1778; Fig. 12.6) which showed the presence of a hierarchical nature in living organisms with the use of scientific method. Soon afterward in 1751, Pierre Louis Maupertuis (1658–1759) argued that natural changes that occur during reproduction could accumulate over several generations and eventually new species emerge, while Georges-Louis Leclerc, Comte de Buffon (1707–1788) proposed that the opposite could also happen, that is, a single "higher" species could degenerate into several "lower" species in the classification hierarchy. The grandfather of Charles Darwin, Erasmus Darwin (1731–1802), was the first to clearly propose that all animals could come from a single microorganism (filament). Jean-Baptiste Lamarck (1744–1829) developed his "transmutation theory" in 1809, which was based on the assumption that animal organs change because of use or disuse and these changes are inherited from parents to children (a theory called later "Lamarckism"). In sharp contrast, the English clergyman William Paley (1743–1805) in his 1802 book *Natural Theology or Evidences of the Existence and Attributes of the Deity* elaborated on the theory of the divine design (Paley 1802). Eventually the theory of evolution of species through natural selection was formulated by Charles Robert Darwin (1809–1882; Fig. 12.1) in his 1859 book *On the Origin of Species* (Darwin 1859). Probably Alfred Russel Wallace (1823–1913) had arrived at a similar theory

Fig. 12.5 Ιεραρχική κλίμακα (scala naturae; hierarchical scale of nature): Drawing from *Retórica cristiana* by Fray Diego Valadés (1579)

more or less simultaneously with Darwin. Thomas Henry Huxley (1825–1895) also known as "Darwin's bulldog" for his support to the theory of Darwin provided evidence that humans and apes shared a common ancestry. This latter caused much criticism and disturbance since it was directly implying that neither humans as a species nor any specific human race have any special place in the universe; this was in sharp contrast with the religious and sociopolitical beliefs and ideas of the time.

Later, in the 1920s and 1930s but especially after the discovery of the DNA by James Watson (1928–) and Francis Crick (1916–2004) in 1953, the accumulated knowledge permitted a resolution of conflicts and discrepancies and the development of a revised and unified theory of evolution that applied generally to all branches of biology.

Essentially it was Charles Darwin who implied for the first time that behavior could obey to the same laws of evolution like biology, thus giving birth to evolutionary psychology. Two of his books concerned animal emotions and psychology (Darwin 1871, 1872).

Fig. 12.6 Carl Linnaeus (1707–1778)

In the 1930s ethology emerged from the works of Nikolaas Tinbergen (1907–1988), Konrad Lorenz (1903–1989), and Karl von Frisch (1886–1982) which were awarded the 1973 Nobel Prize in Physiology or Medicine (Fig. 12.3). In 1975, Edward O. Wilson (1929–) introduced psychobiology (Wilson 1975, 2000, 1978), while mating and sexuality became the focus of evolutionary research during the 1980s and 1990s with the works of Donald Symons (1942–), Leda Cosmides (1957–), and John Tooby (1952–) (Barkow et al. 1995). In the 1970s Robert Ardrey compared human and animal behavior as similar in quality and directly comparable and relevant (Ardrey 2014).

12.3 Ethology

There is a variety of behaviors which are essential for the survival of the individual animal but also for the survival of the species. While usually behaviors serving the survival of the individual serve also the survival of the species, this is not always the case.

As a general rule, it can be said that animals learn to respond to stimuli which are relevant and neglect those which are irrelevant. This prerequisites that it is possible to distinguish between relevant and irrelevant stimuli, and as a consequence, new behaviors are learned and established. One possible mechanism through which these new behaviors are learned is associative learning as shown by Ivan Pavlov

(1849–1936). However, after repetition and by passing the time, responses tend to reduce in intensity with the emerging of habituation toward the stimuli.

Among the abilities and behaviors which are important for survival is the ability to recognize and discriminate the members of one's own species. This is fundamental and functions as a platform for a number of behaviors including collaboration, forming alliances, and mating.

Several animals if not most tend to live in groups of variable sizes. This gives birth to social behavior and requires advanced cognitive abilities. The end result is social life in groups which provides with better chances of survival and development. Living in groups provides better defense against predators, while predators could seek and attack their victims more easily. Searching for food is more efficient since larger areas can be covered and information can be exchanged concerning the location of resources.

Probably learning to identify one's own species only takes place in a very limited period of time. Konrad Lorenz called it "imprinting," after discovering it following extended observations in geese and other birds. He noticed that the young of birds follow their mothers spontaneously almost immediately after they are hatched. He also observed that this response could be triggered by an object other than the mother if it is presented instead of the mother during this critical period and for a few days after hatching. His photo walking and followed by young geese is iconic in the history of science (Fig. 12.7).

Another way of obtaining new behaviors and enriching one's own repertoire is "observational learning." Its simplest form is "imitation" which concerns the exact

Fig. 12.7 The iconic image of Konrad Lorenz walking and followed by young geese

replication of an observed behavior. Usually the observed behavior is carried out by an individual with a higher status in the hierarchy of the group, usually an elder one. In this frame, the behavior is obtained by young low status and inexperienced members of the group, by copying the behaviors of elderly, high status, and experienced members (Horner et al. 2010). However, often there is targeted teaching which demands the "teacher" to deliberately modify its behavior so as to increase the probability the "pupil" will copy and imitate the observed behavior. Indeed the basic principles of teaching do exist in the animal kingdom (Hoppitt et al. 2008).

However, in the frame of group or social life, there is often conflict between members concerning social supremacy. Supremacy gives priority to resources but also to mating and therefore increases further the chances for survival and reproduction. On the other hand, group life has its disadvantages; it facilitates the spreading of disease and parasites, causes continuous conflict, and demands negotiations for the distribution of resources and privileges within the group.

There are also several unanswered questions concerning the existence of sterile subgroups within societies (e.g., in bees) as this seems to contradict essential rules of evolution as we understand it. Also several behaviors including altruism, self-sacrifice, or revenge are difficult to incorporate in a narrow theory of evolution. These behaviors demand to consider evolution not only at the level of the individual animal but also at the level of the species and for a time duration that exceeds the natural life of the individual and spreads across several generations. These proposed mechanisms are complex and currently not well understood; however, it seems that there exist mechanisms that under specific conditions put the interest of the group above the interest of the individual (Cummings et al. 1991).

There is much discussion and controversy on whether there is an optimal group size and how this could be defined. Animal groups tend to increase in size, but after exceeding a certain size, the benefits of social life degrade, and an equilibrium is achieved through the balance of benefits and conflicts (Sibley 1983).

12.4 Evolutionary Psychology

At the core of evolutionary psychology is the assumption that natural selection has provided humans with many psychological adaptations. This has happened with the same mechanisms and processes human anatomy and physiology adapted. In these adaptations the environment and the specific needs that stem from it are the determining factors.

12.4.1 Products of Evolution: Adaptations, Exaptations, By-Products, and Random Variation

The main task of evolutionary psychology is to understand how specific psychological mechanisms developed and exactly how they serve the survival of the species. These psychological mechanisms include also neurocognitive abilities like understanding and interpreting gestures and emotions, discerning kin from nonkin, forming groups and developing cooperation and hierarchy, and identifying and preferring healthier mates, but also they include the involving in conflicts with mates, relatives, and other members of the group.

In general, it is expected that these mechanisms are either innate or easy to learn, and they are spread across cultures in a worldwide fashion. According to George C. Williams (1926–2010), an "adaptation" is characterized by an improbable complexity, species universality, and adaptive functionality (Williams 1966). Behaviors or traits that occur in all human societies and cultures universally around the world are good candidates to be evolutionary adaptations (Brown 1991). These are traits and behaviors related to language, neurocognition, and social and gender skills and roles (Smith 2011; Berent et al. 2008; Chomsky 2005; Sugiyama 2003; Schwartz et al. 2003). It is uncertain whether it is generally obligate or facultative (i.e., resistant or sensitive to typical environmental variation), but it is certain that at least some of them are shaped to a certain degree by the specific contemporary environment (Buss 2005; Barash and Lipton 2001).

Often, behavioral traits are not the products of evolutionary adaptation, but, peculiarly, they constitute by-products of some other behaviors with an adaptation essence. These by-products are called "exaptations" or "spandrels," and they manifest a random variation between individual persons (Buss et al. 1998).

12.4.2 Environment of Evolutionary Adaptedness

The set of recurring selection pressures which cause a specific adaptation to emerge are collectively called "environment of evolutionary adaptedness" (EEA) (Bowlby 1969).

The *Homo* genus appeared 2.5–1.5 million years ago while *Homo sapiens* 1.8–0.2 million years ago (Fig. 12.2). This time period is part of the Pleistocene, which is often colloquially referred to as the Ice Age and lasted from 2.5 million to 12,000 years ago. Its end corresponds to the end of the last glacial period, and its name comes from the Greek *πλείστος* and *καινός* meaning "mostly new." It is divided into four general stages or ages, the Gelasian, Calabrian, Ionian, and Tarantian. During that period, the Earth's climate was characterized by recurrent glacial cycles, and ice came to cover almost one third of the total Earth's surface. A large permafrost zone existed. The entrapment of large quantities of water in the glaciers caused a significant drop in the sea level which at times reached 100 m in comparison to the modern sea level. As a result the coastline was much different than today, and passages existed connecting lands which today are separated by sea (Fig. 12.8). Also the collective memory of cataclysms which survive in ancient myths, from Gilgamesh to Defkalion and to Noah, probably reflect abrupt changes in coastal line in the Mediterranean Basin, in the Black and Caspian Sea, as well as in the Red Sea and the Persian Gulf, because of glacier melting. It is to be noted that at that time, deserts were drier and more extensive.

Fig. 12.8 The coastline during the Pleistocene with sea level 100 m lower than contemporary one. Large strips of land existed where today are sea, and there were passages connecting lands which today are separated by sea. Note the coastline in the Mediterranean Basin, the Black and the Caspian Sea, as well as around the Arabian Peninsula. These areas were probably land at the time and were probably the sites where catastrophic cataclysms occurred when the glaciers melted (image modified from http://www.genesisveracityfoundation.com/Iceage.html after permission). Modified from http://www.genesisveracityfoundation.com/Iceage.html

A major element in evolutionary psychology is the assumption that most of human psychological mechanisms evolved during the Pleistocene and constitute adaptations to survival and reproductive problems caused by the environment of that period. That environment both natural and social was radically different from the modern environment, and the societal structure and the needs were much different from contemporary needs. Humans lived in small hunter-gatherer groups with more stable group features, interpersonal interactions, and identity characteristics, and they were exclusively concerned with food selection and acquisition, selection of territory and physical shelter, as well as avoiding predators and environmental threats (Buss 2011). Differences in gender roles at that time might be behind the higher visuospatial cognitive capacity of males and the higher social cognition of females (Gaulin and McBurney 2003).

As a consequence to the fact that human psychological traits were developed in an environment much different from the modern one and with much different needs, often human psychology exhibits "mismatches" to the modern environment, which, however, is designed and constructed according to the needs and wishes of humanity. These mismatches can take the form of thoughts and biased beliefs but also behaviors (Ohman and Mineka 2001; Pinker 1999; Hagen and Hammerstein 2006). For example, present-day humans are inclined to trace patterns in a series of events even in cases such patterns do not really exist (in a series of random events) and to identify cheating rather than any other irregularity in the events (Gaulin and McBurney 2003).

Maybe several modern conditions including working in large anonymous bureaucratic groups and modern management methods could reflect mismatch and exploitation of instincts (Van Vugt and Ahuja 2011; Van Vugt and Ronay 2014). Another consequence similar to mismatch is the phenomenon of supernormal stimuli. Such a stimulus elicits a response which is far stronger than the same response when elicited by the specific stimulus for which it has originally been evolved. The concept was coined by Niko Tinbergen to refer to nonhuman animal behavior, but later Deirdre Barrett (1954–) carried it to modern human behavior. In this frame, television is a supernormal stimulus for social behavior and attention-grabbing action, junk food for the intake of important nutrients, and pornography of sexual behavior (Barrett 2007, 2010; Hagen and Hammerstein 2006).

12.4.3 Life History Theory

The life history theory starts with the obvious fact that each individual, no matter animal or human, does not have infinite time and energy budgets; on the contrary these are quite finite and precious. Thus, investing effort to solve one problem often precludes the investment in another. In oversimplified terms, the most contrasting investments are the investment in one's bodily growth, safety, and maintenance vs. parenting and kin investment. The first increases the chances of personal survival; the latter increases the reproductive success of genetic relatives. Therefore a trade-off between costs and benefits is always in place, while often investment in complex behaviors increases the success in the solving of more than one adaptive problems (Gadgil and Bossert 1970; Kaplan and Gangestad 2005; Roff 1992; Stearns 1992).

Individual features and characteristics determine the outcome of the trade-off between different investments. These characteristics include total energy, perceived life expectancy, and individual talents and preferences. Individuals might prefer parenting over mating or vice versa, while a perceived short life expectancy might push toward a strategy of immediate expenditure of resources, intense competition, and risk taking for mating (Daly and Wilson 2005a). Biological factors including hormonal might play a role in this trade-off. It is not clear whether it is cause or effect, but it has been reported that male testosterone levels drop with commitment in mateship and levels even fall further after the birth of children (Burnham et al. 2003).

12.4.4 Costly Signaling Theory

Communication is essential in all kinds of social interaction but also for interaction necessary for basic biological functions among members of the same species. Thus individuals also compete on how successfully they will communicate with others in order to be more successful in establishing alliances, achieving social status, and also mating. The reliability of the messages which individuals communicate to others is questionable since there is often an attempt to deceit; by this, the individual might fulfill goals which otherwise would be out of their league. The ways of communication as well as the ability to understand and to trace deceit constitute important adaptations (McAndrew 2002; Miller 2000a; Zahavi 2006).

In this frame, costly signals tend to be honest signals (Zahavi 1975, 2006), since they are demanding in terms of investment and sent only by those who can afford. Activities like fighting or any type of physical contests for males or sex-analogous activities for females serve the purpose to send a honest and reliable message to the opposite sex about the condition of the individual. Complex and difficult to explain behaviors including generosity and altruism could be viewed in this frame (Miller 2007).

The costly signaling theory is linked to life history theory since the quality of the signal which an individual can communicate defines the individual's life history and the strategies he will adopt.

12.4.5 Balancing Selection

Balancing selection refers to a condition when selection does not choose a single solution to a problem and eliminates all others, but instead it permits genetic variation. This leads to the manifestation of different levels of adaptation to a specific environment or the same level of adaptation to different environments (Penke et al. 2007).

One mechanism of balancing selection is through environmental heterogeneity in fitness optima, that is, different environments favor the evolution of different behavioral patterns through a complex selection process including migration (Ebstein 2006; Penke et al. 2007; Chen et al. 1999; Eisenberg et al. 2008).

Another mechanism is frequency-dependent selection. This refers to the situation when two or more different strategies concerning the same adaptation problem are maintained within the same population, and they exist at a particular frequency relative to each other. This means also that the overall fitness of each strategy decreases with its increasing frequency of use in the population. An example of such a strategy is cheating, that is, using deceit to achieve adaptation goals (Mealey 1995).

12.4.6 Mutation Load

Genetic mutations are common and can be neutral or disruptive and could concern any body system or function. On average individual humans carry at least 500 brain-disruptive mutations each (Keller and Miller 2006). Through the process of natural selection, individual mutations or combination patterns could be eliminated through time; however purging is never complete, and mildly harmful mutations could survive for many generations. Most reflect older mutations, inherited from ancestors (Keller and Miller 2006), but the effect of increasing parental age in combination with very low infant mortality and the increasing rate of cesarean sections during the last few decades remain to be seen.

Random mutations and the genetic load they create could be the source of noise or variations in behavior (Buss 2006), and adaptive traits with an important role especially in mating, including emotional stability, conscientiousness, or intelligence, could be disrupted, or alternative facets or traits could manifest (Buss 2006).

12.4.7 Contingent Shifts According to Environmental and Phenotypic Conditions

While natural selection pushes toward the preference for a specific heritable trait which is adaptive for a specific environment, contingent shifts refer to the selection of psychological mechanisms which are flexibly responsive to changes in environmental or cultural conditions (Belsky 1999; Gangestad et al. 2006). This is related to the life history theory since this theory predicts changes in behavior after changes in environment or in personal achievements, e.g., parenting, but also as a response to individual characteristics, e.g., physical size and strength (Tooby and Cosmides 1990; Ishikawa et al. 2001).

12.4.8 Evolution of Emotion

As described in Chap. 1, affects and emotions serve two main aims:

- The first concerns the internal functioning of the individual and provides the individual with fast decisions which serve the survival of the individual but also of the species. Some of these decisions are easy to understand (e.g., fear of animals), but others are incomprehensible at least with a superficial approach (e.g., aesthetics and attraction to the opposite sex). In the same frame, emotions provide feedback concerning the behavior of the individual, and in this way, they enhance the expression of the specific behavior or preclude its future manifestations. For example, sadness constitutes the emotional response to loss, defeat, disappointment, or other adversities. Its adaptive function includes permitting withdrawal to conserve resources, asking for support from significant others and the autonomic arousal which might be present facilitates the search for the lost object or an appropriate substitute.
- The second aim is to communicate the internal emotional state of the individual to others, and this is achieved through facial expressions, gestures, bodily movements and posture, and verbal and nonverbal elements of voice. These ways of communicating emotions vary between cultures, but most of the repertoire is universal for human beings. They constitute a main source for the interaction with others, since the emotions of an individual influence the emotions, thoughts, and behaviors of others, produce positive or negative feedback, and give birth to circles of future interactions and reciprocal influence.

A modern understanding of the issue goes through a basic approach to brain function which could suggest that there are two distinct mental processes: logical thinking and emotions. While emotions are present also in animals, logical thinking is present primarily in humans, while some elements are also evident in the behavior of primates.

Emotional processes are evolutionary older and are characterized by speed and dominance. They lead to fast decision-making, on the basis of predetermined strong assumptions concerning the gross characteristics of the situation. For example, fear is triggered immediately and almost before conscious recognition of the stimuli, and it leads to the fast manifestation of a specific adaptive behavior (fight or flight). A snake will always trigger fear, no matter whether it is poisonous or not. On the contrary, logical thinking is slow, requires the conscious elaboration on the stimuli, and demands concentration and effort, and it is not as strong as emotions are, concerning the effect on behavior. Emotion is biased toward the triggering of those behaviors that serve the survival of the individual and the species, while logical thinking aims toward an "objective" assessment of the situation. In the language of artificial intelligence, the closest description which can be made today is that of "fuzzy" vs. "digital" systems.

The database of assumptions which the emotions use is of unknown origin, probably partially inherited and partially acquired through experience, and possibly it is characteristic of the species. Logical thinking is based mainly on training. Decisions based on emotions are stronger than those based on logical thinking, and when they are in conflict, the person faces a difficult dilemma, since it is very difficult for logical thinking to override emotional pressure.

The two processes, although independent in principle, they interact and influence each other. The emotional status causes bias in logical thinking, and logical analysis triggers emotions depending on the positive or negative outcome. This interaction is likely to happen at multiple levels (e.g., selective memory recall, reinforcement through new analysis, biased selection of possible solutions, etc.).

The interest in emotions from an evolutionary perspective was triggered by the publication of the book *The Expression of the Emotions in Man and Animals* by Charles Darwin in 1872 (Fig. 12.9). In that book, Darwin stresses the universal nature of emotions and the connection of mental states to the neurological organization of movement. Central to his understanding was a shared human and animal ancestry. This was in sharp contrast to the contemporary claims that there were divinely created human muscles to express uniquely human feelings. Darwin's original suggestion was that emotions evolved via natural selection and therefore have cross-culturally universal counterparts; a proposal confirmed almost a century later by the works of Paul Ekman (Ekman 1965, 1980, 1992a, b, 1993, 1994, 2003, 2009, 2016; Ekman and Friesen 1967, 1971; Ekman et al. 1969, 1987). According to Ekman humans share at least five basic emotions: fear, sadness, happiness, anger, and disgust. Furthermore, animals undergo emotions comparable to those of humans. It seems that social interaction based on emotions influences motivation and stimulates the reward systems (Belke and Garland 2007).

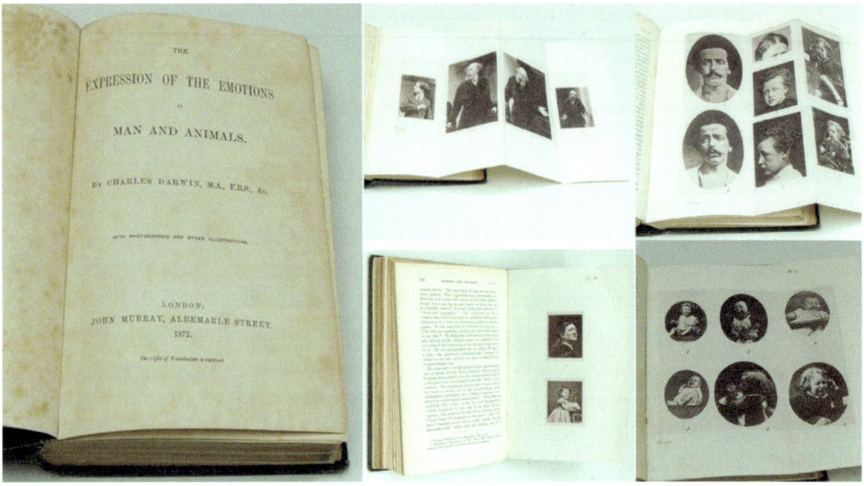

Fig. 12.9 The book *Expression of the Emotions in Man and Animals* by Charles Darwin (1872) (images from Botterweg Auctions Amsterdam, after permission)

12.4.9 Mating and Parenting

Sexual function and mating are important for the survival of the species and more particular for the genes of the individual person. However the strategies to achieve this goal vary considerably among species and genders. According to the r/K selection theory (Huey and Pianka 1977) and the life history theory (Roff 1992; Stearns 1992), some species have many offspring, while on the contrary, others choose to have fewer offspring but invest much more in each one. Humans belong to the second category.

Angus John Bateman (1919–1996) argued that in most species, anisogamy, that is, the fact that males are able to produce millions of sperm cells while females only a relatively small number of eggs, results in different sexual behavior between genders. Females also spend many months in pregnancy. As a consequence, females are the limiting factor of parental investment. Males will compete for this limited female ability to give birth which is combined with higher nurturance of the offspring. Males themselves will overall invest less on the offspring (Bateman 1948). On the contrary, females exercise the ability to have high-quality males (choosiness). Additionally, human females exhibit concealed ovulation ("hidden estrus") which means that it is impossible for the male to know when the female is fertile. Subsequently, frequent mating and stable relationships are necessary to ensure paternity. The erectile ability from the part of the male might provide information concerning his health status, and because of this, it influences the mating choice of the female (Abbot et al. 2011).

This essentially means that females tend to choose males who care for the family and the offspring and its nurture and raising (Barash and Lipton 2001). In 1972

Robert Trivers (1943–) developed the parental investment theory (Trivers and Willard 1973; Trivers 1972) which further built on Bateman's principle. He suggested that there are different levels of parental investment between the sexes. According to his theory, females initially invest more, and this difference leads to different mating and reproductive strategies between males and females and eventually to sexual conflict in the form of sexual dimorphisms in mate choice (differences in behavior between the two sexes beyond the differences in sexual organs), sexual reproductive competition, and courtship displays (Barash and Lipton 2001). There are evolved mechanisms to attract select and secure mates (Workman and Reader 2008; Buss and Barnes 1986; Li et al. 2002; Schmitt and Buss 2001; Buss 1988), but there is also conflict between the sexes (Conroy-Beam et al. 2015; Peters et al. 2002; Botwin et al. 1997). Under specific conditions, females could adopt sexual behavior closer to that to males, but this is rather the exception (Buss 1989).

Same-sex sexual behavior or homosexuality seems impossible under a Darwinian point of view since it is related with low rates of reproduction and constitutes a "Darwinian paradox." However it is not a solely human phenomenon since it has been documented in hundreds of species (Garcia-Cardenas et al. 2015; Gunst et al. 2015; Leca et al. 2015; MacFarlane and Vasey 2016; Triana-Del Rio et al. 2015; Ungerfeld et al. 2014; Vasey et al. 2014). Its frequency is rather low, but the rates seem rather stable in all human populations. The role of environmental factors seems to be weak although there seems to exist some specific environmental conditions that may encourage transient homosexual behavior, e.g., captivity. On the other hand, there are data suggesting the influence of genetic factors, and some studies have pointed out relevant asymmetries in the distribution of both male homosexuality and female fecundity in the parental lines of homosexual vs. heterosexual males.

A number of hypotheses have been proposed to explain this paradox, although none of them has gained the full support of the scientific community. The following are the most important:

1. "Kin selection" which suggests that childless homosexuals might put more effort into helping raise nieces or nephews.
2. "Overdominance" which suggests the presence of a gene which when the person is heterozygous leads to a reproductive advantage (e.g., by increased female fertility or increased sperm motility) but when homozygous leads to homosexuality.
3. The "maternal effects hypothesis" suggests a fetus is influenced by the environment of the mother's womb, resulting in changes that predispose one toward homosexuality.
4. The "sexually antagonistic selection" hypothesis has the highest support from evidence currently. It suggests that there exist genes which are spread throughout the human population and they work by providing the one sex with a specific reproductive advantage while disadvantaging the opposite sex. Such a gene would promote fertility and subsequently reproductory fitness in females but homosexuality in males and vice versa (Camperio Ciani et al. 2008).

5. In animal species, in which the recognition cues of females and males overlap to a certain degree, homosexuality could be a consequence of an adaptive discrimination strategy to avoid the costs of making rejection errors (Engel et al. 2015).

In general it is supposed that the parents' investment depends on the probability of the offspring to survive. Furthermore, according to the Robert Trivers (1943–) and Dan Willard hypothesis (Trivers and Willard 1973), parents in good conditions tend to invest more on sons (who are best able to take advantage of good conditions), while parents in poor conditions tend to invest more in daughters (who are best able to have successful offspring even in poor conditions) (Veller et al. 2016; van Bodegom et al. 2013; Kaptijn et al. 2013; Kolk and Schnettler 2013; Venero Fernandez et al. 2011; Zadzinska et al. 2011; Cameron and Dalerum 2009; Cronk 2007; Koziel and Ulijaszek 2001; Chacon-Puignau and Jaffe 1996; Anderson and Crawford 1993).

According to Buss and Schmitt's sexual strategies theory and strategic interference theory, the differential parental investment observed in males and females led to the evolution of sexually dimorphic behaviors. When the strategies, priorities, and behaviors differ, then conflict between the sexes occurs, and as a result, emotional responses including anger or jealousy emerge (Buss and Schmitt 1993). Females generally react more negatively to emotional infidelity, while males will react more to sexual infidelity, especially also because fatherhood could not be proven until recently (Galperin et al. 2013; Buss 1995; Buss et al. 1992).

The "parental investment theory" is a branch of the "life history theory." It is evident that reproduction in general is costly for both sexes but especially for women. It demands the investment of significant resources, both in the form of time but also of material as well as of social resources (alliances and interactions). As a result the survival and the future reproductive output of the offspring are maximized at the cost of other options the parents have for their overall somatic health, fitness, reduced mating opportunities, and securing survival (Trivers 1972; Clutton-Brock and Vincent 1991; Hamilton 1964a, b). Step parental care could often be problematic because of priorities in favor of the biological child (the Cinderella effect) (Miller 2000b; Daly and Wilson 2005b, 1987 1991).

Evolution concerns the survival and proliferation of genes, and in this sense, the evolutionary success depends on the number of offspring, but this approach cannot explain a number of very frequent and important behaviors, e.g., altruism and self-sacrifice. In 1964, William D. Hamilton (1936–2000) proposed the inclusive fitness theory. In this theory he suggested that behaviors which harm the individual (e.g., self-sacrifice) could be considered to be adaptive in the frame of the group the individual belongs to, and in this way, they increase the chances of survival of the genes of the individual through the survival of family or relatives. Inclusive fitness is the sum of an organism's classical fitness (how many of its own offspring it produces and supports) and the number of equivalents of its own offspring it can add to the population by supporting others (Hamilton 1964a, b, 2001). It also explains our attitude toward more close to us species. This theory proposes an answer to the question concerning "altruism" in the frame of evolution. There is a number of behaviors which are complex and difficult to understand in the frame of the classical evolution theory, and altruism is such an example. To explain their existence, a number of theories have been developed including evolutionary game theory, tit-for-tat reciprocity, and generalized reciprocity (Burt and Trivers 2006; Comins et al. 1980; Hamilton 1970, 1964a; Orlove and Wood 1978; Durand et al. 2011; Ferriere and Michod 2011; Abbot et al. 2011; Michod and Herron 2006; Michod 2006, 1996; Nedelcu and Michod 2006; Michod and Nedelcu 2003).

To further elaborate on the issue of non-selfish behaviors, one should consider the fact that from offspring to the wider family and eventually to wider groups, gene sharing and kinship are not an all or none phenomenon, but on the contrary, they spread on a continuum from close to distant relatives. Kin recognition is a complex ability with unknown characteristics, but it is well known that humans act generally more altruistically to close genetic kin compared to genetic nonkin (Lieberman et al. 2007). But this does not preclude collaboration with nonkin which can be achieved and maintained via mutually beneficial reciprocity. In this frame, natural selection favors strategies which improve the reputation and increase the chances of support from others. These strategies include a repertoire of social behaviors and emotions, including morality, guilt and friendship, as well as the ability to identify non-reciprocators (cheaters) (Fowler 2005).

12.4.10 Evolution of Language

An interesting ability of modern humans is the ability to speak, read, and write. Children learn to speak early without any specific training or intervention, and they are taught to read and write after specific training but again early and rather easily. However reading/writing and maybe speaking were not innate abilities of ancestral humans, although the use of sounds and gestures/behaviors is a common way for animals to communicate and to exchange crucial information.

There are several theories describing evolutionary pathways and mechanisms for the evolution of language, but all remain controversial (Workman and Reader 2008; Fitch 2010; Deacon 1997). The well-documented fact about language is that it is a universal characteristic of the human species and that children are not born with this ability, but seemingly effortlessly and without systematic training they learn to speak and comprehend language between years 1 and 4. These characteristics of language clearly suggest it is a distinctly human psychological adaptation, and it evolved in parallel with the body organs that support it (Pinker and Bloom 1990).

There are different theoretical approaches on the components and the foundation on which language evolved. Steven Pinker suggests the presence of an innate capacity of language in children (Pinker 1994), which is in accord with the ideas of Noam Chomsky (1928–) concerning the innate presence of a basic universal grammar. However, Pinker suggests that the acquisition of language is an adaptation while Chomsky suggests it is a spandrel (Workman and Reader 2008). Pinker argues that language is unique and characterizes only humans, it is an innate human ability rather than an invention, and it is separate from general intelligence and is based on a distinct specialized mental module. On the other hand, Chomsky rejects the radical behaviorism of Skinner which argues that the brain is a "tabula rasa" (blank sheet) and subsequently language is fully acquired and taught. Also he argues that language is unique to humans and different from the ways of communication animals use (Chomsky and McGilvray 2012).

However although human languages share many common elements and structure, the universality of the nature of language has not been proved beyond doubt, and there are reports on the contrary. This debate is still unresolved. Additionally it is wrong to consider human language in the frame of the human species alone, and the idea of Chomsky that human language is completely different from the ways animals use to communicate with each other is an unproven and probably an extreme concept.

Animal communication is not identical with animal language, and it does not imply the presence of a language. However, animals use a variety of sounds or movements to communicate, and often they are complex enough to be considered as a form of language. It is important to note that the higher a species' position in the scale of evolution, the more specialized are the somatic organs which produce the sounds or the movements and of course the more evolved are the neurocognitive function and the content of communication. Almost all animals have specifically developed vocal cords in order to produce specific sounds. Therefore, language should be considered in the frame of extremely advanced general cognitive abilities which make possible the use of already existing organs (cords, fingers) for novel tasks in a creative way, which is easily passed from generation to generation through learning, but it is essential based on the innate ability to express and externalize mental procedures through their reflection on these sounds and movements (Fitch 2010; Deacon 1997). Such an expression could be considered to be a kind of "universal grammar" but in a less narrow definition. Although there was probably a selection pressure concerning language, it is unlikely such a pressure existed for reading/writing (Mabry 1995).

It is obvious that animal "language" lacks key elements which characterize the human language, but this depends on the species and how evolved they are. Some primates are even able to use lexigrams under experimental conditions (Gardner and Gardner 1969; Ward 1983), but normally the animal communication lacks the abstractive nature, the symbolism, the complex content, and the creativity which characterize human language. However it is not uncommon in nature that a quantitative difference, if extreme, could lead to a qualitative difference (Di Vincenzo and Manzi 2013; Traxler et al. 2012; Grodzinsky 2006, 2000; Fitch et al. 2005; Pinker and Jackendoff 2005; Hauser et al. 2002), while the continuum of language evolution is supported by research in primates which showed the presence of at least some human elements in their language (Ward 1983; Gardner and Gardner 1969; Patterson and Linden 1981).

12.4.11 Consciousness

Consciousness is another interesting neurobiological and psychological phenomenon. On one hand, it meets the criteria of species universality and of complexity as determined by George Williams (Nichols and Grantham 2000), and on the other hand, its presence probably increases the overall fitness and survival (Herron and Freeman 2013). It is likely to be the end result of a number of highly adaptive evolutions in brain function (Eccles 1992), and it is not an all-or-none phenomenon. Its main purpose is to put the individual in a perspective concerning the place, the time, and the others, and therefore it is probably present in simpler forms and with somewhat more primitive and incomplete features and functioning also in pre-mammalian species (Baars 1993) with variable complexity and functionality (Gaulin and McBurney 2003).

Self-esteem is not part of consciousness per se, but it reflects self-awareness, that is, cognition about self at a second level. In evolutionary terms, it is essential as an estimation the individual makes concerning its place in social hierarchy. While consciousness positions the individual in terms of place, time, and the others, self-esteem positions the individual among humans and within the scale of social hierarchy. One approach could be that it constitutes a self-assessment in order to choose targets for the allocation of resources. The result is the so-called assortative mating, that is, mating with an individual of the opposite sex with similar qualities.

12.4.12 Personality

The description of temperament, character, and personality is beyond the scope of the present chapter and constitutes a very complex issue (Fountoulakis and Kaprinis 2006; Fountoulakis et al. 2016). Personality reflects individual differences between persons in terms of behavior, and individual behavior in humans manifests significant heterogeneity. These individual differences have been well documented in terms of social behavior, mating and other areas (Sugiyama 2005; Nettle 2006; Ozer and Benet-Martinez 2006; Thornhill and Gangestad 2008), and also in nonhuman species (Wolf et al. 2007; Gosling 2001), and they seem to be heritable to a significant extend (Plomin et al. 2008).

The question whether animals have these characteristics is still a matter of debate, although it is certain that individual differences exist also between individual animals in terms of behavior (D'Eath et al. 2009; Martin and Reale 2008; Reale et al. 2007; King et al. 2006; Whitney 1970). They seem to appear especially in social species, and this might mean that by presenting complex problems, it is the social environment that demands their existence and plays an essential role in manifestation as behaviors (Penke et al. 2007; Perilloux and Buss 2008).

Individual differences constitute an oxymoron for the evolutionary approach because heterogeneity is considered rather as the substrate on which natural selection acts or a starting point rather than the end product of natural selection itself (Gaulin and McBurney 2003). A general concept is that natural selection reduces rather than promotes individual differences. Therefore, the understanding of personality traits in the frame of evolutionary psychology manifests a number of important problems (Buss 1984, 1991, 2009) and has been relatively neglected, with some important exceptions (Buss 1984, 1991, 2009; Sheldon et al. 2007; Segal and MacDonald 1998; MacDonald 1995; Nettle 2006; Wilson 1994; Wilson et al. 1996). On the other hand, when competition exists, individual differences are what matters, and they determine winners and losers. The ability to better monitor and assess these individual differences is also the product of adaptation (Buss 1996). Differences in personality among individuals could be considered as alternative strategies for the solving of adaptive problems which tend to recur (Buss 1996; Denissen and Penke 2008a, b; Hawley 1999; Nettle 2006).

For too long, individual differences were considered to be "noise" rather than "signal" in the big picture of evolution (Tooby and Cosmides 1990), but recent developments challenged this (Nettle 2006; Keller 2007). Apart from the standard life history, costly signaling, balancing selection, and contingent shift theories, a number of additional unique theories were developed specifically to deal with individual differences, such as social contract theory (Cosmides and Tooby 2005), sexual strategies theory (Buss and Schmitt 1993), error management theory (Haselton and Buss 2000), and adaptive cognitive biases (Haselton et al. 2005).

Individual differences are influenced by sex since, for example, it is reported that in the rhesus macaques, males are more aggressive, less socially affiliative, more impulsive, more prone to taking risks, and with higher mortality rates (Mehlman et al. 1994, 1997; Higley and Linnoila 1997; Higley et al. 1991, 1992). In mating, the strategies used differ not only between sexes but also among individuals of the same sex with some pursuing lifelong monogamy while others prefer frequent partner switching (Gangestad and Simpson 1990). Of course mixed strategies also exist. Similar differences exist concerning the pursuit of social status and the preference for a specific rank in social hierarchy (Hawley 1999; Lund et al. 2007b) as well as in the ability to detect deceit and cheaters (Ekman et al. 1999; Buss and Duntley 2008).

According to the life history theory which can provide a conceptual frame for the understanding of personality traits and their evolution (Figueredo et al. 2005; Kaplan and Gangestad 2005; Wolf et al. 2007), individuals make investments to solve adaptation problems, and the optimal trade-off between different allocations of resources depends on individual differences (Daly and Wilson 2005a). In this frame, personality disorders could be considered as behaviors stemming from perceiving different adaptational problems or from the presence of different qualities and resources which dictate different strategies (Daly and Wilson 2005a).

In a similar way, the costly signaling theory (Miller 2007) and the balancing selection with its key components "fitness optima" and "frequency-dependent selection" (Penke et al. 2007) could be also useful. It is reasonable to assume that some environments favor a risk-taking behaviors while others on the contrary favor more cautious harm avoidance behaviors (Camperio Ciani et al. 2007). A specific genetic substrate probably plays a role too (Ebstein 2006; Penke et al. 2007; Chen et al. 1999; Eisenberg et al. 2008). On the other hand, the mutation load theory is rather problematic since it tends to consider individual differences as random noise rather than alternative solution proposals, although there are opinions suggesting a more creative role in general for the mutation load (Buss 2006).

One personality which has been specifically the focus of research is psychopathy which corresponds to aspects of antisocial personality disorder. Frequency-dependent selection has been suggested as a mechanism which can explain the evolution of these traits and their survival in modern humans (Mealey 1995). The core characteristic of this personality is cheating with disregard of social norms and social solidarity. In this way these individuals exploit the strategy of cooperation the majority has. It is more frequent in males, and among other things, it involves the short-term seduction and abandonment of females (Mealey 1995). Another characteristic is the ability to identify potential victims (Buss and Duntley 2008).

More interesting is the way individual differences and personality traits could be understood in the frame of the theory of contingent shifts according to environmental and phenotypic conditions (Belsky 1999; Gangestad et al. 2006). Changes in behavior could come as a response to changing environment such as more aggression and higher risk taking in environment with few resources or mating opportunities or with more co-cooperativeness in dangerous environments. Such changes or facilitation of behaviors and psychological traits can also happen, not because of the environment but because of the emerging individual characteristics, a phenomenon called "reactive heritability" (Tooby and Cosmides 1990). Such an example is body size which tends to determine higher aggression in persons with large body sizes and more pacificity in individuals with small body size (Ishikawa et al. 2001). Another such example are the later-born children which are often more rebellious and less conscientious (Sulloway 1996).

The way individuals copy with the challenges posed by their participation in large groups is also variable and differs significantly from person to person. The position in social hierarchy is usually pursuit with prosocial and coercive strategies, depending on the personality (Hawley 1999). These strategies include deception and manipulation or, on the contrary, communication of positive personal characteristics and also with industriousness (Lund et al. 2007a), and to a significant extent, they correspond to the five-factor personality model (Buss 1992; Costa and McCrae 1985; McCrae et al. 2005). These personality factors were also viewed as motivational forces for the choice of specific strategies (Denissen and Penke 2008a, b).

Interestingly, the five-factor model was developed on the basis of the lexical theory of Gordon Allport (1897–1967) and Henry Odbert (Allport and Odbert 1936). The first such attempt had been made by Franziska Baumgarten-Tramer (1883–1970), who identified 1093 separate words in the German language as reflecting personality traits and mental disorders. This was of course incomplete as Gordon Allport and Henry S. Odbert showed in 1936, when they identified 17,953 such words in the English language and separated them into four categories or "columns." According to them, the first column included 4504 words that reflect personality descriptions. The second column with 4541 words reflected emotions, while the third column with 5226 words reflected social and pragmatic but not psychological evaluations of an individual and its position in society and its hierarchy. The last column with 3682 words included words with miscellaneous meanings and use. This theory is based on the assumption that the natural human language constitutes an important source for the identification of personality traits, because as Raymond Cattell

(1905–1998) suggested, in the course of the evolution of human culture, any behavior or personality concept which would be of importance in human social interaction should have been registered in language since it was the content of communication. The 16 Personality Factor Questionnaire (16PF) (Cattell et al. 1970) and the five-factor personality theory (Costa and McCrae 1985; McCrae et al. 2005) are both products of the lexical hypothesis. The main criticism argues that these imprints in human language are biased lay peoples trivial and superficial descriptions.

12.5 Evolutionary Social Psychology

Evolutionary social psychology is a rather recently developed scientific field and tries to understand the complex area of social behavior in the frame of evolutionary psychology and biology (Santrock 2005; Schaller et al. 2006). Essentially it constitutes the expansion of evolutionary psychology in the social domain.

Social psychology concerns the study of thoughts, feelings, and behaviors in a social environment, that is, in the actual, imagined, or implied presence of others (Allport 1985). Thus, it bridges psychology with sociology and pays attention to the phenomena that occur at the individual but also at the group level (Moscovici and Markova 2006). Its appearance followed the development of sociology in the late nineteenth century and emerged as a new discipline in the early twentieth century although some thoughts in this field existed already in the Arab scholar literature (Gergen 1973). At the core of its existence is the assumption that human behavior and social phenomena can be the focus of scientific research which follows the universal rules of science. One of the pivotal historical cases of social psychology was in the 1960s, the case of the Stanley Milgram (1933–1984) experiments on obedience to authority.

The basic concept in social psychology is "attitude" which is defined as a learned global evaluation of a person, object, place, or issue, and it determines action. It reflects approval or disapproval, favorability or unfavorability, or, in simple words, likes and dislikes (Bem 1970). Attitudes influence behavior, but they are often poor predictors of it. They could be conscious or unconscious (implicit), and they concern most of social interactions. The question whether attitudes are determined genetically or culturally and through learning and to what extent remains unanswered. Another important concept is "persuasion" which refers to the influencing of people by rational or emotive means in order to adopt a specific attitude (Myers 2010).

Social cognition and theory of mind is a field of research concerning the neurocognitive basis of social behavior. This refers to the ability of individuals to process and interpret stimuli relevant to the behavior of others, that is, their intentions, desires, and abilities, but also concerning the prediction of their future behavior. Collectively this is called "attribution" and can be ascribed as an internal (personality, character, etc.) or external (environmental) locus (Reisenzein 2015; Dunfield and Johnson 2015; Schreiber 2012; Seidel et al. 2010; Santiago and Tarantino 2002; Andrews 2001; Klin 2000; Block and Funder 1986; Kruglanski 1986).

A number of biases and errors in the attribution process have been described, and probably all of them are the product of evolution. These biases and errors include the tendency to overestimate the role of personality and underestimate the role of situations, to attribute dispositional causes for successes and failure and blaming victims for their suffering. Also bias is considered the false memory of having predicted events or the overestimation of true predictions after the outcome is known. Confirmation bias leads to search for information or interpretations that confirm preconceptions and disregard the others. All these protect the person from feeling vulnerable and mortal. It seems there is a kind of "white lies" everybody tells himself as a defense mechanism in order to keep psychological well-being, and maybe these "adjustment lies" are not functioning in depression (Andrews 2001). However biases are not identical with errors. Biases can help quick adaptation, but they could constitute errors under specific conditions. This is because the human brain utilizes heuristics in order to arrive at fast decisions to complex and demanding problems. Heuristics are cognitive shortcuts, and the whole procedure is based on the comparison of the situation faced with a prototype situation the people know of. This often demands a simplified and straightforward interpretation of the complex reality and the development of generalized mental representations called schemas. Schemas organize knowledge and guide information processing but often lead to the development of a generalized set of beliefs about groups of people or situations. If this comparison leads to a successful solution, then the bias and the schema lead to adaptation; if not it leads to an error or to a problematic stereotype and prejudice.

Another important concept is that of "social influence" which refers to the persuasive effects people have on each other and includes conformity (act or think like other members of a group), compliance (change in behavior due to a request or suggestion from another person), and obedience (change in behavior as a result of a direct command from another person). An interesting form of social influence is the so-called self-fulfilling prophecy which refers to the situation when a prediction is made; the person's behavior actually causes it to happen, e.g., when expecting hostility from others; the behavior of the person itself actually causes it, while it did not preexist.

Apart from the interaction between individuals and between individuals and groups, groups also interact with each other since they possess a distinct identity, rules to follow, and solidarity among members. A related phenomenon is the behavior of crowds which often leads to deindividuation, a term reflecting a state of altered self-awareness caused by feelings of anonymity (including large crowds, disguise, and online anonymity). This is associated with uninhibited and maybe dangerous behavior.

To put social psychology in an evolutionary perspective is both challenging and rewarding. An essential first observation is that there seems to be a lot of common features that link human cultures from around the world irrespective of how isolated they are from each other and similarities are more than differences (Brown 1991; Rosch 1973). For example, in all human cultures and societies, there are systems to recon kinship and treat individuals according to kinship status (Daly et al. 1997). While this is not the case for the vast majority of mammals, all human societies have some kind of marital bond for the sharing of parenting (Daly and Wilson 1983; Broude 1994; Geary 1998). These cultural similarities exist alongside many cultural variations, which often seem very peculiar like the mating customs of the aboriginal tribe of Tiwi in Australia who manifests an interesting interplay between general human mating preferences and a particular social ecology (Hart and Pillig 1960).

Another example is that historically, in most cultures there is polygyny (one man and more than one wife), while a few permit polyandry (one woman and more than one husband). It is interesting that in spite of this modern societies are monogamous. It is standard procedure that when biologists find variations across species in behavior, they search for correlations with ecological factors (Alcock 2001). In general polyandry, though rare, can be explained by the presence of an environment with limited resources. In such environments (e.g., in the high Himalayas), it needs more than one male to support a female and her offspring successfully. In such places, often brothers marry the same woman (Platek and Shackelford 2007; Salmon and Shackelford 2008). On the contrary, in places rich in resources, the opposite phenomenon, that is, polygyny, is observed. Extreme polygyny is manifested in harems, and they are associated with societies with strict hierarchy in a rich environment (Crook and Crook 1988). However other factors apart from the physical environment play a role in the quality and characteristics of the marital bond. One such factor is the ratio males-to-females which can change because of war, migration, and other similar causes. In excess of females, later marriage, more divorce, and permissive sexual norms are observed. In excess of males, males are committed to more stable monogamy (Gangestad and Buss 1994; Kenrick et al. 2003).

It is interesting to note that in the great apes, all kinds of social organization are seen, from monogamy in gibbons to unimale polygyny in orangutans and gorillas and to multimale polygyny (or polygynandry) in chimpanzees where a group of males defends a group of females and their offspring (Foley and Lewin 2013; Lewin 2009).

In all human cultures, also the presence of status hierarchies is a basic characteristic of society with separations of casts and groups and variable restrictions and rules (Brown 1991). Anthropological and archaeological data suggest that originally, humans were living in small groups of hunter-gatherers with the size of 50–80 individuals on average and the members of each group were biologically related. These groups were characterized by less strict social hierarchy, members knew very well each other, and they were connected with (actual or fictive) bonds of kinship and were occupying large territories in an exclusive way (Barnard 1999; Maryanski and Turner 1992).

Within these groups, the genetic relatedness together with the long-term reciprocal exchanges usually developed trust and cooperation rather than market-like reciprocal exchange (Fiske 1992). Stigmas concerning disability or disease probably reflect fear of threat not only to the psychology of the individual but also to the collective group welfare (Kurzban and Leary 2001; Neuberg et al. 2000).

Also, an unknown individual not belonging to the group was considered to be a potential enemy, and although cordial exchange relationships with other groups were in place, outsiders were always considered to be a threat in terms of stealing, kidnapping females, rape, or homicide (Chagnon 1988; Radcliffe-Brown 1913). Therefore it seems that evolution could have made humans to be cognitively inclined to divide other people into "with us" and "not with us" and to perceive the later as a source of danger and threat (Krebs and Denton 1997; Wilson 1978). However, since outsiders often did not constitute a threat but on the contrary they constituted opportunities for trade and development, a flexible response system would be more adaptive in comparison to a rigid one that would reject all outsiders. Such a flexible system would recognize the true value of the outsider and its place in the complex environment of the group (Kenrick 1994; Kenrick et al. 1994).

12.6 Sociobiology

Sociobiology attempts to predict social behavior by utilizing the tools and theories of evolutionary psychology (Haig 2002). Social life is the preferred mode of living in many species in nature. They prefer to live in groups and manifest characteristics of social life, some of them very complex, but the complexity of human societies is exceptional. This complexity is such that many individuals find it difficult to adjust to its demands and subsequently are concerned more with keeping up with the rules and laws of the society than with survival itself, which these rules are supposed to serve. According to Jean-Jacques Rousseau (1712–1778), "*social institutions are those that best know how to denature man, to take his absolute existence from him in order to give him a relative one and transport the I into the common unity*" (Rousseau 1979).

Through the history of human societies and since the strong prosocial teachings of Socrates (470–399 BC) and Confucius (551–479 BC), a number of theories evolved, which essentially were attacking the very concept of social organization itself, from the philosophical ideas of Jean-Jacques Rousseau to the anarchist political ideology of William Godwin (1756–1836) and Mikhail Alexandrovich Bakunin (1814–1876).

But one of the most fascinating ideas which appeared in the twentieth century was that of the "Noble Savage" which enjoyed wide acceptance, especially in the frame of the civil rights movement after WWII. It referred to an ideal human being, living in peace with nature and his neighbors, with no negative thoughts or feelings. It was preached by Margaret Mead (1901–1978) who claimed that she had found a tribe in Samoa with an ideal psychological and social behavior and lacked things like jealousy or rigid sex roles (Mead 1928). Her reports were proved to be completely false and biased (Shankman 2009; Freeman 1983). However this idea was appealing, since it suggested that humans are inherently good and moral and modern civilization was the cause of all negative things (a convenient external locus). This is of course very close to the ideas of Jean-Jacques Rousseau and his 1762 book *Emil* (Rousseau 1979). Around the same time Margaret Mead was developing her ideas, another concept, that of the "Blank Slate," was widespread. This suggested that all people were able to learn anything with an equal amount of effort and that there were no biological predispositions or inclinations. This reflected a concept of radical equality at all levels and aspects, which took the step from civil rights equality to consider individuals being not only equal but essentially identical in all aspects. This was also suggesting that all individual differences were the product of social forces rather than inherent traits. This has its roots in Aristotle (384–322 BC), but it was shaped with the concept of "tabula rasa" by John Locke (1632–1704), although in a different frame and purpose. However the modern concept has important modern political implications. According to the modern concept of the "Blank Slate," in cases of state failure or in non-state societies, people organize in group for protection, and concepts like revenge and honor become extremely important (Pinker 2002; Rose 2001). This is in sharp contrast to the very influential collectivist anarchism theories developed by Mikhail Alexandrovich Bakunin.

Research has proved that humans have a strong predisposition to learn some behaviors over others and that individual differences do exist (Buss 2001). Therefore, it is pretty evident that as every living organism on Earth has a repertoire of "hard-wired" behaviors and instincts which serve survival, so do humans. For all species, these traits are evolutionary developed and determined. For humans, this evolution took place mainly during the Pleistocene, and contemporary behaviors constitute more abstract and refined versions of basic adaptation strategies (e.g., friendship evolved from sharing of resources) (Kenrick et al. 2003). All aspects of human mental life were developed during that period, and their properties, advantages, and problems can be traced back to then, from differences in cognitive function between sexes to the selective and representative but imprecise way memories are retrieved (Klein et al. 2002).

Social life is characterized by a way of life and behaviors which are often sharply different from a solitude way of life. Thus there is a constant conflict between the needs of the one vs. the needs of the many, and this is a problem evolutionary sociobiology needs to address, but it is often very difficult. The behaviors and restrictions when being part of a group might reduce that reproductive success or induce harm or even death to the individual. Eventually, however, through inclusive fitness, the survival and adaptation of the group and the society as a whole increase. The term "reciprocity" refers to the behaviors with which individuals exchange favors. The issue of altruism and other non-selfish or self-harming behaviors in the frame of inclusive fitness theory has been described above. Issues pertaining mating and polygamy have also been discussed above. The ability for empathy is a rapid way to induce cooperation and to facilitate reciprocal altruism (Burt and Trivers 2006; Dawkins 1976). The same holds for communication and "cheating." Reciprocity is essential for humans; failure to reciprocate reduces, while on the contrary, the reliable reciprocation increases social reputation. Especially for the study of reciprocal systems and cheating, the game theory paradigm known as the "Prisoner's dilemma" tries to model the development of cooperation among independently acting egoists (Barash 2003b).

It has been estimated that in order for the human fetus to be fully developed like the fetuses of other mammals and great apes, a gestation period of 18–21 months would be necessary. The reason why it has been restricted to only 9 months has been considered to be a side effect of upright bipedal walking which leads to a small pelvis size which in turn limits the size of the fetus the human female can carry (Weiner et al. 2008; Charnov and Ernest 2006). This was called the "obstetrical dilemma," a term coined in 1960 by Sherwood Larned Washburn (1911–2000) (Wells et al. 2012; Washburn 1973, 1978, 1982; Washburn and McCown 1972). Another theory suggests that it is the result of limits in the metabolic burden the mother can take in favor of the fetus, and subsequently this puts a limit to how large and energetically expensive a fetus can be (metabolic crossover hypothesis) (Dunsworth et al. 2012). No matter what the cause is, the fact remains that the human newborn is extremely unprepared to survive by itself in nature and demands significant care and parental investment, thus pushing toward stable long-term mating relationships. This makes humans much more monogamous oriented than primates and other mammals.

Observations are consistent with the belief that humans are mildly polygamous by nature and this is a stable cross-cultural characteristic. Serial monogamy is a situation in between monogamy and polygamy (Barash and Lipton 2002; Platek and Shackelford 2007; Salmon and Shackelford 2008). Social norms do not permit departures from monogamy to manifest openly, especially for women. In spite of this, departures from monogamy are present for both sexes universally (Barash and Lipton 2002). This is also true for previously considered monogamous species, as DNA testing proved that 10–80% of offspring come from a father which is not the male social partner of the couple. This might also was the case for prehistoric humans as well, but in modern human societies, the rate is probably below 4% (Bellis et al. 2005). The globalization which started with the colonial era brought a gradual social homogenization, and while before, almost 90% of societies were polygynous, currently the vast majority are monogamous, and monogamy is also a feature, among others, which determines whether a society is modern or antiquated.

Polygyny was socially more dominant, and it is related to a number of differences between sexes. On average, men are physically larger and with a tendency toward competitive and often violent behavior. Females mature sexually earlier than men and often prefer men older than themselves. All these are also found in mammals where polygyny is the rule since through natural selection, males obtain characteristics which help them to compete with other males, with higher strength, status, and chances for success with advancing age, while females reproduce early and more frequently. These correspond to a pattern of "sexual dimorphism" and "sexual bimaturism." In monogamous species like gibbons, males and females are of the same size. In gorillas which live in unimale polygamy, males are 30% bigger (Foley and Lewin 2013; Lewin 2009).

Studies of old societies who accepted polygyny suggest that nearly all women were mated and reproductive. On the contrary, there was significant variability in men, with some men being nonreproductive bachelors, most being monogamous, and a few having a harem. This points to another fact that, although the physical tendency is toward polygamy, humans are clearly capable of monogamy, at least at the social level. Even their physical inclination toward polygamy seems to be less strong in comparison with most mammals. In evolutionary terms, as described above, this could be explained by the fact that the human offspring is helpless at birth and takes several years to grow up, and this is more pronounced in comparison to the offspring of other mammals. Subsequently it needs more postnatal care and parental investment from both parents for several years until it grows up.

On the other hand, the tendency of females to seek multiple sexual partners is difficult to explain. It was also difficult to document until recently, because of the secretive way it was conducted in all species. This tendency occurs in spite of risking even violent behaviors from the side of the male and the high risk of abandonment. The probable explanation is that females try to improve the genetic quality of offspring while at the same time they keep the advantages of stable parenting through social monogamy.

The above are in accord with the presence of a lower threshold for sexual excitement in men which therefore are more susceptible to pornography, prostitution, and paraphilias. They are also more eager to engage in sexual activities with strangers, and they pay more emphasis on the physical attributes of the partner since these reflect fertility status. They pay less attention to intellectual attributes and are jealous and possessive. On the contrary, females seem to be more concern with male access to recourses. The mechanism for these choices is largely unconscious, at least concerning the deeper evolutionary goal.

Men are more inclined to physical aggression and violence than women, and much of their behavior is based on the show-off or physical strength. This is of course culturally enhanced, but it is based on inherent traits. There is no culture in which the sex images are the opposite and women are culturally expected to be more violent than men.

In nature physical aggression and violence is often linked with mating, even in the great apes. While in human society and culture, rape is linked to moral and political issues, in the animal kingdom, rape is probably the strategy followed by the otherwise socially and sexually unsuccessful individuals. In humans, the "domination hypothesis," introduced by Susan Brownmiller (1935–), suggests that rape is not sexually motivated, but instead, it is a conscious process of intimidation by which all men keep all women in a state of fear. This was a highly controversial theory in a clear feminist politico-ideological frame (Lalumiere et al. 2005), and one of the key arguments in favor was that no zoologist ever observed animals raping in their natural habitat. However this is far from true. At that time such evidence was available and in the following decades accumulated (Alcock 2001). Randy Thornhill (1944–) and Greg Palmer argued against this hypothesis and pointed out that the improved understanding of what motivates rape rather than ideological approaches likely helps prevent rape. They argued that sexuality is the motive behind rape since a disproportionate number of victims are very young women and suggested that rape is either an adaptation or a by-product of adaptive traits such as sexual desire and aggressiveness (Thornhill and Palmer 2000; Lalumiere et al. 2005; Figueredo et al. 2011). Furthermore, competition among males is probably more intensive in late adolescence and young adulthood, and this might explain the particularly high crime rates in these age groups (Rose 2001).

Parenting is a complex situation and task and demands a number of somatic but also intellectual abilities from the side of parents. Conflicts between parents but also among parents and offspring are frequent, and from an evolutionary perspective, this is also because their evolutionary goals and agendas are not identical. Parents share only half of their genetic material with each offspring which acts as a limiting factor in parental investment. The area of parent-offspring conflict is predictable and concerns mainly the tendency of the offspring to seeking and obtaining more investment than the parent has chosen to provide. In a reversal of situations, similar conflicts may occur when the offspring shows significantly less than expected inclination toward the parent and preference for other members of the family. The psychological dynamics within a family were the focus of psychological, sociological, and anthropological studies, and the conclusion is that they could be very intense as well as unconscious. Competition with same-sex parent and adjustment to opposite-sex parent can be considered under this view, as they can, even sophisticated psychoanalytic concepts like the "Oedipus concept."

A more complex social and psychological issue is stepparenting. Among animals, stepparenting is rare, and when it occurs, it might hide different aims from the side of the stepparent. When created experimentally, these families are very dysfunctional, and even murders happen. Infanticide and neglect of stepchildren are the rules rather than the exceptions (Power 1975; Hasegawa and Hiraiwa 1980; Barash 2003a).

While most human stepparents are clearly able to function and invest in their stepchildren, which is an important element especially in societies with high divorce rates, in evolutionary psychology, the term "Cinderella effect" (Daly and Wilson 2005b) refers to the alleged higher incidence of child abuse and mistreatment by stepparents in comparison with biological parents. Of course it refers to the famous Cinderella fairytale, and it is believed to be a direct effect of competition concerning mating and parental investment as described above. There is a wide belief that stepfamilies are emotionally conflicted more frequently in comparison to biological families. Non-biological parents have an inherent disinclination to invest in unrelated children. It is both interesting and disturbing that there are some data suggesting that even the accidental injury of children is higher when a stepparent exists but not in single-parent families when one of biological parents has left (Tooley et al. 2006).

One of the sensitive and delicate issues to deal with in a sociobiological perspective is religion and religiosity. Currently there is a controversy whether they are a consequence of evolved psychological adaptations or a by-product of other cognitive adaptations (Beit-Hallahmi 2012; D'Onofrio et al. 1999; Fountoulakis et al. 2008; Rossano 2006). It is to be noted however that primitive religious-like ceremonies have been observed in chimpanzees (Harrod 2014).

An issue which recently started to be the focus of interest and research is the fact that the more the culturally and technologically advanced a society is, the more it demands biologically determined secondary abilities, especially neurocognitive, from its members. This seems to be even more intensified and pronounced with the informatics and electronics revolution of the last few decades (Geary 1995) and creates new challenges since human societies try to be simultaneously competitive but also inclusive and caring.

12.7 Darwinian (Evolutionary) Psychiatry

The Darwinian theory had a major influence on how health and disease are conceptualized and what might constitute the best treatment option. There is a newly emerged field, called "Darwinian medicine." It utilizes a different and novel approach to the consideration of symptoms and diseases and tries to provide with more rational and informed choices for treatment (Sims 2001; Rose 2001).

A number of medical conditions including obesity, anemias, autoimmune diseases, and hypertension were put in an evolutionary frame (Gluckman et al. 2009). A number of questions concerning "why we have this problem" might be answered in an evolutionary frame. For example, why modern humans manifest high rates of obesity? Is this a consequence of generations of humans living in an environment of limited resources and frequent famine which made important the accumulation of reserves? Is primary (essential) hypertension a consequence of adaption to low-salt diet which resulted in hyper-response to salt intake? Is maxillary sinusitis a problem of walking upright since the position of the duct for drainage is perfect for quadrupedal walking, but it is located too high when the head is erect as in bipedal walking? Are fever and diarrhea defense mechanisms to reduce the survival of pathogens inside the body and to eject them?

According to this approach, also mental health and disease could be understood as the end result of the interaction between the organism and the environment, with the addition that abnormal behaviors and symptoms could reflect either an extreme form of otherwise adaptive behaviors or the triggering of them in an out-of-frame or proportion way and under inappropriate conditions.

This is somehow different from the standard approach in clinical psychiatry and psychopathology where connections with recent events and reaction to recent problems in the frame of the individual patient's and core family are considered (proximate mechanisms). In contrast, evolutionary psychiatry stresses that variation (including variation in behavior) is not only normal but evolutionarily necessary and is concerned with the misfit of preexisting normal coping mechanisms which could cause maladjustment or disease if triggered in an inappropriate way or persist for longer than expected and especially in an environment and with social demands extremely different from those encountered by humans during the Pleistocene and the Stone Age (Fabrega 2002).

As psychiatry is part of medicine, so is Darwinian psychiatry a subset of Darwinian medicine. If this is so, it is important to see whether mental symptoms and disorders could be considered in the frame of evolutional adaptations.

The classic psychoanalytic theory of Sigmund Freud (1856–1939) is a very fruitful starting point. Especially his theory on the instincts of Eros (life and survival) vs. Thanatos (death and aggression) and the libidinal investing as well as the Oedipus complex have direct relevance to concepts and theories of evolutionary psychology as described above.

However, psychopathology is quite different from psychology of normal mental functioning. The big question is whether diseases and disorders like schizophrenia, manic depression (bipolar disorder), depression, and others can be analyzed and understood in an evolutionary frame.

The problem is perplexed by the problematic reliability of psychiatric diagnosis and the presence of a "gray zone" between normality and psychopathology. Although it is not standard approach, one could suggest that some psychiatric conditions could be considered to be clearly abnormal and psychotic symptoms are such an example. Experiences similar to psychosis are very rare in the general population, and the average person cannot "understand" how a psychotic person feels. On the other hand, some other conditions like anxiety disorders and depression seem to constitute an extreme version of normal experiences of grief and anxiety, at least in the way the patients experience them and the way the observer perceives. Thus, the inner experience of a depressed patient is quite understood by the average healthy person which in the past had experienced several periods of grief, mourning, and anxiety. A third group of conditions include behaviors and inner experiences which are frequently seen in the general population but in mental patients seem to exist in an unusually accumulated coexistence, and they manifest out of frame concerning the intensity and the environmental stimuli. These conditions include certain personality traits like antisocial behaviors, eccentric dressing and outlook, substance use, body modification, etc. For example, a large proportion of the population has used cannabis at least once, has been involved in minor stealing at least once especially during adolescence, dresses in an odd way occasionally, and has piercing and/or tattoo. When these characteristics accumulate in a single person and to an unusual degree, e.g., daily use of cannabis, habitual delinquency, whole-body cover of tattoos, and extreme piercing with health risks, then it is highly possible the person suffers from some kind of mental disorder.

If one views mental disorders with the above described way, then it is clear that it is highly unlikely to find an evolutionary explanation for the first category. This is reasonable since there is no apparent evolutionary advantage in hearing voices or seeing visions at any age and at any frequency. Such conditions are probably related to a deep primary disruption in the functioning of the brain and subsequently of the mental apparatus. Some delusional ideas, however, could be conceptualized as extreme forms of caution and efforts toward harm avoidance. The only way out could be to find some kind of fitness advantage in the healthy members of families of patients with psychotic disorders. Although there are assumptions that creativity and high intelligence run together with manic depression and schizophrenia in families, this has never been proven. An impressive example of the coexistence of high intelligence and schizophrenia in a family is Albert Einstein (1879–1955) and his son Eduard (1910–1965; Fig. 12.10). Another example of the coexistence of greatness and multiple mental disorders and suicidality is the Hemingway family (Fig. 12.11).

Fig. 12.10 Eduard Einstein (1910–1965)

Fig. 12.11 Members of the Hemingway family c. 1917. Main photo: Clarence (father), Grace (mother), and offspring (Carol, Ernest, Leicester, Ursula, Madelaine "Sunny," and Marceline). Right top: Pauline Pfeiffer (wife). Right bottom: Gregory (son). Left bottom: Margaux (granddaughter). Red dots mark those who died by suicide. The question mark stands for a possible suicide for Marceline

For the other two categories, however, an evolutionary approach is highly possible. Since affect is an adaptation which serves both fast unconscious decisions and communication with peers, mood and anxiety disorders can be conceptualized as an extreme form of communication, a "cry for help." An alternative explanation could be that depressed individuals attempt to preserve their position within the social group but the same time avoid risky and costly behaviors by sending out signals to ask for help (Allen and Badcock 2003). Suicide could be conceptualized as the triggering of a behavior where the individual sees no fitness-enhancing opportunities and death is the only right action since it will conserve resources for kin and enhance inclusive fitness. Substance abuse on the other hand as well as behavioral addictions (e.g., gambling) could be considered to be a by-product of the evolution of the "behavioral and habits formation systems," which permit the individual to acquire new and adaptive behaviors and habits. However the interaction of these systems with the presence of supernormal stimuli (drugs, carbohydrates, games etc.), which have a higher than expected effect both on the attentional grab and also at the biochemical level, leads to the formation of addictions. This is neither an adaptation nor an exaptations (spandrels), but rather a mismatch, that is, a mechanism which had been developed to function in a different environment, is dysfunctional today at least in vulnerable individuals.

On the contrary, the case of antisocial behavior and body modification as well as extreme cases of provocative and bizarre way of dressing probably constitute exaptations (sprandels) since they are based on behaviors which had developed in order to promote fitness but currently are expressed for a different reason. Usually this reason is overt or hidden aggression which takes the form of physical aggression or provocative show-off of a violation of social norms. In most cases there is no gain in terms of better adaptation or fitness; on the contrary the individual could be highly dysfunctional.

There are two significant problems when one tries to put the above in a comprehensive frame for clinical psychiatry. First, in order to have behaviors and experiences "out of proportion" and "out of frame," regardless of their initial evolutionary relevance, there should be some kind of dysfunction of controlling mechanisms. Again the concept of a brain dysfunction at the core of mental disorders seems to be essential even for the evolutionary approach, which, never the less, can explain the source of individual symptoms.

The second problem is political/ideological again, but it is at the root of the whole establishment, and the problem is strengthened by the approach itself. If there is an evolutionary explanation behind many antisocial behaviors, then these constitute essentially normal nonpsychiatric variations rather than abnormal conditions. On the contrary, a more conservative approach would argue that if supernormal stimuli are, at least partially, responsible for many mental problems, then these stimuli are "unnatural" and should be eradicated from human societies (wide suppression policies).

The detailed answer to both these issues is beyond the scope of the present chapter. However, one should have in mind that persons with antisocial behaviors are fully responsible for their acts in front of the law; therefore everybody accepts their "civil right" in the choosing of these behaviors but also in the accepting of their consequences. There is a huge debate whether self-destructive behaviors (including suicide) should be considered as "civil rights" or mental disorders, but the prevailing attitude is in favor of the second.

Concerning supernatural stimuli, an approach in favor of a wide eradication of this kind of stimuli from human societies does not take into consideration the fact that essentially modern societies are based on the presence of supernatural stimuli and demands. This is often the cause of mismatch conditions, where new "health problems" arise, and behaviors are triggered under inappropriate situations while focusing on the wrong target. Such examples are the demands of living in megacities with very complex social roles and dynamics, the cognitive load and demands the modern education and especially modern educational methods including multimedia and the Internet pose on the human mind, and the need for accuracy and precision in occupational tasks. Another interesting feature of modern life is that a significant number of stressors and threats are not visible and material, as they used to be in the past, but, on the contrary, are abstracted, theoretical, complex, and difficult for the average individual to understand in full. For example, during the Pleistocene, the danger was predator animals, while today it is the distant risk of having a disease which will give symptoms sometime in the future, the danger for economic problems, etc. The fact is that while biological evolution is slow (and Darwinian), cultural and societal evolution (which is Lamarckian) is very fast.

12.8 Criticism

While ethology is based on observation, a major problem in evolutionary psychology and sociobiology is that they depend too much on reasoning and less on the analysis of data. Very few theories can be experimentally tested or prospectively studied. Of course they follow the rules of science, but it is doubtful that many of the theories which have been developed could fulfill the falsificationism criterion proposed by Karl Popper (1902–1994) (Popper 1959). Probably while isolated statements do not fulfill the criterion, the whole theoretical framework does, but it is rather underdeveloped and conclusions are difficult.

Apart from the epistemological, there is a number of ideological-political issues which make the application of evolutionary principles to human behavior controversial (Confer et al. 2010). The first is the widespread objection to the Darwinian theory of evolution in general. Then there is the bitter controversy on the importance of nature vs. nurture in the shaping of human behavior. Inevitably these lead to important politico-ideological conflicts, and the evolutionary approach to human behavior has been accused of developing and spreading malevolent political or moral ideas, of justifying existing social hierarchies and "reactionary policies," and of giving support to racist and sexist attitudes.

The obvious problem with this kind of criticism is the well-known "naturalistic fallacy." This biased way of thinking identifies "natural" with "good" (Moore 2004). This is behind lay beliefs that "natural remedies" are good and medications are toxic and that a natural way of life in nature is better for human health. Of course these are in sharp contrast with the improvement of life and the impressive increase in life expectancy as a result of "unnatural" and technological advances during the last few decades. Still, it is a very popular way of viewing things not only among lay people but also among many health and social sciences professionals and philosophers. Darwinism has been a particular problem for this way of viewing things, because it is directly destroying the romantic viewing of nature and the anthropocentric consideration of the universe.

An essential characteristic of politico-ideological criticism is that it mixes epistemology and the results of science with morality and ethics. Is the wolf in the *Little Red Riding Hood* a "bad" and "unmoral" being? Of course the tale is a metaphor; however it reflects our tendency to utilize an anthropomorphic and anthropocentric way when viewing nature. There is nothing moral, good, or bad in the natural relationship between wolves and sheep, the first being the predators and the second being the prey. The description of similar structures and phenomena in human society and related underlying psychological phenomena in the individual human being is neither good nor bad, as long as there is proof that this is the scientific case.

The human society and the human culture were both developed in order to control these behaviors and create a more friendly environment for human beings to live, and both society and culture fight constantly with the most primitive and aggressive aspects of human nature.

As discussed before, there is a long anti-societal and anti-civilization ideology, with deep roots in irrationalism, which however differs as a concept from anti-rationalism and non-rationalism, and in simple words, it interprets the world through wishful thinking (intuition) but simultaneously keeping in touch with reality, e.g., as in the writings of Friedrich Wilhelm Nietzsche (1844–1900). An example of anti-rationalism is the teachings of Saint Augustine of Hippo (354–430). Often irrationalism, non-rationalism, and anti-rationalism coexist in most ideological and philosophical thoughts, since their major concern is not scientific theories and scientific explanation and understanding of the world but the effect such progress might have on various aspects of humanism and especially in the rejection of the perceived special top position humans have in the hierarchical pyramid of the universe.

Last but not least, one should always have in mind that the topics discussed in this chapter are sensitive in terms of humanity and politics and have been used both ways by completely opposing sociopolitical groups. This sensitivity and dangerousness is impressively shown in the life and works of the Nobel Laureate Konrad Lorenz. Lorenz joined the Nazi Party in 1938, and he was also a university chair at the University of Königsberg under the Nazi regime. At the same time, he published articles in accord with Nazi ideology, especially "racial hygiene" couched in pseudoscientific metaphors and with anti-Semitic content. This led to accusations that his scientific work had been contaminated by Nazi sympathies. During the WWII, he served as a military psychologist, conducting racial studies in occupied Poznań under Rudolf Hippius, on the biological characteristics of "German-Polish half-breeds" in order to determine whether they were psychologically and physically fit to be allowed to reproduce humans. His real contribution to this project is unknown, and this 2-year period is not mentioned in his memoirs. Because of the above, in 2015, the University of Salzburg posthumously rescinded an honorary doctorate awarded to him in 1983 and also accused him of using his work to spread "basic elements of the racist ideology of National Socialism" (Burkhardt 2005; Föger and Taschwer 2001; Kalikow 1983; Nisbett 1976). In sharp contrast, Karl von Frisch lost his academic position in 1933 when the Nazi regime passed the Civil Service Law, and he was proved to have 1/8th Jewish ancestry. He had also attracted negative attention for employing Jewish assistants and for practicing "Jewish science." He also worked actively to help Polish scientists who arrested by the Gestapo (Deichmann 1992).

References

Abbot P, Abe J, Alcock J, Alizon S, Alpedrinha JA, Andersson M, Andre JB, van Baalen M, Balloux F, Balshine S, Barton N, Beukeboom LW, Biernaskie JM, Bilde T, Borgia G, Breed M, Brown S, Bshary R, Buckling A, Burley NT, Burton-Chellew MN, Cant MA, Chapuisat M, Charnov EL, Clutton-Brock T, Cockburn A, Cole BJ, Colegrave N, Cosmides L, Couzin ID, Coyne JA, Creel S, Crespi B, Curry RL, Dall SR, Day T, Dickinson JL, Dugatkin LA, El Mouden C, Emlen ST, Evans J, Ferriere R, Field J, Foitzik S, Foster K, Foster WA, Fox CW, Gadau J, Gandon S, Gardner A, Gardner MG, Getty T, Goodisman MA, Grafen A, Grosberg R, Grozinger CM, Gouyon PH, Gwynne D, Harvey PH, Hatchwell BJ, Heinze J, Helantera H, Helms KR, Hill K, Jiricny N, Johnstone RA, Kacelnik A, Kiers ET, Kokko H, Komdeur J, Korb J, Kronauer D, Kummerli R, Lehmann L, Linksvayer TA, Lion S, Lyon B, Marshall JA, McElreath R, Michalakis Y, Michod RE, Mock D, Monnin T, Montgomerie R, Moore AJ, Mueller UG, Noe R, Okasha S, Pamilo P, Parker GA, Pedersen JS, Pen I, Pfennig D, Queller DC, Rankin DJ, Reece SE, Reeve HK, Reuter M, Roberts G, Robson SK, Roze D, Rousset F, Rueppell O, Sachs JL, Santorelli L, Schmid-Hempel P, Schwarz MP, Scott-Phillips T, Shellmann-Sherman J, Sherman PW, Shuker DM, Smith J, Spagna JC, Strassmann B, Suarez AV, Sundstrom L, Taborsky M, Taylor P, Thompson G, Tooby J, Tsutsui ND, Tsuji K, Turillazzi S, Ubeda F, Vargo EL, Voelkl B, Wenseleers T, West SA, West-Eberhard MJ, Westneat DF, Wiernasz DC, Wild G, Wrangham R, Young AJ, Zeh DW, Zeh JA, Zink A (2011) Inclusive fitness theory and eusociality. Nature 471(7339):E1–E4.; author reply E9–10. https://doi.org/10.1038/nature09831

Alcock J (2001) The triumph of sociobiology. Oxford University Press, New York, NY

Allen NB, Badcock PB (2003) The social risk hypothesis of depressed mood: evolutionary, psychosocial, and neurobiological perspectives. Psychol Bull 129(6):887–913. https://doi.org/10.1037/0033-2909.129.6.887

Allport GW (1985) The historical background of social psychology. In: Lindzey G, Aronson E (eds) The handbook of social psychology. McGraw Hill, New York, NY

Allport GW, Odbert HS (1936) Trait-names: a psycho-lexical study. Psychological Review Company, Albany, NY

Anderson JL, Crawford CB (1993) Trivers-willard rules for sex allocation: when do they maximize expected grandchildren in humans? Hum Nat 4(2):137–174. https://doi.org/10.1007/BF02734114

Andrews PW (2001) The psychology of social chess and the evolution of attribution mechanisms: explaining the fundamental attribution error. Evol Hum Behav 22(1):11–29

Ardrey R (2014) The social contract: a personal inquiry into the evolutionary sources of order and disorder. Story Design LTD, Ames, IA

Baars B (1993) Cognitive theory of consciousness. Cambridge University Press, Cambridge

Barash D (2003a) Revolutionary biology: the new, gene-centered view of life. Transaction Publishers, Piscataway, NJ

Barash D (2003b) The survival game: how game theory explains the biology of cooperation and competition. Henry Holt and Company, New York, NY

Barash D, Lipton J (2001) Gender gap: the biology of male-female differences. Transaction Publishers, New Brunswick, NJ
Barash D, Lipton J (2002) The myth of monogamy: fidelity and infidelity in animals and people. Henry Holt, New York, NY
Barkow J, Cosmides L, Tooby J (1995) The adapted mind: evolutionary psychology and the generation of culture. Oxford University Press, Oxford
Barnard A (1999) Modem hunter-gatherers and early symbolic culture. In: Dunbar R, Knight C, Power C (eds) The evolution of culture: an interdisciplinary view. Rutgers University Press, New Brunswick, NJ
Barrett D (2007) The R/Evolutionary science behind our weight and fitness crisis. W.W. Norton, New York, NY
Barrett D (2010) Supernormal stimuli: how primal urges overran their evolutionary purpose. W.W. Norton, New York, NY
Bateman AJ (1948) Intra-sexual selection in Drosophila. Heredity 2(3):349–821
Beit-Hallahmi B (2012) Connecting biological concepts and religious behavior. Behav Brain Sci 35(2):80–81. https://doi.org/10.1017/S0140525X11000938
Belke TW, Garland T Jr (2007) A brief opportunity to run does not function as a reinforcer for mice selected for high daily wheel-running rates. J Exp Anal Behav 88(2):199–213
Bellis MA, Hughes K, Hughes S, Ashton JR (2005) Measuring paternal discrepancy and its public health consequences. J Epidemiol Community Health 59(9):749–754. https://doi.org/10.1136/jech.2005.036517
Belsky J (1999) Modern evolutionary theory and patterns of attachment. In: Cassidy J, Shaver P (eds) Handbook of attachment: theory, research, and clinical applications. Guilford, New York, NY, pp 141–161
Bem D (1970) Beliefs, attitudes, and human affairs. Brooks/Cole, Belmont, CA
Berent I, Lennertz T, Jun J, Moreno MA, Smolensky P (2008) Language universals in human brains. Proc Natl Acad Sci U S A 105(14):5321–5325. https://doi.org/10.1073/pnas.0801469105
Block J, Funder DC (1986) Social roles and social perception: individual differences in attribution and error. J Pers Soc Psychol 51(6):1200–1207
van Bodegom D, Rozing MP, May L, Meij HJ, Thomese F, Zwaan BJ, Westendorp RG (2013) Socioeconomic status determines sex-dependent survival of human offspring. Evol Med Public Health 2013(1):37–45. https://doi.org/10.1093/emph/eot002
Botwin MD, Buss DM, Shackelford TK (1997) Personality and mate preferences: five factors in mate selection and marital satisfaction. J Pers 65(1):107–136
Bowlby J (1969) Attachment. Basic Books, New York, NY
Broude GJ (1994) Marriage, family, and relationships: a cross cultural encyclopedia. ABC-CLIO, Santa Barbara, CA
Brown DE (1991) Human universals. McGraw Hill, New York, NY
Burkhardt RW (2005) Patterns of behavior: Konrad Lorenz, Niko Tinbergen, and the founding of ethology. University of Chicago Press, Chicago, IL
Burnham TC, Chapman JF, Gray PB, McIntyre MH, Lipson SF, Ellison PT (2003) Men in committed, romantic relationships have lower testosterone. Horm Behav 44(2):119–122
Burt A, Trivers R (2006) Genes in conflict: the biology of selfish genetic elements. Belknap Press, Cambridge, MA
Buss DM (1984) Evolutionary biology and personality psychology. Toward a conception of human nature and individual differences. Am Psychol 39(10):1135–1147
Buss DM (1988) From vigilance to violence: tactics of mate retention in American undergraduates. Ethol Sociobiol 9(5):291–317
Buss DM (1989) Conflict between the sexes: strategic interference and the evocation of anger and upset. J Pers Soc Psychol 56(5):735–747
Buss DM (1991) Evolutionary personality psychology. Annu Rev Psychol 42:459–491. https://doi.org/10.1146/annurev.ps.42.020191.002331
Buss DM (1992) Manipulation in close relationships: the five factor model of personality in interactional context. J Pers 60:477–499

Buss DM (1995) Psychological sex differences. Origins through sexual selection. Am Psychol 50(3):164–168. discussion 169–171
Buss DM (1996) Social adaptation and five major factors of personality. In: Wiggins JS (ed) The five-factor model of personality: theoretical perspectives. Guilford, New York, NY, pp 180–207
Buss DM (2001) Human nature and culture: an evolutionary psychological perspective. J Pers 69(6):955–978
Buss D (2005) The handbook of evolutionary psychology. John Wiley & Sons, Inc, Hoboken, NJ
Buss DM (2006) The evolutionary genetics of personality: does mutation load signal relationship load? Behav Brain Sci 29:409
Buss DM (2009) How can evolutionary psychology successfully explain personality and individual differences? Perspect Psychol Sci 4(4):359–366. https://doi.org/10.1111/j.1745-6924.2009.01138.x
Buss DM (2011) Evolutionary psychology. Pearson, Boston, MA
Buss DM, Barnes M (1986) Preferences in human mate selection. J Pers Soc Psychol 50(3):559–570
Buss DM, Duntley JD (2008) Adaptations for exploitation. Group Dynamics 12:53–62
Buss DM, Schmitt DP (1993) Sexual strategies theory: an evolutionary perspective on human mating. Psychol Rev 100(2):204–232
Buss DM, Larsen RJ, Westen D, Semmelroth J (1992) Sex differences in jealousy: evolution, physiology, and psychology. Psychol Sci 3(4):251–255
Buss DM, Haselton MG, Shackelford TK, Bleske AL, Wakefield JC (1998) Adaptations, exaptations, and spandrels. Am Psychol 53(5):533–548
Cameron EZ, Dalerum F (2009) A Trivers-Willard effect in contemporary humans: male-biased sex ratios among billionaires. PLoS One 4(1):e4195. https://doi.org/10.1371/journal.pone.0004195
Camperio Ciani AS, Capiluppi C, Veronese A, Sartori G (2007) The adaptive value of personality differences revealed by small island population dynamics. Eur J Personal 21:3–22
Camperio Ciani A, Cermelli P, Zanzotto G (2008) Sexually antagonistic selection in human male homosexuality. PLoS One 3(6):e2282. https://doi.org/10.1371/journal.pone.0002282
Cattell RB, Eber HW, Tatsuoka MM (1970) Handbook for the Sixteen Personality Factor Questionnaire (16PF). Institute for Personality and Ability Testing, Champaign, IL
Chacon-Puignau GC, Jaffe K (1996) Sex ratio at birth deviations in modern Venezuela: the Trivers-Willard effect. Soc Biol 43(3-4):257–270
Chagnon NA (1988) Life histories, blood revenge, and warfare in a tribal population. Science 239(4843):985–992. https://doi.org/10.1126/science.239.4843.985
Charnov EL, Ernest SK (2006) The offspring-size/clutch-size trade-off in mammals. Am Nat 167(4):578–582. https://doi.org/10.1086/501141
Chen C, Burton M, Greenberger E, Dmitrieva J (1999) Population migration and the variation of dopamine D4 receptor (DRD4) allele frequencies around the globe. Evol Hum Behav Brain Sci 20:309–324
Chomsky N (2005) Universals of human nature. Psychother Psychosom 74(5):263–268. https://doi.org/10.1159/000086316
Chomsky N, McGilvray J (2012) The science of language. Cambridge University Press, Cambridge
Clutton-Brock TH, Vincent AC (1991) Sexual selection and the potential reproductive rates of males and females. Nature 351(6321):58–60. https://doi.org/10.1038/351058a0
Comins HN, Hamilton WD, May RM (1980) Evolutionarily stable dispersal strategies. J Theor Biol 82(2):205–230
Confer JC, Easton JA, Fleischman DS, Goetz CD, Lewis DM, Perilloux C, Buss DM (2010) Evolutionary psychology. Controversies, questions, prospects, and limitations. Am Psychol 65(2):110–126. https://doi.org/10.1037/a0018413
Conroy-Beam D, Buss DM, Pham MN, Shackelford TK (2015) How sexually dimorphic are human mate preferences? Pers Soc Psychol Bull 41(8):1082–1093. https://doi.org/10.1177/0146167215590987
Cosmides L, Tooby J (2005) Neurocognitive adaptations designed for social exchange. In: Buss D (ed) The handbook of evolutionary psychology. Wiley, New York, NY, pp 584–627

Costa PT, McCrae RR (1985) The NEO personality inventory manual. Psychological Assessment Resources, Odessa, FL
Cronk L (2007) Boy or girl: gender preferences from a Darwinian point of view. Reprod Biomed Online 15(Suppl 2):23–32
Crook JH, Crook SJ (1988) Tibetan polyandry: problems of adaptation and fitness. In: Betzig L, Borgerhoff Mulder M, Turke P (eds) Human reproductive behavior: a Darwinian perspective. Cambridge University Press, Cambridge
Cummings M, Zahn-Waxler C, Iannotti R (1991) Altruism and aggression: biological and social origins. Cambridge University Press, Cambridge
D'Eath RB, Roehe R, Turner SP, Ison SH, Farish M, Jack MC, Lawrence AB (2009) Genetics of animal temperament: aggressive behaviour at mixing is genetically associated with the response to handling in pigs. Animal 3(11):1544–1554. https://doi.org/10.1017/S1751731109990528
D'Onofrio BM, Eaves LJ, Murrelle L, Maes HH, Spilka B (1999) Understanding biological and social influences on religious affiliation, attitudes, and behaviors: a behavior genetic perspective. J Pers 67(6):953–984
Daly M, Wilson M (1983) Sex, evolution, and behavior, 2nd edn. Wadsworth, Belmont, CA
Daly M, Wilson M (1987) The Darwinian psychology of discriminative parental solicitude. Nebraska symposium on motivation nebraska symposium on motivation 35:91–144
Daly M, Wilson M (1991) A reply to Gelles: stepchildren are disproportionately abused, and diverse forms of violencecan share causal factors. Hum Nat 2(4):419–426. https://doi.org/10.1007/BF02692199
Daly M, Wilson M (2005a) Carpe diem: adaptation and devaluing the future. Q Rev Biol 80(1):55–60
Daly M, Wilson M (2005b) The "Cinderella effect" is no fairy tale. Trends Cogn Sci 9(11):507–508.; author reply 508–510. https://doi.org/10.1016/j.tics.2005.09.007
Daly M, Salmon C, Wilson MIE (1997) Kinship: the conceptual hole in psychological studies of social cognition and close relationships. In: Simpson J, Kenrick D (eds) Evolutionary social psychology. Lawrence Erlbaum Associates, Mahnaw, NJ, pp 265–296
Darwin C (1859) On the origin of species by means of natural selection, or the preservation of favoured races in the struggle for life. John Murray, London
Darwin C (1871) The descent of man, and selection in relation to sex. John Murray, London
Darwin C (1872) The expression of the emotions in man and animals. John Murray, London
Dawkins R (1976) The selfish gene. Oxford University Press, Oxford
Deacon T (1997) The symbolic species: the co-evolution of language and the brain. W.W. Norton, New York, NY
Deichmann U (1992) Biologists under Hitler: expulsion, careers, research. Harvard University Press, New York, NY
Denissen JJA, Penke L (2008a) Motivational individual reaction norms underlying the five-factor model of personality: first steps towards a theory-based conceptual framework. J Res Pers 42:1285–1302
Denissen JJA, Penke L (2008b) Neuroticism predicts reactions to cues of social inclusion. Eur J Personal 22:497–517
Di Vincenzo F, Manzi G (2013) Social learning and origin of the language faculty by means of natural selection. J Anthropol Sci 91:261–267. https://doi.org/10.4436/jass.91017
Dunbar R, Barret L (2007) The Oxford handbook of evolutionary psychology. Oxford University Press, Oxford
Dunfield KA, Johnson SC (2015) Variability in social reasoning: the influence of attachment security on the attribution of goals. Front Psychol 6:1487. https://doi.org/10.3389/fpsyg.2015.01487
Dunsworth HM, Warrener AG, Deacon T, Ellison PT, Pontzer H (2012) Metabolic hypothesis for human altriciality. Proc Natl Acad Sci U S A 109(38):15212–15216. https://doi.org/10.1073/pnas.1205282109
Durand PM, Rashidi A, Michod RE (2011) How an organism dies affects the fitness of its neighbors. Am Nat 177(2):224–232. https://doi.org/10.1086/657686

Ebstein RP (2006) The molecular genetic architecture of human personality: beyond self-report questionnaires. Mol Psychiatry 11(5):427–445. https://doi.org/10.1038/sj.mp.4001814
Eccles JC (1992) Evolution of consciousness. Proc Natl Acad Sci U S A 89(16):7320–7324
Eisenberg DT, Campbell B, Gray PB, Sorenson MD (2008) Dopamine receptor genetic polymorphisms and body composition in undernourished pastoralists: an exploration of nutrition indices among nomadic and recently settled Ariaal men of northern Kenya. BMC Evol Biol 8:173. https://doi.org/10.1186/1471-2148-8-173
Ekman P (1965) Differential communication of affect by head and body cues. J Pers Soc Psychol 2(5):726–735
Ekman P (1980) Asymmetry in facial expression. Science 209(4458):833–834
Ekman P (1992a) Are there basic emotions? Psychol Rev 99(3):550–553
Ekman P (1992b) Facial expressions of emotion: an old controversy and new findings. Philos Trans R Soc Lond Ser B Biol Sci 335(1273):63–69. https://doi.org/10.1098/rstb.1992.0008
Ekman P (1993) Facial expression and emotion. Am Psychol 48(4):384–392
Ekman P (1994) Strong evidence for universals in facial expressions: a reply to Russell's mistaken critique. Psychol Bull 115(2):268–287
Ekman P (2003) Emotions inside out. 130 years after Darwin's "The expression of the emotions in man and animal". Ann N Y Acad Sci 1000:1–6
Ekman P (2009) Darwin's contributions to our understanding of emotional expressions. Philos Trans R Soc Lond Ser B Biol Sci 364(1535):3449–3451. https://doi.org/10.1098/rstb.2009.0189
Ekman P (2016) What scientists who study emotion agree about. Perspect Psychol Sci 11(1):31–34. https://doi.org/10.1177/1745691615596992
Ekman P, Friesen WV (1967) Head and body cues in the judgment of emotion: a reformulation. Percept Mot Skills 24(3):711–724. https://doi.org/10.2466/pms.1967.24.3.711
Ekman P, Friesen WV (1971) Constants across cultures in the face and emotion. J Pers Soc Psychol 17(2):124–129
Ekman P, Sorenson ER, Friesen WV (1969) Pan-cultural elements in facial displays of emotion. Science 164(3875):86–88
Ekman P, Friesen WV, O'Sullivan M, Chan A, Diacoyanni-Tarlatzis I, Heider K, Krause R, LeCompte WA, Pitcairn T, Ricci-Bitti PE et al (1987) Universals and cultural differences in the judgments of facial expressions of emotion. J Pers Soc Psychol 53(4):712–717
Ekman P, O'Sullivan M, Frank MG (1999) A few can catch a liar. Psychol Sci 10:363–366
Engel KC, Manner L, Ayasse M, Steiger S (2015) Acceptance threshold theory can explain occurrence of homosexual behaviour. Biol Lett 11(1):20140603. https://doi.org/10.1098/rsbl.2014.0603
Fabrega H (2002) Phylogenetic and cultural basis of mental illness. Rutgers University Press, New Brunswick, NJ
Ferriere R, Michod RE (2011) Inclusive fitness in evolution. Nature 471(7339):E6–E8.; author reply E9–10. https://doi.org/10.1038/nature09834
Figueredo AJ, Sefcek JA, Vasquez G, Brumbach BH, King JE, Jacobs WJ (2005) Evolutionary personality psychology. In: Buss D (ed) The handbook of evolutionary psychology. Wiley, New York, NY, pp 851–877
Figueredo A, Gladden P, Hohman Z (2011) The evolutionary psychology of criminal behaviour. In: Roberts SC (ed) Applied evolutionary psychology. Oxford University Press, Oxford
Fiske AP (1992) The four elementary forms of sociality: framework for a unified theory of social relations. Psychol Rev 99(4):689–723
Fitch T (2010) The evolution of language. Cambridge University Press, Cambridge
Fitch WT, Hauser MD, Chomsky N (2005) The evolution of the language faculty: clarifications and implications. Cognition 97(2):179–210.; discussion 211–125. https://doi.org/10.1016/j.cognition.2005.02.005
Föger B, Taschwer K (2001) Die andere Seite des Spiegels: Konrad Lorenz und der Nationalsozialismus. Czernin-Verlag, Wien
Foley R, Lewin R (2013) Principles of Human Evolution. John Wiley & Sons, Hoboken, NJ

Fountoulakis KN, Kaprinis GS (2006) Personality disorders: new data versus old concepts. Curr Opin Psychiatry 19(1):90–94. https://doi.org/10.1097/01.yco.0000196158.98540.b6
Fountoulakis KN, Siamouli M, Magiria S, Kaprinis G (2008) Late-life depression, religiosity, cerebrovascular disease, cognitive impairment and attitudes towards death in the elderly: interpreting the data. Med Hypotheses 70(3):493–496. https://doi.org/10.1016/j.mehy.2007.01.093
Fountoulakis KN, Gonda X, Koufaki I, Hyphantis T, Cloninger CR (2016) The role of temperament in the etiopathogenesis of bipolar spectrum illness. Harv Rev Psychiatry 24(1):36–52. https://doi.org/10.1097/HRP.0000000000000077
Fowler JH (2005) Human cooperation: second-order free-riding problem solved? Nature 437(7058):E8; discussion E8-9. https://doi.org/10.1038/nature04201
Freeman D (1983) Margaret Mead and Samoa. Harvard University Press, London
Gadgil M, Bossert WH (1970) Life historical consequences of natural selection. Am Nat 104:1–24
Galperin A, Haselton MG, Frederick DA, Poore J, von Hippel W, Buss DM, Gonzaga GC (2013) Sexual regret: evidence for evolved sex differences. Arch Sex Behav 42(7):1145–1161. https://doi.org/10.1007/s10508-012-0019-3
Gangestad SW, Buss DM (1994) Pathogen prevalence and human mate preferences. Ethol Sociobiol 14(2):89–96
Gangestad SW, Simpson JA (1990) Toward an evolutionary history of female sociosexual variation. J Pers 58(1):69–96
Gangestad SW, Haselton MG, Buss DM (2006) Evolutionary foundations of cultural variation: evoked culture and mate preferences. Psychol Inq 17:75–95
Garcia-Cardenas N, Olvera-Hernandez S, Gomez-Quintanar BN, Fernandez-Guasti A (2015) Male rats with same sex preference show high experimental anxiety and lack of anxiogenic-like effect of fluoxetine in the plus maze test. Pharmacol Biochem Behav 135:128–135. https://doi.org/10.1016/j.pbb.2015.05.017
Gardner RA, Gardner BT (1969) Teaching sign language to a chimpanzee. Science 165(3894):664–672
Gaulin S, McBurney DH (2003) Evolutionary psychology. Prentice Hall, Upper Saddle River, NJ
Geary DC (1995) Reflections of evolution and culture in children's cognition. Implications for mathematical development and instruction. Am Psychol 50(1):24–37
Geary DC (1998) Male, female: the evolution of human sex differences. American Psychological Association, Washington, DC
Gergen KJ (1973) Social psychology as history. J Pers Soc Psychol 26:309–320
Gluckman P, Beedle A, Hanson M (2009) Principles of evolutionary medicine. Oxford University Press, Oxford
Gosling SD (2001) From mice to men: what can we learn about personality from animal research? Psychol Bull 127(1):45–86
Grodzinsky Y (2000) The neural substrate of the language faculty: suggestions for the future. Brain Lang 71(1):82–84. https://doi.org/10.1006/brln.1999.2219
Grodzinsky Y (2006) The language faculty, Broca's region, and the mirror system. Cortex 42(4):464–468
Gunst N, Leca JB, Vasey PL (2015) Influence of sexual competition and social context on homosexual behavior in adolescent female Japanese macaques. Am J Primatol 77(5):502–515. https://doi.org/10.1002/ajp.22369
Hagen E, Hammerstein P (2006) Game theory and human evolution: a critique of some recent interpretations of experimental games. Theor Popul Biol 69(3):339–348
Haig D (2002) Genomic imprinting and kinship. Rutgers University Press, New Brunswick, NJ
Hamilton WD (1964a) The genetical evolution of social behaviour. I. J Theor Biol 7(1):1–16
Hamilton WD (1964b) The genetical evolution of social behaviour. II. J Theor Biol 7(1):17–52
Hamilton WD (1970) Selfish and spiteful behaviour in an evolutionary model. Nature 228(5277):1218–1220
Hamilton W (2001) Narrow roads of gene land, vol 1, 2 and 3. Oxford University Press, New York, NY
Harrod J (2014) The case for chimpanzee religion. J Stud Relig Nat Cult 8(1):8–45

Hart CW, Pillig AR (1960) The Tiwi of North Australia. Holt, Rinehart, & Winston, New York, NY
Hasegawa T, Hiraiwa M (1980) Social interactions of orphans observed in a free-ranging troop of Japanese monkeys. Folia Primatol 33(1-2):129–158
Haselton MG, Buss DM (2000) Error management theory: a new perspective on biases in cross-sex mind reading. J Pers Soc Psychol 78(1):81–91
Haselton MG, Nettle D, Andrews PW (2005) The evolution of cognitive bias. In: Buss D (ed) The handbook of evolutionary psychology. Wiley, New York, NY, pp 724–746
Hauser MD, Chomsky N, Fitch WT (2002) The faculty of language: what is it, who has it, and how did it evolve? Science 298(5598):1569–1579. https://doi.org/10.1126/science.298.5598.1569
Hawley PH (1999) The ontogenesis of social dominance: a strategy-based evolutionary perspective. Dev Rev 19:97–132
Herron J, Freeman S (2013) Evolutionary Analysis, 5th edn. Pearson, NJ
Higley JD, Linnoila M (1997) A nonhuman primate model of excessive alcohol intake. Personality and neurobiological parallels of type I- and type II-like alcoholism. Recent Dev Alcohol 13:191–219
Higley JD, Suomi SJ, Linnoila M (1991) CSF monoamine metabolite concentrations vary according to age, rearing, and sex, and are influenced by the stressor of social separation in rhesus monkeys. Psychopharmacology 103(4):551–556
Higley JD, Mehlman PT, Taub DM, Higley SB, Suomi SJ, Vickers JH, Linnoila M (1992) Cerebrospinal fluid monoamine and adrenal correlates of aggression in free-ranging rhesus monkeys. Arch Gen Psychiatry 49(6):436–441
Hoppitt WJ, Brown GR, Kendal R, Rendell L, Thornton A, Webster MM, Laland KN (2008) Lessons from animal teaching. Trends Ecol Evol 23(9):486–493. https://doi.org/10.1016/j.tree.2008.05.008
Horner V, Proctor D, Bonnie KE, Whiten A, de Waal FB (2010) Prestige affects cultural learning in chimpanzees. PLoS One 5(5):e10625. https://doi.org/10.1371/journal.pone.0010625
Huey RB, Pianka ER (1977) Natural selection for juvenile lizards mimicking noxious beetles. Science 195(4274):201–203
Ishikawa SS, Raine A, Lencz T, Bihrle S, LaCasse L (2001) Increased height and bulk in antisocial personality disorder and its subtypes. Psychiatry Res 105(3):211–219
Kalikow TJ (1983) Konrad Lorenz's ethological theory: explanation and ideology, 1938–1943. J Hist Biol 16(1):39–73
Kaplan HS, Gangestad SW (2005) Life history theory and evolutionary psychology. In: Buss DM (ed) The handbook of evolutionary psychology. Wiley, New York, NY, pp 68–96
Kaptijn R, Thomese F, Liefbroer AC, Silverstein M (2013) Testing evolutionary theories of discriminative grandparental investment. J Biosoc Sci 45(3):289–310. https://doi.org/10.1017/S0021932012000612
Keller MC (2007) The role of mutations in human mating. In: Geher G, Miller G (eds) Mating intelligence: theoretical, experimental, and differential perspectives. Erlbaum, Mahwah, NJ, pp 173–192
Keller MC, Miller G (2006) Resolving the paradox of common, harmful, heritable mental disorders: which evolutionary genetic models work best? Behav Brain Sci 29(4):385–404. https://doi.org/10.1017/S0140525X06009095. discussion 405–352
Kenrick DT (1994) Evolutionary social psychology: from sexual selection to social cognition. In: Zanna M (ed) Advances in experimental social psychology, vol 26. Academic, San Diego, CA
Kenrick DT, Neuberg SL, Zierk KL, Krones JM (1994) Evolution and social cognition: contrast effects as a function of sex, dominance, and physical attractiveness. Pers Soc Psychol Bull 20:210–217
Kenrick DT, Li NP, Butner J (2003) Dynamical evolutionary psychology: individual decision rules and emergent social norms. Psychol Rev 110(1):3–28
King DA, Schuehle Pfeiffer CE, Randel RD, Welsh TH Jr, Oliphint RA, Baird BE, Curley KO Jr, Vann RC, Hale DS, Savell JW (2006) Influence of animal temperament and stress responsiveness on the carcass quality and beef tenderness of feedlot cattle. Meat Sci 74(3):546–556. https://doi.org/10.1016/j.meatsci.2006.05.004

Klein SB, Cosmides L, Tooby J, Chance S (2002) Decisions and the evolution of memory: multiple systems, multiple functions. Psychol Rev 109(2):306–329

Klin A (2000) Attributing social meaning to ambiguous visual stimuli in higher-functioning autism and Asperger syndrome: the social attribution task. J Child Psychol Psychiatry 41(7):831–846

Kolk M, Schnettler S (2013) Parental status and gender preferences for children: is differential fertility stopping consistent with the trivers-willard hypothesis? J Biosoc Sci 45(5):683–704. https://doi.org/10.1017/S0021932012000557

Koziel S, Ulijaszek SJ (2001) Waiting for Trivers and Willard: do the rich really favor sons? Am J Phys Anthropol 115(1):71–79. https://doi.org/10.1002/ajpa.1058

Krebs DL, Denton K (1997) Social illusions and self-deception: the evolution of biases in person perception. In: Simpson J, Kenrick D (eds) Evolutionary social psychology. Erlbaum, Mahwah, NJ, pp 21–48

Kruglanski AW (1986) Social psychology: attribution. Science 232(4750):665–666. https://doi.org/10.1126/science.232.4750.665

Kurzban R, Leary MR (2001) Evolutionary origins of stigmatization: the functions of social exclusion. Psychol Bull 127(2):187–208

Lalumiere M, Harris G, Quinsey V, Rice M (2005) The causes of rape. American Psychological Association, Washington, DC

Leca JB, Gunst N, Vasey PL (2015) Comparative development of heterosexual and homosexual behaviors in free-ranging female Japanese macaques. Arch Sex Behav 44(5):1215–1231. https://doi.org/10.1007/s10508-014-0437-5

Lewin R (2009) Human evolution: an illustrated introduction, 5th edn. John Wiley & Sons, Hoboken, NY

Li NP, Bailey JM, Kenrick DT, Linsenmeier JAW (2002) The necessities and luxuries of mate preferences: testing the tradeoffs. J Pers Soc Psychol 6(6):947–955

Lieberman D, Tooby J, Cosmides L (2007) The architecture of human kin detection. Nature 445(7129):727–731. https://doi.org/10.1038/nature05510

Lund OCH, Tamnes CK, Moestue C, Buss DM, Vollrath M (2007a) Tactics of hierarchy negotiation. J Res Pers 41:25–44

Lund OCH, Tamnes CK, Moestue C, Buss DM, Vollrath M (2007b) Tactics of hierarchy negotiation. J Res Pers 41:25–44

Mabry JH (1995) Review of Pinker's the language instinct. Anal Verbal Behav 12:87–96

MacDonald K (1995) Evolution, the five factor model, and levels of personality. J Pers 63:525–568

MacFarlane GR, Vasey PL (2016) Promiscuous primates engage in same-sex genital interactions. Behav Process 126:21–26. https://doi.org/10.1016/j.beproc.2016.02.016

Martin JG, Reale D (2008) Animal temperament and human disturbance: implications for the response of wildlife to tourism. Behav Process 77(1):66–72. https://doi.org/10.1016/j.beproc.2007.06.004

Maryanski A, Turner JH (1992) The social cage: human nature and the evolution of society. Stanford University Press, Stanford, CA

Matthews J, Matthews R (2010) Insect behaviour. Springer, New York, NY

McAndrew FT (2002) New evolutionary perspectives on altruism: multilevel selection and costly signaling theories. Curr Dir Psychol Sci 11:79–82

McCrae RR, Costa JPT, Martin TA (2005) The NEO-PI-3: a more readable revised NEO personality inventory. J Pers Assess 84(3):261–270. https://doi.org/10.1207/s15327752jpa8403_05

Mead M (1928) Coming of age in Samoa William Morrow and Co.,

Mealey L (1995) The sociobiology of sociopathy: an integrated evolutionary model. Behav Brain Sci 18:523–599

Mehlman PT, Higley JD, Faucher I, Lilly AA, Taub DM, Vickers J, Suomi SJ, Linnoila M (1994) Low CSF 5-HIAA concentrations and severe aggression and impaired impulse control in non-human primates. Am J Psychiatry 151(10):1485–1491. https://doi.org/10.1176/ajp.151.10.1485

Mehlman PT, Higley JD, Fernald BJ, Sallee FR, Suomi SJ, Linnoila M (1997) CSF 5-HIAA, testosterone, and sociosexual behaviors in free-ranging male rhesus macaques in the mating season. Psychiatry Res 72(2):89–102

Michod RE (1996) Cooperation and conflict in the evolution of individuality. II. Conflict mediation. Proc Biol Sci 263(1372):813–822. https://doi.org/10.1098/rspb.1996.0121
Michod RE (2006) The group covariance effect and fitness trade-offs during evolutionary transitions in individuality. Proc Natl Acad Sci U S A 103(24):9113–9117. https://doi.org/10.1073/pnas.0601080103
Michod RE, Herron MD (2006) Cooperation and conflict during evolutionary transitions in individuality. J Evol Biol 19(5):1406–1409.; discussion 1426–1436. https://doi.org/10.1111/j.1420-9101.2006.01142.x
Michod RE, Nedelcu AM (2003) On the reorganization of fitness during evolutionary transitions in individuality. Integr Comp Biol 43(1):64–73. https://doi.org/10.1093/icb/43.1.64
Miller G (2000a) The mating mind. Penguin, New York, NY
Miller GF (2000b) The mating mind: how sexual choice shaped the evolution of human nature. Anchor Books, New York, NY
Miller GF (2007) Sexual selection for moral virtues. Q Rev Biol 82(2):97–125
Moore G (2004) Principia ethica (original publication 1903). Dover Publications, Mineola, NY
Moscovici S, Markova I (2006) The making of modern social psychology. Polity Press, Cambridge
Myers D (2010) Social psychology, 10th edn. McGraw-Hill, New York, NY
Nedelcu AM, Michod RE (2006) The evolutionary origin of an altruistic gene. Mol Biol Evol 23(8):1460–1464. https://doi.org/10.1093/molbev/msl016
Nettle D (2006) The evolution of personality variation in humans and other animals. Am Psychol 61(6):622–631. https://doi.org/10.1037/0003-066X.61.6.622
Neuberg SL, Smith DM, Asher T (2000) Why people stigmatize: toward a biocultural framework. In: Heatherton T, Kleck R (eds) The social psychology of stigma. The Guilford Press, New York, NY
Nichols S, Grantham T (2000) Adaptive Complexity and Phenomenal Consciousness. Philos Sci 67(4):648–670
Nisbett A (1976) Konrad Lorenz. Littlehampton Book Services Ltd, Worthing
Ohman A, Mineka S (2001) Fears, phobias, and preparedness: toward an evolved module of fear and fear learning. Psychol Rev 108(3):483–522
Orlove MJ, Wood CL (1978) Coefficients of relationship and coefficients of relatedness in kin selection: a covariance form for the RHO formula. J Theor Biol 73(4):679–686
Ozer DJ, Benet-Martinez V (2006) Personality and the prediction of consequential outcomes. Annu Rev Psychol 57:401–421. https://doi.org/10.1146/annurev.psych.57.102904.190127
Paley W (1802) Natural theology or evidences of the existence and attributes of the deity. R. Faulder, London
Patterson F, Linden E (1981) The education of Koko. Holt, New York, NY
Penke L, Denissen JJA, Miller GF (2007) The evolutionary genetics of personality. Eur J Personal 21:549–587
Perilloux C, Buss DM (2008) Breaking up romantic relationships: costs experienced and coping strategies deployed. Evol Psychol 6:164–181
Peters J, Shackelford TK, Buss DM (2002) Understanding domestic violence against women: using evolutionary psychology to extend the feminist functional analysis. Violence Vict 17(2):255–264
Pinker S (1994) The language instinct. William Morrow and Company, New York, NY
Pinker S (1999) How the mind works. WW Norton & Co, New York, NY
Pinker S (2002) The blank slate: the modern denial of human nature. New York, NY, Viking
Pinker S, Bloom P (1990) Natural language and natural selection. Behav Brain Sci 13(4):707–727
Pinker S, Jackendoff R (2005) The faculty of language: what's special about it? Cognition 95(2):201–236. https://doi.org/10.1016/j.cognition.2004.08.004
Platek S, Shackelford T (2007) Female infidelity and paternal uncertainty. Cambridge University Press, Cambridge
Plomin R, DeFries JC, McClearn GE, Rutter M (2008) Behavioral genetics. Freeman, New York, NY
Popper K (1959) The logic of scientific discovery. Basic Books, New York, NY

Power HW (1975) Mountain bluebirds: experimental evidence against altruism. Science 189(4197):142–143. https://doi.org/10.1126/science.189.4197.142
Radcliffe-Brown A (1913) Three tribes of Western Australia. J R Anthropol Inst 43:143–194
Reale D, Reader SM, Sol D, McDougall PT, Dingemanse NJ (2007) Integrating animal temperament within ecology and evolution. Biol Rev Camb Philos Soc 82(2):291–318. https://doi.org/10.1111/j.1469-185X.2007.00010.x
Reisenzein R (2015) On the universality of the attribution-affect model of helping. Int J Psychol 50(4):308–311. https://doi.org/10.1002/ijop.12153
Robson JM (1974) Collected works of John Stuart Mill. University of Toronto Press, Toronto, ON
Roff D (1992) The evolution of life histories: theory and analysis. Chapman & Hall, New York, NY
Rosch EH (1973) Natural categories. Cogn Psychol 4:328–350
Rose S (2001) Revisiting evolutionary psychology and psychiatry. Br J Psychiatry 179:558
Rossano M (2006) The religious mind and the evolution of religion. Rev Gen Psychol 10(4):346–364
Rousseau JJ (1979) Emile, or on education (trans. Allan Bloom). Basic Books, New York, NY
Salmon C, Shackelford T (2008) Family relationships: an evolutionary perspective. Oxford University Press, Oxford
Santiago JH, Tarantino SJ (2002) Individualism and collectivism: cultural orientation in locus of control and moral attribution under conditions of social change. Psychol Rep 91(3 Pt 2):1155–1168. https://doi.org/10.2466/pr0.2002.91.3f.1155
Santrock WJ (2005) A topical approach to life-span development, 3rd edn. McGraw-Hill, New York, NY
Schaller M, Simpson JA, Kenrick DT (2006) Evolution and social psychology. Psychology Press, New York, NY
Schmitt DP, Buss DM (2001) Human mate poaching: tactics and temptations for infiltrating existing relationships. J Pers Soc Psychol 80(6):894–917
Schreiber D (2012) On social attribution: implications of recent cognitive neuroscience research for race, law, and politics. Sci Eng Ethics 18(3):557–566. https://doi.org/10.1007/s11948-012-9381-8
Schwartz DA, Howe CQ, Purves D (2003) The statistical structure of human speech sounds predicts musical universals. J Neurosci 23(18):7160–7168
Segal NL, MacDonald KB (1998) Behavioral genetics and evolutionary psychology: unified perspective on personality research. Hum Biol 70(2):159–184
Seidel EM, Eickhoff SB, Kellermann T, Schneider F, Gur RC, Habel U, Derntl B (2010) Who is to blame? Neural correlates of causal attribution in social situations. Soc Neurosci 5(4):335–350. https://doi.org/10.1080/17470911003615997
Shankman P (2009) The trashing of margaret mead: the anatomy of an anthropological myth. University of Wisconsin Press, Madison, WI
Sheldon KM, Sheldon MS, Nichols CP (2007) Traits and trade-offs are insufficient for evolutionary personality psychology. Am Psychol 62(9):1073–1074. https://doi.org/10.1037/0003-066X.62.9.1073
Sibley RM (1983) Optimal group size is unstable. Anim Behav 31:947–948
Sims CA (2001) Revisiting evolutionary psychology and psychiatry. Br J Psychiatry 179:558–559
Smith EA (2011) Endless forms: human behavioural diversity and evolved universals. Philos Trans R Soc Lond Ser B Biol Sci 366(1563):325–332. https://doi.org/10.1098/rstb.2010.0233
Stearns S (1992) The evolution of life histories. Oxford, Oxford University Press
Sugiyama MS (2003) Cultural variation is part of human nature: literary universals, context-sensitivity, and "Shakespeare in the bush". Hum Nat 14(4):383–396. https://doi.org/10.1007/s12110-003-1012-2
Sugiyama L (2005) Physical attractiveness in adaptationist perspective. In: Buss D (ed) The handbook of evolutionary psychology. Wiley, New York, NY, pp 292–342
Sulloway F (1996) Born to rebel. Pantheon, New York, NY
Thornhill R, Gangestad SW (2008) The evolutionary biology of human female sexuality. Oxford University Press, New York, NY

Thornhill R, Palmer C (2000) A natural history of rape: biological bases of sexual coercion. The MIT Press, Cambridge, MA
Tooby J, Cosmides L (1990) On the universality of human nature and the uniqueness of the individual: the role of genetics and adaptation. J Pers 58(1):17–67
Tooley G, Karakis M, Stokes M, Ozannesmith J (2006) Generalising the Cinderella Effect to unintentional childhood fatalities. Evol Hum Behav Brain Sci 27:224–230
Traxler MJ, Boudewyn M, Loudermilk J (2012) What's special about human language? The contents of the "narrow language faculty" revisited. Lang Ling Compass 6(10):611–621. https://doi.org/10.1002/lnc3.355
Triana-Del Rio R, Tecamachaltzi-Silvaran MB, Diaz-Estrada VX, Herrera-Covarrubias D, Corona-Morales AA, Pfaus JG, Coria-Avila GA (2015) Conditioned same-sex partner preference in male rats is facilitated by oxytocin and dopamine: effect on sexually dimorphic brain nuclei. Behav Brain Res 283:69–77. https://doi.org/10.1016/j.bbr.2015.01.019
Trivers RL (1972) Parental investment and sexual selection. In: Campbell B (ed) Sexual selection and the descent of man, 1871–1971. Aldine-Atherton, Chicago, IL, pp 136–179
Trivers RL, Willard DE (1973) Natural selection of parental ability to vary the sex ratio of offspring. Science 179(4068):90–92
Ungerfeld R, Giriboni J, Freitas-de-Melo A, Lacuesta L (2014) Homosexual behavior in male goats is more frequent during breeding season and in bucks isolated from females. Horm Behav 65(5):516–520. https://doi.org/10.1016/j.yhbeh.2014.04.013
Van Vugt M, Ahuja A (2011) Naturally selected: the evolutionary science of leadership. Harper Business, New York, NY
Van Vugt M, Ronay R (2014) The evolutionary psychology of leadership. Organ Psychol Rev 4:74–95
Varki A, Altheide TK (2005) Comparing the human and chimpanzee genomes: searching for needles in a haystack. Genome Res 15(12):1746–1758. https://doi.org/10.1101/gr.3737405
Vasey PL, Leca JB, Gunst N, VanderLaan DP (2014) Female homosexual behavior and inter-sexual mate competition in Japanese macaques: possible implications for sexual selection theory. Neurosci Biobehav Rev 46(Pt 4):573–578. https://doi.org/10.1016/j.neubiorev.2014.09.002
Veller C, Haig D, Nowak MA (2016) The Trivers-Willard hypothesis: sex ratio or investment? Proc Biol Sci 283:1830. https://doi.org/10.1098/rspb.2016.0126
Venero Fernandez SJ, Medina RS, Britton J, Fogarty AW (2011) The association between living through a prolonged economic depression and the male:female birth ratio—a longitudinal study from Cuba, 1960–2008. Am J Epidemiol 174(12):1327–1331. https://doi.org/10.1093/aje/kwr357
Ward EF (1983) Teaching sign language to a chimpanzee: some historical references. J Exp Anal Behav 40(3):341–342
Washburn SL (1973) The promise of primatology. Am J Phys Anthropol 38(2):177–182. https://doi.org/10.1002/ajpa.1330380206
Washburn SL (1978) The evolution of man. Sci Am 239(3):194–201. 204 passim
Washburn SL (1982) Human evolution. Perspect Biol Med 25(4):583–602
Washburn SL, McCown ER (1972) Evolution of human behavior. Soc Biol 19(2):163–170
Weiner S, Monge J, Mann A (2008) Bipedalism and parturition: an evolutionary imperative for cesarean delivery? Clin Perinatol 35(3):469–478. https://doi.org/10.1016/j.clp.2008.06.003
Wells JC, DeSilva JM, Stock JT (2012) The obstetric dilemma: an ancient game of Russian roulette, or a variable dilemma sensitive to ecology? Am J Phys Anthropol 149(Suppl 55):40–71. https://doi.org/10.1002/ajpa.22160
Whitney G (1970) Timidity and fearfulness of laboratory mice: an illustration of problems in animal temperament. Behav Genet 1(1):77–85
Williams G (1966) Adaptation and natural selection. Princeton University Press, Princeton, NJ
Wilson E (1975) Sociobiology: the new synthesis. Harvard University Press, Cambridge, MA
Wilson E (1978) On human nature. Harvard University Press, Cambridge, MA
Wilson DS (1994) Adaptive genetic variation and human evolutionary psychology. Ethol Sociobiol 15:219–235

Wilson E (2000) Sociobiology: the new synthesis, twenty-fifth anniversary edition new edition edition. Belknap Press, Cambridge, MA

Wilson DS, Near D, Miller RR (1996) Machiavellianism: a synthesis of the evolutionary and psychological literatures. Psychol Bull 119(2):285–299

Wolf M, van Doorn GS, Leimar O, Weissing FJ (2007) Life-history trade-offs favour the evolution of animal personalities. Nature 447(7144):581–584. https://doi.org/10.1038/nature05835

Workman L, Reader W (2008) Evolutionary psychology: an introduction. Cambridge University Press, Cambridge

Zadzinska E, Rosset I, Mikulec A, Domanski C, Pawlowski B (2011) Impact of economic conditions on the secondary sex ratio in a post-communist economy. Homo : internationale Zeitschrift fur die vergleichende. Forsch Mensch 62(3):218–227. https://doi.org/10.1016/j.jchb.2011.03.002

Zahavi A (1975) Mate selection-a selection for a handicap. J Theor Biol 53(1):205–214

Zahavi A (2006) Sexual selection, signal selection and the handicap principle. In: Jamieson BGM (ed) Reproductive biology and phylogeny of birds. Science Publishers, Enfield, NH, pp 143–159

Biological Psychiatry and Psychopharmacology

13

Kostas N. Fountoulakis, Ioannis Nimatoudis, and Stephen M. Stahl

13.1 General Considerations

Biological psychiatry is an approach to psychiatry that aims to understand mental disorders in terms of the biological function of the brain and to establish their biological basis. It is interdisciplinary in its approach and includes basic neuroscience, psychopharmacology, biochemistry, genetics, epigenetics and neuropsychophysiology. Although it has its roots in the text of the Hippocratic school, for the first time the term was coined officially in 1953 (Bennett 1953; Gerard 1955a, b; Rioch 1955; Tourney 1969; van Praag 1971). The difference between biological psychiatry and neurology is that while biological psychiatry deals with the general neurobiological characteristics of diseases of the brain without gross abnormal findings, neurology deals with those disorders with gross and often easily observed pathology in the brain. The two disciplines overlap and also complement each other in the fields of neuropsychiatry and especially in behavioural neurology and psychogeriatrics, particularly when dealing with dementias.

The modern explosive development of the field happened in the 1950s with the introduction of psychotherapeutic medication, that is, antipsychotics and antidepressants. In 1965 the first biological theory concerning a mental disorder was

K. N. Fountoulakis (✉) · I. Nimatoudis
3rd Department of Psychiatry, Faculty of Medicine, School of Health Sciences, Aristotle University of Thessaloniki, Thessaloniki, Greece

S. M. Stahl
Department of Psychiatry, University of California San Diego, San Diego, CA, USA

Department of Psychiatry, Cambridge University, Cambridge, UK

K. N. Fountoulakis, I. Nimatoudis (eds.), *Psychobiology of Behaviour*,
https://doi.org/10.1007/978-3-030-18323-3_13

developed (Schildkraut 1965), and since then this particular approach was proven valuable in the development of novel treatment modalities that changed the lives of hundreds of millions of mental patients worldwide and made the deinstitutionalization effort and the end of the big asylums possible.

The current chapter will attempt a very brief but comprehensive introduction to biological theories on the etiopathogenesis of major mental disorders as well as a brief description of contemporary available psychopharmacological agents.

13.2 Biological Models of Major Mental Disorders

13.2.1 Schizophrenia

Although most of cases of schizophrenia are sporadic, it also runs in families, and the overall genetic component is quite strong with a heritability greater than 60–80% (Hilker et al. 2018; Cardno and Gottesman 2000; Cardno et al. 2012; Kendler 1983; Sullivan et al. 2003). The lifetime risk for schizophrenia in the general population is around 1%, with the lifetime risk of relatives of patients being several times higher: siblings 10%, offspring 13% and parents 6%. The lifetime risk for children with both parents being patients is close to 50% (Gottesman and Shields 1982; Kendler 1993; Kendler et al. 1994).

Thus, and since no gross pathology is identifiable, it is generally accepted that a more subtle and probably nonuniform and complex pathology does exist (Selemon 2001). In the late 1950s and early 1960s, the dopaminergic theory of schizophrenia was developed (Davis et al. 1991; Howes et al. 2015; Carlsson et al. 2001), on the basis of the empirical discovery of antipsychotics and findings suggesting they produce their therapeutic effect by blocking the D2 dopamine receptors (Kapur and Seeman 2001). Following this, the hypothesis was that schizophrenia is a result of abnormal dopamine receptor density (Snyder 1976; Matthysse 1973), but so far there is little direct evidence to support a primary pathology in the dopamine system as the etiological factor in schizophrenia especially since D2 up-regulation could be due to antipsychotic treatment (Silvestri et al. 2000). Subsequently a refined dopamine hypothesis was developed in the 1990s suggesting a subcortical hyperdopaminergic activity being secondary to cortical hypodopaminergic activity, in particular in the frontal regions (Grace 2000; Davis et al. 1991). There is no change in the D2 density in extrastriatal subcortical brain regions (Kambeitz et al. 2014), the data concerning the D1 receptor density are inconclusive (Hirvonen et al. 2006; Kosaka et al. 2010; Abi-Dargham et al. 2012; Okubo et al. 1997a, b; Karlsson et al. 2002), and the data concerning the dopamine transporter density are also negative (Chen et al. 2013; Howes et al. 2012). A recent meta-analysis found a small increase in the density of the D2 and D3 receptors but reported also that heterogeneity between studies was large (Howes et al. 2012). Another meta-analysis reported no change in receptor densities in the thalamus, temporal cortex or substantia nigra (Kambeitz et al. 2014).

However, there seems to be a 14% increased striatal dopamine synthesis and a higher occupancy of D2 receptors by dopamine in patients with schizophrenia especially under stress (Howes et al. 2012; Abi-Dargham et al. 2009; Laruelle et al. 1999; Pogarell et al. 2012; Kegeles et al. 2010; Mizrahi et al. 2014; Catts et al. 2013; Fusar-Poli and Meyer-Lindenberg 2013). On the other hand, it seems that increasing the dopaminergic activity has some beneficial effect on negative symptoms (Laruelle et al. 1999), which is suggestive of regional rather than global changes in dopaminergic activity in the brain (Laruelle 2014).

It has been shown that there is no linear relationship between the D2 occupancy, clinical response and side effects. Response appears with occupancy above 50% and extrapyramidal side effects when it reaches 75% or more (Howes et al. 2009a; Howes and Kapur 2009; Nordstrom et al. 1993). However this has been challenged by the EUFEST study which suggested that for first-generation antipsychotics, extrapyramidal adverse events might appear at lower occupancy than the therapeutic effect (Boter et al. 2009).

It is possible that the alterations in the dopaminergic activity are state rather than trait features and are related with the acute phase and could be consequences of the treatment rather than core features in the etiopathogenesis of schizophrenia (Hietala et al. 1999; Howes and Kapur 2009; Howes et al. 2009b, 2013; Meyer-Lindenberg et al. 2002; Reith et al. 1994; Shotbolt et al. 2011).

A dysfunction in the activity of excitatory amino acids and their receptors has also been proposed. This model is based on animal studies and suggests that glutamate acting on the NMDA receptor in combination with dopaminergic activity on the D2 receptor in combination leads to a balanced intracellular signal at postsynaptic neurons in the striatum. In this way there is an optimal connection and communication between the basal ganglia and thalamus to cortex. In case either activity is dysfunctional and more specifically there is an increase in dopaminergic activity or a decrease in glutamatergic activity, psychosis could emerge (Carlsson and Carlsson 1990a; Carlsson and Carlsson 1990b). Several authors have hypothesized that in schizophrenia, there is a primary glutamatergic dysfunction in the prefrontal cortex, which results in reduced GABAergic activity in the ventral tegmental area (VTA) and eventually to a reduced inhibition of dopaminergic activity in the striatum (Garbutt and Vankammen 1983). Following this reduced tonic inhibition exerted by the thalamic mediodorsal nucleus, the dopaminergic activity increases which is the main cause for the emergence of positive symptoms of schizophrenia (Jones et al. 1988). According to this theory, the dopamine dysfunction seen in schizophrenia could be secondary to a primary glutamatergic dysfunction (McGuire et al. 2008; Stone et al. 2010).

The nature of this glutamatergic dysfunction is unclear but probably is due to a NMDA receptor hypofunction (Stone 2011). Such a deficit has been reported to exist in the left hippocampus in unmedicated patients (Pilowsky et al. 2006). However, the data concerning the existence and exact functionality of this dopamine-glutamate systems interaction are inconclusive, and the effect of glutamate and GABA on dopamine function seems to depend on the specific conditions (Reid et al. 1988; Ishita et al. 1988; Onteniente et al. 1987; Penit-Soria et al. 1987;

Chéramy et al. 1978; Garbutt and Vankammen 1983). Additionally, molecular imaging studies did not confirm the presence of differences in glutamatergic activity between patients with schizophrenia and controls (Kim et al. 2015). An alternative explanation is that NMDA dysfunction makes the dopamine system more sensitive to the effects of psychological stress and causes it to react in an abnormal way (Aalto et al. 2002; Kegeles et al. 2000, 2002, 2010).

Another important element for consideration is the observation that after overstimulation of either NMDA or non-NMDA receptors, excitotoxicity might appear because of the uncontrolled influx of CA2+ ions through the NMDA receptor channels which open (Choi et al. 1988; Rothman and Olney 1987). This could be the cause behind a variety of brain disorders of the neurodegenerative type (Olney 1989). This autoexcitotoxic mechanism could be responsible for the death of postsynaptic cells that house the glutaminergic receptors during early life, leading to schizophrenia (Olney and Farber 1995; Coyle and Puttfarcken 1993).

Metabolic imaging studies suggest the presence of lower metabolism in the frontal lobes, a condition known as 'hypofrontality' (Daniel et al. 1989; Ingvar and Franzen 1974; Liddle et al. 1992; Ragland et al. 1998; Spence et al. 2018; Volz et al. 1997; Weinberger et al. 1986; Yurgelun-Todd et al. 1996).

The overall research findings are split, and while some of them suggest the presence of a neurodevelopmental static encephalopathy (Munn 2000), others point to an ongoing neurodegenerative disorder maybe of the type of acceleration of brain ageing (Schnack et al. 2016). The neurodevelopmental hypothesis is supported by reports on the relationship between obstetric complications and the development of schizophrenia in the offspring (Dalman et al. 1999; Geddes and Lawrie 1995; Hultman et al. 1999; Jones et al. 1998) and also by the finding of ectopic neurons and abnormal cytoarchitecture in the PFC and the entorhinal cortex (Glausier and Lewis 2013; Catts et al. 2013; Eastwood 2004). On the other hand, neurodegenerative theories stress the neurotoxic effect of hyperdopaminergic activity (Lieberman et al. 1990) and the static type of existing encephalopathy, that is, the relative lack of progression in gross brain structural pathology (Csernansky and Bardgett 1998; Pearlson and Marsh 1999).

It is certain that in chronic patients, there is a widespread dysfunction in white matter connections (Kubicki et al. 2007) and significant reductions in neuropil (Eastwood 2004; Glausier and Lewis 2013; Konopaske et al. 2014; Selemon and Goldman-Rakic 1999) and synaptic proteins (Arnold 2006; Kleinman et al. 2011), resulting in fewer synaptic connections in specific brain regions (Glausier and Lewis 2013; Konopaske et al. 2014; Schmidt and Mirnics 2015). This impaired connectivity affects the communication between the default, the affective, the ventral attention, the thalamic and the somatosensory networks and all of them with the frontoparietal network (Dong et al. 2018).

On the other hand, there is much controversy over the presence or not of an enlargement of lateral ventricles (Andreasen et al. 1990; Johnstone et al. 1976; Raz and Raz 1990; Van Horn and McManus 1992; Chua and McKenna 1995; Weinberger et al. 1979) and a reduction in overall brain size (Andreasen et al. 1986; Harvey et al. 1993; Andreasen et al. 1990; DeMyer et al. 1988). Some authors suggest these are localized rather than global findings affecting specific brain areas like the head of the hippocampus (Csernansky et al. 1998) and only in a minority of patients with more severe and chronic form of the disorder.

As already mentioned, mental disorders are not characterized by abnormalities in gross brain morphology, and schizophrenia is not an exception. However, pathological studies suggest the presence of subtle changes in cellular morphology in the prefrontal cortex (PFC), the thalamus and the medial temporal lobe and more specifically a reduction in neuropil volume (Lewis et al. 1999a, b; Selemon and Goldman-Rakic 1999; Arnold 2006). Other studies reported an increase in neuronal density in the PFC without any change in the total number of neurons, which is in accord with neuropil volume reduction (Akbarian et al. 1995; Daviss and Lewis 1995; Pakkenberg 1993; Rajkowska 1997; Rajkowska et al. 1998; Selemon et al. 1998). Such findings however were not found in the hippocampus (Arnold 2006; Glausier and Lewis 2013; Harrison 2004). Some but not all studies report an increase in microglia in the PFC and in the white matter (Bernstein et al. 2015). The finding of ectopic neurons and indications of the presence of some kind of abnormal cytoarchitecture in the PFC and the entorhinal cortex (Glausier and Lewis 2013) provide support for the concept of neurodevelopmental disorder (Catts et al. 2013; Eastwood 2004).

The literature on postmortem studies of NMDA receptors is inconclusive (Humphries et al. 1996; Sokolov 2002; McCullumsmith et al. 2012).

From psychophysiological findings, it is important to note that in almost half of the patients with schizophrenia, there is an intrusion of saccadic movements in smooth pursuit, that is, while following a target across the visual field (Holzman et al. 1997). Successful treatment does not seem to correct this, and it seems also to constitute an endophenotype since it is found also in unaffected relatives of patients (Iacono and Clementz 1993).

In the field of genetics, while the literature is rich, so far it has failed to identify specific genetic loci in a reliable way, since findings are almost never replicated. GWAS are more promising and so far have provided some data supporting the association between the DRD2 (dopamine D2 gene), GRM3 (mGluR3 gene), NRG (neuregulin gene), DTNBP1 (dystrobrevin-binding protein 1 gene) and RGS 4 (regulator of G protein-signaling 4 gene), as well as other genes related to the glutamatergic system. These studies strongly support a polygenic contribution of a very large number of small allelic effects (Schizophrenia Working Group of the Psychiatric Genomics C 2014; Bigdeli et al. 2016; Collier et al. 2016; Devor et al. 2017; Wockner et al. 2014). Epigenetic processes seem also to play a significant role since there seems to exist a differential DNA methylation profile in leucocytes and in the brain in patients with schizophrenia (Kundakovic et al. 2016; Pidsley and Mill 2011).

13.2.2 Bipolar Disorder (Manic Depression)

The first biological theory of mood disorders was the monoamine deficiency hypothesis for depression (Schildkraut 1965; Maas 1975; Van Praag and Leijnse 1963), while later, the cholinergic-noradrenergic imbalance hypothesis included acetylcholine in a broader model for mood disorders (Davidson 1972; Tarsy et al. 1972; Janowsky et al. 1972). More complex models include state changes (depending on the polarity of the mood episode) in the excitatory amino acid function in specific areas of the cortex (Fountoulakis et al. 2008a). However, in spite of decades of extensive research, there is no definite proof for either a deficiency or an excess of either the quantity or the overall functioning of biogenic amines in specific brain structures in bipolar patients (Cannon et al. 2006a, b; Anand et al. 2011; Zavitsanou et al. 2004, 2005; Fountoulakis 2015a). The picture is especially problematic since the only class effect concerning the treatment of bipolar patients exists with antipsychotics exclusively against acute mania, while antidepressants do not seem to be efficacious against bipolar depression (Fountoulakis et al. 2011, 2012 Fountoulakis et al. 2017a, b, c, d; Rosa et al. 2011). A more complex profile for those agents with proven efficacy against bipolar depression seems necessary with an effect on norepinephrine reuptake and 5HT-1A agonism (Fountoulakis et al. 2012, 2015; Wiste et al. 2008; Young et al. 1994).

There are some data suggestive of a possible deficit in GABA and developmental/synaptic neurochemical systems in BD (Torrey et al. 2005; Benes et al. 2000, 2001) with a 60% decrease in NMDA receptor density (Woo et al. 2008b), but other studies failed to confirm any involvement of the GABAergic system (Cotter et al. 2002; Bielau et al. 2007; Bitanihirwe et al. 2010; Woo et al. 2008a). On the other hand, mania has been associated with reduced glutamate levels in the anterior cingulate cortex (ACC) (Moore et al. 2007). There are evidence supporting the decrease in the expression of specific NMDA subunits and associated proteins in the dorsolateral prefrontal cortex (Mueller and Meador-Woodruff 2004) and maybe in regions of the hippocampus (McCullumsmith et al. 2007; Beneyto et al. 2007; Toro and Deakin 2005), as well as in the perirhinal cortex (Beneyto et al. 2007), and the thalamus (Clinton and Meador-Woodruff 2004), but not in the orbitofrontal cortex of bipolar patients (Toro and Deakin 2005). The abnormal composition of NMDA receptors might lead to slower NMDA kinetics and eventually to disorganization (Fountoulakis 2012).

The presence of volumetric changes in the brains of BD patients is present as early as the prodromal phase but seems to increase with the duration of illness (Farrow et al. 2005; Hajek et al. 2005; Hirayasu et al. 1999; Kaur et al. 2005; Lyoo et al. 2006; Lagopoulos et al. 2012; Strakowski et al. 2002). The literature so far suggests that patients with BD manifest an enlargement of the third and lateral ventricles (Soares et al. 2005), white matter hyperintensities (Moore et al. 2001; Silverstone et al. 2003; Marlinge et al. 2014) and reduced grey matter in the prefrontal cortex, the hippocampus and the cerebellum (Moorhead et al. 2007; Soares et al. 2005; Blumberg et al. 2006) as well as a volume reduction in the left cingulate cortex (CC) (Bruno et al. 2006; Lyoo et al. 2006) or the right CC (McDonald et al. 2004). There is a large body of literature pointing to a specific vulnerability of the left anterior CC (ACC) and in particular its subgenual part (sgACC) (Atmaca et al. 2007; Lyoo et al. 2004; Sassi et al. 2004) or left posterior CC (PCC) (Hirayasu et al. 1999; Houenou et al. 2007; Wilke et al.

2004). However some studies report a right or bilateral ACC (Bruno et al. 2004; Cannon et al. 2006a) or a left or bilateral PCC volume decrease (Farrow et al. 2005; Kaur et al. 2005; Lim et al. 1999; Lochhead et al. 2004). While in schizophrenia a loss of brain volume is evident already at onset, in BD this happens latter. This is especially true concerning grey matter, while on the contrary, the loss of white mater volume might happen first and be present already at onset (Berk et al. 2010; Vita et al. 2009; Bora et al. 2010; Strakowski et al. 1993). Therefore, it is possible that white matter pathology is the prominent finding during the early stages (Lim et al. 2013; Lin et al. 2013), while grey matter loss follows years later (Arango et al. 2012; Strakowski et al. 2002). There are evidences that at least three variables might influence the volumetric changes in BD, the presence of cognitive decline (Bruno et al. 2006), response to treatment (Sassi et al. 2004; Cannon et al. 2006a; Atmaca et al. 2007; Bearden et al. 2007) and genetic background (McDonald et al. 2004), but overall the literature on neuroimaging data at different stages of BD is limited, and most of the data are cross-sectional (Lim et al. 2013; Balanza-Martinez et al. 2005) which make generalizable conclusions difficult.

Unfortunately the functional neuroimaging studies are usually restricted to the depressive phase of the illness simply because when patients are in an acute manic phase, it is very difficult to provide the level of collaboration needed to apply this kind of examination. Overall they suggest a state-dependent increased activity in the left dorsal ACC during acute mania (Blumberg et al. 2000) and a decreased activity in the same area during acute bipolar depression (Drevets et al. 1997). Activation studies provided with inconclusive results (Chang et al. 2004; Gruber et al. 2004; Rubinsztein et al. 2001; Benabarre et al. 2005; Altshuler et al. 2005; Blumberg et al. 2003; Roth et al. 2006; Pavuluri et al. 2007; Malhi et al. 2007).

A significant part of the total contribution on the neuropathology of BD is based on the material provided by the Stanley Neuropathology Consortium (Cotter et al. 2002; Knable 1999; Raedler et al. 1999; Torrey et al. 2005; Webster et al. 2005; Zavitsanou et al. 2004, 2005).

The results of these studies are inconclusive since some of them report an approximately 20–30% reduction in the volume in the ACC cortex as well as in the glial cells, with some layers the volume loss being above 60% (Ongur et al. 1998; Bouras et al. 2001; Chana et al. 2003; Torrey et al. 2005; Webster et al. 2005; Savitz et al. 2014), but other studies failed to confirm these findings (Miller et al. 2006; Benes et al. 2000, 2001; Cotter et al. 2001). A meta-analysis reported a 31% decreased density of non-pyramidal neurons bilaterally in layer II of the ACC in BD patients but reported no differences in glia numbers with 2D cell counting but significant glial reduction in layers III, V and VI when using 3D cell counting (Todtenkopf et al. 2005). Proteins related to the number and functioning of synapses such as synaptophysin, complexin II and growth-associated protein-43 (GAP-43) may be reduced in the ACC (Eastwood and Harrison 2001), suggesting a progressive destruction of excitatory rather than inhibitory synapses (Auer et al. 2000), and non-GABAergic cells may be selectively vulnerable to oxidative stress in patients with BD (Buttner et al. 2007).

The role of genetics in the development of BD is well known, and there is significant support in the literature from family, twin and adoption studies. It is also known that the mechanisms through which genetic factors play a role are complex, probably with several of them interacting with the environment, and thus BD is not a Mendelian disease (Andreassen et al. 2013), but instead, it manifests significant genetic heterogeneity. So far one of the greatest problems is that studies suffer from poor replicability and explain only around 70–80% of the genetic variance. Overall mood disorders have a reduced penetrance (less than 100%) which increases with age. Little is known about the role of epigenetics and imprinting in BD (Fountoulakis 2015a).

Interestingly there seems to be a significant genetic overlap between BD and schizophrenia (Fountoulakis et al. 2012; Cross-Disorder Group of the Psychiatric Genomics C et al. 2013; Andreassen et al. 2013).

Family studies indicate a morbid risk of BD in first-degree relatives of bipolar probands between 3% and 8% that is significantly higher in comparison with the general population. Twin studies often pool together unipolar and bipolar disorders and report 2–4 times higher risk for monozygotic twins in comparison with dizygotic. When BD is contrasted with unipolar depression, the genetic load appears to be higher for BD. Again, as in family studies, unipolar depression is the most common mood disorder in monozygotic cotwins of bipolar probands. Concerning adoption studies, only a few exist, and their results are inconclusive (Kelsoe 2009).

A meta-analysis reported that six pathways (corticotropin-releasing hormone signaling, cardiac beta-adrenergic signaling, phospholipase C signaling, glutamate receptor signaling, endothelin 1 signaling and cardiac hypertrophy signaling) and nine genes (CACNA1C, DTNA, FOXP1, GNG2, ITPR2, LSAMP, NPAS3, NCOA2 and NTRK3) were found to relate with BD (Nurnberger Jr. et al. 2014). Another review concluded that the expression profiles of BD-associated genes do not explain the majority of structural abnormalities observed in BD (McCarthy et al. 2014).

A GWAS study in 1461 BD patients reported that after genotyping for 372,193 single nucleotide polymorphisms (SNPs), the strongest results concerned myosin5B (MYO5B) and tetraspanin-8 (TSPAN8) and possibly the epidermal growth factor receptor (EGFR), but the results failed to replicate. Further analysis with the use of controls from the Wellcome Trust Case Control study reported that the results pointed to SNPs related with the voltage-dependent calcium channel, L-type, alpha 1C subunit (CACNA1C) gene (Sklar et al. 2008; Wellcome Trust Case Control 2007). Another GWAS confirmed the involvement of ion channel structural and regulatory genes, including voltage-gated ion channels and the broader ion channel group that comprises both voltage- and ligand-gated channels in the pathogenesis of BD (Askland et al. 2009).

In accord with the above are a number of studies concerning the genes related with NMDA receptors (Mundo et al. 2003; Itokawa et al. 2003; Avramopoulos et al. 2007; Martucci et al. 2006; Fountoulakis et al. 2012). These studies suggested that BD is associated with an abnormal structure of the NMDA receptor because of lower contribution of NR2A subunits, with predominance of more immature forms resulting in disordered receptor properties and slower kinetics. These slower kinetics might make NMDA receptors incapable of dealing with the increased speed of stimuli during manic episodes leading to disorganization (Fountoulakis et al. 2012).

13.2.3 Unipolar Depression

Overall the literature suggests that depressed patients manifest impairments in attention and concentration probably as a non-specific accompanying feature but the contribution of comorbid anxiety is unknown. Failure to activate the left anterior cingulate gyrus (which suppresses interference) accompanied by an overactivation of the dorsolateral prefrontal and the visual cortex probably constitutes the neurobiological substrate of this impairment (Abrams and Taylor 1987; Golinkoff and Sweeney 1989; Watts et al. 1990; Mialet et al. 1996; Austin et al. 1992; Cornblatt et al. 1989; George et al. 1997; Mayberg 1997; Mayberg et al. 1999). The deficits in memory functions are isolated on explicit but not implicit tasks and include short-term memory, recognition and verbal and visual memory, spatial working memory and immediate or delayed recall. Also depressed patients tend to recall more negative memories. A dysfunction in the temporal cortex and the hippocampus is probably behind memory problems (Burt et al. 1995; Cutting 1979; Massman et al. 1992; Richards and Ruff 1989; Heller et al. 1997; Herrmann et al. 2001; Austin et al. 1992; Beats et al. 1996; Sweeney et al. 2000; Shah et al. 1998; Bremner et al. 2000; Mervaala et al. 2000; von Gunten et al. 2000; Sheline et al. 1996). Concerning executive functioning, patients with depression suffer from deficits among others, in set shifting, working memory, inhibition and updating, and these deficits seem to persist even after remission at least in a subgroup of patients and are not related to medication status. A dysfunction in the dorsolateral and the medial prefrontal cortex, the orbitofrontal cortex and the anterior cingulate regions is probably behind these impairments (Channon and Green 1999; Friedman 1964; Silberman et al. 1983; Elliott et al. 1996, 1997; Degl'Innocenti et al. 1998; Trichard et al. 1995; Heller and Nitscke 1997). In general the severity of neurocognitive deficits seems to correlate with the severity of depression. Psychotic features may specifically cause further deterioration in a 'qualitative' way irrespective of the overall symptoms severity (Jeste et al. 1996; Nelson et al. 1998; Basso and Bornstein 1999; Lesser et al. 1991). Successful treatment improves neurocognitive function irrespective of medication status (Thompson 1991; Frith et al. 1983).

In the area of neurotransmitters, the role of monoamines and especially of serotonin in the successful treatment of depression has been solidly proven, but the initially conceived monoamine deficiency hypothesis concerning the etiopathogenesis of depression (Coppen 1967; Maes and Meltzer 1995; Bunney Jr. and Davis 1965; Schildkraut 1965; Delgado 2000; Heninger et al. 1996; Kapur and Mann 1992) should not be considered valid any more, since tryptophan and catecholamine depletion studies provided negative results (Abbott et al. 1992; Delgado et al. 1994; Delgado 2000; Price et al. 1997, 1998; Miller et al. 1996; Hirschfeld 2000; Belmaker and Agam 2008; Homan et al. 2015). The abnormal findings concerning the 5-HT_{1A}

and $5\text{-}HT_2$ receptors are probably due to the long-term treatment with antidepressants, while the initial findings concerning the serotonin transporters failed to replicate (Parsey et al. 1998; Meyer et al. 1999; Meltzer et al. 1999). The literature is inconsistent concerning noradrenergic receptors and overall the dopaminergic, cholinergic, GABAergic and glutamatergic systems, but they are much less researched in comparison to the serotonergic.

Stress could be the intermediate factor between depression and pathological findings, possibly through the development of hypercortisolaemia in combination with a dysfunction in neuroprotective systems, especially BDNF (Duman et al. 1997; Fuchs and Flugge 1998; Sapolsky 2000; Coplan et al. 2014; Lyons et al. 2001; Sheline et al. 2003). Stress might also dysregulate several neuroendocrine systems, with a specific effect on thyroid function (Haggerty Jr. et al. 1993; Staner et al. 1992; Legros et al. 1985; Rao et al. 1996). Especially early severe stress, e.g. maternal deprivation, might predispose for the development of adult depression (Rots et al. 1996; Smith et al. 1997; Suchecki et al. 1993; Zhang et al. 2002).

The brains of severely depressed and maybe refractory patients manifest local volumetric deficits and more specifically ventricular enlargement, cortical atrophy and sulcal widening especially in the prefrontal cortex, the amygdala and the hippocampus (Drevets et al. 1997; Siegle et al. 2012; Shah et al. 1998, 1999; Vakili et al. 2000; Mervaala et al. 2000; Sheline et al. 1996, 1998; McEwen 2000; Sheline et al. 2003; Fujita et al. 2000; Pearlson et al. 1997; Botteron and Figiel 1997; Duman and Charney 1999; Bremner et al. 2000). They seem to be reversible with successful treatment suggesting the loss of volume is mainly due to neuropil volume reduction rather than loss of neurons (Korte et al. 1998; van Winkel et al. 2014; Fujita et al. 2000).

Functional brain imaging in patients with depression during the resting state shows that there is an abnormally increased connectivity in the brain's default mode network and a decrease in the connectivity in the central executive network suggesting a dysregulation in the recruitment as well as the deactivation of regions and circuits responsible for the control of emotions. This includes a reduction in the activity of the DLPFC, in combination with an increase in the activity in the VMPFC as well as in the amygdala (Gudayol-Ferre et al. 2015; Posner et al. 2016; Watters et al. 2018; Biver et al. 1994; Baxter Jr. et al. 1989; Galynker et al. 1998; Rive et al. 2013; Drevets et al. 1992; Greicius et al. 2007; Mayberg et al. 2000, 2005; Motzkin et al. 2015; McGrath et al. 2013; Zhang et al. 2016; Brody et al. 2001).

Pathological studies report the presence of structural disturbances in the parahippocampal cortical regions with malformations in the entorhinal lamination, reduction in neuronal size and alterations in the prefrontal regions and the temporal lobe (Duman et al. 1997; Fuchs and Flugge 1998; Rajkowska et al. 1999, 2001; Sapolsky 2000; Bernstein et al. 1998; Altshuler et al. 1990; Goodwin 1997; Rajkowska 1997).

A number of psychophysiological abnormalities pointing to ANS dysfunction have been reported, but the most important findings concern sleep disturbances and more specifically the premature loss of slow-wave deep sleep which is a consequence of reduced REM latency (<65 min) and reduced SWS, particularly during the first sleep cycle, the increase in nocturnal arousal and awakenings (sleep fragmentation), the reduction in total sleep time and the increased phasic REM sleep. Increased REM density seems to be specific for depression (Benca et al. 1992; Brunello et al. 2000; Lauer et al. 1991; Reynolds 3rd and Kupfer 1987; Riemann et al. 2001).

The genetics of unipolar depression are extremely complex and based on a mixture of a great number of genes of both large and small effect, which are transmitted in a variety of ways (Ament et al. 2015). The concordance rate of mood disorders for MZ twins is around 70–80% suggesting the role of epigenetic and maybe environment-gene interactions. Overall genetic data collectively suggest that genes explain approximately 30% of the aetiology of major depression (Wray and Gottesman 2012). In general it seems that early-onset depressions are more heritable. Unfortunately although a number of genes have been proposed as candidate genetic markers, including the SERT gene (SLC6A4), the results are inconclusive and rather negative so far (Culverhouse et al. 2018). GWAS studies seem promising but more studies of this kind are needed.

13.2.4 Alzheimer's Disease (AD)

AD is a neurodegenerative disorder, with a complex aetiology, especially concerning sporadic cases. Risk factors include ageing, gender and genetic and environmental factors, but the overall prediction power is relatively low. Neuropathological and imaging data point to the hippocampus as the early location of anatomical change (Jack Jr. et al. 1999).

Imaging studies have revealed marked atrophy (30–40% in comparison with controls) in the hippocampus (Kesslak et al. 1991), the parahippocampal gyrus (Ikeda et al. 1994), the amygdala (Lehericy et al. 1994), the subiculum and the perihippocampal clefts (De Leon et al. 1992). This loss of brain volume is reported to correlate with the severity of dementia (Fox et al. 1999). Functional imaging data suggest a pattern of hypometabolism which is most prominent in the temporoparietal and frontal cortical association areas while the primary visual and sensory motor cortices are typically spared (Haxby et al. 1986; McGeer et al. 1986a, b, c; Hirsch et al. 1997; Bartenstein et al. 1997; Weber et al. 1997; Holman et al. 1992) and suggested that hypoperfusion in the temporoparietal areas correlated with the global severity of the disease (Imran et al. 1999; Robert et al. 1992; Ashford et al. 2000).

At the pathological level, the disease is characterized by significant neuronal loss which is more striking with a reduction of up to 40% in the size of the hippocampus, amygdala, thalamus and anterior temporal lobe (Pantel et al. 1997). It is also characterized by the accumulation of amyloid-β (Aβ) protein in the brain. This protein can be found both in the brain parenchyma and on the walls of the leptomeningeal and parenchymal vessels. The core protein of amyloid plaques, Aβ, is a 4-kDa peptide, derived from the much larger beta-precursor protein (βPP). It is coded in chromosome 21, exists in several isoforms, is synthesized as an integral membrane molecule (Kang et al. 1987; Ashall and Goate 1994) and is secreted as a product of proteolytic cleavage of βPP (Selkoe 1994). An additional pathological characteristic is the presence of neurofibrillary changes. Amyloid accumulation and neurofibrillary changes could be diffuse or could form neuritic plaques and neurofibrillary tangles, which are both considered to be hallmark lesions of AD in which the post-mortem diagnosis is based on them (Fox et al. 1985; Khachaturian 1985; Mirra et al. 1991; Braak and Braak 1991). Neuritic plaques have a complex structure and include abnormal neurites with paired helical filaments and activated glial cells (Wisniewski et al. 1989, 1990; Wisniewski and Wegiel 1995). The neurofibrillary tangles consist of abnormal accumulations of abnormally phosphorylated tau within the perikaryal cytoplasm of certain neurons (Perl 2010).

The mechanisms behind the development of the typical AD neuropathology are not well understood. Overall the data suggest that a low-grade chronic inflammatory process is active in AD, with activated microglia being the main factor. Centrally located Aβ aggregates are surrounded by astrocytes. However, microglia and astrocytes are associated with neuritic plaques but are lacking from most diffuse plaques. The fibril formation is influenced by Aβ peptide concentration, pH, length of Aβ peptide, as well as by interaction with other proteins (Selkoe 1994; Kisilevsky and Fraser 1997). The processes involved include neurotoxicity of fibrillar Aβ, oxidative injury and immune activation. Thus, it seems that activated glial cells produce substances contributing to the production and the fibrillization of Aβ as well as to neuronal injury (Akiyama et al. 1991; Eikelenboom et al. 1998; McGeer et al. 1994; Overmyer et al. 1999). A number of interleukins are present in neuritic plaques, including interleukin 1 alpha (IL-1α), interleukin 1 beta (IL-1β), interleukin 6 (IL-6) and tumour necrosis factor (TNF) (Eikelenboom et al. 1994; McGeer and McGeer 1999) which all have been shown promote an altered processing of βPP leading to the production of Aβ (Griffin et al. 2006; Sheng et al. 1996; Wood et al. 1993; Del Bo et al. 1995; Forloni et al. 1992; Ohyagi and Tabira 1993; Vasilakos et al. 1994), which Aβ in turn, in a vicious cycle, stimulates glial cells to produce cytokines and basic fibroblast growth factor (BFGF) (Araujo and Cotman 1992; Meda et al. 1995). A number of studies implicated IL-1 in memory deficits and reduced long-term potentiation in the ageing rat hippocampus (Lynch 1998) and in hippocampal cell damage (Araujo 1992). Activated microglia secretes a number of additional neurotoxic factors, including eicosanoids, free radicals, nitric oxide, proteases, protease inhibitors and excitotoxins (Eikelenboom et al. 1994; Espey et al. 1997; Giulian et al. 1995; McGeer et al. 1989).

Since ApoE ε4 allele is considered an important risk factor for the development of AD, its possible relationship with Aβ aggregation has been studied (Corder et al. 1993; Saunders et al. 1993). ApoE is involved in the transport and cellular uptake of lipid complexes via the low-density lipoprotein receptor (LDL-R) and the low-density lipoprotein receptor-related protein (LRP) receptor (Mahley 1988; Rebeck et al. 1995). It binds to Aβ and is co-localized with it in plaques (Strittmatter et al. 1993b; Namba et al. 1991), an observation which leads to the suggestion that there is a possible LRP-mediated uptake of ApoE/Aβ complexes and the risk for the development of AD associated with the ApoE genotype could be related to the clearance of the Aβ from the neuropil (Rebeck et al. 1995),.

The earliest and most consistent neurochemical abnormalities to appear in the brains of AD concern the cholinergic system (Mesulam et al. 1983). The decline in cholinergic transmission is not uniform throughout the brain. There is a significant reduction in the basal forebrain cholinergic system (30–90% loss of neurons), but typically there is no loss in other cholinergic-rich areas, particularly the brainstem (Mesulam 1996). This loss of neurons results in a dramatic reduction of choline acetyltransferase (ChAT) in the temporal lobes with a less marked reduction in frontal lobes. Biopsy studies suggest that the decline in cortical ChAT activity occurs within the first year after the onset of clinical signs (Davis et al. 1999; Bowen et al. 1983). Postmortem and PET studies reported the presence of an accompanying reduction in the density of presynaptic nicotinic receptors in the cortex and parahippocampal gyrus (Perry et al. 1995; Nordberg 1996). It is important to note that muscarinic activity remains relatively intact.

The basis for the selective vulnerability of basal forebrain cholinergic neurons, as opposed to other populations of cholinergic neurons, remains unclear. The data so far point to the depletion of neurons in the basal nucleus of Meynert as the core issue concerning the declining cholinergic activity (Ezrin-Waters and Resch 1986; Ferreira-Vieira et al. 2016).

There is the hypothesis that cholinergic degeneration is, in part, the result of impaired nerve growth factor (NGF) signaling and transport to the basal forebrain cholinergic neurons (Cuello et al. 2010; Tuszynski et al. 2015). Other studies proposed that there is a decrease in the number of neurons expressing tropomyosin receptor kinase A (TrkA) in the nucleus of Meynert, again as a result of cholinergic neuronal loss, and since TrkA interacts with Aβ, the decrease leads to apoptosis and the formation of amyloid plaques (Boissiere et al. 1997; Zheng et al. 2015). Another finding is that acetylcholinesterase AChE activity is reduced in AD (Herholz et al. 2000; Kuhl et al. 1999; Geula and Mesulam 1996) and that this reduction correlates well with reductions in regional cerebral blood flow or glucose metabolism, particularly in temporolateral regions of the cortex (Herholz et al. 2000). It is not clear whether this reduction is part of a vicious cycle of Aβ and P-tau dysregulation (Garcia-Ayllon et al. 2011).

The progressive loss of noradrenergic neurons in the locus coeruleus has been documented using postmortem as well as biopsy tissue samples (Mann 1998), but there is little relationship between the extent of loss of noradrenaline in the locus coeruleus and the cognitive deficits seen in AD (Palmer et al. 1987a, b; Palmer and DeKosky 1993). Interestingly, the loss of noradrenergic innervation greatly exacerbates AD pathogenesis and progression (Gannon et al. 2015). There are some data suggesting that the norepinephrine system mediates the effect of chronic stress in the pathogenesis of dementia (Ross et al. 2015).

There are several reports indicating widespread serotonergic dysfunctions early in AD, although the overall picture is less consistent (Bowen et al. 1983; Palmer et al. 1987a). A significant reduction equal to 30–40% in the number of 5-HT-positive neurons in the median and dorsal raphe nuclei is reported (Chen et al. 2000; Aletrino et al. 1992; Halliday et al. 1992; Kovacs et al. 2003; Yamamoto and Hirano 1985; Zweig et al. 1988). There is little evidence that changes in serotonergic transmission contribute significantly to the neurocognitive deficits seen in AD and probably play a role in the development of affective and behavioural symptomatology that accompany AD (Court and Perry 1991). Postmortem analyses reveal significant reductions in 5-HT levels in the frontal (Arai et al. 1984; D'Amato et al. 1987; Palmer et al. 1987a) and temporal cortex (Palmer et al. 1987c), the hippocampus (Cross et al. 1984), the hypothalamus (Sparks et al. 1988), the basal nucleus of Meynert (Sparks et al. 1992) and the basal ganglia (Sparks et al. 1988).

Changes in the dopaminergic systems in AD are more variable from those seen in the cholinergic and noradrenergic systems, and whether there is a role for dopamine in the etiopathogenesis of AD remains unclear (Martorana and Koch 2014; Attems et al. 2007; Portet et al. 2009; Trillo et al. 2013).

It is well established that GABA neurons are spared in AD (Nagga et al. 1999; Howell et al. 2000), but it seems that the α1 subunit-containing GABA-A receptors appear to be vulnerable as the disease progresses (Mizukami et al. 1998a, b). The roles of these variable responses of selected GABA subunits in the process of the disease are not clear (Li et al. 2016), but it is possible they are implicated in the development of behavioural symptoms in AD patients (Solas et al. 2015).

Glutamatergic neurons are probably those most vulnerable to degeneration in AD. Loss of such neurons is observed in the entorhinal cortex and the hippocampus (Braak and Braak 1991; Hyman et al. 1984; Pearson et al. 1985; Greenamyre et al. 1985, 1987), with a subsequent decrease in glutamic acid levels and loss of glutamate terminals in these regions (Cowburn et al. 1990; Hyman et al. 1987; Procter et al. 1989), whereas a minority of studies report stable levels (Geddes et al. 1986; Monaghan et al. 1987). Thus, it is believed that the specific loss of glutamatergic neurons plays a significant role in the pathological manifestations of AD (Greenamyre and Young 1989; Myhrer et al. 2003; Myhrer 1993; Palmer and Gershon 1990; Danysz and Parsons 2012); however it is well known that glutamate plays a role in neuronal death in general, due to excitotoxicity, through a cascade involving increased calcium entry into the cell (Blass and Gibson 1991; Gooch and Stennett 1996; Kornhuber and Weller 1997). Therefore, such a role might not be exclusive to AD (Choi 1988).

The reduction of NMDA receptors could be the cause behind the memory decline in AD, since glutamatergic hypoactivity in AD patients postmortem correlated well with the severity of cognitive deficits (Greenamyre and Maragos 1993; Sumpter et al. 1986). Additionally, βPP plays an important role in regulating glutamate levels in the synapse, and Aβ protein can enhance the neurotoxicity of glutamate (Mattson et al. 1999; Blanchard et al. 1997; Fernandez-Tome et al. 2004).

There is a number of genetic markers, with some data in support of their involvement in the development of AD. They include fully penetrant mutations in βPP (Mann 1988; Ponte et al. 1988; Tanzi et al. 1987; Wisniewski et al. 1985), Presenilin 1 (Doan et al. 1996; Lehmann et al. 1997; Li and Greenwald 1998), Presenilin 2 (Citron et al. 1997; Levy-Lahad et al. 1995a, b) and the varepsilon4 allele of Apolipoprotein E (Namba et al. 1991; Saunders et al. 1993; Strittmatter et al. 1993a). There are more than 20 additional genetic risk loci identified in genome-wide association studies and massive parallel resequencing efforts (Van Cauwenberghe et al. 2016).

13.2.5 Contemporary Neurobiological Understanding of Substance Abuse

The positive reinforcing properties of substances in nondependent subjects are well known (Di Chiara and Imperato 1988; Fibiger 1978; Robinson and Berridge 1993; Wise 1978, 1988; Wise and Bozarth 1987), and it is also known that the compulsive craving of substances comes from changes in the reward circuitry (Dackis and Gold 1985; Frank et al. 1992; Kokkinidis and McCarter 1990; Kokkinidis et al. 1980; Koob and Bloom 1988; Leith and Barrett 1976; Solomon and Corbit 1973), which concern probably a desensitization of the mechanisms of reward and a potentiation of the rewarding effects of lateral hypothalamic and related brain stimulation (Frank et al. 1992, 1988; Kokkinidis and McCarter 1990; Kokkinidis et al. 1980; Leith and Barrett 1976; Schulteis et al. 1995; Wise 1996; Wise and Munn 1995).

Sensitization refers to the phenomenon that after repeated abuse of an addictive substance, the brain function is biased in the opposite direction from that of the acute substance state (Koob et al. 1989; Solomon and Corbit 1973). This is especially well documented for psychostimulants. Such phenomena are the reward-specific effects of alcohol and nicotine (George et al. 1998; Nurmi et al. 1996), amphetamine (Lett 1989; Lorrain et al. 2000; Piazza et al. 1990), cocaine (Horger et al. 1990; Lett 1989; Shippenberg and Heidbreder 1995) and morphine (Lett 1989; Shippenberg and Heidbreder 1995).

While tolerance usually lasts only a few days and gradually disappears after cessation of substance administration, sensitization is very long lasting. It is for these reasons that this specific between-session sensitization constitutes the strongest and longer-lasting positive reinforcement feature of addiction (Robinson and Berridge 1993). A phenomenon of cross-sensitization among different substances also exists (Horger et al. 1992; Itzhak and Martin 1999)

Trigger zones are areas of specific importance for each substance to act and produce its abusive potential. The nucleus accumbens (NAc) is the primary trigger zone for amphetamine (Carr and White 1983; Hoebel et al. 1983) and one of the two sites for nicotine (Pontieri et al. 1996; Reid et al. 2000; Toth et al. 1993; Vidal 1994; Shoaib et al. 1997), cocaine (de Wit and Wise 1977; Risner and Jones 1980), phencyclidine and opiates (Goeders et al. 1984; Olds 1982; Ettenberg et al. 1982; Pettit et al. 1984). Amphetamine and cocaine act via their effect on the dopamine transporter (Pierce and Kalivas 1997; Reith et al. 1986; Heikkila et al. 1975; Hurd et al. 1989; Pettit and Justice 1989; Wise et al. 1995), while phencyclidine (PCP) and morphine act on glutamate or opioid receptors (Chaudieu et al. 1989; Ohmori et al. 1992). Tetrahydrocannabinol (THC) which is the major psychoactive component of marijuana and hashish elevates NAc dopamine levels (Chen et al. 1990; 1993; Tanda 1997; Ton et al. 1988) and so does ethanol (Weiss et al. 1993; Gonzales and Weiss 1998; Imperato and Di Chiara 1986).

A second trigger zone for opiates is the ventral tegmental area (VTA) (Bozarth and Wise 1981; Welzl et al. 1989) where μ-opioid receptors increase the mesocorticolimbic dopamine output by inhibiting GABAergic activity (Dilts and Kalivas 1989; Gysling and Wang 1983; Westfall et al. 1989; Clarke and Pert 1985; Nisell et al. 1994b; Johnson and North 1992; Johnson and Pillai 1990). THC has also actions in VTA (Chen et al. 1990; French 1997; Melis et al. 2000; Tanda 1997) and the same holds true for alcohol (Gessa et al. 1985; Brodie et al. 1990; Gatto et al. 1994; Rodd-Henricks et al. 2000). Alcohol modulates GABA-gated chloride and glutamate-gated sodium and calcium conductances (Fadda and Rossetti 1998; Lovinger 1997) leading to the inhibition of GABAergic interneurons activity and subsequently to an increase in the dopamine cell firing and dopamine release within the VTA (Bailey et al. 1998; Brodie et al. 1999; Yim et al. 1998). Phencyclidine has an action similar to alcohol (French and Ceci 1990). Nicotine acts in the VTA directly to acetylcholine nicotinic receptors on dopamine cells to increase their firing frequency (Mereu et al. 1987; Clarke and Pert 1985; Benwell et al. 1993; Grenhoff et al. 1986; Nisell et al. 1994a).

An additional trigger zone for cocaine and phencyclidine is the medial prefrontal cortex (PFC), where addictive substances exert their effect through the NMDA receptor and probably on dopamine through the blockade of the noradrenaline transporter (Carlezon Jr. et al. 1996; Goeders et al. 1986; Goeders and Smith 1986). It has been shown that drug craving is associated with metabolic activation of the anterior cingulate gyrus and the amygdala (Childress et al. 1999; Grant et al. 1996; Kilts et al. 2001). This probably suggests the presence of a transition from dopamine- to glutamate-dependent behaviours with the amygdala mediating the communication between the PFC and the NAc during the conditioning of behaviour (Brown and Fibiger 1993; Everitt et al. 1991; Grimm and See 2000; Berke and Hyman 2000; Robinson and Becker 1986; Wise and Rompre 1989).

Dopamine release and cell firing in the trigger zones are increased by the presentation of novel and motivationally relevant common environmental stimuli (Schultz 1997; Berridge and Robinson 1998), leading to the recruitment of cortically derived memories, cognitive strategies and motor output. In this way, dopamine mediates the initiating and establishing of neuroplastic changes coupled with developing behavioural strategies and habit formation, which are all necessary to adapt to novel stimuli. One characteristic of the activation of dopamine transmission by common environmental stimuli is that dopamine release diminishes with repeated exposure to the same stimulus as the organism establishes an adaptive behavioural response (Cabib and Puglisi-Allegra 1996; Deutch et al. 1985). According to this model, while dopamine contributes and is probably necessary for the establishment of neuroplastic changes that mediate behavioural adaptation and habit formation, it might not have a role in the expression of those behaviours after the habit has been established. After this, it seems that the stimulus elicits the behaviour via interactions among the limbic cortex, thalamus and basal ganglia, with less or not at all involvement of the regional dopaminergic neurotransmission (Graybiel 1998; Jog et al. 1999). It is interesting to note that studies in animal models have confirmed the role of increasing dopamine release in the trigger zones after natural rewarding or aversive stimuli (Berridge and Robinson 1998; Cabib and Puglisi-Allegra 1996; Doherty and Gratton 1997; Kalivas and Duffy 1995; Mermelstein and Becker 1995; Mitchell and Gratton 1991; Taber and Fibiger 1997). The important observation from these studies is that common environmental stimuli cause a dopamine release which is of substantially less magnitude and duration than the pharmacologic release elicited by most substances of abuse (supernatural stimuli) (Kalivas et al. 1998; Kuczenski and Segal 1999; Marshall et al. 2002; Tanda 1997; Yim et al. 1998).

Relapse often occurs on re-exposure to substances of abuse but also to cues associated with substance administration. This points to the important role for associative learning (O'Brien et al. 1992) which represents enduring or permanent alterations in patterns of synaptic connectivity and function (Berke and Hyman 2000; Quinn and Harden 2013). It seems that associative learning mechanisms interact with neurobiological mechanisms, and therefore the expression of tolerance may be context-dependent (Cepeda-Benito et al. 1999), while the threshold of learning could be modified by homeostatic responses.

Clinical data suggest that in addicted humans, late relapses appear to involve associative learning, as they often occur after encounters with cues previously associated with substance use (Shiffman 1996). These conditioned responses persist far longer than withdrawal symptoms (O'Brien et al. 1992; Berke and Hyman 2000). Thus, the persistence of substance addiction reflects the persistence of the memory for this learned experience, involving a variety of brain areas and neurochemical mechanisms (Hyman 1996; Bourtchuladze et al. 1994; Frey et al. 1996; Nguyen et al. 1994; Nguyen et al. 1992; Castellanos-Ryan et al. 2011) including genetic ones (Berke and Hyman 2000; Cole et al. 1992; Fosnaugh et al. 2002; Brakeman et al. 1997; Cole et al. 1989; Lyford et al. 1995).

There are strong data from twin and adoption studies suggesting that genetic factors explain a significant proportion of the variability of substance use disorders (Bierut et al. 1998; Bolos et al. 1990; Cadoret et al. 1986, 1995, 1996; Kendler et al. 2000; 2007; Kendler and Prescott 1998a, b; Lin et al. 1996; Merikangas et al. 1998; Sabol et al. 1999; Uhl et al. 2001; Haile et al. 2007). They play a stronger role for substance dependence and problematic use rather than substance use per se and probably with a significant interaction with malleable environmental risk factors (Boomsma et al. 1994; Cutrona et al. 1994; Grove et al. 1990; McGue et al. 2000; True et al. 1997; Demers et al. 2014; Volkow and Muenke 2012).

13.3 Psychopharmacology

13.3.1 Lithium

Lithium is a rather rare chemical element with atomic number 3 and its symbol is 'Li'. It belongs to the alkali metal group; it is the lightest metal and the least dense solid element. Two stable lithium isotopes can be found in nature. It is soft, silver-white and highly reactive and inflammable.

Although trace amounts exist in all organisms, there are no known physiological functions for lithium and live organisms can survive without it. In spite of this, lithium has been used as medication already since the late nineteenth century to cure a variety of diseases but especially uric arthritis. It was also widely used in beverages in the late nineteenth and early twentieth century. However it was proven to be toxic if taken in high dosages, and its use was abandoned in the early twentieth century (Marmol 2008; Shorter 2009; Johnson and Amdisen 1983; Strobusch and Jefferson 1980).

After the WWII, in 1949 the Australian John Cade (1912–1980) reported positive results from the treatment of ten acutely manic patients (Cade 1949, 2000); however 2 years later, he reported the first death because of lithium toxicity in a patient whose bipolar illness otherwise responded extremely well to treatment. During the 1950s several researchers studied lithium and its usefulness in BD (Noack and Trautner 1951). However the important contribution that made the difference came from Mogens Schou (1918–2005) who randomized acutely manic patients to lithium or placebo, and in 1954 he published the results which made a significant impact to undertake a randomized controlled trial of lithium in acute mania (Bech 2006; Schou et al. 1954). Poul Christian Baastrup (1918–2002) demonstrated in 1964 the efficacy of lithium for the maintenance phase of bipolar disorder (Baastrup 1964). In the US, in 1960 Samuel Gershon published the first North American paper on lithium (Gershon and Yuwiler 1960). Further studies established lithium and robustly linked it to the treatment of all phases of (Schou et al. 1970; Angst et al. 1969, 1970; Baastrup et al. 1970; Baastrup and Schou 1967; Bech 2006; Schioldann 1999, 2006, 2011; Johnstone et al. 1988; Mitchell and Hadzi-Pavlovic 2000). Latter Fred Goodwin suggested it could be also useful in the treatment of depression as add-on to antidepressants (Goodwin et al. 1969a, b, 1972, 2003; Goodwin 2002; Goodwin and Zis 1979). The recommended serum lithium levels were determined with certainty in 1976 (Bech et al. 1976).

The specific biochemical mechanism of lithium action in bipolar disorder is unknown. Treatment with lithium demands regular serum level tests and monitoring of thyroid and kidney function. Dehydration can result in increasing lithium levels. Serum lithium concentrations are recommended to be in the 0.4–1.2 mmol/l range (lower end of the range for maintenance therapy and the elderly, higher end for children) on samples taken 12 h after the preceding dose (Amdisen 1977; Chen et al. 2004; Solomon et al. 1996; Perlis et al. 2002).

The adverse effects of lithium include leukocytosis, polyuria and polydipsia, dry mouth, hand tremor, headache, neurocognitive problems, confusion, muscle weakness, ECG changes, nausea, vomiting, diarrhoea or constipation, muscle twitch, vertigo, EPS, euthyroid goitre, hypothyroidism, acne, hair loss and hair thinning, renal toxicity and renal interstitial fibrosis, seizures, coma, hallucinations, erythema multiforme, Brugada syndrome, sinus node dysfunction, pseudotumour cerebri, increased intracranial pressure and papilledema and weight gain or loss. Lithium is also a teratogen, causing birth defects in a small number of newborn babies, including Ebstein's anomaly (Shepard et al. 2002). Most adverse effects are dose-dependent.

Lithium toxicity manifestations include nausea, vomiting, diarrhoea, asthenia, ataxia, confusion, lethargy, polyuria, seizures and coma, coarse tremor, muscle twitching, convulsions and renal failure. Several authors have described 'syndrome of irreversible lithium-effectuated neurotoxicity' (SILENT), associated with episodes of acute lithium toxicity or long-term treatment within the appropriate dosage range. Symptoms are said to include cerebellar dysfunction (Ikeda et al. 2010; Porto et al. 2009; Adityanjee et al. 2005; Adityanjee 1989, 1987). Unfortunately, in long-term use, toxic effects might be induced even at therapeutic plasma levels (Fountoulakis et al. 2008b).

13.3.2 Antipsychotics

Antipsychotics or neuroleptics (also called previously major tranquilizers) were developed initially for the treatment of schizophrenia and psychotic symptoms, but today their usefulness has been proved in the treatment among others of bipolar disorder and refractory unipolar depression (Fountoulakis 2015b).

Chlorpromazine was the first to be discovered in 1952 and initially was developed as an anaesthetic agent for general surgical use. The French Henri Laborit (1914–1995) reported that chlorpromazine was inducing indifference towards traumatic events in otherwise mentally healthy persons. Jean Delay (1907–1987) and Pierre Deniker (1917–1998) were the first to use it as monotherapy in agitated psychosis (Delay and Deniker 1955, 1956; Delay et al. 1952, 1956).

Antipsychotics are grouped into the first-generation antipsychotics (FGAs), also called typical antipsychotics, and the second-generation agents (SGAs), also called atypical antipsychotics. The common pharmacodynamics property of all antipsychotic agents is dopamine D2 receptors blockade, and it is believed that they act their therapeutic effect through this pathway. Most antipsychotics also affect a number of other neurotransmitters.

The most frequent adverse events for FGAs are extrapyramidal symptoms (EPS) and hyperprolactinaemia, while weight gain and metabolic abnormalities are caused mainly by SGAs. Other adverse effects include sedation, headaches, dizziness, diarrhoea, sexual dysfunction, osteoporosis, orthostatic hypotension, anticholinergic side effects, memory problems, angle-closure glaucoma, blurred vision, constipation, dry mouth or hypersalivation, agranulocytosis, leukopenia and neutropenia and QT prolongation. Tardive dyskinesia and neuroleptic malignant syndrome are the most severe adverse events.

13.3.3 Antidepressants

Antidepressants are agents used for the treatment of depression but also of anxiety disorders, obsessive compulsive disorder, eating disorders, chronic pain, neuropathic pain and other neurological and psychiatric conditions.

In 1951, research on the two new antituberculosis agents, isoniazid and iproniazid, suggested the two agents also possessed some psychotropic properties (Selikoff and Robitzek 1952; Robitzek et al. 1952; Selifoff et al. 1952). Following these reports, in 1952 Max Lurie (born 1920) and Harry Salzer (born 1906) reported that isoniazid improved depression in two thirds of their patients. They also introduced the term antidepressant (Salzer and Lurie 1953). A year before, in France, Jean Delay with the resident Jean-Francois Buisson reported the positive effect of isoniazid on depressed patients, but they published these results years later (Delay and Buisson 1958). Nathan Kline supported the use of iproniazid as an antidepressant, but eventually in 1961, it was withdrawn from the market because of lethal hepatotoxicity (Lopez-Munoz et al. 2007).

In 1957 the Swiss psychiatrist Roland Kuhn (1912–2005) discovered the first tricyclic antidepressant in the process of improvement of the efficacy of chlorpromazine in conjunction with the Geigy Pharmaceutical Company. He also coined the term 'thymoleptic' (Kuhn 1958, 1957). In 1988, fluoxetine, the first SSRI, was introduced. It was developed at Eli Lilly and Company in the early 1970s by Bryan Molloy, Klaus Schmiegel, David Wong and others. In spite of a long-lasting recent debate, the efficacy of antidepressants in the treatment of unipolar depression is no longer a matter of dispute (Fountoulakis et al. 2013; Fountoulakis and Moller 2012; Sartorius et al. 2007).

Currently there are several classes of antidepressants including the selective serotonin reuptake inhibitors (SSRIs), the serotonin-norepinephrine reuptake inhibitors (SNRIs), the tricyclic antidepressants (TCAs) and the monoamine oxidase inhibitors (MAOIs).

The main neurotransmitter pathway through which antidepressants seem to exert their beneficial effect is that of serotonin. Pure noradrenergic action is unlikely to be sufficient to produce an antidepressant effect; however double-acting agents (which affect both serotonin and noradrenaline pathways) might be more efficacious in comparison with purely serotonergic agents but also with more adverse effects.

The most common adverse effects include nausea, increased appetite and weight gain, loss of sexual desire and other sexual problems (e.g. erectile dysfunction and decreased ability to achieve orgasm), fatigue and drowsiness, insomnia, dry mouth, blurred vision, constipation, dizziness, agitation, irritability, anxiety, sexual problems and hyperprolactinaemia. Serotonergic syndrome is a potentially lethal event. Treatment with antidepressants also might induce suicidal thoughts, but no completed suicide has been attributed to treatment with antidepressants. Some agents after abrupt stop of treatment might cause withdrawal symptoms which persist for no more than 1–2 weeks.

Although the teratogenic risk is low with antidepressants, SSRI use in pregnancy has been associated with an increased risk of spontaneous abortion, preterm birth and low birth weight (Malm 2012; Rahimi et al. 2006).

The usefulness of antipsychotics in the treatment of bipolar depression is a matter of continuous debate (Pacchiarotti et al. 2013). It is interesting that some data suggest that norepinephrine activity is necessary for an antidepressant to act in bipolar depression; still this very activity increases the risk for the patients to switch to mania or hypomania (Fountoulakis et al. 2012).

13.3.4 Valproate

Sodium valproate is the sodium salt of valproic acid. It is an anticonvulsant efficacious in the treatment of epilepsy (all partial and generalized seizures including absence seizures) as well as in the prevention of migraine headaches. It was first synthesized in 1882 by B.S. Burton as an analogue of valeric acid which can be found naturally in valerian and was used in the cosmetics industry. Valeric acid appears similar in structure to GABA but lacks the alcohol and amine functional groups that contribute to the biological activities of the GABA. In 1962 Pierre Eymard accidentally discovered its anticonvulsant properties (Meunier et al. 1963). Valproic acid was approved as an antiepileptic for the first time in 1967 in France. Later it has been proven to be efficacious in the treatment during all phases of bipolar disorder (Fountoulakis et al. 2017a, c)

Its mechanism of action includes weak blocking of sodium ion channels and weak inhibition of enzymes that deactivate GABA (e.g. GABA transaminase). It is unclear whether it also stimulates GABA synthesis.

Adverse effects include tiredness, tremor, nausea, vomiting, sedation and gastrointestinal symptoms as well as reversible hair loss in about 10% of patients. Also some patients experience vision problems, endocrinological disorder (increased testosterone production in females and menstrual irregularities), memory problems, weight gain, infections, drowsiness and headache, liver damage, polycystic ovaries, movement disorders (even hallucinations, anxiety and confusion), swollen pancreas, low body temperature and potentially life-threatening blood abnormalities (e.g. low platelet count). Valproate has the highest risk of birth defects of any of the commonly used antiepileptic drugs during pregnancy (Cummings et al. 2011). Overdose results in tremor, respiratory depression, coma and metabolic acidosis and eventually can result to death. Serum or plasma levels of valproic acid concentrations should be in the range of 50–150 mg/l for the treatment of BD.

13.3.5 Carbamazepine

Carbamazepine is an antiepileptic efficacious against partial seizures, generalized tonic-clonic seizures and mixed seizures and also useful for the treatment of trigeminal neuralgia. It was discovered in 1953 and was first marketed as a drug to treat trigeminal neuralgia in 1962 and as an antiepileptic in the UK since 1965 and the USA since 1974. In 1971 the first studies concerning bipolar disorder appeared in Japan (Okuma and Kishimoto 1998), and in the next few years, it has been proven to be efficacious in the treatment of bipolar disorder (Fountoulakis et al. 2017a, c).

The mechanism of action of carbamazepine includes the stabilization of the inactivated state of sodium channels, thus reducing the excitability of the neurons. It has also been shown to bind to GABA receptors (Granger et al. 1995).

The most common adverse effects with carbamazepine treatment may include drowsiness, dizziness, headaches and migraines, motor coordination impairment, nausea, vomiting, constipation, cardiac arrhythmias, blurry or double vision, aplastic anaemia or agranulocytosis and a dangerous or even fatal skin reaction (Stevens-Johnson syndrome and toxic epidermal necrolysis). It can also exacerbate preexisting hypothyroidism. It can cause syndrome of inappropriate antidiuretic hormone, and it can aggravate juvenile myoclonic epilepsy and other types of epilepsy, especially absence seizures (Liu et al. 2006).

Among the agents used in the treatment of BD, carbamazepine is the drug most potent to interactions with other medication. It is a CYP450 inducer, and thus it might increase the metabolism and elimination of many agents, including warfarin, lamotrigine, phenytoin, theophylline, valproic acid, benzodiazepines and some antipsychotics. It also reduces the effectiveness of birth control pills, thus leading to unexpected pregnancies. Other agents, like erythromycin, cimetidine, valproic acid, valnoctamide and calcium channel blockers as well as grapefruit juice, decrease carbamazepine metabolism and increase its serum availability often to toxic levels. As a drug that induces cytochrome P450 enzymes, it accelerates elimination of many benzodiazepines and decreases their action.

Carbamazepine is teratogenic and is associated among others with the development of spina bifida (Jentink et al. 2010), neurodevelopmental problems and delays (Cummings et al. 2011), craniofacial defects, cardiovascular malformations and hypospadias.

13.3.6 Lamotrigine

Lamotrigine is an antiepileptic efficacious in the treatment of focal seizures, primary and secondary tonic-clonic seizures and seizures associated with Lennox-Gastaut syndrome. It was marketed for the first time in 1994. It is chemically different to other antiepileptics. It has been proven efficacious in the prevention of depressive episodes during maintenance treatment of bipolar disorder (Fountoulakis et al. 2017a, c)

Lamotrigine is a triazine derivate that inhibits voltage-sensitive sodium channels, leading to stabilization of neuronal membranes. It also blocks calcium channels and has weak 5-HT3 receptor inhibition. Probably other actions also exist since lamotrigine exerts a variety of effects and adverse events which cannot be explained by its above pharmacodynamics properties alone (Rogawski and Loscher 2004a, b; Lees and Leach 1993). It is metabolized by hepatic glucuronidation.

Its adverse effects include life-threatening skin reactions, including Stevens-Johnson syndrome, DRESS syndrome and toxic epidermal necrolysis. Since December 2010, lamotrigine carries an FDA black box warning for aseptic meningitis. Other adverse events include loss of balance or coordination, double vision, blurred vision, dizziness, drowsiness, insomnia, anxiety, vivid dreams or nightmares, dry mouth, mouth ulcers, memory and cognitive problems, runny nose, cough, indigestion, abdominal pain, weight loss, missed or painful menstrual periods, vaginitis and leukopenia.

Certain contraceptives decrease serum levels of lamotrigine (Reimers et al. 2005). It has low teratogenic action; however if used during the first trimester, it may increase the risk for cleft lip and palate malformation in newborns.

Lamotrigine has fewer drug interactions than other antiepileptics; however caution is needed when co-administered with hepatic enzyme-inducing medications (Anderson 1998).

References

Aalto S, Hirvonen J, Kajander J, Scheinin H, Nagren K, Vilkman H, Gustafsson L, Syvalahti E, Hietala J (2002) Ketamine does not decrease striatal dopamine D2 receptor binding in man. Psychopharmacology (Berl) 164(4):401–406. https://doi.org/10.1007/s00213-002-1236-6

Abbott FV, Etienne P, Franklin KB, Morgan MJ, Sewitch MJ, Young SN (1992) Acute tryptophan depletion blocks morphine analgesia in the cold-pressor test in humans. Psychopharmacology (Berl) 108(1-2):60–66. https://doi.org/10.1007/bf02245286

Abi-Dargham A, van de Giessen E, Slifstein M, Kegeles LS, Laruelle M (2009) Baseline and amphetamine-stimulated dopamine activity are related in drug-naive schizophrenic subjects. Biol Psychiatry 65(12):1091–1093. https://doi.org/10.1016/j.biopsych.2008.12.007

Abi-Dargham A, Xu X, Thompson JL, Gil R, Kegeles LS, Urban N, Narendran R, Hwang DR, Laruelle M, Slifstein M (2012) Increased prefrontal cortical D(1) receptors in drug naive patients with schizophrenia: a PET study with [(1)(1)C]NNC112. J Psychopharmacol 26(6):794–805. https://doi.org/10.1177/0269881111409265

Abrams R, Taylor MA (1987) Cognitive dysfunction in melancholia. Psychol Med 17(2):359–362. https://doi.org/10.1017/s0033291700024909

Adityanjee (1987) The syndrome of irreversible lithium effectuated neurotoxicity. J Neurol Neurosurg Psychiatry 50(9):1246–1247

Adityanjee (1989) The syndrome of irreversible lithium-effectuated neurotoxicity (SILENT). Pharmacopsychiatry 22(2):81–83. https://doi.org/10.1055/s-2007-1014583

Adityanjee, Munshi KR, Thampy A (2005) The syndrome of irreversible lithium-effectuated neurotoxicity. Clin Neuropharmacol 28(1):38–49

Akbarian S, Kim JJ, Potkin SG, Hagman JO, Tafazzoli A, Bunney WE, Jones EG (1995) Gene-expression for glutamic-acid decarboxylase is reduced without loss of neurons in prefrontal cortex of schizophrenics. Arch Gen Psychiatry 52(4):258–266. https://doi.org/10.1001/archpsyc.1995.03950160008002

Akiyama H, Kawamata T, Dedhar S, McGeer PL (1991) Immunohistochemical localization of vitronectin, its receptor and beta-3 integrin in Alzheimer brain tissue. J Neuroimmunol 32(1):19–28

Aletrino MA, Vogels OJ, Van Domburg PH, Ten Donkelaar HJ (1992) Cell loss in the nucleus raphes dorsalis in Alzheimer's disease. Neurobiol Aging 13(4):461–468

Altshuler LL, Casanova MF, Goldberg TE, Kleinman JE (1990) The hippocampus and parahippocampus in schizophrenic, suicide, and control brains. Arch Gen Psychiatry 47(11):1029–1034. https://doi.org/10.1001/archpsyc.1990.01810230045008

Altshuler LL, Bookheimer SY, Townsend J, Proenza MA, Eisenberger N, Sabb F, Mintz J, Cohen MS (2005) Blunted activation in orbitofrontal cortex during mania: a functional magnetic resonance imaging study. Biol Psychiatry 58(10):763–769. https://doi.org/10.1016/j.biopsych.2005.09.012

Amdisen A (1977) Serum level monitoring and clinical pharmacokinetics of lithium. Clin Pharmacokinet 2(2):73–92

Ament SA, Szelinger S, Glusman G, Ashworth J, Hou L, Akula N, Shekhtman T, Badner JA, Brunkow ME, Mauldin DE, Stittrich AB, Rouleau K, Detera-Wadleigh SD, Nurnberger JI Jr, Edenberg HJ, Gershon ES, Schork N, Bipolar Genome S, Price ND, Gelinas R, Hood L, Craig D, McMahon FJ, Kelsoe JR, Roach JC (2015) Rare variants in neuronal excitability genes influence risk for bipolar disorder. Proc Natl Acad Sci U S A 112(11):3576–3581. https://doi.org/10.1073/pnas.1424958112

Anand A, Barkay G, Dzemidzic M, Albrecht D, Karne H, Zheng QH, Hutchins GD, Normandin MD, Yoder KK (2011) Striatal dopamine transporter availability in unmedicated bipolar disorder. Bipolar Disord 13(4):406–413. https://doi.org/10.1111/j.1399-5618.2011.00936.x

Anderson GD (1998) A mechanistic approach to antiepileptic drug interactions. Ann Pharmacother 32(5):554–563. https://doi.org/10.1345/aph.17332

Andreasen N, Nasrallah HA, Dunn V, Olson SC, Grove WM, Ehrhardt JC, Coffman JA, Crossett JH (1986) Structural abnormalities in the frontal system in schizophrenia. A magnetic resonance imaging study. Arch Gen Psychiatry 43(2):136–144. https://doi.org/10.1001/archpsyc.1986.01800020042006

Andreasen NC, Swayze VW 2nd, Flaum M, Yates WR, Arndt S, McChesney C (1990) Ventricular enlargement in schizophrenia evaluated with computed tomographic scanning. Effects of gender, age, and stage of illness. Arch Gen Psychiatry 47(11):1008–1015. https://doi.org/10.1001/archpsyc.1990.01810230024005

Andreassen OA, Thompson WK, Schork AJ, Ripke S, Mattingsdal M, Kelsoe JR, Kendler KS, O'Donovan MC, Rujescu D, Werge T, Sklar P, Psychiatric Genomics C, Bipolar D, Schizophrenia Working G, Roddey JC, Chen CH, McEvoy L, Desikan RS, Djurovic S, Dale AM (2013) Improved detection of common variants associated with schizophrenia and bipolar disorder using pleiotropy-informed conditional false discovery rate. PLoS Genet 9(4):e1003455. https://doi.org/10.1371/journal.pgen.1003455

Angst J, Grof P, Schou M (1969) Lithium. Lancet 1(7605):1097

Angst J, Weis P, Grof P, Baastrup PC, Schou M (1970) Lithium prophylaxis in recurrent affective disorders. Br J Psychiatry 116(535):604–614. https://doi.org/10.1192/bjp.116.535.604

Arai H, Kosaka K, Iizuka R (1984) Changes of biogenic amines and their metabolites in postmortem brains from patients with Alzheimer-type dementia. J Neurochem 43(2):388–393

Arango C, Rapado-Castro M, Reig S, Castro-Fornieles J, Gonzalez-Pinto A, Otero S, Baeza I, Moreno C, Graell M, Janssen J, Parellada M, Moreno D, Bargallo N, Desco M (2012) Progressive brain changes in children and adolescents with first-episode psychosis. Arch Gen Psychiatry 69(1):16–26. https://doi.org/10.1001/archgenpsychiatry.2011.150

Araujo DM (1992) Contrasting effects of specific lymphokines on the survival of hippocampal neurons in culture. Adv Behav Biol. https://doi.org/10.1007/978-1-4615-3432-7_9

Araujo DM, Cotman CW (1992) Beta-amyloid stimulates glial cells in vitro to produce growth factors that accumulate in senile plaques in Alzheimer's disease. Brain Res 569(1):141–145

Arnold SE (2006) Cellular and molecular neuropathology of the parahippocampal region in schizophrenia. Ann N Y Acad Sci 911(1):275–292. https://doi.org/10.1111/j.1749-6632.2000.tb06732.x

Ashall F, Goate AM (1994) Role of the beta-amyloid precursor protein in Alzheimer's disease. Trends Biochem Sci 19(1):42–46

Ashford JW, Shih WJ, Coupal J, Shetty R, Schneider A, Cool C, Aleem A, Kiefer VH, Mendiondo MS, Schmitt FA (2000) Single SPECT measures of cerebral cortical perfusion reflect time-index estimation of dementia severity in Alzheimer's disease. J Nucl Med 41(1):57–64

Askland K, Read C, Moore J (2009) Pathways-based analyses of whole-genome association study data in bipolar disorder reveal genes mediating ion channel activity and synaptic neurotransmission. Hum Genet 125(1):63–79. https://doi.org/10.1007/s00439-008-0600-y

Atmaca M, Ozdemir H, Cetinkaya S, Parmaksiz S, Belli H, Poyraz AK, Tezcan E, Ogur E (2007) Cingulate gyrus volumetry in drug free bipolar patients and patients treated with valproate or valproate and quetiapine. J Psychiatr Res 41(10):821–827. https://doi.org/10.1016/j.jpsychires.2006.07.006

Attems J, Quass M, Jellinger KA (2007) Tau and alpha-synuclein brainstem pathology in Alzheimer disease: relation with extrapyramidal signs. Acta Neuropathol 113(1):53–62. https://doi.org/10.1007/s00401-006-0146-9

Auer DP, Putz B, Kraft E, Lipinski B, Schill J, Holsboer F (2000) Reduced glutamate in the anterior cingulate cortex in depression: an in vivo proton magnetic resonance spectroscopy study. Biol Psychiatry 47(4):305–313

Austin MP, Ross M, Murray C, O'Carroll RE, Ebmeier KP, Goodwin GM (1992) Cognitive function in major depression. J Affect Disord 25(1):21–29. https://doi.org/10.1016/0165-0327(92)90089-o

Avramopoulos D, Lasseter VK, Fallin MD, Wolyniec PS, McGrath JA, Nestadt G, Valle D, Pulver AE (2007) Stage II follow-up on a linkage scan for bipolar disorder in the Ashkenazim provides suggestive evidence for chromosome 12p and the GRIN2B gene. Genet Med 9(11):745–751. https://doi.org/10.1097/GIM.0b013e318159a37c

Baastrup PC (1964) The use of lithium in manic-depressive psychosis. Compr Psychiatry 5(6):396–408

Baastrup PC, Schou M (1967) Lithium as a prophylactic agents. Its effect against recurrent depressions and manic-depressive psychosis. Arch Gen Psychiatry 16(2):162–172

Baastrup PC, Poulsen JC, Schou M, Thomsen K, Amdisen A (1970) Prophylactic lithium: double blind discontinuation in manic-depressive and recurrent-depressive disorders. Lancet 2(7668):326–330

Bailey CP, Manley SJ, Watson WP, Wonnacott S, Molleman A, Little HJ (1998) Chronic ethanol administration alters activity in ventral tegmental area neurons after cessation of withdrawal hyperexcitability. Brain Res 803(1-2):144–152. https://doi.org/10.1016/s0006-8993(98)00654-4

Balanza-Martinez V, Tabares-Seisdedos R, Selva-Vera G, Martinez-Aran A, Torrent C, Salazar-Fraile J, Leal-Cercos C, Vieta E, Gomez-Beneyto M (2005) Persistent cognitive dysfunctions in bipolar I disorder and schizophrenic patients: a 3-year follow-up study. Psychother Psychosom 74(2):113–119. https://doi.org/10.1159/000083170

Bartenstein P, Minoshima S, Hirsch C, Buch K, Willoch F, Mosch D, Schad D, Schwaiger M, Kurz A (1997) Quantitative assessment of cerebral blood flow in patients with Alzheimer's disease by SPECT. J Nucl Med 38(7):1095–1101

Basso MR, Bornstein RA (1999) Neuropsychological deficits in psychotic versus nonpsychotic unipolar depression. Neuropsychology 13(1):69–75. https://doi.org/10.1037//0894-4105.13.1.69

Baxter LR Jr, Schwartz JM, Phelps ME, Mazziotta JC, Guze BH, Selin CE, Gerner RH, Sumida RM (1989) Reduction of prefrontal cortex glucose metabolism common to three types of depression. Arch Gen Psychiatry 46(3):243–250. https://doi.org/10.1001/archpsyc.1989.01810030049007

Bearden CE, Thompson PM, Dalwani M, Hayashi KM, Lee AD, Nicoletti M, Trakhtenbroit M, Glahn DC, Brambilla P, Sassi RB, Mallinger AG, Frank E, Kupfer DJ, Soares JC (2007) Greater cortical gray matter density in lithium-treated patients with bipolar disorder. Biol Psychiatry 62(1):7–16. https://doi.org/10.1016/j.biopsych.2006.10.027

Beats BC, Sahakian BJ, Levy R (1996) Cognitive performance in tests sensitive to frontal lobe dysfunction in the elderly depressed. Psychol Med 26(3):591–603. https://doi.org/10.1017/s0033291700035662

Bech P (2006) The full story of lithium. A tribute to Mogens Schou (1918–2005). Psychother Psychosom 75(5):265–269. https://doi.org/10.1159/000093947

Bech P, Vendsborg PB, Rafaelsen OJ (1976) Lithium maintenance treatment of manic-melancholic patients: its role in the daily routine. Acta Psychiatr Scand 53(1):70–81

Belmaker RH, Agam G (2008) Major depressive disorder. N Engl J Med 358(1):55–68. https://doi.org/10.1056/NEJMra073096

Benabarre A, Vieta E, Martinez-Aran A, Garcia-Garcia M, Martin F, Lomena F, Torrent C, Sanchez-Moreno J, Colom F, Reinares M, Brugue E, Valdes M (2005) Neuropsychological disturbances and cerebral blood flow in bipolar disorder. Aust N Z J Psychiatry 39(4):227–234. https://doi.org/10.1080/j.1440-1614.2004.01558.x

Benca RM, Obermeyer WH, Thisted RA, Gillin JC (1992) Sleep and psychiatric disorders. A meta-analysis. Arch Gen Psychiatry 49(8):651–668.; discussion 669–670. https://doi.org/10.1001/archpsyc.1992.01820080059010

Benes FM, Todtenkopf MS, Logiotatos P, Williams M (2000) Glutamate decarboxylase(65)-immunoreactive terminals in cingulate and prefrontal cortices of schizophrenic and bipolar brain. J Chem Neuroanat 20(3-4):259–269

Benes FM, Vincent SL, Todtenkopf M (2001) The density of pyramidal and nonpyramidal neurons in anterior cingulate cortex of schizophrenic and bipolar subjects. Biol Psychiatry 50(6):395–406

Beneyto M, Kristiansen LV, Oni-Orisan A, McCullumsmith RE, Meador-Woodruff JH (2007) Abnormal glutamate receptor expression in the medial temporal lobe in schizophrenia and mood disorders. Neuropsychopharmacology 32(9):1888–1902. https://doi.org/10.1038/sj.npp.1301312

Bennett AE (1953) Biological psychiatry. Am J Psychiatry 110(4):244–252. https://doi.org/10.1176/ajp.110.4.244
Benwell ME, Balfour DJ, Lucchi HM (1993) Influence of tetrodotoxin and calcium on changes in extracellular dopamine levels evoked by systemic nicotine. Psychopharmacology (Berl) 112(4):467–474. https://doi.org/10.1007/bf02244896
Berk M, Hallam K, Malhi GS, Henry L, Hasty M, Macneil C, Yucel M, Pantelis C, Murphy B, Vieta E, Dodd S, McGorry PD (2010) Evidence and implications for early intervention in bipolar disorder. J Ment Health 19(2):113–126. https://doi.org/10.3109/09638230903469111
Berke JD, Hyman SE (2000) Addiction, dopamine, and the molecular mechanisms of memory. Neuron 25(3):515–532. https://doi.org/10.1016/s0896-6273(00)81056-9
Bernstein HG, Krell D, Baumann B, Danos P, Falkai P, Diekmann S, Henning H, Bogerts B (1998) Morphometric studies of the entorhinal cortex in neuropsychiatric patients and controls: clusters of heterotopically displaced lamina II neurons are not indicative of schizophrenia. Schizophr Res 33(3):125–132. https://doi.org/10.1016/S0920-9964(98)00071-1
Bernstein HG, Steiner J, Guest PC, Dobrowolny H, Bogerts B (2015) Glial cells as key players in schizophrenia pathology: recent insights and concepts of therapy. Schizophr Res 161(1):4–18. https://doi.org/10.1016/j.schres.2014.03.035
Berridge KC, Robinson TE (1998) What is the role of dopamine in reward: hedonic impact, reward learning, or incentive salience? Brain Res Rev 28(3):309–369. https://doi.org/10.1016/S0165-0173(98)00019-8
Bielau H, Steiner J, Mawrin C, Trubner K, Brisch R, Meyer-Lotz G, Brodhun M, Dobrowolny H, Baumann B, Gos T, Bernstein HG, Bogerts B (2007) Dysregulation of GABAergic neurotransmission in mood disorders: a postmortem study. Ann N Y Acad Sci 1096:157–169. https://doi.org/10.1196/annals.1397.081
Bierut LJ, Dinwiddie SH, Begleiter H, Crowe RR, Hesselbrock V, Nurnberger JI Jr, Porjesz B, Schuckit MA, Reich T (1998) Familial transmission of substance dependence: alcohol, marijuana, cocaine, and habitual smoking: a report from the collaborative study on the genetics of alcoholism. Arch Gen Psychiatry 55(11):982–988. https://doi.org/10.1001/archpsyc.55.11.982
Bigdeli TB, Ripke S, Bacanu SA, Lee SH, Wray NR, Gejman PV, Rietschel M, Cichon S, St Clair D, Corvin A, Kirov G, McQuillin A, Gurling H, Rujescu D, Andreassen OA, Werge T, Blackwood DH, Pato CN, Pato MT, Malhotra AK, O'Donovan MC, Kendler KS, Fanous AH, Schizophrenia Working Group of the Psychiatric Genomics C (2016) Genome-wide association study reveals greater polygenic loading for schizophrenia in cases with a family history of illness. Am J Med Genet B Neuropsychiatr Genet 171B(2):276–289. https://doi.org/10.1002/ajmg.b.32402
Bitanihirwe BK, Lim MP, Woo TU (2010) N-methyl-D-aspartate receptor expression in parvalbumin-containing inhibitory neurons in the prefrontal cortex in bipolar disorder. Bipolar Disord 12(1):95–101. https://doi.org/10.1111/j.1399-5618.2009.00785.x
Biver F, Goldman S, Delvenne V, Luxen A, De Maertelaer V, Hubain P, Mendlewicz J, Lotstra F (1994) Frontal and parietal metabolic disturbances in unipolar depression. Biol Psychiatry 36(6):381–388
Blanchard BJ, Konopka G, Russell M, Ingram VM (1997) Mechanism and prevention of neurotoxicity caused by beta-amyloid peptides: relation to Alzheimer's disease. Brain Res 776(1-2):40–50
Blass JP, Gibson GE (1991) The role of oxidative abnormalities in the pathophysiology of Alzheimer's disease. Rev Neurol 147(6–7):513–525
Blumberg HP, Stern E, Martinez D, Ricketts S, de Asis J, White T, Epstein J, McBride PA, Eidelberg D, Kocsis JH, Silbersweig DA (2000) Increased anterior cingulate and caudate activity in bipolar mania. Biol Psychiatry 48(11):1045–1052
Blumberg HP, Leung HC, Skudlarski P, Lacadie CM, Fredericks CA, Harris BC, Charney DS, Gore JC, Krystal JH, Peterson BS (2003) A functional magnetic resonance imaging study of bipolar disorder: state- and trait-related dysfunction in ventral prefrontal cortices. Arch Gen Psychiatry 60(6):601–609. https://doi.org/10.1001/archpsyc.60.6.601

Blumberg HP, Krystal JH, Bansal R, Martin A, Dziura J, Durkin K, Martin L, Gerard E, Charney DS, Peterson BS (2006) Age, rapid-cycling, and pharmacotherapy effects on ventral prefrontal cortex in bipolar disorder: a cross-sectional study. Biol Psychiatry 59(7):611–618. https://doi.org/10.1016/j.biopsych.2005.08.031

Boissiere F, Hunot S, Faucheux B, Hersh LB, Agid Y, Hirsch EC (1997) Trk neurotrophin receptors in cholinergic neurons of patients with Alzheimer's disease. Dement Geriatr Cogn Disord 8(1):1–8. https://doi.org/10.1159/000106594

Bolos AM, Dean M, Lucasderse S, Ramsburg M, Brown GL, Goldman D (1990) Population and pedigree studies reveal a lack of association between the dopamine-D2 receptor gene and alcoholism. JAMA J Am Med Assoc 264(24):3156–3160. https://doi.org/10.1001/jama.264.24.3156

Boomsma DI, Koopmans JR, Vandoornen LJP, Orlebeke JF (1994) Genetic and social influences on starting to smoke – a study of Dutch adolescent twins and their parents. Addiction 89(2):219–226. https://doi.org/10.1111/j.1360-0443.1994.tb00881.x

Bora E, Fornito A, Yucel M, Pantelis C (2010) Voxelwise meta-analysis of gray matter abnormalities in bipolar disorder. Biol Psychiatry 67(11):1097–1105. https://doi.org/10.1016/j.biopsych.2010.01.020

Boter H, Peuskens J, Libiger J, Fleischhacker WW, Davidson M, Galderisi S, Kahn RS, Group ES (2009) Effectiveness of antipsychotics in first-episode schizophrenia and schizophreniform disorder on response and remission: an open randomized clinical trial (EUFEST). Schizophr Res 115(2-3):97–103. https://doi.org/10.1016/j.schres.2009.09.019

Botteron KN, Figiel GS (1997) The neuromorphometry of affective disorders. In: Krishnan KRR, Doraiswamy PM (eds) Brain imaging in clinical psychiatry. Marcel Dekker, New York, NY, pp 145–184

Bouras C, Kovari E, Hof PR, Riederer BM, Giannakopoulos P (2001) Anterior cingulate cortex pathology in schizophrenia and bipolar disorder. Acta Neuropathol 102(4):373–379

Bourtchuladze R, Frenguelli B, Blendy J, Cioffi D, Schutz G, Silva AJ (1994) Deficient long-term memory in mice with a targeted mutation of the cAMP-responsive element-binding protein. Cell 79(1):59–68. https://doi.org/10.1016/0092-8674(94)90400-6

Bowen DM, Allen SJ, Benton JS, Goodhardt MJ, Haan EA, Palmer AM, Sims NR, Smith CC, Spillane JA, Esiri MM, Neary D, Snowdon JS, Wilcock GK, Davison AN (1983) Biochemical assessment of serotonergic and cholinergic dysfunction and cerebral atrophy in Alzheimer's disease. J Neurochem 41(1):266–272

Bozarth MA, Wise RA (1981) Intracranial self-administration of morphine into the ventral tegmental area in rats. Life Sci 28(5):551–555. https://doi.org/10.1016/0024-3205(81)90148-x

Braak H, Braak E (1991) Neuropathological stageing of Alzheimer-related changes. Acta Neuropathol 82(4):239–259

Brakeman PR, Lanahan AA, O'Brien R, Roche K, Barnes CA, Huganir RL, Worley PF (1997) Homer: a protein that selectively binds metabotropic glutamate receptors. Nature 386(6622):284–288. https://doi.org/10.1038/386284a0

Bremner JD, Narayan M, Anderson ER, Staib LH, Miller HL, Charney DS (2000) Hippocampal volume reduction in major depression. Am J Psychiatry 157(1):115–118. https://doi.org/10.1176/ajp.157.1.115

Brodie MS, Shefner SA, Dunwiddie TV (1990) Ethanol increases the firing rate of dopamine neurons of the rat ventral tegmental area in vitro. Brain Res 508(1):65–69. https://doi.org/10.1016/0006-8993(90)91118-z

Brodie MS, Pesold C, Appel SB (1999) Ethanol directly excites dopaminergic ventral tegmental area reward neurons. Alcohol Clin Exp Res 23(11):1848–1852. https://doi.org/10.1097/00000374-199911000-00019

Brody AL, Saxena S, Mandelkern MA, Fairbanks LA, Ho ML, Baxter LR (2001) Brain metabolic changes associated with symptom factor improvement in major depressive disorder. Biol Psychiatry 50(3):171–178

Brown EE, Fibiger HC (1993) Differential effects of excitotoxic lesions of the amygdala on cocaine-induced conditioned locomotion and conditioned place preference. Psychopharmacology (Berl) 113(1):123–130. https://doi.org/10.1007/bf02244344

Brunello N, Armitage R, Feinberg I, Holsboer-Trachsler E, Leger D, Linkowski P, Mendelson WB, Racagni G, Saletu B, Sharpley AL, Turek F, Van Cauter E, Mendlewicz J (2000) Depression and sleep disorders: clinical relevance, economic burden and pharmacological treatment. Neuropsychobiology 42(3):107–119. https://doi.org/10.1159/000026680
Bruno SD, Barker GJ, Cercignani M, Symms M, Ron MA (2004) A study of bipolar disorder using magnetization transfer imaging and voxel-based morphometry. Brain 127(Pt 11):2433–2440. https://doi.org/10.1093/brain/awh274
Bruno SD, Papadopoulou K, Cercignani M, Cipolotti L, Ron MA (2006) Structural brain correlates of IQ changes in bipolar disorder. Psychol Med 36(5):609–618. https://doi.org/10.1017/S0033291706007112
Bunney WE Jr, Davis JM (1965) Norepinephrine in depressive reactions. A review. Arch Gen Psychiatry 13(6):483–494. https://doi.org/10.1001/archpsyc.1965.01730060001001
Burt DB, Zembar MJ, Niederehe G (1995) Depression and memory impairment: a meta-analysis of the association, its pattern, and specificity. Psychol Bull 117(2):285–305. https://doi.org/10.1037//0033-2909.117.2.285
Buttner N, Bhattacharyya S, Walsh J, Benes FM (2007) DNA fragmentation is increased in non-GABAergic neurons in bipolar disorder but not in schizophrenia. Schizophr Res 93(1-3):33–41. https://doi.org/10.1016/j.schres.2007.01.030
Cabib S, Puglisi-Allegra S (1996) Different effects of repeated stressful experiences on mesocortical and mesolimbic dopamine metabolism. Neuroscience 73(2):375–380. https://doi.org/10.1016/0306-4522(96)00750-6
Cade JF (1949) Lithium salts in the treatment of psychotic excitement. Med J Aust 2(10):349–352
Cade JF (2000) Lithium salts in the treatment of psychotic excitement. 1949. Bull World Health Organ 78(4):518–520
Cadoret RJ, Troughton E, O'Gorman TW, Heywood E (1986) An adoption study of genetic and environmental factors in drug abuse. Arch Gen Psychiatry 43(12):1131–1136. https://doi.org/10.1001/archpsyc.1986.01800120017004
Cadoret RJ, Yates WR, Troughton E, Woodworth G, Stewart MA (1995) Adoption study demonstrating two genetic pathways to drug abuse. Arch Gen Psychiatry 52(1):42–52. https://doi.org/10.1001/archpsyc.1995.03950130042005
Cadoret RJ, Yates WR, Troughton E, Woodworth G, Stewart MA (1996) An adoption study of drug abuse/dependency in females. Compr Psychiatry 37(2):88–94. https://doi.org/10.1016/s0010-440x(96)90567-2
Cannon DM, Carson RE, Nugent AC, Eckelman WC, Kiesewetter DO, Williams J, Rollis D, Drevets M, Gandhi S, Solorio G, Drevets WC (2006a) Reduced muscarinic type 2 receptor binding in subjects with bipolar disorder. Arch Gen Psychiatry 63(7):741–747. https://doi.org/10.1001/archpsyc.63.7.741
Cannon DM, Ichise M, Fromm SJ, Nugent AC, Rollis D, Gandhi SK, Klaver JM, Charney DS, Manji HK, Drevets WC (2006b) Serotonin transporter binding in bipolar disorder assessed using [11C]DASB and positron emission tomography. Biol Psychiatry 60(3):207–217. https://doi.org/10.1016/j.biopsych.2006.05.005
Cardno AG, Gottesman II (2000) Twin studies of schizophrenia: from bow-and-arrow concordances to star wars Mx and functional genomics. Am J Med Genet 97(1):12–17. https://doi.org/10.1002/(sici)1096-8628(200021)97:1<12::aid-ajmg3>3.3.co;2-l
Cardno AG, Rijsdijk FV, West RM, Gottesman II, Craddock N, Murray RM, McGuffin P (2012) A twin study of schizoaffective-mania, schizoaffective-depression, and other psychotic syndromes. Am J Med Genet B Neuropsychiatr Genet 159B(2):172–182. https://doi.org/10.1002/ajmg.b.32011
Carlezon WA Jr, Wise RA, Carlezon WA Jr (1996) Microinjections of phencyclidine (PCP) and related drugs into nucleus accumbens shell potentiate medial forebrain bundle brain stimulation reward. Psychopharmacology (Berl) 128(4):413–420. https://doi.org/10.1007/s002130050151
Carlsson M, Carlsson A (1990a) Interactions between glutamatergic and monoaminergic systems within the basal ganglia-implications for schizophrenia and Parkinson's disease. Trends Neurosci 13(7):272–276. https://doi.org/10.1016/0166-2236(90)90108-m

Carlsson M, Carlsson A (1990b) Schizophrenia: a subcortical neurotransmitter imbalance syndrome? Schizophr Bull 16(3):425–432. https://doi.org/10.1093/schbul/16.3.425
Carlsson A, Waters N, Holm-Waters S, Tedroff J, Nilsson M, Carlsson ML (2001) Interactions between monoamines, glutamate, and GABA in schizophrenia: new evidence. Annu Rev Pharmacol Toxicol 41(1):237–260. https://doi.org/10.1146/annurev.pharmtox.41.1.237
Carr GD, White NM (1983) Conditioned place preference from intra-accumbens but not intra-caudate amphetamine injections. Life Sci 33(25):2551–2557. https://doi.org/10.1016/0024-3205(83)90165-0
Castellanos-Ryan N, Rubia K, Conrod PJ (2011) Response inhibition and reward response bias mediate the predictive relationships between impulsivity and sensation seeking and common and unique variance in conduct disorder and substance misuse. Alcohol Clin Exp Res 35(1):140–155. https://doi.org/10.1111/j.1530-0277.2010.01331.x
Catts VS, Fung SJ, Long LE, Joshi D, Vercammen A, Allen KM, Fillman SG, Rothmond DA, Sinclair D, Tiwari Y, Tsai SY, Weickert TW, Shannon Weickert C (2013) Rethinking schizophrenia in the context of normal neurodevelopment. Front Cell Neurosci 7:60. https://doi.org/10.3389/fncel.2013.00060
Cepeda-Benito A, Tiffany ST, Cox LS (1999) Context-specific morphine tolerance on the paw-pressure and tail-shock vocalization tests: evidence of associative tolerance without conditioned compensatory responding. Psychopharmacology 145(4):426–432. https://doi.org/10.1007/s002130051077
Chana G, Landau S, Beasley C, Everall IP, Cotter D (2003) Two-dimensional assessment of cytoarchitecture in the anterior cingulate cortex in major depressive disorder, bipolar disorder, and schizophrenia: evidence for decreased neuronal somal size and increased neuronal density. Biol Psychiatry 53(12):1086–1098
Chang K, Adleman NE, Dienes K, Simeonova DI, Menon V, Reiss A (2004) Anomalous prefrontal-subcortical activation in familial pediatric bipolar disorder: a functional magnetic resonance imaging investigation. Arch Gen Psychiatry 61(8):781–792. https://doi.org/10.1001/archpsyc.61.8.781
Channon S, Green PS (1999) Executive function in depression: the role of performance strategies in aiding depressed and non-depressed participants. J Neurol Neurosurg Psychiatry 66(2):162–171. https://doi.org/10.1136/jnnp.66.2.162
Chaudieu I, Vignon J, Chicheportiche M, Kamenka JM, Trouiller G, Chicheportiche R (1989) Role of the aromatic group in the inhibition of phencyclidine binding and dopamine uptake by PCP analogs. Pharmacol Biochem Behav 32(3):699–705. https://doi.org/10.1016/0091-3057(89)90020-8
Chen J, Paredes W, Li J, Smith D, Lowinson J, Gardner EL (1990) Δ9-Tetrahydrocannabinol produces naloxone-blockable enhancement of presynaptic basal dopamine efflux in nucleus accumbens of conscious, freely-moving rats as measured by intracerebral microdialysis. Psychopharmacology (Berl) 102(2):156–162. https://doi.org/10.1007/bf02245916
Chen J, Marmur R, Pulles A, Paredes W, Gardner EL (1993) Ventral tegmental microinjection of Δ9-tetrahydrocannabinol enhances ventral tegmental somatodendritic dopamine levels but not forebrain dopamine levels: evidence for local neural action by marijuana's psychoactive ingredient. Brain Res 621(1):65–70. https://doi.org/10.1016/0006-8993(93)90298-2
Chen CP, Eastwood SL, Hope T, McDonald B, Francis PT, Esiri MM (2000) Immunocytochemical study of the dorsal and median raphe nuclei in patients with Alzheimer's disease prospectively assessed for behavioural changes. Neuropathol Appl Neurobiol 26(4):347–355
Chen KP, Shen WW, Lu ML (2004) Implication of serum concentration monitoring in patients with lithium intoxication. Psychiatry Clin Neurosci 58(1):25–29
Chen KC, Yang YK, Howes O, Lee IH, Landau S, Yeh TL, Chiu NT, Chen PS, Lu RB, David AS, Bramon E (2013) Striatal dopamine transporter availability in drug-naive patients with schizophrenia: a case-control SPECT study with [(99m)Tc]-TRODAT-1 and a meta-analysis. Schizophr Bull 39(2):378–386. https://doi.org/10.1093/schbul/sbr163
Chéramy A, Nieoullon A, Glowinski J (1978) Gabaergic processes involved in the control of dopamine release from nigrostriatal dopaminergic neurons in the cat. Eur J Pharmacol 48(3):281–295. https://doi.org/10.1016/0014-2999(78)90087-0

Childress AR, Mozley PD, McElgin W, Fitzgerald J, Reivich M, O'Brien CP (1999) Limbic activation during cue-induced cocaine craving. Am J Psychiatry 156(1):11–18. https://doi.org/10.1176/ajp.156.1.11
Choi DW (1988) Glutamate neurotoxicity and diseases of the nervous system. Neuron 1(8):623–634
Choi DW, Yokoyama M, Koh J (1988) Zinc neurotoxicity in cortical cell culture. Neuroscience 24(1):67–79. https://doi.org/10.1016/0306-4522(88)90312-0
Chua SE, McKenna PJ (1995) Schizophrenia—a brain disease? A critical review of structural and functional cerebral abnormality in the disorder. Br J Psychiatry 166(5):563–582. https://doi.org/10.1192/bjp.166.5.563
Citron M, Westaway D, Xia W, Carlson G, Diehl T, Levesque G, Johnson-wood K, Lee M, Seubert P, Davis A, Kholodenko D, Motter R, Sherrington R, Perry B, Yao H, Strome R, Lieberburg I, Rommens J, Kim S, Schenk D, Fraser P, St George Hyslop P, Selkoe DJ (1997) Mutant presenilins of Alzheimer's disease increase production of 42-residue amyloid β-protein in both transfected cells and transgenic mice. Nat Med 3(1):67–72. https://doi.org/10.1038/nm0197-67
Clarke PB, Pert A (1985) Autoradiographic evidence for nicotine receptors on nigrostriatal and mesolimbic dopaminergic neurons. Brain Res 348(2):355–358. https://doi.org/10.1016/0006-8993(85)90456-1
Clinton SM, Meador-Woodruff JH (2004) Abnormalities of the NMDA receptor and associated intracellular molecules in the thalamus in schizophrenia and bipolar disorder. Neuropsychopharmacology 29(7):1353–1362. https://doi.org/10.1038/sj.npp.1300451
Cole AJ, Saffen DW, Baraban JM, Worley PF (1989) Rapid increase of an immediate early gene messenger RNA in hippocampal neurons by synaptic NMDA receptor activation. Nature 340(6233):474–476. https://doi.org/10.1038/340474a0
Cole AJ, Bhat RV, Patt C, Worley PF, Baraban JM (1992) D1Dopamine receptor activation of multiple transcription factor genes in rat striatum. J Neurochem 58(4):1420–1426. https://doi.org/10.1111/j.1471-4159.1992.tb11358.x
Collier DA, Eastwood BJ, Malki K, Mokrab Y (2016) Advances in the genetics of schizophrenia: toward a network and pathway view for drug discovery. Ann N Y Acad Sci 1366(1):61–75. https://doi.org/10.1111/nyas.13066
Coplan JD, Fathy HM, Jackowski AP, Tang CY, Perera TD, Mathew SJ, Martinez J, Abdallah CG, Dwork AJ, Pantol G, Carpenter D, Gorman JM, Nemeroff CB, Owens MJ, Kaffman A, Kaufman J (2014) Early life stress and macaque amygdala hypertrophy: preliminary evidence for a role for the serotonin transporter gene. Front Behav Neurosci 8:342. https://doi.org/10.3389/fnbeh.2014.00342
Coppen A (1967) The biochemistry of affective disorders. Br J Psychiatry 113(504):1237–1264. https://doi.org/10.1192/bjp.113.504.1237
Corder EH, Saunders AM, Strittmatter WJ, Schmechel DE, Gaskell PC, Small GW, Roses AD, Haines JL, Pericak-Vance MA (1993) Gene dose of apolipoprotein E type 4 allele and the risk of Alzheimer's disease in late onset families. Science 261(5123):921–923. https://doi.org/10.1126/science.8346443
Cornblatt BA, Lenzenweger MF, Erlenmeyer-Kimling L (1989) The continuous performance test, identical pairs version: II. Contrasting attentional profiles in schizophrenic and depressed patients. Psychiatry Res 29(1):65–85. https://doi.org/10.1016/0165-1781(89)90188-1
Cotter D, Mackay D, Landau S, Kerwin R, Everall I (2001) Reduced glial cell density and neuronal size in the anterior cingulate cortex in major depressive disorder. Arch Gen Psychiatry 58(6):545–553
Cotter D, Landau S, Beasley C, Stevenson R, Chana G, MacMillan L, Everall I (2002) The density and spatial distribution of GABAergic neurons, labelled using calcium binding proteins, in the anterior cingulate cortex in major depressive disorder, bipolar disorder, and schizophrenia. Biol Psychiatry 51(5):377–386
Court JA, Perry EK (1991) Dementia: the neurochemical basis of putative transmitter orientated therapy. Pharmacol Ther 52(3):423–443
Cowburn RF, Hardy JA, Roberts PJ (1990) Glutamatergic neurotransmission in Alzheimer's disease. Biochem Soc Trans 18(3):390–392

Coyle JT, Puttfarcken P (1993) Oxidative stress, glutamate, and neurodegenerative disorders. Science 262(5134):689–695. https://doi.org/10.1126/science.7901908

Cross AJ, Crow TJ, Ferrier IN, Johnson JA, Bloom SR, Corsellis JA (1984) Serotonin receptor changes in dementia of the Alzheimer type. J Neurochem 43(6):1574–1581

Cross-Disorder Group of the Psychiatric Genomics C, Lee SH, Ripke S, Neale BM, Faraone SV, Purcell SM, Perlis RH, Mowry BJ, Thapar A, Goddard ME, Witte JS, Absher D, Agartz I, Akil H, Amin F, Andreassen OA, Anjorin A, Anney R, Anttila V, Arking DE, Asherson P, Azevedo MH, Backlund L, Badner JA, Bailey AJ, Banaschewski T, Barchas JD, Barnes MR, Barrett TB, Bass N, Battaglia A, Bauer M, Bayes M, Bellivier F, Bergen SE, Berrettini W, Betancur C, Bettecken T, Biederman J, Binder EB, Black DW, Blackwood DH, Bloss CS, Boehnke M, Boomsma DI, Breen G, Breuer R, Bruggeman R, Cormican P, Buccola NG, Buitelaar JK, Bunney WE, Buxbaum JD, Byerley WF, Byrne EM, Caesar S, Cahn W, Cantor RM, Casas M, Chakravarti A, Chambert K, Choudhury K, Cichon S, Cloninger CR, Collier DA, Cook EH, Coon H, Cormand B, Corvin A, Coryell WH, Craig DW, Craig IW, Crosbie J, Cuccaro ML, Curtis D, Czamara D, Datta S, Dawson G, Day R, De Geus EJ, Degenhardt F, Djurovic S, Donohoe GJ, Doyle AE, Duan J, Dudbridge F, Duketis E, Ebstein RP, Edenberg HJ, Elia J, Ennis S, Etain B, Fanous A, Farmer AE, Ferrier IN, Flickinger M, Fombonne E, Foroud T, Frank J, Franke B, Fraser C, Freedman R, Freimer NB, Freitag CM, Friedl M, Frisen L, Gallagher L, Gejman PV, Georgieva L, Gershon ES, Geschwind DH, Giegling I, Gill M, Gordon SD, Gordon-Smith K, Green EK, Greenwood TA, Grice DE, Gross M, Grozeva D, Guan W, Gurling H, De Haan L, Haines JL, Hakonarson H, Hallmayer J, Hamilton SP, Hamshere ML, Hansen TF, Hartmann AM, Hautzinger M, Heath AC, Henders AK, Herms S, Hickie IB, Hipolito M, Hoefels S, Holmans PA, Holsboer F, Hoogendijk WJ, Hottenga JJ, Hultman CM, Hus V, Ingason A, Ising M, Jamain S, Jones EG, Jones I, Jones L, Tzeng JY, Kahler AK, Kahn RS, Kandaswamy R, Keller MC, Kennedy JL, Kenny E, Kent L, Kim Y, Kirov GK, Klauck SM, Klei L, Knowles JA, Kohli MA, Koller DL, Konte B, Korszun A, Krabbendam L, Krasucki R, Kuntsi J, Kwan P, Landen M, Langstrom N, Lathrop M, Lawrence J, Lawson WB, Leboyer M, Ledbetter DH, Lee PH, Lencz T, Lesch KP, Levinson DF, Lewis CM, Li J, Lichtenstein P, Lieberman JA, Lin DY, Linszen DH, Liu C, Lohoff FW, Loo SK, Lord C, Lowe JK, Lucae S, DJ MI, Madden PA, Maestrini E, Magnusson PK, Mahon PB, Maier W, Malhotra AK, Mane SM, Martin CL, Martin NG, Mattheisen M, Matthews K, Mattingsdal M, McCarroll SA, McGhee KA, McGough JJ, McGrath PJ, McGuffin P, McInnis MG, McIntosh A, McKinney R, McLean AW, McMahon FJ, McMahon WM, McQuillin A, Medeiros H, Medland SE, Meier S, Melle I, Meng F, Meyer J, Middeldorp CM, Middleton L, Milanova V, Miranda A, Monaco AP, Montgomery GW, Moran JL, Moreno-De-Luca D, Morken G, Morris DW, Morrow EM, Moskvina V, Muglia P, Muhleisen TW, Muir WJ, Muller-Myhsok B, Murtha M, Myers RM, Myin-Germeys I, Neale MC, Nelson SF, Nievergelt CM, Nikolov I, Nimgaonkar V, Nolen WA, Nothen MM, Nurnberger JI, Nwulia EA, Nyholt DR, O'Dushlaine C, Oades RD, Olincy A, Oliveira G, Olsen L, Ophoff RA, Osby U, Owen MJ, Palotie A, Parr JR, Paterson AD, Pato CN, Pato MT, Penninx BW, Pergadia ML, Pericak-Vance MA, Pickard BS, Pimm J, Piven J, Posthuma D, Potash JB, Poustka F, Propping P, Puri V, Quested DJ, Quinn EM, Ramos-Quiroga JA, Rasmussen HB, Raychaudhuri S, Rehnstrom K, Reif A, Ribases M, Rice JP, Rietschel M, Roeder K, Roeyers H, Rossin L, Rothenberger A, Rouleau G, Ruderfer D, Rujescu D, Sanders AR, Sanders SJ, Santangelo SL, Sergeant JA, Schachar R, Schalling M, Schatzberg AF, Scheftner WA, Schellenberg GD, Scherer SW, Schork NJ, Schulze TG, Schumacher J, Schwarz M, Scolnick E, Scott LJ, Shi J, Shilling PD, Shyn SI, Silverman JM, Slager SL, Smalley SL, Smit JH, Smith EN, Sonuga-Barke EJ, St Clair D, State M, Steffens M, Steinhausen HC, Strauss JS, Strohmaier J, Stroup TS, Sutcliffe JS, Szatmari P, Szelinger S, Thirumalai S, Thompson RC, Todorov AA, Tozzi F, Treutlein J, Uhr M, van den Oord EJ, Van Grootheest G, Van Os J, Vicente AM, Vieland VJ, Vincent JB, Visscher PM, Walsh CA, Wassink TH, Watson SJ, Weissman MM, Werge T, Wienker TF, Wijsman EM, Willemsen G, Williams N, Willsey AJ, Witt SH, Xu W, Young AH, Yu TW, Zammit S, Zandi PP, Zhang P, Zitman FG, Zollner S, Devlin B, Kelsoe JR, Sklar P, Daly MJ, O'Donovan MC, Craddock N, Sullivan PF, Smoller JW, Kendler KS, Wray NR, International Inflammatory Bowel Disease

Genetics C (2013) Genetic relationship between five psychiatric disorders estimated from genome-wide SNPs. Nat Genet 45(9):984–994. https://doi.org/10.1038/ng.2711

Csernansky JG, Bardgett ME (1998) Limbic-cortical neuronal damage and the pathophysiology of schizophrenia. Schizophr Bull 24(2):231–248. https://doi.org/10.1093/oxfordjournals.schbul.a033323

Csernansky JG, Joshi S, Wang L, Haller JW, Gado M, Miller JP, Grenander U, Miller MI (1998) Hippocampal morphometry in schizophrenia by high dimensional brain mapping. Proc Natl Acad Sci U S A 95(19):11406–11411. https://doi.org/10.1073/pnas.95.19.11406

Cuello AC, Bruno MA, Allard S, Leon W, Iulita MF (2010) Cholinergic involvement in Alzheimer's disease. A link with NGF maturation and degradation. J Mol Neurosci 40(1-2):230–235. https://doi.org/10.1007/s12031-009-9238-z

Culverhouse RC, Saccone NL, Horton AC, Ma Y, Anstey KJ, Banaschewski T, Burmeister M, Cohen-Woods S, Etain B, Fisher HL, Goldman N, Guillaume S, Horwood J, Juhasz G, Lester KJ, Mandelli L, Middeldorp CM, Olie E, Villafuerte S, Air TM, Araya R, Bowes L, Burns R, Byrne EM, Coffey C, Coventry WL, Gawronski KAB, Glei D, Hatzimanolis A, Hottenga JJ, Jaussent I, Jawahar C, Jennen-Steinmetz C, Kramer JR, Lajnef M, Little K, Zu Schwabedissen HM, Nauck M, Nederhof E, Petschner P, Peyrot WJ, Schwahn C, Sinnamon G, Stacey D, Tian Y, Toben C, Van der Auwera S, Wainwright N, Wang JC, Willemsen G, Anderson IM, Arolt V, Aslund C, Bagdy G, Baune BT, Bellivier F, Boomsma DI, Courtet P, Dannlowski U, de Geus EJC, Deakin JFW, Easteal S, Eley T, Fergusson DM, Goate AM, Gonda X, Grabe HJ, Holzman C, Johnson EO, Kennedy M, Laucht M, Martin NG, Munafo MR, Nilsson KW, Oldehinkel AJ, Olsson CA, Ormel J, Otte C, Patton GC, Penninx B, Ritchie K, Sarchiapone M, Scheid JM, Serretti A, Smit JH, Stefanis NC, Surtees PG, Volzke H, Weinstein M, Whooley M, Nurnberger JI Jr, Breslau N, Bierut LJ (2018) Collaborative meta-analysis finds no evidence of a strong interaction between stress and 5-HTTLPR genotype contributing to the development of depression. Mol Psychiatry 23(1):133–142. https://doi.org/10.1038/mp.2017.44

Cummings C, Stewart M, Stevenson M, Morrow J, Nelson J (2011) Neurodevelopment of children exposed in utero to lamotrigine, sodium valproate and carbamazepine. Arch Dis Child 96(7):643–647. https://doi.org/10.1136/adc.2009.176990

Cutrona CE, Cadoret RJ, Suhr JA, Richards CC, Troughton E, Schutte K, Woodworth G (1994) Interpersonal variables in the prediction of alcoholism among adoptees – evidence for gene-environment interactions. Compr Psychiatry 35(3):171–179. https://doi.org/10.1016/0010-440x(94)90188-0

Cutting J (1979) Memory in functional psychosis. J Neurol Neurosurg Psychiatry 42(11):1031–1037. https://doi.org/10.1136/jnnp.42.11.1031

D'Amato RJ, Zweig RM, Whitehouse PJ, Wenk GL, Singer HS, Mayeux R, Price DL, Snyder SH (1987) Aminergic systems in Alzheimer's disease and Parkinson's disease. Ann Neurol 22(2):229–236. https://doi.org/10.1002/ana.410220207

Dackis CA, Gold MS (1985) New concepts in cocaine addiction: the dopamine depletion hypothesis. Neurosci Biobehav Rev 9(3):469–477. https://doi.org/10.1016/0149-7634(85)90022-3

Dalman C, Allebeck P, Cullberg J, Grunewald C, Koster M (1999) Obstetric complications and the risk of schizophrenia: a longitudinal study of a national birth cohort. Arch Gen Psychiatry 56(3):234–240. https://doi.org/10.1001/archpsyc.56.3.234

Daniel DG, Jones DW, Coppola R, Goldberg TE, Bigelow LB, Weinberger DR (1989) Effect of amphetamine on cerebral blood flow (XE-133 dynamic spect) in schizophrenia. Biol Psychiatry 25(7):A157. https://doi.org/10.1016/0006-3223(89)91803-9

Danysz W, Parsons CG (2012) Alzheimer's disease, beta-amyloid, glutamate, NMDA receptors and memantine—searching for the connections. Br J Pharmacol 167(2):324–352. https://doi.org/10.1111/j.1476-5381.2012.02057.x

Davidson J (1972) Cholinergic-adrenergic hypothesis of mania and depression. Lancet 2(7789):1249

Davis KL, Kahn RS, Ko G, Davidson M (1991) Dopamine in schizophrenia: a review and reconceptualization. Am J Psychiatry 148(11):1474–1486. https://doi.org/10.1176/ajp.148.11.1474

Davis KL, Mohs RC, Marin DB, Purohit DP, Perl DP, Lantz M, Austin G, Haroutunian V (1999) Neuropeptide abnormalities in patients with early Alzheimer disease. Arch Gen Psychiatry 56(11):981–987. https://doi.org/10.1001/archpsyc.56.11.981
Daviss SR, Lewis DA (1995) Local circuit neurons of the prefrontal cortex in schizophrenia: selective increase in the density of calbindin-immunoreactive neurons. Psychiatry Res 59(1-2):81–96. https://doi.org/10.1016/0165-1781(95)02720-3
De Leon MJ, George AE, Golomb J, Convit A, De SS (1992) Hippocampal atrophy. Behav Pharmacol 3(Suppl):31. https://doi.org/10.1097/00008877-199204001-00090
Degl'Innocenti A, Agren H, Backman L (1998) Executive deficits in major depression. Acta Psychiatr Scand 97(3):182–188. https://doi.org/10.1111/j.1600-0447.1998.tb09985.x
Del Bo R, Angeretti N, Lucca E, De Simoni MG, Forloni G (1995) Reciprocal control of inflammatory cytokines, IL-1 and IL-6, and beta-amyloid production in cultures. Neurosci Lett 188(1):70–74
Delay J, Buisson JF (1958) Psychic action of isoniazid in the treatment of depressive states. J Clin Exp Psychopathol 19(2, Suppl. 1):51–55
Delay J, Deniker P (1955) Neuroleptic effects of chlorpromazine in therapeutics of neuropsychiatry. J Clin Exp Psychopathol 16(2):104–112
Delay J, Deniker P (1956) Chlorpromazine and neuroleptic treatments in psychiatry. J Clin Exp Psychopathol 17(1):19–24
Delay J, Deniker P, Harl JM (1952) Therapeutic use in psychiatry of phenothiazine of central elective action (4560 RP). Ann Med Psychol (Paris) 110(21):112–117
Delay J, Deniker P, Ropert R (1956) Four years of experience with chlorpromazine in therapy of psychoses. Presse Med 64(22):493–496
Delgado PL (2000) Depression: the case for a monoamine deficiency. J Clin Psychiatry 61(Suppl 6):7–11
Delgado PL, Price LH, Miller HL, Salomon RM, Aghajanian GK, Heninger GR, Charney DS (1994) Serotonin and the neurobiology of depression. Effects of tryptophan depletion in drug-free depressed patients. Arch Gen Psychiatry 51(11):865–874. https://doi.org/10.1001/archpsyc.1994.03950110025005
Demers CH, Bogdan R, Agrawal A (2014) The genetics, neurogenetics and pharmacogenetics of addiction. Curr Behav Neurosci Rep 1(1):33–44. https://doi.org/10.1007/s40473-013-0004-8
DeMyer MK, Gilmor RL, Hendrie HC, DeMyer WE, Augustyn GT, Jackson RK (1988) Magnetic resonance brain images in schizophrenic and normal subjects: influence of diagnosis and education. Schizophr Bull 14(1):21–37. https://doi.org/10.1093/schbul/14.1.21
Deutch AY, Tam S-Y, Roth RH (1985) Footshock and conditioned stress increase 3, 4-dihydroxyphenylacetic acid (DOPAC) in the ventral tegmental area but not substantia nigra. Brain Res 333(1):143–146. https://doi.org/10.1016/0006-8993(85)90134-9
Devor A, Andreassen OA, Wang Y, Maki-Marttunen T, Smeland OB, Fan CC, Schork AJ, Holland D, Thompson WK, Witoelar A, Chen CH, Desikan RS, McEvoy LK, Djurovic S, Greengard P, Svenningsson P, Einevoll GT, Dale AM (2017) Genetic evidence for role of integration of fast and slow neurotransmission in schizophrenia. Mol Psychiatry 22(6):792–801. https://doi.org/10.1038/mp.2017.33
Di Chiara G, Imperato A (1988) Drugs abused by humans preferentially increase synaptic dopamine concentrations in the mesolimbic system of freely moving rats. Proc Natl Acad Sci U S A 85(14):5274–5278. https://doi.org/10.1073/pnas.85.14.5274
Dilts RP, Kalivas PW (1989) Autoradiographic localization of μ-opioid and neurotensin receptors within the mesolimbic dopamine system. Brain Res 488(1-2):311–327. https://doi.org/10.1016/0006-8993(89)90723-3
Doan A, Thinakaran G, Borchelt DR, Slunt HH, Ratovitsky T, Podlisny M, Selkoe DJ, Seeger M, Gandy SE, Price DL, Sisodia SS (1996) Protein topology of presenilin 1. Neuron 17(5):1023–1030. https://doi.org/10.1016/s0896-6273(00)80232-9
Doherty MD, Gratton A (1997) NMDA receptors in nucleus accumbens modulate stress-induced dopamine release in nucleus accumbens and ventral tegmental area. Synapse 26(3):225–234. https://doi.org/10.1002/(SICI)1098-2396(199707)26:3<225::AID-SYN4>3.0.CO;2-9

Dong D, Wang Y, Chang X, Luo C, Yao D (2018) Dysfunction of large-scale brain networks in schizophrenia: a meta-analysis of resting-state functional connectivity. Schizophr Bull 44(1):168–181. https://doi.org/10.1093/schbul/sbx034
Drevets WC, Videen TO, Price JL, Preskorn SH, Carmichael ST, Raichle ME (1992) A functional anatomical study of unipolar depression. J Neurosci 12(9):3628–3641
Drevets WC, Price JL, Simpson JR Jr, Todd RD, Reich T, Vannier M, Raichle ME (1997) Subgenual prefrontal cortex abnormalities in mood disorders. Nature 386(6627):824–827. https://doi.org/10.1038/386824a0
Duman RS, Charney DS (1999) Cell atrophy and loss in major depression. Biol Psychiatry 45(9):1083–1084
Duman RS, Heninger GR, Nestler EJ (1997) A molecular and cellular theory of depression. Arch Gen Psychiatry 54(7):597–606. https://doi.org/10.1001/archpsyc.1997.01830190015002
Eastwood SL (2004) The synaptic pathology of schizophrenia: is aberrant neurodevelopment and plasticity to blame? Int Rev Neurobiol. https://doi.org/10.1016/s0074-7742(04)59003-7
Eastwood SL, Harrison PJ (2001) Synaptic pathology in the anterior cingulate cortex in schizophrenia and mood disorders. A review and a Western blot study of synaptophysin, GAP-43 and the complexins. Brain Res Bull 55(5):569–578
Eikelenboom P, Zhan SS, van Gool WA, Allsop D (1994) Inflammatory mechanisms in Alzheimer's disease. Trends Pharmacol Sci 15(12):447–450
Eikelenboom P, Rozemuller JM, van Muiswinkel FL (1998) Inflammation and Alzheimer's disease: relationships between pathogenic mechanisms and clinical expression. Exp Neurol 154(1):89–98. https://doi.org/10.1006/exnr.1998.6920
Elliott R, Sahakian BJ, McKay AP, Herrod JJ, Robbins TW, Paykel ES (1996) Neuropsychological impairments in unipolar depression: the influence of perceived failure on subsequent performance. Psychol Med 26(5):975–989. https://doi.org/10.1017/s0033291700035303
Elliott R, Baker SC, Rogers RD, OLeary DA, Paykel ES, Frith CD, Dolan RJ, Sahakian BJ (1997) Prefrontal dysfunction in depressed patients performing a complex planning task: a study using positron emission tomography. Psychol Med 27(4):931–942. https://doi.org/10.1017/S0033291797005187
Espey MG, Chernyshev ON, Reinhard JF Jr, Namboodiri MA, Colton CA (1997) Activated human microglia produce the excitotoxin quinolinic acid. Neuroreport 8(2):431–434. https://doi.org/10.1097/00001756-199701200-00011
Ettenberg A, Pettit HO, Bloom FE, Koob GF (1982) Heroin and cocaine intravenous self-administration in rats: mediation by separate neural systems. Psychopharmacology (Berl) 78(3):204–209. https://doi.org/10.1007/bf00428151
Everitt BJ, Morris KA, Obrien A, Robbins TW (1991) The basolateral amygdala ventral striatal system and conditioned place preference – further evidence of limbic striatal interactions underlying reward-related processes. Neuroscience 42(1):1–18. https://doi.org/10.1016/0306-4522(91)90145-E
Ezrin-Waters C, Resch L (1986) The nucleus basalis of Meynert. Can J Neurol Sci 13(1):8–14. https://doi.org/10.1017/s0317167100035721
Fadda F, Rossetti ZL (1998) Chronic ethanol consumption: from neuroadaptation to neurodegeneration. Prog Neurobiol 56(4):385–431. https://doi.org/10.1016/s0301-0082(98)00032-x
Farrow TF, Whitford TJ, Williams LM, Gomes L, Harris AW (2005) Diagnosis-related regional gray matter loss over two years in first episode schizophrenia and bipolar disorder. Biol Psychiatry 58(9):713–723. https://doi.org/10.1016/j.biopsych.2005.04.033
Fernandez-Tome P, Brera B, Arevalo MA, de Ceballos ML (2004) Beta-amyloid25-35 inhibits glutamate uptake in cultured neurons and astrocytes: modulation of uptake as a survival mechanism. Neurobiol Dis 15(3):580–589. https://doi.org/10.1016/j.nbd.2003.12.006
Ferreira-Vieira TH, Guimaraes IM, Silva FR, Ribeiro FM (2016) Alzheimer's disease: targeting the cholinergic system. Curr Neuropharmacol 14(1):101–115
Fibiger HC (1978) Drugs and reinforcement mechanisms: a critical review of the catecholamine theory. Annu Rev Pharmacol Toxicol 18(1):37–56. https://doi.org/10.1146/annurev.pa.18.040178.000345

Forloni G, Demicheli F, Giorgi S, Bendotti C, Angeretti N (1992) Expression of amyloid precursor protein mRNAs in endothelial, neuronal and glial cells: modulation by interleukin-1. Brain Res Mol Brain Res 16(1-2):128–134

Fosnaugh JS, Bhat RV, Yamagata K, Worley PF, Baraban JM (2002) Activation of arc, a putative "effector" immediate early gene, by cocaine in rat brain. J Neurochem 64(5):2377–2380. https://doi.org/10.1046/j.1471-4159.1995.64052377.x

Fountoulakis KN (2012) The possible involvement of NMDA glutamate receptor in the etiopathogenesis of bipolar disorder. Curr Pharm Des 18(12):1605–1608

Fountoulakis K (2015a) Aetiopathogenesis of bipolar disorder. In: Fountoulakis K (ed) Bipolar disorder: an evidence-based guide to manic depression. Springer-Verlag, Berlin, pp 389–419. https://doi.org/10.1007/978-3-642-37216-2

Fountoulakis K (2015b) Biological therapies. In: Fountoulakis K (ed) Bipolar disorder: an evidence-based guide to manic depression. Springer-Verlag, Berlin, pp 461–625. https://doi.org/10.1007/978-3-642-37216-2

Fountoulakis KN, Moller HJ (2012) Antidepressant drugs and the response in the placebo group: the real problem lies in our understanding of the issue. J Psychopharmacol 26(5):744–750. https://doi.org/10.1177/0269881111421969

Fountoulakis KN, Giannakopoulos P, Kovari E, Bouras C (2008a) Assessing the role of cingulate cortex in bipolar disorder: neuropathological, structural and functional imaging data. Brain Res Rev 59(1):9–21. https://doi.org/10.1016/j.brainresrev.2008.04.005

Fountoulakis KN, Vieta E, Bouras C, Notaridis G, Giannakopoulos P, Kaprinis G, Akiskal H (2008b) A systematic review of existing data on long-term lithium therapy: neuroprotective or neurotoxic? Int J Neuropsychopharmacol 11(2):269–287. https://doi.org/10.1017/S1461145707007821

Fountoulakis KN, Gonda X, Vieta E, Rihmer Z (2011) Class effect of pharmacotherapy in bipolar disorder: fact or misbelief? Ann Gen Psychiatry 10(1):8. https://doi.org/10.1186/1744-859X-10-8

Fountoulakis KN, Kelsoe JR, Akiskal H (2012) Receptor targets for antidepressant therapy in bipolar disorder: an overview. J Affect Disord 138(3):222–238. https://doi.org/10.1016/j.jad.2011.04.043

Fountoulakis KN, Veroniki AA, Siamouli M, Moller HJ (2013) No role for initial severity on the efficacy of antidepressants: results of a multi-meta-analysis. Ann Gen Psychiatry 12(1):26. https://doi.org/10.1186/1744-859X-12-26

Fountoulakis KN, Gazouli M, Kelsoe J, Akiskal H (2015) The pharmacodynamic properties of lurasidone and their role in its antidepressant efficacy in bipolar disorder. Eur Neuropsychopharmacol 25(3):335–342. https://doi.org/10.1016/j.euroneuro.2014.11.010

Fountoulakis KN, Grunze H, Vieta E, Young A, Yatham L, Blier P, Kasper S, Moeller HJ (2017a) The International College of Neuro-Psychopharmacology (CINP) treatment guidelines for bipolar disorder in adults (CINP-BD-2017), part 3: the clinical guidelines. Int J Neuropsychopharmacol 20(2):180–195. https://doi.org/10.1093/ijnp/pyw109

Fountoulakis KN, Vieta E, Young A, Yatham L, Grunze H, Blier P, Moeller HJ, Kasper S (2017b) The International College of Neuropsychopharmacology (CINP) treatment guidelines for bipolar disorder in adults (CINP-BD-2017), part 4: unmet needs in the treatment of bipolar disorder and recommendations for future research. Int J Neuropsychopharmacol 20(2):196–205. https://doi.org/10.1093/ijnp/pyw072

Fountoulakis KN, Yatham L, Grunze H, Vieta E, Young A, Blier P, Kasper S, Moeller HJ (2017c) The International College of Neuro-Psychopharmacology (CINP) treatment guidelines for bipolar disorder in adults (CINP-BD-2017), part 2: review, grading of the evidence, and a precise algorithm. Int J Neuropsychopharmacol 20(2):121–179. https://doi.org/10.1093/ijnp/pyw100

Fountoulakis KN, Young A, Yatham L, Grunze H, Vieta E, Blier P, Moeller HJ, Kasper S (2017d) The International College of Neuropsychopharmacology (CINP) treatment guidelines for bipolar disorder in adults (CINP-BD-2017), part 1: background and methods of the development of guidelines. Int J Neuropsychopharmacol 20(2):98–120. https://doi.org/10.1093/ijnp/pyw091

Fox JH, Penn R, Clasen R, Martin E, Wilson R, Savoy S (1985) Pathological diagnosis in clinically typical Alzheimer's disease. N Engl J Med 313(22):1419–1420

Fox NC, Warrington EK, Rossor MN (1999) Serial magnetic resonance imaging of cerebral atrophy in preclinical Alzheimer's disease. Lancet 353(9170):2125. https://doi.org/10.1016/S0140-6736(99)00496-1

Frank RA, Martz S, Pommering T (1988) The effect of chronic cocaine on self-stimulation train-duration thresholds. Pharmacol Biochem Behav 29(4):755–758. https://doi.org/10.1016/0091-3057(88)90199-2

Frank RA, Manderscheid PZ, Panicker S, Williams HP, Kokoris D (1992) Cocaine euphoria, dysphoria, and tolerance assessed using drug-induced changes in brain-stimulation reward. Pharmacol Biochem Behav 42(4):771–779. https://doi.org/10.1016/0091-3057(92)90028-E

French ED (1997) Δ9-Tetrahydrocannabinol excites rat VTA dopamine neurons through activation of cannabinoid CB1 but not opioid receptors. Neurosci Lett 226(3):159–162. https://doi.org/10.1016/s0304-3940(97)00278-4

French ED, Ceci A (1990) Non-competitive N-methyl-D-aspartate antagonists are potent activators of ventral tegmental A10 dopamine neurons. Neurosci Lett 119(2):159–162. https://doi.org/10.1016/0304-3940(90)90823-r

Frey U, Frey S, Schollmeier F, Krug M (1996) Influence of actinomycin D, a RNA synthesis inhibitor, on long-term potentiation in rat hippocampal neurons in vivo and in vitro. J Physiol 490(Pt 3):703–711. https://doi.org/10.1113/jphysiol.1996.sp021179

Friedman AS (1964) Minimal effects of severe depression on cognitive functioning. J Abnorm Psychol 69(3):237–243. https://doi.org/10.1037/h0048608

Frith CD, Stevens M, Johnstone EC, Deakin JFW, Lawler P, Crow TJ (1983) Effects of ECT and depression on various aspects of memory. Br J Psychiatry 142:610–617. https://doi.org/10.1192/bjp.142.6.610

Fuchs E, Flugge G (1998) Stress, glucocorticoids and structural plasticity of the hippocampus. Neurosci Biobehav Rev 23(2):295–300. https://doi.org/10.1016/s0149-7634(98)00031-1

Fujita M, Charney DS, Innis RB (2000) Imaging serotonergic neurotransmission in depression: hippocampal pathophysiology may mirror global brain alterations. Biol Psychiatry 48(8):801–812. https://doi.org/10.1016/S0006-3223(00)00960-4

Fusar-Poli P, Meyer-Lindenberg A (2013) Striatal presynaptic dopamine in schizophrenia, part II: meta-analysis of [(18)F/(11)C]-DOPA PET studies. Schizophr Bull 39(1):33–42. https://doi.org/10.1093/schbul/sbr180

Galynker II, Cai J, Ongseng F, Finestone H, Dutta E, Serseni D (1998) Hypofrontality and negative symptoms in major depressive disorder. J Nucl Med 39(4):608–612

Gannon M, Che P, Chen Y, Jiao K, Roberson ED, Wang Q (2015) Noradrenergic dysfunction in Alzheimer's disease. Front Neurosci 9:220. https://doi.org/10.3389/fnins.2015.00220

Garbutt JC, Vankammen DP (1983) The interaction between gaba and dopamine – implications for schizophrenia. Schizophr Bull 9(3):336–353. https://doi.org/10.1093/schbul/9.3.336

Garcia-Ayllon MS, Small DH, Avila J, Saez-Valero J (2011) Revisiting the role of acetylcholinesterase in Alzheimer's disease: cross-talk with P-tau and beta-amyloid. Front Mol Neurosci 4:22. https://doi.org/10.3389/fnmol.2011.00022

Gatto GJ, Mcbride WJ, Murphy JM, Lumeng L, Li TK (1994) Ethanol self-infusion into the ventral tegmental area by alcohol-preferring rats. Alcohol 11(6):557–564. https://doi.org/10.1016/0741-8329(94)90083-3

Geddes JR, Lawrie SM (1995) Obstetric complications and schizophrenia: a meta-analysis. Br J Psychiatry 167(6):786–793. https://doi.org/10.1192/bjp.167.6.786

Geddes JW, Chang-Chui H, Cooper SM, Lott IT, Cotman CW (1986) Density and distribution of NMDA receptors in the human hippocampus in Alzheimer's disease. Brain Res 399(1):156–161

George MS, Ketter TA, Parekh PI, Rosinsky N, Ring HA, Pazzaglia PJ, Marangell LB, Callahan AM, Post RM (1997) Blunted left cingulate activation in mood disorder subjects during a response interference task (the Stroop). J Neuropsychiatry Clin Neurosci 9(1):55–63. https://doi.org/10.1176/jnp.9.1.55

George TP, Verrico CD, Roth RH (1998) Effects of repeated nicotine pre-treatment on mesoprefrontal dopaminergic and behavioral responses to acute footshock stress. Brain Res 801(1-2):36–49. https://doi.org/10.1016/s0006-8993(98)00537-x

Gerard RW (1955a) The biological roots of psychiatry. Am J Psychiatry 112(2):81–90. https://doi.org/10.1176/ajp.112.2.81

Gerard RW (1955b) Biological roots of psychiatry. Science 122(3162):225–230

Gershon S, Yuwiler A (1960) Lithium ion: a specific psychopharmacological approach to the treatment of mania. J Neuropsychiatr 1:229–241

Gessa GL, Muntoni F, Collu M, Vargiu L, Mereu G (1985) Low doses of ethanol activate dopaminergic neurons in the ventral tegmental area. Brain Res 348(1):201–203. https://doi.org/10.1016/0006-8993(85)90381-6

Geula C, Mesulam MM (1996) Systematic regional variations in the loss of cortical cholinergic fibers in Alzheimer's disease. Cereb Cortex 6(2):165–177

Giulian D, Haverkamp LJ, Li J, Karshin WL, Yu J, Tom D, Li X, Kirkpatrick JB (1995) Senile plaques stimulate microglia to release a neurotoxin found in Alzheimer brain. Neurochem Int 27(1):119–137. https://doi.org/10.1016/0197-0186(95)00067-I

Glausier JR, Lewis DA (2013) Dendritic spine pathology in schizophrenia. Neuroscience 251:90–107. https://doi.org/10.1016/j.neuroscience.2012.04.044

Goeders NE, Smith JE (1986) Reinforcing properties of cocaine in the medial prefrontal cortex – primary action on presynaptic dopaminergic terminals. Pharmacol Biochem Behav 25(1):191–199. https://doi.org/10.1016/0091-3057(86)90252-2

Goeders NE, Lane JD, Smith JE (1984) Self-administration of methionine enkephalin into the nucleus accumbens. Pharmacol Biochem Behav 20(3):451–455. https://doi.org/10.1016/0091-3057(84)90284-3

Goeders NE, Dworkin SI, Smith JE (1986) Neuropharmacological assessment of cocaine self-administration into the medial prefrontal cortex. Pharmacol Biochem Behav 24(5):1429–1440. https://doi.org/10.1016/0091-3057(86)90206-6

Golinkoff M, Sweeney JA (1989) Cognitive impairments in depression. J Affect Disord 17(2):105–112. https://doi.org/10.1016/0165-0327(89)90032-3

Gonzales RA, Weiss F (1998) Suppression of ethanol-reinforced behavior by naltrexone is associated with attenuation of the ethanol-induced increase in dialysate dopamine levels in the nucleus accumbens. J Neurosci 18(24):10663–10671

Gooch MD, Stennett DJ (1996) Molecular basis of Alzheimer's disease. Am J Health Syst Pharm 53(13):1545–1557. quiz 1603–1544

Goodwin GM (1997) Neuropsychological and neuroimaging evidence for the involvement of the frontal lobes in depression. J Psychopharmacol 11(2):115–122. https://doi.org/10.1177/026988119701100204

Goodwin FK (2002) Rationale for long-term treatment of bipolar disorder and evidence for long-term lithium treatment. J Clin Psychiatry 63(Suppl 10):5–12

Goodwin FK, Zis AP (1979) Lithium in the treatment of mania: comparisons with neuroleptics. Arch Gen Psychiatry 36(8 Spec):840–844

Goodwin FK, Murphy DL, Bunney WE Jr (1969a) Lithium. Lancet 2(7613):212–213

Goodwin FK, Murphy DL, Bunney WE Jr (1969b) Lithium-carbonate treatment in depression and mania. A longitudinal double-blind study. Arch Gen Psychiatry 21(4):486–496

Goodwin FK, Murphy DL, Dunner DL, Bunney WE Jr (1972) Lithium response in unipolar versus bipolar depression. Am J Psychiatry 129(1):44–47. https://doi.org/10.1176/ajp.129.1.44

Goodwin FK, Fireman B, Simon GE, Hunkeler EM, Lee J, Revicki D (2003) Suicide risk in bipolar disorder during treatment with lithium and divalproex. JAMA 290(11):1467–1473. https://doi.org/10.1001/jama.290.11.1467

Gottesman II, Shields J (1982) Schizophrenia: the epigenetic puzzle. Cambridge University Press, Cambridge

Grace AA (2000) Gating of information flow within the limbic system and the pathophysiology of schizophrenia. Brain Res Brain Res Rev 31(2-3):330–341

Granger P, Biton B, Faure C, Vige X, Depoortere H, Graham D, Langer SZ, Scatton B, Avenet P (1995) Modulation of the gamma-aminobutyric acid type A receptor by the antiepileptic drugs carbamazepine and phenytoin. Mol Pharmacol 47(6):1189–1196

Grant S, London ED, Newlin DB, Villemagne VL, Liu X, Contoreggi C, Phillips RL, Kimes AS, Margolin A (1996) Activation of memory circuits during cue-elicited cocaine craving. Proc Natl Acad Sci U S A 93(21):12040–12045. https://doi.org/10.1073/pnas.93.21.12040

Graybiel AM (1998) The basal ganglia and chunking of action repertoires. Neurobiol Learn Mem 70(1-2):119–136. https://doi.org/10.1006/nlme.1998.3843

Greenamyre JT, Maragos WF (1993) Neurotransmitter receptors in Alzheimer disease. Cerebrovasc Brain Metab Rev 5(2):61–94

Greenamyre JT, Young AB (1989) Synaptic localization of striatal NMDA, quisqualate and kainate receptors. Neurosci Lett 101(2):133–137

Greenamyre JT, Penney JB, Young AB, D'Amato CJ, Hicks SP, Shoulson I (1985) Alterations in L-glutamate binding in Alzheimer's and Huntington's diseases. Science 227(4693):1496–1499

Greenamyre JT, Penney JB, D'Amato CJ, Young AB (1987) Dementia of the Alzheimer's type: changes in hippocampal L-[3H]glutamate binding. J Neurochem 48(2):543–551

Greicius MD, Flores BH, Menon V, Glover GH, Solvason HB, Kenna H, Reiss AL, Schatzberg AF (2007) Resting-state functional connectivity in major depression: abnormally increased contributions from subgenual cingulate cortex and thalamus. Biol Psychiatry 62(5):429–437. https://doi.org/10.1016/j.biopsych.2006.09.020

Grenhoff J, Aston-Jones G, Svensson TH (1986) Nicotinic effects on the firing pattern of midbrain dopamine neurons. Acta Physiol Scand 128(3):351–358. https://doi.org/10.1111/j.1748-1716.1986.tb07988.x

Griffin WST, Sheng JG, Royston MC, Gentleman SM, McKenzie JE, Graham DI, Roberts GW, Mrak RE (2006) Glial-neuronal interactions in Alzheimer's disease: the potential role of a 'cytokine cycle' in disease progression. Brain Pathol 8(1):65–72. https://doi.org/10.1111/j.1750-3639.1998.tb00136.x

Grimm JW, See RE (2000) Dissociation of primary and secondary reward-relevant limbic nuclei in an animal model of relapse. Neuropsychopharmacology 22(5):473–479. https://doi.org/10.1016/S0893-133X(99)00157-8

Grove WM, Eckert ED, Heston L, Bouchard TJ Jr, Segal N, Lykken DT (1990) Heritability of substance abuse and antisocial behavior: a study of monozygotic twins reared apart. Biol Psychiatry 27(12):1293–1304. https://doi.org/10.1016/0006-3223(90)90500-2

Gruber SA, Rogowska J, Yurgelun-Todd DA (2004) Decreased activation of the anterior cingulate in bipolar patients: an fMRI study. J Affect Disord 82(2):191–201. https://doi.org/10.1016/j.jad.2003.10.010

Gudayol-Ferre E, Pero-Cebollero M, Gonzalez-Garrido AA, Guardia-Olmos J (2015) Changes in brain connectivity related to the treatment of depression measured through fMRI: a systematic review. Front Hum Neurosci 9:582. https://doi.org/10.3389/fnhum.2015.00582

von Gunten A, Fox NC, Cipolotti L, Ron MA (2000) A volumetric study of hippocampus and amygdala in depressed patients with subjective memory problems. J Neuropsychiatry Clin Neurosci 12(4):493–498. https://doi.org/10.1176/jnp.12.4.493

Gysling K, Wang RY (1983) Morphine-induced activation of A10 dopamine neurons in the rat. Brain Res 277(1):119–127. https://doi.org/10.1016/0006-8993(83)90913-7

Haggerty JJ Jr, Stern RA, Mason GA, Beckwith J, Morey CE, Prange AJ Jr (1993) Subclinical hypothyroidism: a modifiable risk factor for depression? Am J Psychiatry 150(3):508–510. https://doi.org/10.1176/ajp.150.3.508

Haile CN, Kosten TR, Kosten TA (2007) Genetics of dopamine and its contribution to cocaine addiction. Behav Genet 37(1):119–145. https://doi.org/10.1007/s10519-006-9115-2

Hajek T, Carrey N, Alda M (2005) Neuroanatomical abnormalities as risk factors for bipolar disorder. Bipolar Disord 7(5):393–403. https://doi.org/10.1111/j.1399-5618.2005.00238.x

Halliday GM, McCann HL, Pamphlett R, Brooks WS, Creasey H, McCusker E, Cotton RG, Broe GA, Harper CG (1992) Brain stem serotonin-synthesizing neurons in Alzheimer's disease: a clinicopathological correlation. Acta Neuropathol 84(6):638–650

Harrison PJ (2004) The hippocampus in schizophrenia: a review of the neuropathological evidence and its pathophysiological implications. Psychopharmacology (Berl) 174(1):151–162. https://doi.org/10.1007/s00213-003-1761-y

Harvey I, Ron MA, Du Boulay G, Wicks D, Lewis SW, Murray RM (1993) Reduction of cortical volume in schizophrenia on magnetic resonance imaging. Psychol Med 23(3):591–604. https://doi.org/10.1017/s003329170002537x

Haxby JV, Grady CL, Duara R, Schlageter N, Berg G, Rapoport SI (1986) Neocortical metabolic abnormalities precede nonmemory cognitive defects in early Alzheimer's-type dementia. Arch Neurol 43(9):882–885. https://doi.org/10.1001/archneur.1986.00520090022010

Heikkila RE, Orlansky H, Cohen G (1975) Studies on the distinction between uptake inhibition and release of [3H]dopamine in rat brain tissue slices. Biochem Pharmacol 24(8):847–852. https://doi.org/10.1016/0006-2952(75)90152-5

Heller W, Nitscke JB (1997) regional brain activity in emotion: a framework for understanding cognition in depression. Cognit Emot 11(5-6):637–661. https://doi.org/10.1080/0269993973 79845a

Heller W, Nitschke JB, Etienne MA, Miller GA (1997) Patterns of regional brain activity differentiate types of anxiety. J Abnorm Psychol 106(3):376–385. https://doi.org/10.1037//0021-843x.106.3.376

Heninger GR, Delgado PL, Charney DS (1996) The revised monoamine theory of depression: a modulatory role for monoamines, based on new findings from monoamine depletion experiments in humans. Pharmacopsychiatry 29(1):2–11. https://doi.org/10.1055/s-2007-979535

Herholz K, Bauer B, Wienhard K, Kracht L, Mielke R, Lenz MO, Strotmann T, Heiss WD (2000) In-vivo measurements of regional acetylcholine esterase activity in degenerative dementia: comparison with blood flow and glucose metabolism. J Neural Transm 107(12):1457–1468. https://doi.org/10.1007/s007020070009

Herrmann M, Rotte M, Grubich C, Ebert AD, Schiltz K, Munte TF, Heinze HJ (2001) Control of semantic interference in episodic memory retrieval is associated with an anterior cingulate-prefrontal activation pattern. Hum Brain Mapp 13(2):94–103. https://doi.org/10.1002/hbm.1027

Hietala J, Syvalahti E, Vilkman H, Vuorio K, Rakkolainen V, Bergman J, Haaparanta M, Solin O, Kuoppamaki M, Eronen E, Ruotsalainen U, Salokangas RK (1999) Depressive symptoms and presynaptic dopamine function in neuroleptic-naive schizophrenia. Schizophr Res 35(1):41–50. https://doi.org/10.1016/s0920-9964(98)00113-3

Hilker R, Helenius D, Fagerlund B, Skytthe A, Christensen K, Werge TM, Nordentoft M, Glenthoj B (2018) Heritability of schizophrenia and schizophrenia spectrum based on the Nationwide Danish Twin Register. Biol Psychiatry 83(6):492–498. https://doi.org/10.1016/j.biopsych.2017.08.017

Hirayasu Y, Shenton ME, Salisbury DF, Kwon JS, Wible CG, Fischer IA, Yurgelun-Todd D, Zarate C, Kikinis R, Jolesz FA, McCarley RW (1999) Subgenual cingulate cortex volume in first-episode psychosis. Am J Psychiatry 156(7):1091–1093. https://doi.org/10.1176/ajp.156.7.1091

Hirsch C, Bartenstein P, Minoshima S, Mosch D, Willoch F, Buch K, Schad D, Schwaiger M, Kurz A (1997) Reduction of regional cerebral blood flow and cognitive impairment in patients with Alzheimer's disease: evaluation of an observer-independent analytic approach. Dement Geriatr Cogn Disord 8(2):98–104. https://doi.org/10.1159/000106613

Hirschfeld RM (2000) History and evolution of the monoamine hypothesis of depression. J Clin Psychiatry 61(Suppl 6):4–6

Hirvonen J, van Erp TG, Huttunen J, Aalto S, Nagren K, Huttunen M, Lonnqvist J, Kaprio J, Cannon TD, Hietala J (2006) Brain dopamine d1 receptors in twins discordant for schizophrenia. Am J Psychiatry 163(10):1747–1753. https://doi.org/10.1176/ajp.2006.163.10.1747

Hoebel BG, Monaco AP, Hernandez L, Aulisi EF, Stanley BG, Lenard L (1983) Self-injection of amphetamine directly into the brain. Psychopharmacology 81(2):158–163. https://doi.org/10.1007/Bf00429012

Holman BL, Johnson KA, Gerada B, Carvalho PA, Satlin A (1992) The scintigraphic appearance of Alzheimer's disease: a prospective study using technetium-99m-HMPAO SPECT. J Nucl Med 33(2):181–185
Holzman PS, Levy DL, Matthysse SW, Abel LA (1997) Smooth pursuit eye tracking in twins. A critical commentary. Arch Gen Psychiatry 54(5):429–431
Homan P, Neumeister A, Nugent AC, Charney DS, Drevets WC, Hasler G (2015) Serotonin versus catecholamine deficiency: behavioral and neural effects of experimental depletion in remitted depression. Transl Psychiatry 5(3):e532. https://doi.org/10.1038/tp.2015.25
Horger BA, Shelton K, Schenk S (1990) Preexposure sensitizes rats to the rewarding effects of cocaine. Pharmacol Biochem Behav 37(4):707–711. https://doi.org/10.1016/0091-3057(90)90552-s
Horger BA, Giles MK, Schenk S (1992) Preexposure to amphetamine and nicotine predisposes rats to self-administer a low dose of cocaine. Psychopharmacology (Berl) 107(2-3):271–276. https://doi.org/10.1007/bf02245147
Houenou J, Wessa M, Douaud G, Leboyer M, Chanraud S, Perrin M, Poupon C, Martinot JL, Paillere-Martinot ML (2007) Increased white matter connectivity in euthymic bipolar patients: diffusion tensor tractography between the subgenual cingulate and the amygdalo-hippocampal complex. Mol Psychiatry 12(11):1001–1010. https://doi.org/10.1038/sj.mp.4002010
Howell O, Atack JR, Dewar D, McKernan RM, Sur C (2000) Density and pharmacology of alpha5 subunit-containing GABA(A) receptors are preserved in hippocampus of Alzheimer's disease patients. Neuroscience 98(4):669–675
Howes OD, Kapur S (2009) The dopamine hypothesis of schizophrenia: version III—the final common pathway. Schizophr Bull 35(3):549–562. https://doi.org/10.1093/schbul/sbp006
Howes OD, Egerton A, Allan V, McGuire P, Stokes P, Kapur S (2009a) Mechanisms underlying psychosis and antipsychotic treatment response in schizophrenia: insights from PET and SPECT imaging. Curr Pharm Des 15(22):2550–2559. https://doi.org/10.2174/138161209788957528
Howes OD, Montgomery AJ, Asselin MC, Murray RM, Valli I, Tabraham P, Bramon-Bosch E, Valmaggia L, Johns L, Broome M, McGuire PK, Grasby PM (2009b) Elevated striatal dopamine function linked to prodromal signs of schizophrenia. Arch Gen Psychiatry 66(1):13–20. https://doi.org/10.1001/archgenpsychiatry.2008.514
Howes OD, Kambeitz J, Kim E, Stahl D, Slifstein M, Abi-Dargham A, Kapur S (2012) The nature of dopamine dysfunction in schizophrenia and what this means for treatment. Arch Gen Psychiatry 69(8):776–786. https://doi.org/10.1001/archgenpsychiatry.2012.169
Howes OD, Williams M, Ibrahim K, Leung G, Egerton A, McGuire PK, Turkheimer F (2013) Midbrain dopamine function in schizophrenia and depression: a post-mortem and positron emission tomographic imaging study. Brain 136(Pt 11):3242–3251. https://doi.org/10.1093/brain/awt264
Howes O, McCutcheon R, Stone J (2015) Glutamate and dopamine in schizophrenia: an update for the 21st century. J Psychopharmacol 29(2):97–115. https://doi.org/10.1177/0269881114563634
Hultman CM, Sparen P, Takei N, Murray RM, Cnattingius S, Geddes J (1999) Prenatal and perinatal risk factors for schizophrenia, affective psychosis, and reactive psychosis of early onset: case-control study prenatal and perinatal risk factors for early onset schizophrenia, affective psychosis, and reactive psychosis. BMJ 318(7181):421–426. https://doi.org/10.1136/bmj.318.7181.421
Humphries C, Mortimer A, Hirsch S, de Belleroche J (1996) NMDA receptor mRNA correlation with antemortem cognitive impairment in schizophrenia. Neuroreport 7(12):2051–2055. https://doi.org/10.1097/00001756-199608120-00040
Hurd YL, Weiss F, Koob GF, And NE, Ungerstedt U (1989) Cocaine reinforcement and extracellular dopamine overflow in rat nucleus accumbens: an in vivo microdialysis study. Brain Res 498(1):199–203. https://doi.org/10.1016/0006-8993(89)90422-8
Hyman SE (1996) Shaking out the cause of addiction. Science 273(5275):611–612. https://doi.org/10.1126/science.273.5275.611

Hyman BT, Van Hoesen GW, Damasio AR, Barnes CL (1984) Alzheimer's disease: cell-specific pathology isolates the hippocampal formation. Science 225(4667):1168–1170

Hyman BT, Van Hoesen GW, Damasio AR (1987) Alzheimer's disease: glutamate depletion in the hippocampal perforant pathway zone. Ann Neurol 22(1):37–40. https://doi.org/10.1002/ana.410220110

Iacono WG, Clementz BA (1993) A strategy for elucidating genetic influences on complex psychopathological syndromes (with special reference to ocular motor functioning and schizophrenia). Prog Exp Pers Psychopathol Res 16:11–65

Ikeda M, Tanabe H, Nakagawa Y, Kazui H, Oi H, Yamazaki H, Harada K, Nishimura T (1994) MRI-based quantitative assessment of the hippocampal region in very mild to moderate Alzheimer's disease. Neuroradiology 36(1):7–10. https://doi.org/10.1007/bf00599184

Ikeda Y, Kameyama M, Narita K, Takei Y, Suda M, Aoyama Y, Yuuki N, Sakurai N, Fukuda M, Mikuni M, Amanuma M (2010) Total and regional brain volume reductions due to the Syndrome of Irreversible Lithium-Effectuated Neurotoxicity (SILENT): a voxel-based morphometric study. Prog Neuropsychopharmacol Biol Psychiatry 34(1):244–246. https://doi.org/10.1016/j.pnpbp.2009.10.010

Imperato A, Di Chiara G (1986) Preferential stimulation of dopamine release in the nucleus accumbens of freely moving rats by ethanol. J Pharmacol Exp Ther 239(1):219–228

Imran MB, Kawashima R, Awata S, Sato K, Kinomura S, Ono S, Sato M, Fukuda H (1999) Tc-99m HMPAO SPECT in the evaluation of Alzheimer's disease: correlation between neuropsychiatric evaluation and CBF images. J Neurol Neurosurg Psychiatry 66(2):228–232. https://doi.org/10.1136/jnnp.66.2.228

Ingvar DH, Franzen G (1974) Distribution of cerebral activity in chronic schizophrenia. Lancet 2(7895):1484–1486. https://doi.org/10.1016/s0140-6736(74)90221-9

Ishita S, Negishi K, Teranishi T, Shimada Y, Kato S (1988) GABAergic inhibition on dopamine cells of the fish retina: a [3H]dopamine release study with isolated cell fractions. J Neurochem 50(1):1–6. https://doi.org/10.1111/j.1471-4159.1988.tb13221.x

Itokawa M, Yamada K, Iwayama-Shigeno Y, Ishitsuka Y, Detera-Wadleigh S, Yoshikawa T (2003) Genetic analysis of a functional GRIN2A promoter (GT)n repeat in bipolar disorder pedigrees in humans. Neurosci Lett 345(1):53–56

Itzhak Y, Martin JL (1999) Effects of cocaine, nicotine, dizocipline and alcohol on mice locomotor activity: cocaine–alcohol cross-sensitization involves upregulation of striatal dopamine transporter binding sites. Brain Res 818(2):204–211. https://doi.org/10.1016/s0006-8993(98)01260-8

Jack CR Jr, Petersen RC, Xu YC, O'Brien PC, Smith GE, Ivnik RJ, Boeve BF, Waring SC, Tangalos EG, Kokmen E (1999) Prediction of AD with MRI-based hippocampal volume in mild cognitive impairment. Neurology 52(7):1397–1403. https://doi.org/10.1212/wnl.52.7.1397

Janowsky DS, El-Yousef MK, Davis JM, Sekerke HJ (1972) A cholinergic-adrenergic hypothesis of mania and depression. Lancet 2(7778):632–635

Jentink J, Dolk H, Loane MA, Morris JK, Wellesley D, Garne E, de Jong-van den Berg L, EASW Group (2010) Intrauterine exposure to carbamazepine and specific congenital malformations: systematic review and case-control study. BMJ 341:c6581. https://doi.org/10.1136/bmj.c6581

Jeste DV, Heaton SC, Paulsen JS, Ercoli L, Harris J, Heaton RK (1996) Clinical and neuropsychological comparison of psychotic depression with nonpsychotic depression and schizophrenia. Am J Psychiatry 153(4):490–496. https://doi.org/10.1176/ajp.153.4.490

Jog MS, Kubota Y, Connolly CI, Hillegaart V, Graybiel AM (1999) Building neural representations of habits. Science 286(5445):1745–1749. https://doi.org/10.1126/science.286.5445.1745

Johnson FN, Amdisen A (1983) The first era of lithium in medicine. An historical note. Pharmacopsychiatria 16(2):61–63. https://doi.org/10.1055/s-2007-1017450

Johnson SW, North RA (1992) Opioids excite dopamine neurons by hyperpolarization of local interneurons. J Neurosci 12(2):483–488

Johnson SM, Pillai NP (1990) Hyperpolarization of myenteric neurons by opioids does not involve cyclic adenosine-3′,5′-monophosphate. Neuroscience 36(2):299–304. https://doi.org/10.1016/0306-4522(90)90427-6

Johnstone EC, Crow TJ, Frith CD, Husband J, Kreel L (1976) Cerebral ventricular size and cognitive impairment in chronic schizophrenia. Lancet 2(7992):924–926. https://doi.org/10.1016/s0140-6736(76)90890-4

Johnstone EC, Crow TJ, Frith CD, Owens DG (1988) The Northwick Park "functional" psychosis study: diagnosis and treatment response. Lancet 2(8603):119–125

Jones MW, Kilpatrick IC, Phillipson OT (1988) Dopamine function in the prefrontal cortex of the rat is sensitive to a reduction of tonic Gaba-mediated inhibition in the thalamic mediodorsal nucleus. Exp Brain Res 69(3):623–634. https://doi.org/10.1007/bf00247314

Jones PB, Rantakallio P, Hartikainen AL, Isohanni M, Sipila P (1998) Schizophrenia as a long-term outcome of pregnancy, delivery, and perinatal complications: a 28-year follow-up of the 1966 north Finland general population birth cohort. Am J Psychiatry 155(3):355–364. https://doi.org/10.1176/ajp.155.3.355

Kalivas PW, Duffy P (1995) Selective activation of dopamine transmission in the shell of the nucleus accumbens by stress. Brain Res 675(1-2):325–328. https://doi.org/10.1016/0006-8993(95)00013-g

Kalivas PW, Duffy P, White SR (1998) MDMA elicits behavioral and neurochemical sensitization in rats. Neuropsychopharmacology 18(6):469–479. https://doi.org/10.1016/S0893-133X(97)00195-4

Kambeitz J, Abi-Dargham A, Kapur S, Howes OD (2014) Alterations in cortical and extrastriatal subcortical dopamine function in schizophrenia: systematic review and meta-analysis of imaging studies. Br J Psychiatry 204(6):420–429. https://doi.org/10.1192/bjp.bp.113.132308

Kang J, Lemaire HG, Unterbeck A, Salbaum JM, Masters CL, Grzeschik KH, Multhaup G, Beyreuther K, Muller-Hill B (1987) The precursor of Alzheimer's disease amyloid A4 protein resembles a cell-surface receptor. Nature 325(6106):733–736. https://doi.org/10.1038/325733a0

Kapur S, Mann JJ (1992) Role of the dopaminergic system in depression. Biol Psychiatry 32(1):1–17. https://doi.org/10.1016/0006-3223(92)90137-o

Kapur S, Seeman P (2001) Does fast dissociation from the dopamine d(2) receptor explain the action of atypical antipsychotics? A new hypothesis. Am J Psychiatry 158(3):360–369. https://doi.org/10.1176/appi.ajp.158.3.360

Karlsson P, Farde L, Halldin C, Sedvall G (2002) PET study of D(1) dopamine receptor binding in neuroleptic-naive patients with schizophrenia. Am J Psychiatry 159(5):761–767. https://doi.org/10.1176/appi.ajp.159.5.761

Kaur S, Sassi RB, Axelson D, Nicoletti M, Brambilla P, Monkul ES, Hatch JP, Keshavan MS, Ryan N, Birmaher B, Soares JC (2005) Cingulate cortex anatomical abnormalities in children and adolescents with bipolar disorder. Am J Psychiatry 162(9):1637–1643. https://doi.org/10.1176/appi.ajp.162.9.1637

Kegeles LS, Abi-Dargham A, Zea-Ponce Y, Rodenhiser-Hill J, Mann JJ, Van Heertum RL, Cooper TB, Carlsson A, Laruelle M (2000) Modulation of amphetamine-induced striatal dopamine release by ketamine in humans: implications for schizophrenia. Biol Psychiatry 48(7):627–640. https://doi.org/10.1016/s0006-3223(00)00976-8

Kegeles LS, Martinez D, Kochan LD, Hwang DR, Huang Y, Mawlawi O, Suckow RF, Van Heertum RL, Laruelle M (2002) NMDA antagonist effects on striatal dopamine release: positron emission tomography studies in humans. Synapse 43(1):19–29. https://doi.org/10.1002/syn.10010

Kegeles LS, Abi-Dargham A, Frankle WG, Gil R, Cooper TB, Slifstein M, Hwang DR, Huang Y, Haber SN, Laruelle M (2010) Increased synaptic dopamine function in associative regions of the striatum in schizophrenia. Arch Gen Psychiatry 67(3):231–239. https://doi.org/10.1001/archgenpsychiatry.2010.10

Kelsoe J (2009) Mood disorders: genetics. In: Sadock B, Sadock V, Ruiz P (eds) Kaplan & Sadock's comprehensive textbook of psychiatry, 9th edn. Lippincott Williams & Wilkins, Philadelphia, PA

Kendler KS (1983) Overview: a current perspective on twin studies of schizophrenia. Am J Psychiatry 140(11):1413–1425. https://doi.org/10.1176/ajp.140.11.1413

Kendler KS (1993) The roscommon family study. Arch Gen Psychiatry 50(8):645. https://doi.org/10.1001/archpsyc.1993.01820200059006

Kendler KS, Prescott CA (1998a) Cannabis use, abuse, and dependence in a population-based sample of female twins. Am J Psychiatry 155(8):1016–1022. https://doi.org/10.1176/ajp.155.8.1016

Kendler KS, Prescott CA (1998b) Cocaine use, abuse and dependence in a population-based sample of female twins. Br J Psychiatry 173(4):345–350. https://doi.org/10.1192/bjp.173.4.345

Kendler KS, Gruenberg AM, Kinney DK (1994) Independent diagnoses of adoptees and relatives as defined by Dsm-Iii in the provincial and National Samples of the Danish Adoption Study of Schizophrenia. Arch Gen Psychiatry 51(6):456–468. https://doi.org/10.1001/archpsyc.1994.03950060020002

Kendler KS, Karkowski LM, Neale MC, Prescott CA (2000) Illicit psychoactive substance use, heavy use, abuse, and dependence in a US population-based sample of male twins. Arch Gen Psychiatry 57(3):261–269. https://doi.org/10.1001/archpsyc.57.3.261

Kendler KS, Karkowski L, Prescott CA (2007) Hallucinogen, opiate, sedative and stimulant use and abuse in a population-based sample of female twins. Acta Psychiatr Scand 99(5):368–376. https://doi.org/10.1111/j.1600-0447.1999.tb07243.x

Kesslak JP, Nalcioglu O, Cotman CW (1991) Quantification of magnetic resonance scans for hippocampal and parahippocampal atrophy in Alzheimer's disease. Neurology 41(1):51–54. https://doi.org/10.1212/wnl.41.1.51

Khachaturian ZS (1985) Diagnosis of Alzheimer's disease. Arch Neurol 42(11):1097–1105

Kilts CD, Schweitzer JB, Quinn CK, Gross RE, Faber TL, Muhammad F, Ely TD, Hoffman JM, Drexler KP (2001) Neural activity related to drug craving in cocaine addiction. Arch Gen Psychiatry 58(4):334–341. https://doi.org/10.1001/archpsyc.58.4.334

Kim J, Horti AG, Mathews WB, Pogorelov V, Valentine H, Brasic JR, Holt DP, Ravert HT, Dannals RF, Zhou L, Jedynak B, Kamiya A, Pletnikov MV, Wong DF (2015) Quantitative multi-modal brain autoradiography of glutamatergic, dopaminergic, cannabinoid, and nicotinic receptors in mutant disrupted-in-schizophrenia-1 (DISC1) mice. Mol Imaging Biol 17(3):355–363. https://doi.org/10.1007/s11307-014-0786-4

Kisilevsky R, Fraser PE (1997) A beta amyloidogenesis: unique, or variation on a systemic theme? Crit Rev Biochem Mol Biol 32(5):361–404. https://doi.org/10.3109/10409239709082674

Kleinman JE, Law AJ, Lipska BK, Hyde TM, Ellis JK, Harrison PJ, Weinberger DR (2011) Genetic neuropathology of schizophrenia: new approaches to an old question and new uses for postmortem human brains. Biol Psychiatry 69(2):140–145. https://doi.org/10.1016/j.biopsych.2010.10.032

Knable MB (1999) Schizophrenia and bipolar disorder: findings from studies of the Stanley Foundation Brain Collection. Schizophr Res 39(2):149–152. discussion 163

Kokkinidis L, McCarter BD (1990) Postcocaine depression and sensitization of brain-stimulation reward: analysis of reinforcement and performance effects. Pharmacol Biochem Behav 36(3):463–471. https://doi.org/10.1016/0091-3057(90)90242-a

Kokkinidis L, Zacharko RM, Predy PA (1980) Post-amphetamine depression of self-stimulation responding from the substantia nigra: reversal by tricyclic antidepressants. Pharmacol Biochem Behav 13(3):379–383. https://doi.org/10.1016/0091-3057(80)90242-7

Konopaske GT, Lange N, Coyle JT, Benes FM (2014) Prefrontal cortical dendritic spine pathology in schizophrenia and bipolar disorder. JAMA Psychiat 71(12):1323–1331. https://doi.org/10.1001/jamapsychiatry.2014.1582

Koob GF, Bloom FE (1988) Cellular and molecular mechanisms of drug dependence. Science 242(4879):715–723. https://doi.org/10.1126/science.2903550

Koob GF, Stinus L, Le Moal M, Bloom FE (1989) Opponent process theory of motivation: neurobiological evidence from studies of opiate dependence. Neurosci Biobehav Rev 13(2-3):135–140. https://doi.org/10.1016/s0149-7634(89)80022-3

Kornhuber J, Weller M (1997) Psychotogenicity and N-methyl-D-aspartate receptor antagonism: implications for neuroprotective pharmacotherapy. Biol Psychiatry 41(2):135–144. https://doi.org/10.1016/S0006-3223(96)00047-9

Korte M, Kang H, Bonhoeffer T, Schuman E (1998) A role for BDNF in the late-phase of hippocampal long-term potentiation. Neuropharmacology 37(4-5):553–559. https://doi.org/10.1016/s0028-3908(98)00035-5

Kosaka J, Takahashi H, Ito H, Takano A, Fujimura Y, Matsumoto R, Nozaki S, Yasuno F, Okubo Y, Kishimoto T, Suhara T (2010) Decreased binding of [11C]NNC112 and [11C]SCH23390 in patients with chronic schizophrenia. Life Sci 86(21-22):814–818. https://doi.org/10.1016/j.lfs.2010.03.018

Kovacs GG, Kloppel S, Fischer I, Dorner S, Lindeck-Pozza E, Birner P, Botefur IC, Pilz P, Volk B, Budka H (2003) Nucleus-specific alteration of raphe neurons in human neurodegenerative disorders. Neuroreport 14(1):73–76. https://doi.org/10.1097/01.wnr.0000050301.92401.0c

Kubicki M, McCarley R, Westin CF, Park HJ, Maier S, Kikinis R, Jolesz FA, Shenton ME (2007) A review of diffusion tensor imaging studies in schizophrenia. J Psychiatr Res 41(1-2):15–30. https://doi.org/10.1016/j.jpsychires.2005.05.005

Kuczenski R, Segal DS (1999) Dynamic changes in sensitivity occur during the acute response to cocaine and methylphenidate. Psychopharmacology (Berl) 147(1):96–103. https://doi.org/10.1007/s002130051147

Kuhl DE, Koeppe RA, Minoshima S, Snyder SE, Ficaro EP, Foster NL, Frey KA, Kilbourn MR (1999) In vivo mapping of cerebral acetylcholinesterase activity in aging and Alzheimer's disease. Neurology 52(4):691–699

Kuhn R (1957) Treatment of depressive states with an iminodibenzyl derivative (G 22355). Schweiz Med Wochenschr 87(35–36):1135–1140

Kuhn R (1958) The treatment of depressive states with G 22355 (imipramine hydrochloride). Am J Psychiatry 115(5):459–464. https://doi.org/10.1176/ajp.115.5.459

Kundakovic M, Peter C, Roussos P, Akbarian S (2016) Epigenetic approaches to define the molecular and genetic risk architectures of schizophrenia. Neurobiol Schizophrenia. https://doi.org/10.1016/b978-0-12-801829-3.00013-6

Lagopoulos J, Hermens DF, Naismith SL, Scott EM, Hickie IB (2012) Frontal lobe changes occur early in the course of affective disorders in young people. BMC Psychiatry 12:4. https://doi.org/10.1186/1471-244X-12-4

Laruelle M (2014) Schizophrenia: from dopaminergic to glutamatergic interventions. Curr Opin Pharmacol 14:97–102. https://doi.org/10.1016/j.coph.2014.01.001

Laruelle M, Abi-Dargham A, Gil R, Kegeles L, Innis R (1999) Increased dopamine transmission in schizophrenia: relationship to illness phases. Biol Psychiatry 46(1):56–72. https://doi.org/10.1016/s0006-3223(99)00067-0

Lauer CJ, Riemann D, Wiegand M, Berger M (1991) From early to late adulthood. Changes in EEG sleep of depressed patients and healthy volunteers. Biol Psychiatry 29(10):979–993. https://doi.org/10.1016/0006-3223(91)90355-p

Lees G, Leach MJ (1993) Studies on the mechanism of action of the novel anticonvulsant lamotrigine (Lamictal) using primary neurological cultures from rat cortex. Brain Res 612(1-2):190–199

Legros S, Mendlewicz J, Wybran J (1985) Immunoglobulins, autoantibodies and other serum protein fractions in psychiatric disorders. Eur Arch Psychiatry Neurol Sci 235(1):9–11

Lehericy S, Baulac M, Chiras J, Pierot L, Martin N, Pillon B, Deweer B, Dubois B, Marsault C (1994) Amygdalohippocampal MR volume measurements in the early stages of Alzheimer disease. AJNR Am J Neuroradiol 15(5):929–937

Lehmann S, Chiesa R, Harris DA (1997) Evidence for a six-transmembrane domain structure of presenilin 1. J Biol Chem 272(18):12047–12051. https://doi.org/10.1074/jbc.272.18.12047

Leith NJ, Barrett RJ (1976) Amphetamine and the reward system: evidence for tolerance and post-drug depression. Psychopharmacologia 46(1):19–25. https://doi.org/10.1007/bf00421544

Lesser IM, Miller BL, Boone KB, Hill-Gutierrez E, Mehringer CM, Wong K, Mena I (1991) Brain injury and cognitive function in late-onset psychotic depression. J Neuropsychiatry Clin Neurosci 3(1):33–40. https://doi.org/10.1176/jnp.3.1.33

Lett BT (1989) Repeated exposures intensify rather than diminish the rewarding effects of amphetamine, morphine, and cocaine. Psychopharmacology 98(3):357–362. https://doi.org/10.1007/Bf00451687

Levy-Lahad E, Wasco W, Poorkaj P, Romano DM, Oshima J, Pettingell WH, Yu CE, Jondro PD, Schmidt SD, Wang K et al (1995a) Candidate gene for the chromosome 1 familial Alzheimer's disease locus. Science 269(5226):973–977. https://doi.org/10.1126/science.7638622

Levy-Lahad E, Wijsman E, Nemens E, Anderson L, Goddard K, Weber J, Bird T, Schellenberg G (1995b) A familial Alzheimer's disease locus on chromosome 1. Science 269(5226):970–973. https://doi.org/10.1126/science.7638621

Lewis DA, Pierri JN, Volk DW, Melchitzky DS, Woo TU (1999a) Altered GABA neurotransmission and prefrontal cortical dysfunction in schizophrenia. Biol Psychiatry 46(5):616–626. https://doi.org/10.1016/s0006-3223(99)00061-x

Lewis R, Kapur S, Jones C, DaSilva J, Brown GM, Wilson AA, Houle S, Zipursky RB (1999b) Serotonin 5-HT2 receptors in schizophrenia: a PET study using [18F]setoperone in neuroleptic-naive patients and normal subjects. Am J Psychiatry 156(1):72–78. https://doi.org/10.1176/ajp.156.1.72

Li X, Greenwald I (1998) Additional evidence for an eight-transmembrane-domain topology for Caenorhabditis elegans and human presenilins. Proc Natl Acad Sci U S A 95(12):7109–7114. https://doi.org/10.1073/pnas.95.12.7109

Li Y, Sun H, Chen Z, Xu H, Bu G, Zheng H (2016) Implications of GABAergic neurotransmission in Alzheimer's disease. Front Aging Neurosci 8:31. https://doi.org/10.3389/fnagi.2016.00031

Liddle PF, Frith CD, Friston KJ (1992) Supervisory mental processes in schizophrenia – a study using pet. Schizophr Res 6(2):148–148. https://doi.org/10.1016/0920-9964(92)90221-P

Lieberman JA, Kinon BJ, Loebel AD (1990) Dopaminergic mechanisms in idiopathic and drug-induced psychoses. Schizophr Bull 16(1):97–110. https://doi.org/10.1093/schbul/16.1.97

Lim KO, Rosenbloom MJ, Faustman WO, Sullivan EV, Pfefferbaum A (1999) Cortical gray matter deficit in patients with bipolar disorder. Schizophr Res 40(3):219–227

Lim CS, Baldessarini RJ, Vieta E, Yucel M, Bora E, Sim K (2013) Longitudinal neuroimaging and neuropsychological changes in bipolar disorder patients: review of the evidence. Neurosci Biobehav Rev 37(3):418–435. https://doi.org/10.1016/j.neubiorev.2013.01.003

Lin N, Eisen SA, Scherrer JF, Goldberg J, True WR, Lyons MJ, Tsuang MT (1996) The influence of familial and non-familial factors on the association between major depression and substance abuse dependence in 1874 monozygotic male twin pairs. Drug Alcohol Depend 43(1-2):49–55. https://doi.org/10.1016/S0376-8716(96)01287-2

Lin A, Reniers RL, Wood SJ (2013) Clinical staging in severe mental disorder: evidence from neurocognition and neuroimaging. Br J Psychiatry Suppl 54:s11–s17. https://doi.org/10.1192/bjp.bp.112.119156

Liu L, Zheng T, Morris MJ, Wallengren C, Clarke AL, Reid CA, Petrou S, O'Brien TJ (2006) The mechanism of carbamazepine aggravation of absence seizures. J Pharmacol Exp Ther 319(2):790–798. https://doi.org/10.1124/jpet.106.104968

Lochhead RA, Parsey RV, Oquendo MA, Mann JJ (2004) Regional brain gray matter volume differences in patients with bipolar disorder as assessed by optimized voxel-based morphometry. Biol Psychiatry 55(12):1154–1162. https://doi.org/10.1016/j.biopsych.2004.02.026

Lopez-Munoz F, Alamo C, Juckel G, Assion HJ (2007) Half a century of antidepressant drugs: on the clinical introduction of monoamine oxidase inhibitors, tricyclics, and tetracyclics. Part I: monoamine oxidase inhibitors. J Clin Psychopharmacol 27(6):555–559. https://doi.org/10.1097/jcp.0b013e3181bb617

Lorrain DS, Arnold GM, Vezina P (2000) Previous exposure to amphetamine increases incentive to obtain the drug: long-lasting effects revealed by the progressive ratio schedule. Behav Brain Res 107(1-2):9–19. https://doi.org/10.1016/s0166-4328(99)00109-6

Lovinger DM (1997) Alcohols and neurotransmitter gated ion channels: past, present and future. Naunyn Schmiedebergs Arch Pharmacol 356(3):267–282. https://doi.org/10.1007/pl00005051
Lyford GL, Yamagata K, Kaufmann WE, Barnes CA, Sanders LK, Copeland NG, Gilbert DJ, Jenkins NA, Lanahan AA, Worley PF (1995) Arc, a growth-factor and activity-regulated gene, encodes a novel cytoskeleton-associated protein that is enriched in neuronal dendrites. Neuron 14(2):433–445. https://doi.org/10.1016/0896-6273(95)90299-6
Lynch M (1998) Age-related impairment in long-term potentiation in hippocampus: a role for the cytokine, interleukin-1Î²? Prog Neurobiol 56(5):571–589. https://doi.org/10.1016/s0301-0082(98)00054-9
Lyons DM, Yang C, Sawyer-Glover AM, Moseley ME, Schatzberg AF (2001) Early life stress and inherited variation in monkey hippocampal volumes. Arch Gen Psychiatry 58(12):1145–1151. https://doi.org/10.1001/archpsyc.58.12.1145
Lyoo IK, Kim MJ, Stoll AL, Demopulos CM, Parow AM, Dager SR, Friedman SD, Dunner DL, Renshaw PF (2004) Frontal lobe gray matter density decreases in bipolar I disorder. Biol Psychiatry 55(6):648–651. https://doi.org/10.1016/j.biopsych.2003.10.017
Lyoo IK, Sung YH, Dager SR, Friedman SD, Lee JY, Kim SJ, Kim N, Dunner DL, Renshaw PF (2006) Regional cerebral cortical thinning in bipolar disorder. Bipolar Disord 8(1):65–74. https://doi.org/10.1111/j.1399-5618.2006.00284.x
Maas JW (1975) Biogenic amines and depression. Biochemical and pharmacological separation of two types of depression. Arch Gen Psychiatry 32(11):1357–1361
Maes M, Meltzer HY (1995) The serotonin hypothesis of major depression. In: Bloom FE, Kupfer DJ (eds) Psychopharmacology: the fourth generation of progress. Raven Press, New York, NY, pp 933–944
Mahley RW (1988) Apolipoprotein E: cholesterol transport protein with expanding role in cell biology. Science 240(4852):622–630. https://doi.org/10.1126/science.3283935
Malhi GS, Lagopoulos J, Owen AM, Ivanovski B, Shnier R, Sachdev P (2007) Reduced activation to implicit affect induction in euthymic bipolar patients: an fMRI study. J Affect Disord 97(1-3):109–122. https://doi.org/10.1016/j.jad.2006.06.005
Malm H (2012) Prenatal exposure to selective serotonin reuptake inhibitors and infant outcome. Ther Drug Monit 34(6):607–614. https://doi.org/10.1097/FTD.0b013e31826d07ea
Mann DMA (1988) The pathological association between down syndrome and Alzheimer-disease. Mech Ageing Dev 43(2):99–136. https://doi.org/10.1016/0047-6374(88)90041-3
Mann DMA (1998) Neuropathological and neurochemical aspects of Alzheimer's disease. In: Iversen LL, Iversen SD, Snyder SH (eds) Psychopharmacology of the aging nervous system. Plenum, New York, NY, pp 1–67
Marlinge E, Bellivier F, Houenou J (2014) White matter alterations in bipolar disorder: potential for drug discovery and development. Bipolar Disord 16(2):97–112. https://doi.org/10.1111/bdi.12135
Marmol F (2008) Lithium: bipolar disorder and neurodegenerative diseases Possible cellular mechanisms of the therapeutic effects of lithium. Prog Neuropsychopharmacol Biol Psychiatry 32(8):1761–1771. https://doi.org/10.1016/j.pnpbp.2008.08.012
Marshall DL, Redfern PH, Wonnacott S (2002) Presynaptic nicotinic modulation of dopamine release in the three ascending pathways studied by in vivo microdialysis: comparison of naive and chronic nicotine-treated rats. J Neurochem 68(4):1511–1519. https://doi.org/10.1046/j.1471-4159.1997.68041511.x
Martorana A, Koch G (2014) Is dopamine involved in Alzheimer's disease? Front Aging Neurosci 6:252. https://doi.org/10.3389/fnagi.2014.00252
Martucci L, Wong AH, De Luca V, Likhodi O, Wong GW, King N, Kennedy JL (2006) N-methyl-D-aspartate receptor NR2B subunit gene GRIN2B in schizophrenia and bipolar disorder: polymorphisms and mRNA levels. Schizophr Res 84(2-3):214–221. https://doi.org/10.1016/j.schres.2006.02.001
Massman PJ, Delis DC, Butters N, Dupont RM, Gillin JC (1992) The subcortical dysfunction hypothesis of memory deficits in depression: neuropsychological validation in a subgroup of patients. J Clin Exp Neuropsychol 14(5):687–706. https://doi.org/10.1080/01688639208402856

Matthysse S (1973) Antipsychotic drug actions: a clue to the neuropathology of schizophrenia? Fed Proc 32(2):200–205

Mattson MP, Guo ZH, Geiger JD (1999) Secreted form of amyloid precursor protein enhances basal glucose and glutamate transport and protects against oxidative impairment of glucose and glutamate transport in synaptosomes by a cyclic GMP-mediated mechanism. J Neurochem 73(2):532–537

Mayberg HS (1997) Limbic-cortical dysregulation: a proposed model of depression. J Neuropsychiatry Clin Neurosci 9(3):471–481. https://doi.org/10.1176/jnp.9.3.471

Mayberg HS, Liotti M, Brannan SK, McGinnis S, Mahurin RK, Jerabek PA, Silva JA, Tekell JL, Martin CC, Lancaster JL, Fox PT (1999) Reciprocal limbic-cortical function and negative mood: converging PET findings in depression and normal sadness. Am J Psychiatry 156(5):675–682. https://doi.org/10.1176/ajp.156.5.675

Mayberg HS, Brannan SK, Tekell JL, Silva JA, Mahurin RK, McGinnis S, Jerabek PA (2000) Regional metabolic effects of fluoxetine in major depression: serial changes and relationship to clinical response. Biol Psychiatry 48(8):830–843

Mayberg HS, Lozano AM, Voon V, McNeely HE, Seminowicz D, Hamani C, Schwalb JM, Kennedy SH (2005) Deep brain stimulation for treatment-resistant depression. Neuron 45(5):651–660. https://doi.org/10.1016/j.neuron.2005.02.014

McCarthy MJ, Liang S, Spadoni AD, Kelsoe JR, Simmons AN (2014) Whole brain expression of bipolar disorder associated genes: structural and genetic analyses. PLoS One 9(6):e100204. https://doi.org/10.1371/journal.pone.0100204

McCullumsmith RE, Kristiansen LV, Beneyto M, Scarr E, Dean B, Meador-Woodruff JH (2007) Decreased NR1, NR2A, and SAP102 transcript expression in the hippocampus in bipolar disorder. Brain Res 1127(1):108–118. https://doi.org/10.1016/j.brainres.2006.09.011

McCullumsmith RE, Hammond J, Funk A, Meador-Woodruff JH (2012) Recent advances in targeting the ionotropic glutamate receptors in treating schizophrenia. Curr Pharm Biotechnol 13(8):1535–1542

McDonald C, Bullmore ET, Sham PC, Chitnis X, Wickham H, Bramon E, Murray RM (2004) Association of genetic risks for schizophrenia and bipolar disorder with specific and generic brain structural endophenotypes. Arch Gen Psychiatry 61(10):974–984. https://doi.org/10.1001/archpsyc.61.10.974

McEwen BS (2000) Effects of adverse experiences for brain structure and function. Biol Psychiatry 48(8):721–731. https://doi.org/10.1016/s0006-3223(00)00964-1

McGeer EG, McGeer PL (1999) Brain inflammation in Alzheimer disease and the therapeutic implications. Curr Pharm Des 5(10):821–836

McGeer EG, McGeer PL, Kamo H, Tago H, Harrop R (1986a) Cortical metabolism, acetylcholinesterase staining and pathological changes in Alzheimer's disease. Can J Neurol Sci 13(4 Suppl):511–516

McGeer PL, Kamo H, Harrop R, Li DK, Tuokko H, McGeer EG, Adam MJ, Ammann W, Beattie BL, Calne DB et al (1986b) Positron emission tomography in patients with clinically diagnosed Alzheimer's disease. Can Med Assoc J 134(6):597–607

McGeer PL, McGeer EG, Kamo H, Wong K (1986c) Positron emission tomography and the possible origins of cytopathology in Alzheimer's disease. Prog Neuropsychopharmacol Biol Psychiatry 10(3-5):501–518

McGeer PL, Akiyama H, Itagaki S, McGeer EG (1989) Immune system response in Alzheimer's disease. Can J Neurol Sci 16(4 Suppl):516–527. https://doi.org/10.1017/s0317167100029863

McGeer PL, Rogers J, McGeer EG (1994) Neuroimmune mechanisms in Alzheimer disease pathogenesis. Alzheimer Dis Assoc Disord 8(3):149–158

McGrath CL, Kelley ME, Holtzheimer PE, Dunlop BW, Craighead WE, Franco AR, Craddock RC, Mayberg HS (2013) Toward a neuroimaging treatment selection biomarker for major depressive disorder. JAMA Psychiat 70(8):821–829. https://doi.org/10.1001/jamapsychiatry.2013.143

McGue M, Elkins I, Iacono WG (2000) Genetic and environmental influences on adolescent substance use and abuse. Am J Med Genet 96(5):671–677. https://doi.org/10.1002/1096-8628(20001009)96:5<671::Aid-Ajmg14>3.0.Co;2-W

McGuire P, Howes OD, Stone J, Fusar-Poli P (2008) Functional neuroimaging in schizophrenia: diagnosis and drug discovery. Trends Pharmacol Sci 29(2):91–98. https://doi.org/10.1016/j.tips.2007.11.005

Meda L, Cassatella MA, Szendrei GI, Otvos L Jr, Baron P, Villalba M, Ferrari D, Rossi F (1995) Activation of microglial cells by beta-amyloid protein and interferon-gamma. Nature 374(6523):647–650. https://doi.org/10.1038/374647a0

Melis M, Gessa GL, Diana M (2000) Different mechanisms for dopaminergic excitation induced by opiates and cannabinoids in the rat midbrain. Prog Neuropsychopharmacol Biol Psychiatry 24(6):993–1006. https://doi.org/10.1016/s0278-5846(00)00119-6

Meltzer CC, Price JC, Mathis CA, Greer PJ, Cantwell MN, Houck PR, Mulsant BH, Ben-Eliezer D, Lopresti B, DeKosky ST, Reynolds CF 3rd (1999) PET imaging of serotonin type 2A receptors in late-life neuropsychiatric disorders. Am J Psychiatry 156(12):1871–1878. https://doi.org/10.1176/ajp.156.12.1871

Mereu G, Yoon KW, Boi V, Gessa GL, Naes L, Westfall TC (1987) Preferential stimulation of ventral tegmental area dopaminergic neurons by nicotine. Eur J Pharmacol 141(3):395–399. https://doi.org/10.1016/0014-2999(87)90556-5

Merikangas KR, Stolar M, Stevens DE, Goulet J, Preisig MA, Fenton B, Zhang HP, O'Malley SS, Rounsaville BJ (1998) Familial transmission of substance use disorders. Arch Gen Psychiatry 55(11):973–979. https://doi.org/10.1001/archpsyc.55.11.973

Mermelstein PG, Becker JB (1995) Increased extracellular dopamine in the nucleus accumbens and striatum of the female rat during paced copulatory behavior. Behav Neurosci 109(2):354–365. https://doi.org/10.1037//0735-7044.109.2.354

Mervaala E, Fohr J, Kononen M, Valkonen-Korhonen M, Vainio P, Partanen K, Partanen J, Tiihonen J, Viinamaki H, Karjalainen AK, Lehtonen J (2000) Quantitative MRI of the hippocampus and amygdala in severe depression. Psychol Med 30(1):117–125. https://doi.org/10.1017/s0033291799001567

Mesulam MM (1996) The systems-level organization of cholinergic innervation in the human cerebral cortex and its alterations in Alzheimer's disease. Prog Brain Res 109:285–297

Mesulam MM, Mufson EJ, Levey AI, Wainer BH (1983) Cholinergic innervation of cortex by the basal forebrain: cytochemistry and cortical connections of the septal area, diagonal band nuclei, nucleus basalis (substantia innominata), and hypothalamus in the rhesus monkey. J Comp Neurol 214(2):170–197. https://doi.org/10.1002/cne.902140206

Meunier H, Carraz G, Neunier Y, Eymard P, Aimard M (1963) Pharmacodynamic properties of N-dipropylacetic acid. Therapie 18:435–438

Meyer JH, Kapur S, Houle S, DaSilva J, Owczarek B, Brown GM, Wilson AA, Kennedy SH (1999) Prefrontal cortex 5-HT2 receptors in depression: an [18F]setoperone PET imaging study. Am J Psychiatry 156(7):1029–1034. https://doi.org/10.1176/ajp.156.7.1029

Meyer-Lindenberg A, Miletich RS, Kohn PD, Esposito G, Carson RE, Quarantelli M, Weinberger DR, Berman KF (2002) Reduced prefrontal activity predicts exaggerated striatal dopaminergic function in schizophrenia. Nat Neurosci 5(3):267–271. https://doi.org/10.1038/nn804

Mialet JP, Pope HG, Yurgelun-Todd D (1996) Impaired attention in depressive states: a non-specific deficit? Psychol Med 26(5):1009–1020. https://doi.org/10.1017/s0033291700035339

Miller HL, Delgado PL, Salomon RM, Heninger GR, Charney DS (1996) Effects of alpha-methyl-para-tyrosine (AMPT) in drug-free depressed patients. Neuropsychopharmacology 14(3):151–157. https://doi.org/10.1016/0893-133X(95)00072-L

Miller CL, Llenos IC, Dulay JR, Weis S (2006) Upregulation of the initiating step of the kynurenine pathway in postmortem anterior cingulate cortex from individuals with schizophrenia and bipolar disorder. Brain Res 1073-1074:25–37. https://doi.org/10.1016/j.brainres.2005.12.056

Mirra SS, Heyman A, McKeel D, Sumi SM, Crain BJ, Brownlee LM, Vogel FS, Hughes JP, van Belle G, Berg L (1991) The Consortium to Establish a Registry for Alzheimer's Disease (CERAD). Part II standardization of the neuropathologic assessment of Alzheimer's disease. Neurology 41(4):479–486

Mitchell JB, Gratton A (1991) Opioid modulation and sensitization of dopamine release elicited by sexually relevant stimuli: a high speed chronoamperometric study in freely behaving rats. Brain Res 551(1-2):20–27. https://doi.org/10.1016/0006-8993(91)90908-e

Mitchell PB, Hadzi-Pavlovic D (2000) Lithium treatment for bipolar disorder. Bull World Health Organ 78(4):515–517

Mizrahi R, Kenk M, Suridjan I, Boileau I, George TP, McKenzie K, Wilson AA, Houle S, Rusjan P (2014) Stress-induced dopamine response in subjects at clinical high risk for schizophrenia with and without concurrent cannabis use. Neuropsychopharmacology 39(6):1479–1489. https://doi.org/10.1038/npp.2013.347

Mizukami K, Grayson DR, Ikonomovic MD, Sheffield R, Armstrong DM (1998a) GABAA receptor beta 2 and beta 3 subunits mRNA in the hippocampal formation of aged human brain with Alzheimer-related neuropathology. Brain Res Mol Brain Res 56(1-2):268–272

Mizukami K, Ikonomovic MD, Grayson DR, Sheffield R, Armstrong DM (1998b) Immunohistochemical study of GABAA receptor alpha1 subunit in the hippocampal formation of aged brains with Alzheimer-related neuropathologic changes. Brain Res 799(1):148–155

Monaghan DT, Geddes JW, Yao D, Chung C, Cotman CW (1987) [3H]TCP binding sites in Alzheimer's disease. Neurosci Lett 73(2):197–200

Moore PB, Shepherd DJ, Eccleston D, Macmillan IC, Goswami U, McAllister VL, Ferrier IN (2001) Cerebral white matter lesions in bipolar affective disorder: relationship to outcome. Br J Psychiatry 178:172–176

Moore CM, Biederman J, Wozniak J, Mick E, Aleardi M, Wardrop M, Dougherty M, Harpold T, Hammerness P, Randall E, Lyoo IK, Renshaw PF (2007) Mania, glutamate/glutamine and risperidone in pediatric bipolar disorder: a proton magnetic resonance spectroscopy study of the anterior cingulate cortex. J Affect Disord 99(1-3):19–25. https://doi.org/10.1016/j.jad.2006.08.023

Moorhead TW, McKirdy J, Sussmann JE, Hall J, Lawrie SM, Johnstone EC, McIntosh AM (2007) Progressive gray matter loss in patients with bipolar disorder. Biol Psychiatry 62(8):894–900. https://doi.org/10.1016/j.biopsych.2007.03.005

Motzkin JC, Philippi CL, Wolf RC, Baskaya MK, Koenigs M (2015) Ventromedial prefrontal cortex is critical for the regulation of amygdala activity in humans. Biol Psychiatry 77(3):276–284. https://doi.org/10.1016/j.biopsych.2014.02.014

Mueller HT, Meador-Woodruff JH (2004) NR3A NMDA receptor subunit mRNA expression in schizophrenia, depression and bipolar disorder. Schizophr Res 71(2-3):361–370. https://doi.org/10.1016/j.schres.2004.02.016

Mundo E, Tharmalingham S, Neves-Pereira M, Dalton EJ, Macciardi F, Parikh SV, Bolonna A, Kerwin RW, Arranz MJ, Makoff AJ, Kennedy JL (2003) Evidence that the N-methyl-D-aspartate subunit 1 receptor gene (GRIN1) confers susceptibility to bipolar disorder. Mol Psychiatry 8(2):241–245. https://doi.org/10.1038/sj.mp.4001218

Munn NA (2000) Microglia dysfunction in schizophrenia: an integrative theory. Med Hypotheses 54(2):198–202. https://doi.org/10.1054/mehy.1999.0018

Myhrer T (1993) Animal models of Alzheimer's disease: glutamatergic denervation as an alternative approach to cholinergic denervation. Neurosci Biobehav Rev 17(2):195–202

Myhrer T, Danscher G, Fonnum F (2003) Degenerative patterns following denervation of temporal structures in a rat model of mnemonic dysfunction. Brain Res 967(1-2):293–300

Nagga K, Bogdanovic N, Marcusson J (1999) GABA transporters (GAT-1) in Alzheimer's disease. J Neural Transm 106(11-12):1141–1149. https://doi.org/10.1007/s007020050230

Namba Y, Tomonaga M, Kawasaki H, Otomo E, Ikeda K (1991) Apolipoprotein E immunoreactivity in cerebral amyloid deposits and neurofibrillary tangles in Alzheimer's disease and kuru plaque amyloid in Creutzfeldt-Jakob disease. Brain Res 541(1):163–166. https://doi.org/10.1016/0006-8993(91)91092-f

Nelson EB, Sax KW, Strakowski SM (1998) Attentional performance in patients with psychotic and nonpsychotic major depression and schizophrenia. Am J Psychiatry 155(1):137–139. https://doi.org/10.1176/ajp.155.1.137

Nguyen TV, Kosofsky BE, Birnbaum R, Cohen BM, Hyman SE (1992) Differential expression of c-fos and zif268 in rat striatum after haloperidol, clozapine, and amphetamine. Proc Natl Acad Sci U S A 89(10):4270–4274. https://doi.org/10.1073/pnas.89.10.4270

Nguyen PV, Abel T, Kandel ER (1994) Requirement of a critical period of transcription for induction of a late phase of LTP. Science 265(5175):1104–1107. https://doi.org/10.1126/science.8066450

Nisell M, Nomikos GG, Svensson TH (1994a) Infusion of nicotine in the ventral tegmental area or the nucleus accumbens of the rat differentially affects accumbal dopamine release. Pharmacol Toxicol 75(6):348–352. https://doi.org/10.1111/j.1600-0773.1994.tb00373.x

Nisell M, Nomikos GG, Svensson TH (1994b) Systemic nicotine-induced dopamine release in the rat nucleus accumbens is regulated by nicotinic receptors in the ventral tegmental area. Synapse 16(1):36–44. https://doi.org/10.1002/syn.890160105

Noack CH, Trautner EM (1951) The lithium treatment of maniacal psychosis. Med J Aust 2(7):219–222

Nordberg A (1996) Pharmacological treatment of cognitive dysfunction in dementia disorders. Acta Neurol Scand Suppl 168:87–92

Nordstrom AL, Farde L, Wiesel FA, Forslund K, Pauli S, Halldin C, Uppfeldt G (1993) Central D2-dopamine receptor occupancy in relation to antipsychotic drug effects: a double-blind PET study of schizophrenic patients. Biol Psychiatry 33(4):227–235. https://doi.org/10.1016/0006-3223(93)90288-O

Nurmi M, Ashizawa T, Sinclair JD, Kiianmaa K (1996) Effect of prior ethanol experience on dopamine overflow in accumbens of AA and ANA rats. Eur J Pharmacol 315(3):277–283. https://doi.org/10.1016/s0014-2999(96)00650-4

Nurnberger JI Jr, Koller DL, Jung J, Edenberg HJ, Foroud T, Guella I, Vawter MP, Kelsoe JR, Psychiatric Genomics Consortium Bipolar Group (2014) Identification of pathways for bipolar disorder: a meta-analysis. JAMA Psychiat 71(6):657–664. https://doi.org/10.1001/jamapsychiatry.2014.176

O'Brien CP, Childress AR, McLellan AT, Ehrman R (1992) Classical conditioning in drug-dependent humans. Ann N Y Acad Sci 654:400–415. https://doi.org/10.1111/j.1749-6632.1992.tb25984.x

Ohmori T, Koyama T, Nakamura F, Wang P, Yamashita I (1992) Effect of phencyclidine on spontaneous and N-methyl-D-aspartate (NMDA)-induced efflux of dopamine from superfused slices of rat striatum. Neuropharmacology 31(5):461–467. https://doi.org/10.1016/0028-3908(92)90084-3

Ohyagi Y, Tabira T (1993) Effect of growth factors and cytokines on expression of amyloid beta protein precursor mRNAs in cultured neural cells. Brain Res Mol Brain Res 18(1-2):127–132

Okubo Y, Suhara T, Sudo Y, Toru M (1997a) Possible role of dopamine D1 receptors in schizophrenia. Mol Psychiatry 2(4):291–292

Okubo Y, Suhara T, Suzuki K, Kobayashi K, Inoue O, Terasaki O, Someya Y, Sassa T, Sudo Y, Matsushima E, Iyo M, Tateno Y, Toru M (1997b) Decreased prefrontal dopamine D1 receptors in schizophrenia revealed by PET. Nature 385(6617):634–636. https://doi.org/10.1038/385634a0

Okuma T, Kishimoto A (1998) A history of investigation on the mood stabilizing effect of carbamazepine in Japan. Psychiatry Clin Neurosci 52(1):3–12. https://doi.org/10.1111/j.1440-1819.1998.tb00966.x

Olds ME (1982) Reinforcing effects of morphine in the nucleus accumbens. Brain Res 237(2):429–440. https://doi.org/10.1016/0006-8993(82)90454-1

Olney JW (1989) Excitatory amino acids and neuropsychiatric disorders. Biol Psychiatry 26(5):505–525. https://doi.org/10.1016/0006-3223(89)90072-3

Olney JW, Farber NB (1995) Glutamate receptor dysfunction and schizophrenia. Arch Gen Psychiatry 52(12):998–1007. https://doi.org/10.1001/archpsyc.1995.03950240016004

Ongur D, Drevets WC, Price JL (1998) Glial reduction in the subgenual prefrontal cortex in mood disorders. Proc Natl Acad Sci U S A 95(22):13290–13295

Onteniente B, Simon H, Taghzouti K, Geffard M, Lemoal M, Calas A (1987) Dopamine Gaba interactions in the nucleus-accumbens and lateral septum of the rat. Brain Res 421(1–2):391–396. https://doi.org/10.1016/0006-8993(87)91315-1

Overmyer M, Helisalmi S, Soininen H, Laakso M, Riekkinen P Sr, Alafuzoff I (1999) Astrogliosis and the ApoE genotype. an immunohistochemical study of postmortem human brain tissue. Dement Geriatr Cogn Disord 10(4):252–257. https://doi.org/10.1159/000017128

Pacchiarotti I, Bond DJ, Baldessarini RJ, Nolen WA, Grunze H, Licht RW, Post RM, Berk M, Goodwin GM, Sachs GS, Tondo L, Findling RL, Youngstrom EA, Tohen M, Undurraga J, Gonzalez-Pinto A, Goldberg JF, Yildiz A, Altshuler LL, Calabrese JR, Mitchell PB, Thase ME, Koukopoulos A, Colom F, Frye MA, Malhi GS, Fountoulakis KN, Vazquez G, Perlis RH, Ketter TA, Cassidy F, Akiskal H, Azorin JM, Valenti M, Mazzei DH, Lafer B, Kato T, Mazzarini L, Martinez-Aran A, Parker G, Souery D, Ozerdem A, McElroy SL, Girardi P, Bauer M, Yatham LN, Zarate CA, Nierenberg AA, Birmaher B, Kanba S, El-Mallakh RS, Serretti A, Rihmer Z, Young AH, Kotzalidis GD, MacQueen GM, Bowden CL, Ghaemi SN, Lopez-Jaramillo C, Rybakowski J, Ha K, Perugi G, Kasper S, Amsterdam JD, Hirschfeld RM, Kapczinski F, Vieta E (2013) The International Society for Bipolar Disorders (ISBD) task force report on antidepressant use in bipolar disorders. Am J Psychiatry 170(11):1249–1262. https://doi.org/10.1176/appi.ajp.2013.13020185

Pakkenberg B (1993) Total nerve cell number in neocortex in chronic schizophrenics and controls estimated using optical disectors. Biol Psychiatry 34(11):768–772. https://doi.org/10.1016/0006-3223(93)90065-l

Palmer AM, DeKosky ST (1993) Monoamine neurons in aging and Alzheimer's disease. J Neural Transm Gen Sect 91(2-3):135–159

Palmer AM, Gershon S (1990) Is the neuronal basis of Alzheimer's disease cholinergic or glutamatergic? FASEB J 4(10):2745–2752

Palmer AM, Francis PT, Benton JS, Sims NR, Mann DM, Neary D, Snowden JS, Bowen DM (1987a) Presynaptic serotonergic dysfunction in patients with Alzheimer's disease. J Neurochem 48(1):8–15

Palmer AM, Francis PT, Bowen DM, Benton JS, Neary D, Mann DM, Snowden JS (1987b) Catecholaminergic neurones assessed ante-mortem in Alzheimer's disease. Brain Res 414(2):365–375

Palmer AM, Wilcock GK, Esiri MM, Francis PT, Bowen DM (1987c) Monoaminergic innervation of the frontal and temporal lobes in Alzheimer's disease. Brain Res 401(2):231–238

Pantel J, Schroder J, Schad LR, Friedlinger M, Knopp MV, Schmitt R, Geissler M, Bluml S, Essig M, Sauer H (1997) Quantitative magnetic resonance imaging and neuropsychological functions in dementia of the Alzheimer type. Psychol Med 27(1):221–229

Parsey RV, Hwang D, Simpson N, Kegeles L, Anjivel S, Zea-Ponce Y, Lombardo I, Popilskis S, Van Heertum R, Mann JJ, Laruelle M (1998) Kinetic derivation of serotonin 5HT-1A receptor binding potential with [C-11]carbonyl-WAY 100635 and competition studies with endogenous serotonin. J Nucl Med 39(5):167p–167p

Pavuluri MN, O'Connor MM, Harral E, Sweeney JA (2007) Affective neural circuitry during facial emotion processing in pediatric bipolar disorder. Biol Psychiatry 62(2):158–167. https://doi.org/10.1016/j.biopsych.2006.07.011

Pearlson GD, Marsh L (1999) Structural brain imaging in schizophrenia: a selective review. Biol Psychiatry 46(5):627–649. https://doi.org/10.1016/s0006-3223(99)00071-2

Pearlson GD, Barta PE, Powers RE, Menon RR, Richards SS, Aylward EH, Federman EB, Chase GA, Petty RG, Tien AY (1997) Ziskind-Somerfeld Research Award 1996. Medial and superior temporal gyral volumes and cerebral asymmetry in schizophrenia versus bipolar disorder. Biol Psychiatry 41(1):1–14. https://doi.org/10.1016/s0006-3223(96)00373-3

Pearson RC, Esiri MM, Hiorns RW, Wilcock GK, Powell TP (1985) Anatomical correlates of the distribution of the pathological changes in the neocortex in Alzheimer disease. Proc Natl Acad Sci U S A 82(13):4531–4534

Penit-Soria J, Audinat E, Crepel F (1987) Excitation of rat prefrontal cortical neurons by dopamine: an in vitro electrophysiological study. Brain Res 425(2):263–274. https://doi.org/10.1016/0006-8993(87)90509-9

Perl DP (2010) Neuropathology of Alzheimer's disease. Mount Sinai J Med 77(1):32–42. https://doi.org/10.1002/msj.20157

Perlis RH, Sachs GS, Lafer B, Otto MW, Faraone SV, Kane JM, Rosenbaum JF (2002) Effect of abrupt change from standard to low serum levels of lithium: a reanalysis of double-blind

lithium maintenance data. Am J Psychiatry 159(7):1155–1159. https://doi.org/10.1176/appi.ajp.159.7.1155
Perry EK, Morris CM, Court JA, Cheng A, Fairbairn AF, McKeith IG, Irving D, Brown A, Perry RH (1995) Alteration in nicotine binding sites in Parkinson's disease, Lewy body dementia and Alzheimer's disease: possible index of early neuropathology. Neuroscience 64(2):385–395
Pettit HO, Justice JB (1989) Dopamine in the nucleus accumbens during cocaine self-administration as studied by in vivo microdialysis. Pharmacol Biochem Behav 34(4):899–904. https://doi.org/10.1016/0091-3057(89)90291-8
Pettit HO, Ettenberg A, Bloom FE, Koob GF (1984) Destruction of dopamine in the nucleus accumbens selectively attenuates cocaine but not heroin self-administration in rats. Psychopharmacology (Berl) 84(2):167–173. https://doi.org/10.1007/bf00427441
Piazza PV, Deminiere JM, le Moal M, Simon H (1990) Stress- and pharmacologically-induced behavioral sensitization increases vulnerability to acquisition of amphetamine self-administration. Brain Res 514(1):22–26. https://doi.org/10.1016/0006-8993(90)90431-a
Pidsley R, Mill J (2011) Epigenetic studies of psychosis: current findings, methodological approaches, and implications for postmortem research. Biol Psychiatry 69(2):146–156. https://doi.org/10.1016/j.biopsych.2010.03.029
Pierce RC, Kalivas PW (1997) A circuitry model of the expression of behavioral sensitization to amphetamine-like psychostimulants. Brain Res Brain Res Rev 25(2):192–216. https://doi.org/10.1016/s0165-0173(97)00021-0
Pilowsky LS, Bressan RA, Stone JM, Erlandsson K, Mulligan RS, Krystal JH, Ell PJ (2006) First in vivo evidence of an NMDA receptor deficit in medication-free schizophrenic patients. Mol Psychiatry 11(2):118–119. https://doi.org/10.1038/sj.mp.4001751
Pogarell O, Koch W, Karch S, Dehning S, Muller N, Tatsch K, Poepperl G, Moller HJ (2012) Dopaminergic neurotransmission in patients with schizophrenia in relation to positive and negative symptoms. Pharmacopsychiatry 45(Suppl 1):S36–S41. https://doi.org/10.1055/s-0032-1306313
Ponte P, Gonzalez-DeWhitt P, Schilling J, Miller J, Hsu D, Greenberg B, Davis K, Wallace W, Lieberburg I, Fuller F (1988) A new A4 amyloid mRNA contains a domain homologous to serine proteinase inhibitors. Nature 331(6156):525–527. https://doi.org/10.1038/331525a0
Pontieri FE, Tanda G, Orzi F, Di Chiara G (1996) Effects of nicotine on the nucleus accumbens and similarity to those of addictive drugs. Nature 382(6588):255–257. https://doi.org/10.1038/382255a0
Portet F, Scarmeas N, Cosentino S, Helzner EP, Stern Y (2009) Extrapyramidal signs before and after diagnosis of incident Alzheimer disease in a prospective population study. Arch Neurol 66(9):1120–1126. https://doi.org/10.1001/archneurol.2009.196
Porto FH, Leite MA, Fontenelle LF, Marrocos RP, Szczerback NF, de Freitas MR (2009) The Syndrome of Irreversible Lithium-Effectuated Neurotoxicity (SILENT): one-year follow-up of a single case. J Neurol Sci 277(1-2):172–173. https://doi.org/10.1016/j.jns.2008.10.010
Posner J, Cha J, Wang Z, Talati A, Warner V, Gerber A, Peterson BS, Weissman M (2016) Increased default mode network connectivity in individuals at high familial risk for depression. Neuropsychopharmacology 41(7):1759–1767. https://doi.org/10.1038/npp.2015.342
van Praag HM (1971) The position of biological psychiatry among the psychiatric disciplines. Compr Psychiatry 12(1):1–7
Price LH, Malison RT, McDougle CJ, McCance-Katz EF, Owen KR, Heninger GR (1997) Neurobiology of tryptophan depletion in depression: effects of m-chlorophenylpiperazine (mCPP). Neuropsychopharmacology 17(5):342–350. https://doi.org/10.1016/S0893-133X(97)00084-5
Price LH, Malison RT, McDougle CJ, Pelton GH, Heninger GR (1998) The neurobiology of tryptophan depletion in depression: effects of intravenous tryptophan infusion. Biol Psychiatry 43(5):339–347. https://doi.org/10.1016/s0006-3223(97)00284-9
Procter AW, Wong EH, Stratmann GC, Lowe SL, Bowen DM (1989) Reduced glycine stimulation of [3H]MK-801 binding in Alzheimer's disease. J Neurochem 53(3):698–704

Quinn PD, Harden KP (2013) Differential changes in impulsivity and sensation seeking and the escalation of substance use from adolescence to early adulthood. Dev Psychopathol 25(1):223–239. https://doi.org/10.1017/S0954579412000284

Raedler TJ, Knable MB, Lafargue T, Urbina RA, Egan MF, Pickar D, Weinberger DR (1999) In vivo determination of striatal dopamine D2 receptor occupancy in patients treated with olanzapine. Psychiatry Res 90(2):81–90

Ragland JD, Gur RC, Glahn DC, Censits DM, Smith RJ, Lazarev MG, Alavi A, Gur RE (1998) Frontotemporal cerebral blood flow change during executive and declarative memory tasks in schizophrenia: a positron emission tomography study. Neuropsychology 12(3):399–413. https://doi.org/10.1037//0894-4105.12.3.399

Rahimi R, Nikfar S, Abdollahi M (2006) Pregnancy outcomes following exposure to serotonin reuptake inhibitors: a meta-analysis of clinical trials. Reprod Toxicol 22(4):571–575. https://doi.org/10.1016/j.reprotox.2006.03.019

Rajkowska G (1997) Morphometric methods for studying the prefrontal cortex in suicide victims and psychiatric patients. Ann N Y Acad Sci 836(1 Neurobiology):253–268. https://doi.org/10.1111/j.1749-6632.1997.tb52364.x

Rajkowska G, Selemon LD, Goldman-Rakic PS (1998) Neuronal and glial somal size in the prefrontal cortex: a postmortem morphometric study of schizophrenia and Huntington disease. Arch Gen Psychiatry 55(3):215–224. https://doi.org/10.1001/archpsyc.55.3.215

Rajkowska G, Miguel-Hidalgo JJ, Wei J, Dilley G, Pittman SD, Meltzer HY, Overholser JC, Roth BL, Stockmeier CA (1999) Morphometric evidence for neuronal and glial prefrontal cell pathology in major depression. See accompanying Editorial, in this issue. Biol Psychiatry 45(9):1085–1098. https://doi.org/10.1016/s0006-3223(99)00041-4

Rajkowska G, Halaris A, Selemon LD (2001) Reductions in neuronal and glial density characterize the dorsolateral prefrontal cortex in bipolar disorder. Biol Psychiatry 49(9):741–752. https://doi.org/10.1016/s0006-3223(01)01080-0

Rao ML, Ruhrmann S, Retey B, Liappis N, Fuger J, Kraemer M, Kasper S, Moller HJ (1996) Low plasma thyroid indices of depressed patients are attenuated by antidepressant drugs and influence treatment outcome. Pharmacopsychiatry 29(5):180–186. https://doi.org/10.1055/s-2007-979568

Raz S, Raz N (1990) Structural brain abnormalities in the major psychoses: a quantitative review of the evidence from computerized imaging. Psychol Bull 108(1):93–108. https://doi.org/10.1037//0033-2909.108.1.93

Rebeck GW, Harr SD, Strickland DK, Hyman BT (1995) Multiple, diverse senile plaque-associated proteins are ligands of an apolipoprotein E receptor, the alpha 2-macroglobulin receptor/low-density-lipoprotein receptor-related protein. Ann Neurol 37(2):211–217. https://doi.org/10.1002/ana.410370212

Reid M, Herrera-Marschitz M, Hökfelt T, Terenius L, Ungerstedt U (1988) Differential modulation of striatal dopamine release by intranigral injection of γ-aminobutyric acid (GABA), dynorphin A and substance P. Eur J Pharmacol 147(3):411–420. https://doi.org/10.1016/0014-2999(88)90176-8

Reid MS, Fox L, Ho LB, Berger SP (2000) Nicotine stimulation of extracellular glutamate levels in the nucleus accumbens: neuropharmacological characterization. Synapse 35(2):129–136. https://doi.org/10.1002/(SICI)1098-2396(200002)35:2<129::AID-SYN5>3.0.CO;2-D

Reimers A, Helde G, Brodtkorb E (2005) Ethinyl estradiol, not progestogens, reduces lamotrigine serum concentrations. Epilepsia 46(9):1414–1417. https://doi.org/10.1111/j.1528-1167.2005.10105.x

Reith ME, Meisler BE, Sershen H, Lajtha A (1986) Structural requirements for cocaine congeners to interact with dopamine and serotonin uptake sites in mouse brain and to induce stereotyped behavior. Biochem Pharmacol 35(7):1123–1129. https://doi.org/10.1016/0006-2952(86)90148-6

Reith J, Benkelfat C, Sherwin A, Yasuhara Y, Kuwabara H, Andermann F, Bachneff S, Cumming P, Diksic M, Dyve SE, Etienne P, Evans AC, Lal S, Shevell M, Savard G, Wong DF, Chouinard G, Gjedde A (1994) Elevated dopa decarboxylase activity in living brain

of patients with psychosis. Proc Natl Acad Sci U S A 91(24):11651–11654. https://doi.org/10.1073/pnas.91.24.11651
Reynolds CF III, Kupfer DJ (1987) Sleep research in affective illness: state of the art circa 1987. Sleep 10(3):199–215. https://doi.org/10.1093/sleep/10.3.199
Richards PM, Ruff RM (1989) Motivational effects on neuropsychological functioning: comparison of depressed versus nondepressed individuals. J Consult Clin Psychol 57(3):396–402. https://doi.org/10.1037//0022-006x.57.3.396
Riemann D, Berger M, Voderholzer U (2001) Sleep and depression—results from psychobiological studies: an overview. Biol Psychol 57(1-3):67–103. https://doi.org/10.1016/s0301-0511(01)00090-4
Rioch DM (1955) Psychiatry as a biological science. Psychiatry 18(4):313–321
Risner ME, Jones BE (1980) Intravenous self-administration of cocaine and norcocaine by dogs. Psychopharmacology (Berl) 71(1):83–89. https://doi.org/10.1007/bf00433258
Rive MM, van Rooijen G, Veltman DJ, Phillips ML, Schene AH, Ruhe HG (2013) Neural correlates of dysfunctional emotion regulation in major depressive disorder. A systematic review of neuroimaging studies. Neurosci Biobehav Rev 37(10 Pt 2):2529–2553. https://doi.org/10.1016/j.neubiorev.2013.07.018
Robert P, Migneco O, Darcourt J, Ricq O, Aubin V, Bonhomme P, Pringuey D, Lapulus F, Darcourt G (1992) Correlation between ^{99m}Tc – HMPAO brain uptake and severity of dementia in Alzheimer's disease: assessment using an automatized technique. Dement Geriatr Cogn Disord 3(1):15–20. https://doi.org/10.1159/000106988
Robinson TE, Becker JB (1986) Enduring changes in brain and behavior produced by chronic amphetamine administration: a review and evaluation of animal models of amphetamine psychosis. Brain Res 396(2):157–198. https://doi.org/10.1016/s0006-8993(86)80193-7
Robinson TE, Berridge KC (1993) The neural basis of drug craving: an incentive-sensitization theory of addiction. Brain Res Brain Res Rev 18(3):247–291. https://doi.org/10.1016/0165-0173(93)90013-p
Robitzek EH, Selikoff IJ, Ornstein GG (1952) Chemotherapy of human tuberculosis with hydrazine derivatives of isonicotinic acid; preliminary report of representative cases. Q Bull Sea View Hosp 13(1):27–51
Rodd-Henricks ZA, McKinzie DL, Crile RS, Murphy JM, McBride WJ (2000) Regional heterogeneity for the intracranial self-administration of ethanol within the ventral tegmental area of female Wistar rats. Psychopharmacology 149(3):217–224. https://doi.org/10.1007/s002139900347
Rogawski MA, Loscher W (2004a) The neurobiology of antiepileptic drugs. Nat Rev Neurosci 5(7):553–564. https://doi.org/10.1038/nrn1430
Rogawski MA, Loscher W (2004b) The neurobiology of antiepileptic drugs for the treatment of nonepileptic conditions. Nat Med 10(7):685–692. https://doi.org/10.1038/nm1074
Rosa AR, Fountoulakis K, Siamouli M, Gonda X, Vieta E (2011) Is anticonvulsant treatment of mania a class effect? Data from randomized clinical trials. CNS Neurosci Ther 17(3):167–177. https://doi.org/10.1111/j.1755-5949.2009.00089.x
Ross JA, McGonigle P, Van Bockstaele EJ (2015) Locus Coeruleus, norepinephrine and Abeta peptides in Alzheimer's disease. Neurobiol Stress 2:73–84. https://doi.org/10.1016/j.ynstr.2015.09.002
Roth RM, Koven NS, Randolph JJ, Flashman LA, Pixley HS, Ricketts SM, Wishart HA, Saykin AJ (2006) Functional magnetic resonance imaging of executive control in bipolar disorder. Neuroreport 17(11):1085–1089. https://doi.org/10.1097/01.wnr.0000227979.06013.57
Rothman SM, Olney JW (1987) Excitotoxity and the NMDA receptor. Trends Neurosci 10(7):299–302. https://doi.org/10.1016/0166-2236(87)90177-9
Rots NY, de Jong J, Workel JO, Levine S, Cools AR, DeKloet ER (1996) Neonatal maternally deprived rats have as adults elevated basal pituitary-adrenal activity and enhanced susceptibility to apomorphine. J Neuroendocrinol 8(7):501–506. https://doi.org/10.1046/j.1365-2826.1996.04843.x

Rubinsztein JS, Fletcher PC, Rogers RD, Ho LW, Aigbirhio FI, Paykel ES, Robbins TW, Sahakian BJ (2001) Decision-making in mania: a PET study. Brain 124(Pt 12):2550–2563
Sabol SZ, Nelson ML, Fisher C, Gunzerath L, Brody CL, Hu S, Sirota LA, Marcus SE, Greenberg BD, Lucas FR, Benjamin J, Murphy DL, Hamer DH (1999) A genetic association for cigarette smoking behavior. Health Psychol 18(1):7–13. https://doi.org/10.1037//0278-6133.18.1.7
Salzer HM, Lurie ML (1953) Anxiety and depressive states treated with isonicotinyl hydrazide (isoniazid). AMA Arch Neurol Psychiatry 70(3):317–324
Sapolsky RM (2000) Glucocorticoids and hippocampal atrophy in neuropsychiatric disorders. Arch Gen Psychiatry 57(10):925–935. https://doi.org/10.1001/archpsyc.57.10.925
Sartorius N, Baghai TC, Baldwin DS, Barrett B, Brand U, Fleischhacker W, Goodwin G, Grunze H, Knapp M, Leonard BE, Lieberman J, Nakane Y, Pinder RM, Schatzberg AF, Svestka J, Baumann P, Ghalib K, Markowitz JC, Padberg F, Fink M, Furukawa T, Fountoulakis KN, Jensen P, Kanba S, Riecher-Rossler A (2007) Antidepressant medications and other treatments of depressive disorders: a CINP Task Force report based on a review of evidence. Int J Neuropsychopharmacol 10(Suppl 1):S1–S207. https://doi.org/10.1017/S1461145707008255
Sassi RB, Brambilla P, Hatch JP, Nicoletti MA, Mallinger AG, Frank E, Kupfer DJ, Keshavan MS, Soares JC (2004) Reduced left anterior cingulate volumes in untreated bipolar patients. Biol Psychiatry 56(7):467–475. https://doi.org/10.1016/j.biopsych.2004.07.005
Saunders AM, Strittmatter WJ, Schmechel D, George-Hyslop PH, Pericak-Vance MA, Joo SH, Rosi BL, Gusella JF, Crapper-MacLachlan DR, Alberts MJ et al (1993) Association of apolipoprotein E allele epsilon 4 with late-onset familial and sporadic Alzheimer's disease. Neurology 43(8):1467–1472. https://doi.org/10.1212/wnl.43.8.1467
Savitz JB, Price JL, Drevets WC (2014) Neuropathological and neuromorphometric abnormalities in bipolar disorder: view from the medial prefrontal cortical network. Neurosci Biobehav Rev 42:132–147. https://doi.org/10.1016/j.neubiorev.2014.02.008
Schildkraut JJ (1965) The catecholamine hypothesis of affective disorders: a review of supporting evidence. Am J Psychiatry 122(5):509–522. https://doi.org/10.1176/ajp.122.5.509
Schioldann J (1999) John Cade's seminal lithium paper turns fifty. Acta Psychiatr Scand 100(6):403–405
Schioldann J (2006) Mogens Abelin Schou (1918–2005) – half a century with lithium – obituary. Hist Psychiatry 17(2):247–252. https://doi.org/10.1177/0957154x06061602
Schioldann J (2011) On periodical depressions and their pathogenesis' by Carl Lange (1886). Hist Psychiatry 22(85 Pt 1):108–130. https://doi.org/10.1177/0957154X10396807
Schizophrenia Working Group of the Psychiatric Genomics C (2014) Biological insights from 108 schizophrenia-associated genetic loci. Nature 511(7510):421–427. https://doi.org/10.1038/nature13595
Schmidt MJ, Mirnics K (2015) Neurodevelopment, GABA system dysfunction, and schizophrenia. Neuropsychopharmacology 40(1):190–206. https://doi.org/10.1038/npp.2014.95
Schnack HG, van Haren NE, Nieuwenhuis M, Hulshoff Pol HE, Cahn W, Kahn RS (2016) Accelerated brain aging in schizophrenia: a longitudinal pattern recognition study. Am J Psychiatry 173(6):607–616. https://doi.org/10.1176/appi.ajp.2015.15070922
Schou M, Juel-Nielsen N, Stromgren E, Voldby H (1954) The treatment of manic psychoses by the administration of lithium salts. J Neurol Neurosurg Psychiatry 17(4):250–260
Schou M, Baastrup PC, Grof P, Weis P, Angst J (1970) Pharmacological and clinical problems of lithium prophylaxis. Br J Psychiatry 116(535):615–619
Schulteis G, Markou A, Cole M, Koob GF (1995) Decreased brain reward produced by ethanol withdrawal. Proc Natl Acad Sci U S A 92(13):5880–5884. https://doi.org/10.1073/pnas.92.13.5880
Schultz W (1997) The phasic reward signal of primate dopamine neurons. Adv Pharmacol. https://doi.org/10.1016/s1054-3589(08)60841-8
Selemon LD (2001) Regionally diverse cortical pathology in schizophrenia: clues to the etiology of the disease. Schizophr Bull 27(3):349–377. https://doi.org/10.1093/oxfordjournals.schbul.a006881

Selemon LD, Goldman-Rakic PS (1999) The reduced neuropil hypothesis: a circuit based model of schizophrenia. Biol Psychiatry 45(1):17–25. https://doi.org/10.1016/s0006-3223(98)00281-9

Selemon LD, Rajkowska G, Goldman-Rakic PS (1998) Elevated neuronal density in prefrontal area 46 in brains from schizophrenic patients: application of a three-dimensional, stereologic counting method. J Comp Neurol 392(3):402–412. https://doi.org/10.1002/(sici)1096-9861(19980316)392:3<402::aid-cne9>3.0.co;2-5

Selifoff IJ, Robitzek EH, Ornstein GG (1952) Toxicity of hydrazine derivatives of isonicotinic acid in the chemotherapy of human tuberculosis; a preliminary report. Q Bull Sea View Hosp 13(1):17–26

Selikoff IJ, Robitzek EH (1952) Tuberculosis chemotherapy with hydrazine derivatives of isonicotinic acid. Dis Chest 21(4):385–438

Selkoe DJ (1994) Normal and abnormal biology of the beta-amyloid precursor protein. Annu Rev Neurosci 17:489–517. https://doi.org/10.1146/annurev.ne.17.030194.002421

Shah PJ, Ebmeier KP, Glabus MF, Goodwin GM (1998) Cortical grey matter reductions associated with treatment-resistant chronic unipolar depression. Controlled magnetic resonance imaging study. Br J Psychiatry 172(6):527–532. https://doi.org/10.1192/bjp.172.6.527

Shah PJ, O'Carroll RE, Rogers A, Moffoot AP, Ebmeier KP (1999) Abnormal response to negative feedback in depression. Psychol Med 29(1):63–72. https://doi.org/10.1017/s0033291798007880

Sheline YI, Wang PW, Gado MH, Csernansky JG, Vannier MW (1996) Hippocampal atrophy in recurrent major depression. Proc Natl Acad Sci U S A 93(9):3908–3913. https://doi.org/10.1073/pnas.93.9.3908

Sheline YI, Gado MH, Price JL (1998) Amygdala core nuclei volumes are decreased in recurrent major depression. Neuroreport 9(9):2023–2028. https://doi.org/10.1097/00001756-199806220-00021

Sheline YI, Gado MH, Kraemer HC (2003) Untreated depression and hippocampal volume loss. Am J Psychiatry 160(8):1516–1518. https://doi.org/10.1176/appi.ajp.160.8.1516

Sheng JG, Ito K, Skinner RD, Mrak RE, Rovnaghi CR, VanEldik LJ, Griffin WST (1996) In vivo and in vitro evidence supporting a role for the inflammatory cytokine interleukin-1 as a driving force in Alzheimer pathogenesis. Neurobiol Aging 17(5):761–766. https://doi.org/10.1016/0197-4580(96)00104-2

Shepard TH, Brent RL, Friedman JM, Jones KL, Miller RK, Moore CA, Polifka JE (2002) Update on new developments in the study of human teratogens. Teratology 65(4):153–161. https://doi.org/10.1002/tera.10032

Shiffman S (1996) Addiction versus stages of change models in predicting smoking cessation – "addiction versus stages of change models" vs "addiction and stages of change models" – comment. Addiction 91(9):1289–1290. https://doi.org/10.1111/j.1360-0443.1996.tb03614.x

Shippenberg TS, Heidbreder C (1995) The δ-opioid receptor antagonist naltrindole prevents sensitization to the conditioned rewarding effects of cocaine. Eur J Pharmacol 280(1):55–61. https://doi.org/10.1016/0014-2999(95)00185-n

Shoaib M, Schindler CW, Goldberg SR, Pauly JR (1997) Behavioural and biochemical adaptations to nicotine in rats: influence of MK801, an NMDA receptor antagonist. Psychopharmacology (Berl) 134(2):121–130. https://doi.org/10.1007/s002130050433

Shorter E (2009) The history of lithium therapy. Bipolar Disord 11(Suppl 2):4–9. https://doi.org/10.1111/j.1399-5618.2009.00706.x

Shotbolt P, Stokes PR, Owens SF, Toulopoulou T, Picchioni MM, Bose SK, Murray RM, Howes OD (2011) Striatal dopamine synthesis capacity in twins discordant for schizophrenia. Psychol Med 41(11):2331–2338. https://doi.org/10.1017/S0033291711000341

Siegle GJ, Thompson WK, Collier A, Berman SR, Feldmiller J, Thase ME, Friedman ES (2012) Toward clinically useful neuroimaging in depression treatment: prognostic utility of subgenual cingulate activity for determining depression outcome in cognitive therapy across studies, scanners, and patient characteristics. Arch Gen Psychiatry 69(9):913–924. https://doi.org/10.1001/archgenpsychiatry.2012.65

Silberman EK, Weingartner H, Post RM (1983) Thinking disorder in depression. Arch Gen Psychiatry 40(7):775–780. https://doi.org/10.1001/archpsyc.1983.01790060073009

Silverstone T, McPherson H, Li Q, Doyle T (2003) Deep white matter hyperintensities in patients with bipolar depression, unipolar depression and age-matched control subjects. Bipolar Disord 5(1):53–57

Silvestri S, Seeman MV, Negrete JC, Houle S, Shammi CM, Remington GJ, Kapur S, Zipursky RB, Wilson AA, Christensen BK, Seeman P (2000) Increased dopamine D2 receptor binding after long-term treatment with antipsychotics in humans: a clinical PET study. Psychopharmacology (Berl) 152(2):174–180. https://doi.org/10.1007/s002130000532

Sklar P, Smoller JW, Fan J, Ferreira MA, Perlis RH, Chambert K, Nimgaonkar VL, McQueen MB, Faraone SV, Kirby A, de Bakker PI, Ogdie MN, Thase ME, Sachs GS, Todd-Brown K, Gabriel SB, Sougnez C, Gates C, Blumenstiel B, Defelice M, Ardlie KG, Franklin J, Muir WJ, McGhee KA, MacIntyre DJ, McLean A, VanBeck M, McQuillin A, Bass NJ, Robinson M, Lawrence J, Anjorin A, Curtis D, Scolnick EM, Daly MJ, Blackwood DH, Gurling HM, Purcell SM (2008) Whole-genome association study of bipolar disorder. Mol Psychiatry 13(6):558–569. https://doi.org/10.1038/sj.mp.4002151

Smith MA, Kim SY, van Oers HJ, Levine S (1997) Maternal deprivation and stress induce immediate early genes in the infant rat brain. Endocrinology 138(11):4622–4628. https://doi.org/10.1210/endo.138.11.5529

Snyder SH (1976) The dopamine hypothesis of schizophrenia: focus on the dopamine receptor. Am J Psychiatry 133(2):197–202. https://doi.org/10.1176/ajp.133.2.197

Soares JC, Kochunov P, Monkul ES, Nicoletti MA, Brambilla P, Sassi RB, Mallinger AG, Frank E, Kupfer DJ, Lancaster J, Fox P (2005) Structural brain changes in bipolar disorder using deformation field morphometry. Neuroreport 16(6):541–544

Sokolov BP (2002) Expression of NMDAR1, GluR1, GluR7, and KA1 glutamate receptor mRNAs is decreased in frontal cortex of "neuroleptic-free" schizophrenics: evidence on reversible up-regulation by typical neuroleptics. J Neurochem 71(6):2454–2464. https://doi.org/10.1046/j.1471-4159.1998.71062454.x

Solas M, Puerta E, Ramirez MJ (2015) Treatment options in Alzheimer's disease: the GABA Story. Curr Pharm Des 21(34):4960–4971

Solomon RL, Corbit JD (1973) An opponent-process theory of motivation. II. Cigarette addiction. J Abnorm Psychol 81(2):158–171. https://doi.org/10.1037/h0034534

Solomon DA, Ristow WR, Keller MB, Kane JM, Gelenberg AJ, Rosenbaum JF, Warshaw MG (1996) Serum lithium levels and psychosocial function in patients with bipolar I disorder. Am J Psychiatry 153(10):1301–1307. https://doi.org/10.1176/ajp.153.10.1301

Sparks DL, DeKosky ST, Markesbery WR (1988) Alzheimer's disease. Aminergic-cholinergic alterations in hypothalamus. Arch Neurol 45(9):994–999

Sparks DL, Hunsaker JC 3rd, Slevin JT, DeKosky ST, Kryscio RJ, Markesbery WR (1992) Monoaminergic and cholinergic synaptic markers in the nucleus basalis of Meynert (nbM): normal age-related changes and the effect of heart disease and Alzheimer's disease. Ann Neurol 31(6):611–620. https://doi.org/10.1002/ana.410310608

Spence SA, Hirsch SR, Brooks DJ, Grasby PM (2018) Prefrontal cortex activity in people with schizophrenia and control subjects. Br J Psychiatry 172(04):316–323. https://doi.org/10.1192/bjp.172.4.316

Staner L, De La Fuente JM, Kerkhofs M, Linkowski P, Mendlewicz J (1992) Biological and clinical features of recurrent brief depression: a comparison with major depressed and healthy subjects. J Affect Disord 26(4):241–245

Stone JM (2011) Glutamatergic antipsychotic drugs: a new dawn in the treatment of schizophrenia? Ther Adv Psychopharmacol 1(1):5–18. https://doi.org/10.1177/2045125311400779

Stone JM, Howes OD, Egerton A, Kambeitz J, Allen P, Lythgoe DJ, O'Gorman RL, McLean MA, Barker GJ, McGuire P (2010) Altered relationship between hippocampal glutamate levels and striatal dopamine function in subjects at ultra high risk of psychosis. Biol Psychiatry 68(7):599–602. https://doi.org/10.1016/j.biopsych.2010.05.034

Strakowski SM, Wilson DR, Tohen M, Woods BT, Douglass AW, Stoll AL (1993) Structural brain abnormalities in first-episode mania. Biol Psychiatry 33(8-9):602–609

Strakowski SM, DelBello MP, Zimmerman ME, Getz GE, Mills NP, Ret J, Shear P, Adler CM (2002) Ventricular and periventricular structural volumes in first- versus multiple-episode bipolar disorder. Am J Psychiatry 159(11):1841–1847. https://doi.org/10.1176/appi.ajp.159.11.1841
Strittmatter WJ, Saunders AM, Schmechel D, Pericak-Vance M, Enghild J, Salvesen GS, Roses AD (1993a) Apolipoprotein E: high-avidity binding to beta-amyloid and increased frequency of type 4 allele in late-onset familial Alzheimer disease. Proc Natl Acad Sci U S A 90(5):1977–1981. https://doi.org/10.1073/pnas.90.5.1977
Strittmatter WJ, Weisgraber KH, Huang DY, Dong LM, Salvesen GS, Pericak-Vance M, Schmechel D, Saunders AM, Goldgaber D, Roses AD (1993b) Binding of human apolipoprotein E to synthetic amyloid beta peptide: isoform-specific effects and implications for late-onset Alzheimer disease. Proc Natl Acad Sci U S A 90(17):8098–8102. https://doi.org/10.1073/pnas.90.17.8098
Strobusch AD, Jefferson JW (1980) The checkered history of lithium in medicine. Pharm Hist 22(2):72–76
Suchecki D, Mozaffarian D, Gross G, Rosenfeld P, Levine S (1993) Effects of maternal deprivation on the ACTH stress response in the infant rat. Neuroendocrinology 57(2):204–212. https://doi.org/10.1159/000126361
Sullivan PF, Kendler KS, Neale MC (2003) Schizophrenia as a complex trait: evidence from a meta-analysis of twin studies. Arch Gen Psychiatry 60(12):1187–1192. https://doi.org/10.1001/archpsyc.60.12.1187
Sumpter PQ, Mann DM, Davies CA, Neary D, Snowden JS, Yates PO (1986) A quantitative study of the ultrastructure of pyramidal neurons of the cerebral cortex in Alzheimer's disease in relationship to the degree of dementia. Neuropathol Appl Neurobiol 12(3):321–329
Sweeney JA, Kmiec JA, Kupfer DJ (2000) Neuropsychologic impairments in bipolar and unipolar mood disorders on the CANTAB neurocognitive battery. Biol Psychiatry 48(7):674–684. https://doi.org/10.1016/s0006-3223(00)00910-0
Taber MT, Fibiger HC (1997) Feeding-evoked dopamine release in the nucleus, accumbens: regulation by glutamatergic mechanisms. Neuroscience 76(4):1105–1112. https://doi.org/10.1016/s0306-4522(96)00450-2
Tanda G (1997) Cannabinoid and heroin activation of mesolimbic dopamine transmission by a common μ1 opioid receptor mechanism. Science 276(5321):2048–2050. https://doi.org/10.1126/science.276.5321.2048
Tanzi RE, Gusella JF, Watkins PC, Bruns GA, St George-Hyslop P, Van Keuren ML, Patterson D, Pagan S, Kurnit DM, Neve RL (1987) Amyloid beta protein gene: cDNA, mRNA distribution, and genetic linkage near the Alzheimer locus. Science 235(4791):880–884. https://doi.org/10.1126/science.2949367
Tarsy D, Leopold N, Sax DS (1972) Cholinergic-adrenergic hypothesis of mania and depression. Lancet 2(7787):1153
Thompson PJ (1991) Antidepressants and memory – a review. Hum Psychopharmacol Clin Exp 6(2):79–90. https://doi.org/10.1002/hup.470060202
Todtenkopf MS, Vincent SL, Benes FM (2005) A cross-study meta-analysis and three-dimensional comparison of cell counting in the anterior cingulate cortex of schizophrenic and bipolar brain. Schizophr Res 73(1):79–89. https://doi.org/10.1016/j.schres.2004.08.018
Ton JMNC, Gerhardt GA, Friedemann M, Etgen AM, Rose GM, Sharpless NS, Gardner EL (1988) The effects of Δ9-tetrahydrocannabinol on potassium-evoked release of dopamine in the rat caudate nucleus: an in vivo electrochemical and in vivo microdialysis study. Brain Res 451(1-2):59–68. https://doi.org/10.1016/0006-8993(88)90749-4
Toro C, Deakin JF (2005) NMDA receptor subunit NRI and postsynaptic protein PSD-95 in hippocampus and orbitofrontal cortex in schizophrenia and mood disorder. Schizophr Res 80(2-3):323–330. https://doi.org/10.1016/j.schres.2005.07.003
Torrey EF, Barci BM, Webster MJ, Bartko JJ, Meador-Woodruff JH, Knable MB (2005) Neurochemical markers for schizophrenia, bipolar disorder, and major depression in postmortem brains. Biol Psychiatry 57(3):252–260. https://doi.org/10.1016/j.biopsych.2004.10.019

Toth E, Vizi ES, Lajtha A (1993) Effect of nicotine on levels of extracellular amino acids in regions of the rat brain in vivo. Neuropharmacology 32(8):827–832. https://doi.org/10.1016/0028-3908(93)90192-6

Tourney G (1969) History of biological psychiatry in America. Am J Psychiatry 126(1):29–42. https://doi.org/10.1176/ajp.126.1.29

Trichard C, Martinot JL, Alagille M, Masure MC, Hardy P, Ginestet D, Feline A (1995) Time course of prefrontal lobe dysfunction in severely depressed in-patients: a longitudinal neuropsychological study. Psychol Med 25(1):79–85. https://doi.org/10.1017/s0033291700028105

Trillo L, Das D, Hsieh W, Medina B, Moghadam S, Lin B, Dang V, Sanchez MM, De Miguel Z, Ashford JW, Salehi A (2013) Ascending monoaminergic systems alterations in Alzheimer's disease. translating basic science into clinical care. Neurosci Biobehav Rev 37(8):1363–1379. https://doi.org/10.1016/j.neubiorev.2013.05.008

True WR, Heath AC, Scherrer JF, Waterman B, Goldberg J, Lin N, Eisen SA, Lyons MJ, Tsuang MT (1997) Genetic and environmental contributions to smoking. Addiction 92(10):1277–1287. https://doi.org/10.1046/j.1360-0443.1997.921012775.x

Tuszynski MH, Yang JH, Barba D, HS U, Bakay RA, Pay MM, Masliah E, Conner JM, Kobalka P, Roy S, Nagahara AH (2015) Nerve growth factor gene therapy: activation of neuronal responses in Alzheimer disease. JAMA Neurol 72(10):1139–1147. https://doi.org/10.1001/jamaneurol.2015.1807

Uhl GR, Liu QR, Walther D, Hess J, Naiman D (2001) Polysubstance abuse-vulnerability genes: genome scans for association, using 1,004 subjects and 1,494 single-nucleotide polymorphisms. Am J Hum Genet 69(6):1290–1300. https://doi.org/10.1086/324467

Vakili K, Pillay SS, Lafer B, Fava M, Renshaw PF, Bonello-Cintron CM, Yurgelun-Todd DA (2000) Hippocampal volume in primary unipolar major depression: a magnetic resonance imaging study. Biol Psychiatry 47(12):1087–1090. https://doi.org/10.1016/s0006-3223(99)00296-6

Van Cauwenberghe C, Van Broeckhoven C, Sleegers K (2016) The genetic landscape of Alzheimer disease: clinical implications and perspectives. Genet Med 18(5):421–430. https://doi.org/10.1038/gim.2015.117

Van Horn JD, McManus IC (1992) Ventricular enlargement in schizophrenia. A meta-analysis of studies of the ventricle:brain ratio (VBR). Br J Psychiatry 160(5):687–697. https://doi.org/10.1192/bjp.160.5.687

Van Praag M, Leijnse B (1963) Die bedeutung dermonoamineoxydashemmung als antidepressives prinzip I. Psychopharmacologia 4:1–14

Vasilakos JP, Carroll RT, Emmerling MR, Doyle PD, Davis RE, Kim KS, Shivers BD (1994) Interleukin-1 beta dissociates beta-amyloid precursor protein and beta-amyloid peptide secretion. FEBS Lett 354(3):289–292

Vidal C (1994) Nicotinic potentiation of glutamatergic synapses in the prefrontal cortex – new insight into the analysis of the role of nicotinic receptors in cognitive functions. Drug Dev Res 31(2):120–126. https://doi.org/10.1002/ddr.430310206

Vita A, De Peri L, Sacchetti E (2009) Gray matter, white matter, brain, and intracranial volumes in first-episode bipolar disorder: a meta-analysis of magnetic resonance imaging studies. Bipolar Disord 11(8):807–814. https://doi.org/10.1111/j.1399-5618.2009.00759.x

Volkow ND, Muenke M (2012) The genetics of addiction. Hum Genet 131(6):773–777. https://doi.org/10.1007/s00439-012-1173-3

Volz HP, Gaser C, Hager F, Rzanny R, Mentzel HJ, Kreitschmann-Andermahr I, Kaiser WA, Sauer H (1997) Brain activation during cognitive stimulation with the Wisconsin Card Sorting Test—a functional MRI study on healthy volunteers and schizophrenics. Psychiatry Res 75(3):145–157. https://doi.org/10.1016/s0925-4927(97)00053-x

Watters AJ, Korgaonkar MS, Carpenter JS, Harris AWF, Gross JJ, Williams LM (2018) Profiling risk for depressive disorder by circuit, behavior and self-report measures of emotion function. J Affect Disord 227:595–602. https://doi.org/10.1016/j.jad.2017.11.067

Watts FN, Dalgleish T, Bourke P, Healy D (1990) Memory deficit in clinical depression: processing resources and the structure of materials. Psychol Med 20(2):345–349. https://doi.org/10.1017/s0033291700017657
Weber W, Bartenstein P, Gross MW, Kinzel D, Daschner H, Feldmann HJ, Reidel G, Ziegler SI, Lumenta C, Molls M, Schwaiger M (1997) Fluorine-18-FDG PET and iodine-123-IMT SPECT in the evaluation of brain tumors. J Nucl Med 38(5):802–808
Webster MJ, O'Grady J, Kleinman JE, Weickert CS (2005) Glial fibrillary acidic protein mRNA levels in the cingulate cortex of individuals with depression, bipolar disorder and schizophrenia. Neuroscience 133(2):453–461. https://doi.org/10.1016/j.neuroscience.2005.02.037
Weinberger DR, Torrey EF, Neophytides AN, Wyatt RJ (1979) Lateral cerebral ventricular enlargement in chronic schizophrenia. Arch Gen Psychiatry 36(7):735–739. https://doi.org/10.1001/archpsyc.1979.01780070013001
Weinberger DR, Berman KF, Zec RF (1986) Physiologic dysfunction of dorsolateral prefrontal cortex in schizophrenia. I. Regional cerebral blood flow evidence. Arch Gen Psychiatry 43(2):114–124. https://doi.org/10.1001/archpsyc.1986.01800020020004
Weiss F, Lorang MT, Bloom FE, Koob GF (1993) Oral alcohol self-administration stimulates dopamine release in the rat nucleus accumbens: genetic and motivational determinants. J Pharmacol Exp Ther 267(1):250–258
Wellcome Trust Case Control C (2007) Genome-wide association study of 14,000 cases of seven common diseases and 3,000 shared controls. Nature 447(7145):661–678. https://doi.org/10.1038/nature05911
Welzl H, Kuhn G, Huston JP (1989) Self-administration of small amounts of morphine through glass micropipettes into the ventral tegmental area of the rat. Neuropharmacology 28(10):1017. https://doi.org/10.1016/0028-3908(89)90112-3
Westfall TC, Mereu G, Vickery L, Perry H, Naes L, Yoon K-WP (1989) Regulation by nicotine of midbrain dopamine neurons. In: Nicotinic receptors in the CNS their role in synaptic transmission. Elsevier, Amsterdam. https://doi.org/10.1016/s0079-6123(08)62477-2
Wilke M, Kowatch RA, DelBello MP, Mills NP, Holland SK (2004) Voxel-based morphometry in adolescents with bipolar disorder: first results. Psychiatry Res 131(1):57–69. https://doi.org/10.1016/j.pscychresns.2004.01.004
van Winkel M, Peeters F, van Winkel R, Kenis G, Collip D, Geschwind N, Jacobs N, Derom C, Thiery E, van Os J, Myin-Germeys I, Wichers M (2014) Impact of variation in the BDNF gene on social stress sensitivity and the buffering impact of positive emotions: replication and extension of a gene-environment interaction. Eur Neuropsychopharmacol 24(6):930–938. https://doi.org/10.1016/j.euroneuro.2014.02.005
Wise RA (1978) Catecholamine theories of reward: a critical review. Brain Res 152(2):215–247. https://doi.org/10.1016/0006-8993(78)90253-6
Wise RA (1988) The neurobiology of craving: implications for the understanding and treatment of addiction. J Abnorm Psychol 97(2):118–132. https://doi.org/10.1037//0021-843x.97.2.118
Wise RA (1996) Addictive drugs and brain stimulation reward. Annu Rev Neurosci 19(1):319–340. https://doi.org/10.1146/annurev.ne.19.030196.001535
Wise RA, Bozarth MA (1987) A psychomotor stimulant theory of addiction. Psychol Rev 94(4):469–492. https://doi.org/10.1037//0033-295x.94.4.469
Wise RA, Munn E (1995) Withdrawal from chronic amphetamine elevates baseline intracranial self-stimulation thresholds. Psychopharmacology (Berl) 117(2):130–136. https://doi.org/10.1007/bf02245178
Wise RA, Rompre PP (1989) Brain dopamine and reward. Annu Rev Psychol 40(1):191–225. https://doi.org/10.1146/annurev.ps.40.020189.001203
Wise RA, Leone P, Rivest R, Leeb K (1995) Elevations of nucleus accumbens dopamine and DOPAC levels during intravenous heroin self-administration. Synapse 21(2):140–148. https://doi.org/10.1002/syn.890210207

Wisniewski HM, Wegiel J (1995) The neuropathology of Alzheimer's disease. Neuroimaging Clin N Am 5(1):45–57
Wisniewski KE, Dalton AJ, McLachlan C, Wen GY, Wisniewski HM (1985) Alzheimer's disease in Down's syndrome: clinicopathologic studies. Neurology 35(7):957–961. https://doi.org/10.1212/wnl.35.7.957
Wisniewski HM, Wegiel J, Wang KC, Kujawa M, Lach B (1989) Ultrastructural studies of the cells forming amyloid fibers in classical plaques. Can J Neurol Sci 16(4 Suppl):535–542
Wisniewski HM, Vorbrodt AW, Wegiel J, Morys J, Lossinsky AS (1990) Ultrastructure of the cells forming amyloid fibers in Alzheimer disease and scrapie. Am J Med Genet Suppl 7:287–297
Wiste AK, Arango V, Ellis SP, Mann JJ, Underwood MD (2008) Norepinephrine and serotonin imbalance in the locus coeruleus in bipolar disorder. Bipolar Disord 10(3):349–359. https://doi.org/10.1111/j.1399-5618.2007.00528.x
de Wit H, Wise RA (1977) Blockade of cocaine reinforcement in rats with the dopamine receptor blocker pimozide, but not with the noradrenergic blockers phentolamine or phenoxybenzamine. Can J Psychol 31(4):195–203. https://doi.org/10.1037/h0081662
Wockner LF, Noble EP, Lawford BR, Young RM, Morris CP, Whitehall VL, Voisey J (2014) Genome-wide DNA methylation analysis of human brain tissue from schizophrenia patients. Transl Psychiatry 4(1):e339. https://doi.org/10.1038/tp.2013.111
Woo TU, Kim AM, Viscidi E (2008a) Disease-specific alterations in glutamatergic neurotransmission on inhibitory interneurons in the prefrontal cortex in schizophrenia. Brain Res 1218:267–277. https://doi.org/10.1016/j.brainres.2008.03.092
Woo TU, Shrestha K, Lamb D, Minns MM, Benes FM (2008b) N-methyl-D-aspartate receptor and calbindin-containing neurons in the anterior cingulate cortex in schizophrenia and bipolar disorder. Biol Psychiatry 64(9):803–809. https://doi.org/10.1016/j.biopsych.2008.04.034
Wood JA, Wood PL, Ryan R, Graff-Radford NR, Pilapil C, Robitaille Y, Quirion R (1993) Cytokine indices in Alzheimer's temporal cortex: no changes in mature IL-1β or IL-1RA but increases in the associated acute phase proteins IL-6, α2-macroglobulin and C-reactive protein. Brain Res 629(2):245–252. https://doi.org/10.1016/0006-8993(93)91327-o
Wray NR, Gottesman II (2012) Using summary data from the Danish national registers to estimate heritabilities for schizophrenia, bipolar disorder, and major depressive disorder. Front Genet 3:118. https://doi.org/10.3389/fgene.2012.00118
Yamamoto T, Hirano A (1985) Nucleus raphe dorsalis in Alzheimer's disease: neurofibrillary tangles and loss of large neurons. Ann Neurol 17(6):573–577. https://doi.org/10.1002/ana.410170608
Yim HJ, Schallert T, Randall PK, Gonzales RA (1998) Comparison of local and systemic ethanol effects on extracellular dopamine concentration in rat nucleus accumbens by microdialysis. Alcohol Clin Exp Res 22(2):367–374. https://doi.org/10.1111/j.1530-0277.1998.tb03662.x
Young LT, Warsh JJ, Kish SJ, Shannak K, Hornykeiwicz O (1994) Reduced brain 5-HT and elevated NE turnover and metabolites in bipolar affective disorder. Biol Psychiatry 35(2):121–127
Yurgelun-Todd DA, Waternaux CM, Cohen BM, Gruber SA, English CD, Renshaw PF (1996) Functional magnetic resonance imaging of schizophrenic patients and comparison subjects during word production. Am J Psychiatry 153(2):200–205. https://doi.org/10.1176/ajp.153.2.200
Zavitsanou K, Katsifis A, Mattner F, Huang XF (2004) Investigation of m1/m4 muscarinic receptors in the anterior cingulate cortex in schizophrenia, bipolar disorder, and major depression disorder. Neuropsychopharmacology 29(3):619–625. https://doi.org/10.1038/sj.npp.1300367
Zavitsanou K, Katsifis A, Yu Y, Huang XF (2005) M2/M4 muscarinic receptor binding in the anterior cingulate cortex in schizophrenia and mood disorders. Brain Res Bull 65(5):397–403. https://doi.org/10.1016/j.brainresbull.2005.02.007
Zhang LX, Levine S, Dent G, Zhan Y, Xing G, Okimoto D, Kathleen Gordon M, Post RM, Smith MA (2002) Maternal deprivation increases cell death in the infant rat brain. Brain Res Dev Brain Res 133(1):1–11. https://doi.org/10.1016/s0926-6410(01)00118-5

Zhang K, Zhu Y, Zhu Y, Wu S, Liu H, Zhang W, Xu C, Zhang H, Hayashi T, Tian M (2016) Molecular, functional, and structural imaging of major depressive disorder. Neurosci Bull 32(3):273–285. https://doi.org/10.1007/s12264-016-0030-0

Zheng C, Geetha T, Gearing M, Babu JR (2015) Amyloid beta-abrogated TrkA ubiquitination in PC12 cells analogous to Alzheimer's disease. J Neurochem 133(6):919–925. https://doi.org/10.1111/jnc.13076

Zweig RM, Ross CA, Hedreen JC, Steele C, Cardillo JE, Whitehouse PJ, Folstein MF, Price DL (1988) The neuropathology of aminergic nuclei in Alzheimer's disease. Ann Neurol 24(2):233–242. https://doi.org/10.1002/ana.410240210

MIX
Papier aus verantwortungsvollen Quellen
Paper from responsible sources
FSC® C105338

If you have any concerns about our products,
you can contact us on
ProductSafety@springernature.com

In case Publisher is established outside the EU,
the EU authorized representative is:
Springer Nature Customer Service Center GmbH
Europaplatz 3, 69115 Heidelberg, Germany

Printed by Libri Plureos GmbH
in Hamburg, Germany